1st International Conference on Hydromechatronics and Advanced Robot Control Technology (HARCT 2025)

Journal of Physics: Conference Series Volume 3024

Guilin, China
28-30 March 2025

ISBN: 979-8-3313-2160-4
ISSN: 1742-6588

Printed from e-media with permission by:

Curran Associates, Inc.
57 Morehouse Lane
Red Hook, NY 12571

Some format issues inherent in the e-media version may also appear in this print version.

Institute of Physics
Dirac House, Temple Back
Bristol BS1 6BE UK

Phone: 44 1 17 929 7481
Fax: 44 1 17 920 0979

techtracking@iop.org

Additional copies of this publication are available from:

Curran Associates, Inc.
57 Morehouse Lane
Red Hook, NY 12571 USA
Phone: 845-758-0400
Fax: 845-758-2633
Email: curran@proceedings.com
Web: www.proceedings.com

TABLE OF CONTENTS

Journal of Physics: Conference Series **3024** (2025) 011001 doi:10.1088/1742-6596/3024/1/011001

Preface

It is our great pleasure to present the proceedings of International Conference on Hydromechatronics and Advanced Robot Control Technology (HARCT), which held in Guilin, China, from March 28 to 30, 2025. This inaugural conference marks a significant milestone in the convergence of hydromechatronics and advanced robot control technologies, bringing together experts and scholars from diverse backgrounds to share cutting-edge research and innovations. The papers presented in this proceedings reflect the latest developments in hydromechatronics, including fluid-driven actuators, intelligent control systems, and their applications in robotics. These contributions not only highlight the current research status but also pave the way for future research directions in this rapidly evolving field.

HARCT2025 is sponsored by Guilin University of Aerospace Technology, Guangxi University, Guilin University of Electronic Technology, Guilin University of Technology, and is co-sponsored by Beihang University, Central South University, Beijing Institute of Technology, Hunan University, South China University of Technology, Guangxi Mechanical Engineering Society, Guilin University of Aerospace Technology, Guangxi Key Laboratory of Special Engineering Equipment and Control, Guangxi Key Lab of Manufacturing System and Advanced Manufacturing Technology, Fluid Transmission and Intelligent Control Branch of Guangxi Mechanical Engineering Society, Research and Development Center for Micro Special Motors, China Aerospace Science and Technology Corporation. We are deeply grateful for their generous support. Their contributions have been instrumental in making this conference a success.

We extend our gratitude to all authors, reviewers, and participants for their invaluable contributions to this conference. Their collective efforts have enriched our understanding of the complex interactions between hydromechatronics and advanced robot control, fostering a collaborative environment that will continue to drive innovation and progress in these critical areas.We also extend our special thanks to our guest editors, Prof. Liang Yan and Prof. Deqiang He, whose expertise and dedication have been instrumental in ensuring the quality and relevance of the papers presented.

We hope that this collection of papers will serve as a valuable resource for researchers, engineers, and practitioners seeking to advance the frontiers of hydromechatronics and robotics. As we look to the future, we are confident that the insights and discoveries presented here will play a pivotal role in shaping the next generation of intelligent systems and technologies.

Journal of Physics: Conference Series **3024** (2025) 011001 doi:10.1088/1742-6596/3024/1/011001

Committee

Conference General Chair

Jinzhao Wu, Guilin University Of Aerospace Technology, China

Conference Co-Chair

Lieping Zhang, Guilin University Of Aerospace Technology, China

Honorary Chair

Yangzeng Xu, Lanzhou University of Technology, China

Program Chair

Deqiang He, Guangxi University, China

Technical Chair

Liang Yan, Beihang University, China
Deqing Huang, Southwest Jiaotong University, China
Shuai Mo, Guangxi University, China

Publicity Chair

Zhonggang Xiong, University Of Aerospace Technology, China

Technical Program Committee

Xueshan Gao, Beijing Institute of Technology, China
Xingjian Wang, Beihang University, China
Wensi Ding, South China University of Technology, China
Jizhuang Fan, Harbin Institute of Technology, China
Mingxing Lin, Shandong University, China
Ming Yan, Southern University of Science and Technology , China
Fanglan Ma, Guangxi University, China
Tianliang Lin, Huaqiao University, China
Pingyu Zhu, Guangzhou University, China
Wei Li, Jiangsu University, China
Ji Wang, Ningbo University, China
Zhong Liu, Guilin University Of Aerospace Technology, China
Xingguo Han, Guilin University Of Aerospace Technology, China
Rongjiang Tang, Guilin University of Electronic Technology, China
Zhixian Zhong, Guilin University of Technology, China
Shuyi Yang, Hunan University of Science and Technology, China
Qi Chen, University of Shanghai for Science and Technology, China
Min Fu, Chongqing University of Technology, China
Yong Chen, Chengdu Fluid Dynamics Innovation Center, China

Journal of Physics: Conference Series **3024** (2025) 011001 doi:10.1088/1742-6596/3024/1/011001

Yong Guo, Hunan University of Science and Technology, China
Yanbao Guo, China University of Petroleum, China
Laikuang Lin, Central South University, China
Özlem Boydak, Istanbul Medeniyet University, Turkey
Songbao Luo, Aviation Industry Corporation of China Beijing Precision Engineering Institute for Aircraft Industry, China
Hugo Wai Leung MAK, The Chinese University of Hong Kong & The Hong Kong University of Science and Technology, China

Journal of Physics: Conference Series **3024** (2025) 011002 doi:10.1088/1742-6596/3024/1/011002

Peer Review Statement

All papers published in this volume have been reviewed through processes administered by the Editors. Reviews were conducted by expert referees to the professional and scientific standards expected of a proceedings journal published by IOP Publishing.

- **Type of peer review:** Single Anonymous
- **Conference submission management system:** Morressier
- **Number of submissions received:** 81
- **Number of submissions sent for review:** 81
- **Number of submissions accepted:** 58
- **Acceptance Rate (Submissions Accepted / Submissions Received × 100):** 71.6
- **Average number of reviews per paper:** 1.02
- **Total number of reviewers involved:** 8
- **Contact person for queries:**
  Name: Jinzhao Wu
  Email: harct2025@163.com
  Affiliation: Guilin University Of Aerospace Technology

Journal of Physics: Conference Series **3024** (2025) 012001 doi:10.1088/1742-6596/3024/1/012001

Turnout Switch Machine Fault Diagnosis Based on 1DCBA

Zheyu Yuan, Baofu Qin, Deqiang He*, Yanbo Wang, Jiayang Zhao

Guangxi Key Laboratory of Manufacturing System & Advanced Manufacturing Technology, School of Mechanical Engineering of Guangxi University, Nanning 530004, China

*Corresponding author's e-mail: hdqlqy@gxu.edu.cn

Abstract. Turnout switch machine fault data is characterized by limited quantity and imbalanced class distribution. This limits the use of deep learning in fault diagnosis applications. Therefore, this paper presents a fault diagnosis method for turnout switch machines based on the 1DCBA model. Firstly, to mitigate the problem of insufficient fault data, the KMeans-SMOTE is employed to oversample the current data, effectively expanding the dataset. Secondly, a 1D-CNN is utilized to capture local features, and it is combined with a BiLSTM to exploit the temporal dependencies within the data. Finally, an attention mechanism is introduced to improve the model's classification ability and fault diagnosis performance by allowing it to focus on the most critical features. The results of experiments indicate that the suggested approach attains a fault diagnosis accuracy of 100%.

1. Introduction

Urban rail transit has emerged as an essential transportation mode in major cities, owing to its safety, efficiency, and high passenger capacity. [1]. As rail transit expands rapidly and passenger flow increases sharply, the safety of train operations has become a growing concern[2]. Faults in turnout switch machines account for over 40% of all signaling equipment failures[3]. Therefore, conducting in-depth research into fault diagnosis technologies for turnout switch machines not only facilitates timely detection and accurate identification of potential faults, effectively ensuring the operational safety of trains, but also significantly reduces maintenance costs, optimizes railway transportation organization, and consequently improves the overall operational efficiency and service quality of railway transportation.

Researchers have extensively investigated fault diagnosis methods for turnout switch machines, primarily emphasizing machine learning and deep learning approaches. Fault diagnosis methods based on machine learning first extract features from turnout switch machine signals and then classify these features. Wang et al.[4]segment the power signals of turnout switch machines and extract time-domain features, proposing a fault diagnosis method based on the Cost-Sensitive Extreme Learning Machine. Sun et al.[5] computed the Wavelet Packet Energy Entropy and Multi-scale Fractional Permutation Entropy (MFPE) from turnout switch machine signals to construct a more discriminative feature set. Subsequently, an optimized Support Vector Machine (SVM)was employed for fault classification, significantly enhancing the accuracy of the diagnostic results. Although traditional machine learning methods have yielded good diagnostic results, the feature extraction process is complicated and heavily dependent on expert knowledge.

Deep learning, with its deeper network structure, can automatically learn and extract features without the need for manual selection of specific information. Chen et al.[6]proposed a fault diagnosis method

Journal of Physics: Conference Series **3024** (2025) 012001 doi:10.1088/1742-6596/3024/1/012001

for metro turnout switch machines using a convolutional autoencoder, which converts current data into 2D images for fault detection via the autoencoder. Huang et al.[7] introduced a fault diagnosis method for turnout switch machines that leverages Convolutional Neural Networks (CNN) and self-generated samples, transforming current signals into images for fault recognition through CNN. Li et al.[8] proposed an enhanced autoencoder to reduce noise and extract deep features, integrating data augmentation techniques to perform fault diagnosis for turnout switch machines. Although these methods are highly adaptable and valuable, they have limitations in capturing temporal information and addressing data imbalance. Thus, it is essential to develop a diagnostic method that retains temporal information while effectively handling data imbalance.

To address this, this paper proposes a One-Dimensional CNN integrated with Bidirectional Long Short-Term Memory and Attention mechanism (1DCBA) fault diagnosis algorithm aimed at improving the accuracy of turnout switch machine fault diagnosis. The method leverages the KMeans-Synthetic Minority Over-sampling Technique (KMeans-SMOTE) to solve the data imbalance issue while combining 1D-CNN and BiLSTM to enhance feature extraction capabilities. The 1D-CNN extracts local features, while the BiLSTM captures temporal dependencies through bidirectional learning. The attention mechanism then weights and fuses the BiLSTM outputs, optimizing the extraction of temporal information. Experimental results demonstrate that this method performs excellently in terms of fault diagnosis accuracy and loss values, effectively addressing data imbalance and the loss of temporal information.

2. Analysis of the Turnout Switch Machine Current Curve

2.1. Analysis of the Normal Current Curve in Turnout Switch Machines

The ZYJ series turnout switch machines employ an electro-hydraulic drive principle. Figure 1 illustrates the specific structural schematic of the ZYJ7 model turnout switch machines selected in this study.

No.	Component Name	No.	Component Name
1	Locking Rod Set	10	Oil Cylinder
2	Inertial Wheel	11	Locking Block
3	Motor	12	Secondary Action Valve
4	Oil Injection Hole	13	Locking Pillar
5	Overflow Valve	14	Hollow Cylinder Set
6	Oil Pump	15	Action Plate
7	Oil Indicator	16	Roller
8	Electrical Connection Set	17	Observation Window
9	Protective Tube	18	Circuit Breaker

Figure 1. ZYJ7 turnout switch machines

During train operations, the signal monitoring system continuously collects critical operational parameters such as power and current from the turnout switch machines, generating corresponding data records. Based on the operational mechanism of turnout switch machines, the entire operational cycle can typically be divided into three distinct stages: unlocking, switching, and locking. Figure 2 illustrates the three-phase current curves of the ZYJ7 turnout switch machine during a complete normal operation cycle. In this figure, the horizontal axis indicates the sequence of sampling points collected at a sampling frequency of 50 Hz, while the vertical axis represents measured current values in amperes (A). Through this illustration, the dynamic characteristics of current variations at each operational stage can be clearly observed, providing crucial data foundations for subsequent fault diagnosis analysis.

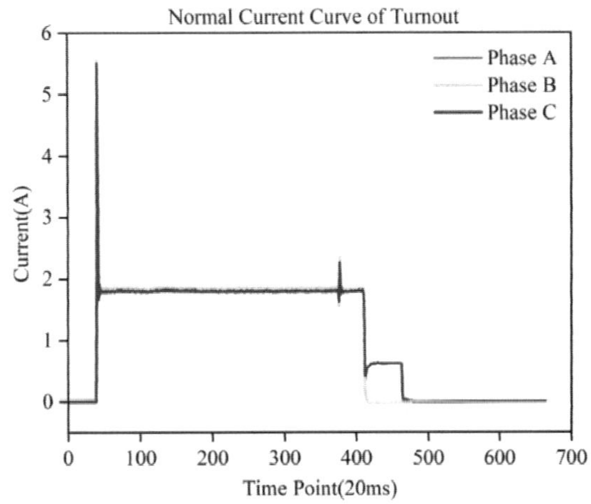

Figure 2. Three-phase action current curve of ZYJ7 switch

2.2. Investigation of Typical Fault Types in Turnouts

To obtain a diverse set of real-world operational data, this study collected and accumulated current signal data of turnout switch machines from signal monitoring systems across multiple railway stations. Through systematic data cleaning, classification, and analysis, several current curve samples containing fault characteristics were identified. As a result, six representative fault types of the ZYJ7 turnout switch machine were extracted and denoted as F1 through F6. These typical fault categories reflect distinct abnormal operating states under various conditions, as detailed in Table 1, and provide a solid data foundation for the subsequent development of fault identification and classification models.

Table 1. Typical faults of ZYJ7 witch

Fault Type	Fault Cause
F1	A-phase action circuit fault
F2	DQJ coil does not rotate
F3	Diode open circuit
F4	Jamming fault
F5	Stuck gap fault
F6	Repeatedly manipulating switches

3. Algorithm Theory

3.1. Data Enhancement Algorithm

In practical engineering applications, imbalanced datasets are a common and challenging issue. For instance, in this study, the three-phase current data collected from turnout switch machines under normal operating conditions are significantly more abundant than those corresponding to various fault states, resulting in a typical class imbalance. Such an uneven data distribution causes diagnostic models to become biased toward the majority class during training, thereby reducing their ability to accurately identify minority-class faults and ultimately compromising both diagnostic accuracy and model robustness.

KMeans-SMOTE is a sample augmentation method that combines KMeans clustering and SMOTE. First, the KMeans algorithm is used to divide the data into K clusters based on similarity. Then, the SMOTE algorithm generates synthetic samples in the neighborhood of minority class samples. Finally, the generated samples are merged with the original dataset, balancing the class distribution within the dataset. This method retains the local structure of the data and effectively addresses the issue of class imbalance. Figure 3 illustrates the principle of KMeans-SMOTE.

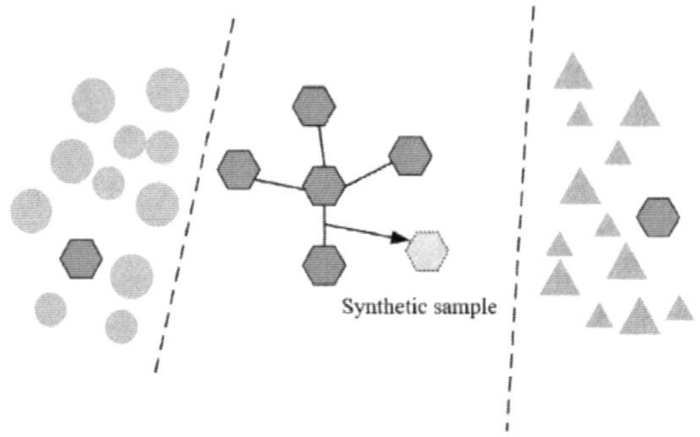

Figure 3. KMeans-SMOTE Synthetic Sample Generation Diagram

3.2. 1DCBA

Figure 4 depicts the 1DCBA-based fault diagnosis model for turnout switch machines, integrating 1D-CNN, BiLSTM, and Attention Mechanism to diagnose faults via hierarchical feature extraction and temporal modeling.

CNN is a supervised learning model with multiple layers consisting of neurons that have trainable weights and biases. Each neuron receives input data and extracts local features through convolution operations; after processing with activation functions, these features form feature maps that are subsequently used for classification tasks. When processing sequential data, BiLSTM, an improved version of the traditional LSTM network, introduce bidirectional information flow to simultaneously capture dependencies from both past and future data. While traditional LSTM relies solely on past information for prediction, BiLSTM utilizes both forward and backward propagation techniques to fully integrate historical data and future context, thereby achieving higher accuracy in predicting fault patterns in sequential data.

The attention mechanism dynamically assigns weights to select important features. In sequential data, certain time steps carry more diagnostic significance than others. By computing the 'importance' weight for each time step, the attention mechanism directs the model's focus to key moments or features, thereby improving the fault diagnosis accuracy.

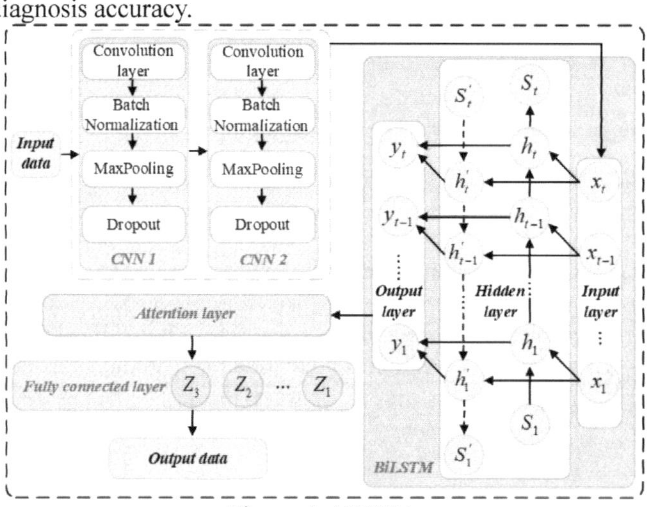

Figure 4. 1DCBA

4. Experiments

4.1. KMeans-SMOTE-based Data Augmentation Method for Turnout Data

The data used in this study were collected from Nanning Metro Line 4, China. The dataset consists of

200 normal samples and 50 samples for each fault type, as presented in Table 2. Different fault types have varying durations from initiation to failure, leading to discrepancies in the length of the data for each type. Specifically, the "No Fault" samples have a length of 665, while the lengths for F1 to F6 are 500, 335, 650, 895, 610, and 660, respectively. Therefore, data preprocessing is required. Firstly, to ensure the validity of the data, all leading and trailing zero values in the samples are removed. Since single-phase current data cannot fully capture fault information, the three-phase current data of each sample is fused using a weighted average method. The weights for phases A, B, and C are set to 3.34:3.33:3.33, respectively. Secondly, the KMeans-SMOTE method is applied to the fused data for oversampling, increasing each fault type to 200 samples, thereby enhancing the data volume and balancing the dataset. Finally, the data are aligned and processed using the Z-score normalization method to eliminate differences in scales, thereby enhancing the stability and performance of model training. The preprocessed data length is 618 for all samples.

Table 2. Data Summary for Fault Classification

Faulty modes	Label	Sample	Sample Length
No Fault	0	200	665
F1	1	50	500
F2	2	50	335
F3	3	50	650
F4	4	50	895
F5	5	50	610
F6	6	50	660

4.2. Fault Diagnosis Approach Using the 1DCBA Model

The model was constructed in a Python 3.8 environment, utilizing Keras as the deep learning framework. The Adam optimizer was chosen, and its learning rate was configured as 0.0001. Table 3 presents the details of the model developed in this study. Figure 5 illustrates the impact of different Dropout parameters on training accuracy. When Dropout = 0.5, the model converges quickly during training and maintains stability, achieving accuracy close to 1.0.

Table 3. 1DCBA structure

Layer	Setting	Input Shape	Output Shape
InputLayer	----	(----,618,1)	(----,618,1)
Conv1D	(32, kernel size=5)	(----,618,1)	(----,618,32)
BatchNormalization	----	(----,618,32)	(----,618,32)
MaxPooling1D	(pool size=2)	(----,618,32)	(----,309,32)
Dropout	0.5	(----,309,32)	(----,309,32)
Conv1D	(64, kernel size=5)	(----,309,32)	(----,305,64)
BatchNormalization	----	(----,305,64)	(----,305,64)
MaxPooling1D	(pool size=2)	(----,305,64)	(----,152,64)
Dropout	0.5	(----,152,64)	(----,152, 64)
Bidirectional (LSTM)	64, return	(----,152,64)	(----,152,128)
Dropout	0.5	(----,152,128)	(----,152,128)
Attention	----	(----,152,128)	(----,152,128)
Flatten	----	(----,152,128)	(----, 19456)
Dense	128	(----,19456)	(----,128)
Dropout	0.5	(----,128)	(----,128)
Dense (Softmax)	7	(----,128)	(----,7)

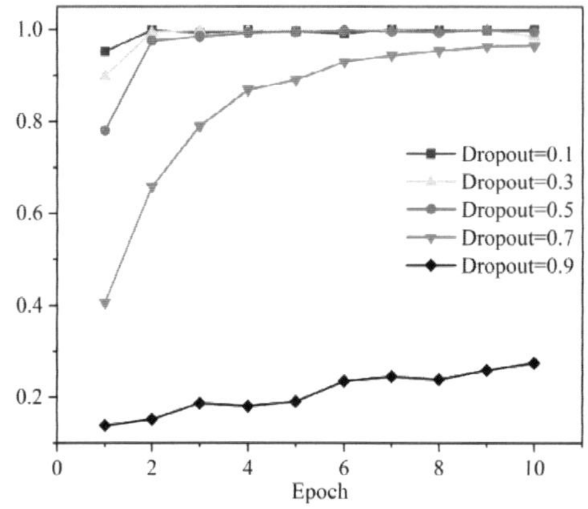

Figure 5. Effect of Different Dropout Rates on Training Accuracy

4.3. Insights from Experimental Analysis

To assess the performance of the 1DCBA model and analyze the effect of imbalanced datasets on diagnostic accuracy, experiments were conducted, as shown in Figures 6(a) and (b). Among them, the 1DCBA is denoted by the red line. UnB-1DCBA represents the 1DCBA model without KMeans-SMOTE preprocessing, indicated by the green line. CNN is represented by the blue line. The experiment was repeated 50 times, and the averaged results were utilized for analytical purposes.

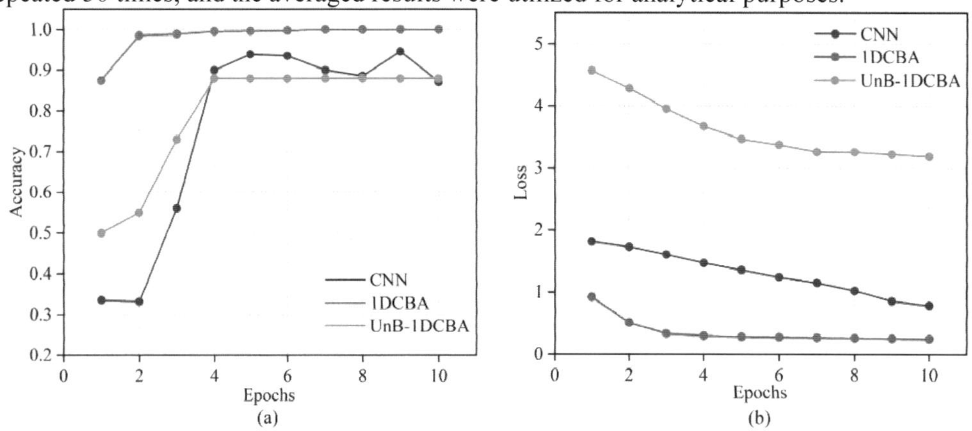

Figure 6. (a)Accuracy Curve; (b)Loss Curve

The experimental outcomes show that the 1DCBA model exhibits remarkable performance regarding both accuracy and loss convergence speed. The model's accuracy approaches 100%, and the loss decreases swiftly, maintaining a low value throughout the training process. Under the influence of imbalanced data, the accuracy of UnB-1DCBA reached 88%. This indicates the significant effect of the KMeans-SMOTE preprocessing. The traditional CNN model, lacking temporal feature extraction and attention mechanisms, achieved an accuracy of 87%. This demonstrates the powerful feature extraction capability of the 1DCBA model, which can improve the accuracy of turnout switch machine fault diagnosis. Figure 7 (a), (b), and (c) present the confusion matrices for 1DCBA, CNN, and UnB-1DCBA, respectively. Each row represents the actual fault categories of the turnout switch machine, while each column represents the fault categories predicted by the model. Each value within the matrix indicates the number of samples of a true fault type diagnosed as the corresponding predicted category. It can be observed that for the CNN and UnB-1DCBA models, the F3 fault is almost unclassifiable. The method proposed in this paper accurately classifies all fault types, demonstrating its effectiveness.

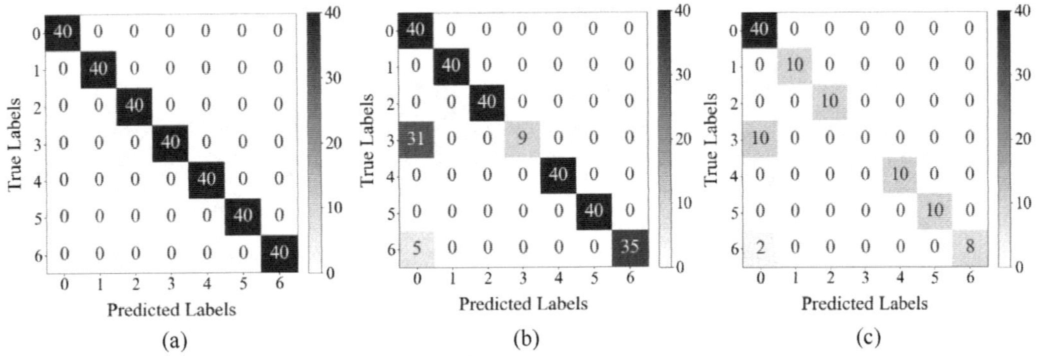

Figure 7. Confusion Matrices

5. Conclusion

To address the issue of low fault diagnosis accuracy in current turnout switch machine fault diagnosis methods caused by insufficient operational data and severe data imbalance, this paper proposes an intelligent fault diagnosis approach for turnout switch machines based on the 1DCBA model. Specifically, for the ZYJ7 turnout switch machine, time-domain features from three-phase current signals are extracted and subsequently fused through a weighted averaging method, thereby enhancing the overall feature representation capability of the model. Moreover, to mitigate the negative effects resulting from imbalanced class distributions, the KMeans-SMOTE algorithm is employed to oversample minority-class fault data, reducing the model bias caused by data imbalance and improving diagnostic accuracy. Finally, a hybrid deep learning model termed 1DCBA, integrating 1D-CNN, BiLSTM, and the Attention mechanism, is developed to further enhance the depth and precision of feature learning, significantly improving the fault diagnosis performance. Experimental results confirm that the proposed approach demonstrates excellent performance and robust generalization ability in fault diagnosis tasks of turnout switch machines.

Acknowledgments

The research was supported by the National Natural Science Foundation of China [Grant Number U22A2053], Major Talent Project in Guangxi Zhuang Autonomous Region, the Major Science and Technology Project of Guangxi Province of China [Grant Number Guike AA23073019], Innovation Project of Guangxi Graduate Education [Grant Number YCBZ2024013] and Guangxi Manufacturing Systems and Advanced Manufacturing Technology Key Laboratory Director Fund [Grant Number 23-050-44-S015].

References

[1] Chen, C., Xu, Z., Mei, M., Huang, K., & Lo, S. M. (2024). Fault Diagnosis Scheme for Railway Switch Machine Using Multi-Sensor Fusion Tensor Machine. Computers, Materials & Continua, 79(3).

[2] Lao, Z., He, D., Wei, Z., Shang, H., Jin, Z., Miao, J., & Ren, C. (2023). Intelligent fault diagnosis for rail transit switch machine based on adaptive feature selection and improved LightGBM. Engineering Failure Analysis, 148, 107219.

[3] Sun, Y., Cao, Y., & Li, P. (2022). Contactless fault diagnosis for railway point machines based on multi-scale fractional wavelet packet energy entropy and synchronous optimization strategy. IEEE Transactions on Vehicular Technology, 71(6), 5906-5914.

[4] Wang, Z., Wang, N., Zhang, H., Jia, L., Qin, Y., Zuo, Y., ... & Dong, H. (2023). Segmentalized mRMR features and cost-sensitive ELM with fixed inputs for fault diagnosis of high-speed railway turnouts. IEEE Transactions on Intelligent Transportation Systems, 24(5), 4975-4987.

[5] Sun, Y., Cao, Y., & Li, P. (2022). Contactless fault diagnosis for railway point machines based on multi-scale fractional wavelet packet energy entropy and synchronous optimization strategy.

IEEE Transactions on Vehicular Technology, 71(6), 5906-5914.

[6] Chen, C., Li, X., Huang, K., Xu, Z., & Mei, M. (2023). A convolutional autoencoder based fault detection method for metro railway turnout. CMES-Computer Modeling in Engineering and Sciences, 136(1), 471-485.

[7] Huang, S., Yang, L., Zhang, F., Chen, W., & Wu, Z. (2020). Turnout fault diagnosis based on CNNs with self-generated samples. Journal of Transportation Engineering, Part A: Systems, 146(9), 04020105.

[8] Li, M., Hei, X., Ji, W., Zhu, L., Wang, Y., & Qiu, Y. (2022). A Fault-Diagnosis Method for Railway Turnout Systems Based on Improved Autoencoder and Data Augmentation. Sensors, 22(23), 9438.

Research on Intelligent Additive Manufacturing of Mullite Whiskers

Wenkai Wen[1], Xiaohua Gu[1,2,3,4*], Canyan Long[2,3], Heng Li[5], Zhuyu Chen[2], Qingyong Su[2]

1. School of Artificial Intelligence, Guilin University of Aerospace Technology, Guilin, 541004, P.R. China

2. School of Energy and Building Environment, Guilin University of Aerospace Technology, Guilin 541004, P.R. China

3. Guangxi Key Laboratory of Special Engineering Equipment and Control, Guilin University of Aerospace Technology, Guilin 541004, P.R. China

4. University Engineering Research Center of Green Upgrade Key Technology for Energy Industry, Guilin 541004, P.R. China

5. School of Mechanical Engineering, Guangxi University, Guangxi, 530004, P. R. China

*Corresponding author's e-mail: 2022049@guat.edu.cn; Tel.: +86-18115098088

Abstract: Mullite whiskers exhibit exceptional properties such as mechanical durability, oxidative stability, low thermal expansion coefficient, high-temperature tolerance, and excellent thermal shock resistance, making them widely used in aerospace, mechanical manufacturing, and other fields. This study investigates the synergistic effects of thermal processing parameters and molybdenum oxide catalyst concentration on mullite whisker's evolution. We propose a predictive model for mullite whisker preparation based on integrating Particle Swarm Optimization (PSO) and Backpropagation Neural Network (BPNN). The model employs a BPNN optimized by the PSO algorithm, leveraging historical experimental data to predict the influence of varying temperatures and molybdenum oxide concentrations on whisker growth without additional experiments. This approach significantly reduces experimental time and costs. Comparative analysis with other nonlinear regression models demonstrates that the proposed model exhibits high accuracy and generalization capability in predicting the impact of temperature and molybdenum oxide concentration on mullite whisker growth. Thus, it represents an effective method for predicting the growth of mullite whiskers.

1. Introduction

Mullite, recognized for its superior properties as a high-temperature structural ceramic material, finds extensive applications in aerospace, mechanical manufacturing, and other fields[1]. The crystal structure of mullite comprises aluminium-oxygen tetrahedra [AlO4] and silicon-oxygen tetrahedra [SiO4], which are arranged along the c-axis to form double chains. These double chains are interconnected by aluminium-oxygen octahedra [AlO6]. In an environment conducive to free growth, mullite tends to grow preferentially along the c-axis, enabling the synthesis of needle-like mullite whiskers by controlling the synthesis conditions. Due to their small diameter, these whiskers are less likely to contain defects commonly found in larger crystals, and their atoms are highly ordered. As a result, the strength of mullite whiskers approaches the theoretical value of a perfect crystal.

Currently, there are lots of methods for preparing mullite whiskers. A common approach is the high-energy ball milling method based on solid-state reactions, which involves uniformly mixing raw materials followed by direct sintering. However, this method requires temperatures as high as 1200°C or above [2]. Another method is the catalytic approach based on gas-solid reactions, where catalysts are added to generate gaseous substances during the reaction, promoting the anisotropic growth of mullite whiskers. This method also necessitates high temperatures, typically no less than 1200°C and even up to 1400°C, to produce mullite whiskers with a high aspect ratio [3]. Alternatively, based on liquid-solid reactions, the molten salt method involves adding salts soluble in water at room temperature and melting at high temperatures. This method enables the synthesis of mullite whiskers at temperatures below 1200°C. The molten salt exists in a liquid state within the reaction system at certain temperatures, accelerating the reaction of raw materials and reducing the synthesis temperature [4][5][6].

Among the above methods, the temperature required for synthesizing mullite whiskers through the molten salt remains significantly lower than conventional techniques. This study will use the molten salt method to prepare mullite whiskers. During the preparation process, the size of the mullite whiskers varies depending on the reaction conditions, and a specific relationship exists between these conditions and the whisker growth. However, searching for this relationship through repeated trials is time-consuming and laborious, leading to significant material waste. To address this issue, we propose establishing a mathematical model to find the relationship between reaction conditions and whisker size. Artificial neural networks, as a novel machine learning model, can map any input-output relationships and can be utilized to explore the correlation between the preparation conditions of mullite whiskers and their resulting size.

Therefore, we construct an artificial neural network model based on previous experimental data and use this data to predict the mullite whiskers' length and aspect ratio under various temperatures and molybdenum catalyst concentrations. Our goal is to improve production efficiency and reduce the experimental cost.

2. Preliminaries

2.1. BP Neural Network

The Backpropagation Neural Network (BPNN) is an artificial neural network model widely used for pattern recognition, classification, and function approximation[7]. Trained through a learning algorithm, it possesses activation functions and nonlinear processing capabilities, making it suitable for solving complex problems. As illustrated in Figure 1, the network consists of multiple neurons arranged in layers, encompassing an input layer, one or more intermediate hidden layer(s), and an output layer. The input layer, the first layer of the network, is in charge of acquiring incoming information from external data sources, such as temperature and molybdenum oxide catalyst concentration, in this study. The intermediate layer(s) (hidden layer(s)) execute feature extraction and representation learning, mapping input data to higher-level feature representations to enhance the network's ability for pattern recognition and complex function approximation. In this context, the output layer generates the ultimate output results: mullite whiskers' length and aspect ratio. During forward propagation, input signals propagate sequentially through the neural network's hierarchical architecture, traversing from the initial input layer, passing through one or more hidden layers for feature transformation, and ultimately reaching the terminal output layer for final prediction generation. When the output results fail to meet the desired criteria, the network enters the backpropagation process, which consists of two main steps: first, comparing the predicted values with the measured actual values to calculate the error generated during forward propagation; second, propagating this error backwards through the network to generate weight adjustments. This iterative process continues until the error converges to the desired accuracy. At this point, the model completes its learning.

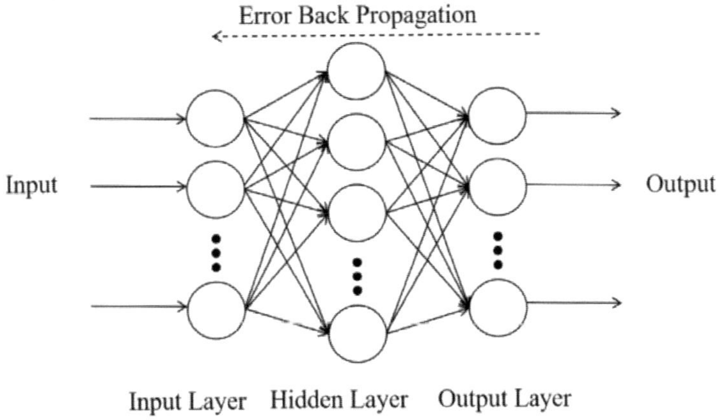

Figure 1. The BPNN model

2.2. *Particle Swarm Optimization (PSO)*

The PSO algorithm is an intelligent optimization algorithm that updates the position and velocity of particles by tracking individual optimal particles and group optimal particles to acquire the optimal solution. During each iteration, particles update their states by tracking two 'extremum values' (p_{best}, g_{best}). After finding these optimal values, the specific updated formulas are shown in equations (1) and (2).

$$v_i^{t+1} = \omega v_i^t + c_1 r_1 (p_{i,best}^t - x_i^t) + c_2 r_2 (g_{best}^t - x_i^t) \tag{1}$$

$$x_i^{t+1} = x_i^t + v_i^{t+1} \tag{2}$$

Where: v_i^t represents the velocity of the particle i at the t-th iteration; x_i^t denotes the position of the particle i at the t-th iteration; ω is the inertia weight; c_1 and c_2 are acceleration coefficients; r_1 and r_2 are two stochastic numbers with values in the range [0, 1]. $p_{i,best}^t$ signifies the individual best position found by particle i after the t-th iteration, while g_{best}^t represents the global best position found after the t-th iteration.

Neural network parameters, such as initial weights and threshold values, are typically assigned through randomized initialization manners. Such stochastic initialization methods can easily result in training instability and suboptimal convergence performance during the network optimization phase, significantly affecting the model's prediction accuracy. The PSO algorithm was systematically implemented to optimize initial synaptic weights and neuronal threshold values to mitigate potential biases induced by stochastic initialisation parameters on model performance. This computational strategy significantly enhances both training convergence rate and predictive accuracy in the BP neural network architectures while maintaining algorithmic stability. In this paper, particles in the PSO algorithm represent the BP neural network's initial weights and threshold values. Through iterative optimization, superior initial weights and threshold values are obtained, making the optimized BP NN more accurate in predictions on the sample data.

3. Proposed method

3.1. *Dataset preparation and preprocessing*

All experiments were conducted using Matlab 2023b on the PC to verify the proposed algorithm's efficiency. Forty-two samples from the previously prepared mullite whiskers were selected as training and testing samples for the PSO-BPNN. The inputs were temperature and the concentration of the molybdenum oxide catalyst, and the outputs were the whiskers' length and aspect ratio, as shown in Table 1.

Journal of Physics: Conference Series **3024** (2025) 012002 doi:10.1088/1742-6596/3024/1/012002

Thirty samples were stochasticly partitioned from them as the training set using the *randperm* function, with the remaining twelve samples reserved as the test set. The *mapminmax* function was employed to normalize the data to ensure data consistency and better network convergence, thus preventing the differences among features from exerting unnecessary impacts on the training process.

Table 1 Experiment datasets

Temperatures (°C)	Catalyst Concentration	Length (μm)	Aspect Ratio
700	0. 05	8. 9798	19. 5212
	0. 07	9. 8838	21. 0292
	0. 09	9. 4392	20. 5200
	0. 11	9. 0960	20. 2133
	0. 13	9. 0045	19. 5750
	0. 15	8. 9470	19. 4500
750	0. 05	9. 1531	19. 8981
	0. 07	10. 4216	21. 2685
	0. 09	9. 9129	21. 0913
	0. 11	9. 1275	20. 2833
	0. 13	8. 8565	19. 6811
	0. 15	8. 8221	19. 6047
800	0. 05	9. 4470	20. 9933
	0. 07	10. 7415	21. 9214
	0. 09	9. 8231	21. 8292
	0. 11	9. 5128	20. 6800
	0. 13	9. 0496	20. 1101
	0. 15	8. 9425	19. 8722
850	0. 05	10. 6918	21. 3836
	0. 07	12. 1477	22. 4957
	0. 09	10. 7106	21. 8583
	0. 11	9. 7737	21. 1016
	0. 13	9. 5317	20. 5150
	0. 15	9. 3726	19. 9417
900	0. 05	9. 5527	20. 7667
	0. 07	11. 5395	22. 1914
	0. 09	10. 7567	21. 5133
	0. 11	9. 5740	20. 3867
	0. 13	9. 4909	20. 1933
	0. 15	9. 3920	19.5667
950	0. 05	9. 5508	20. 7625
	0. 07	11. 2285	22. 0167
	0. 09	10. 2320	21. 3167
	0. 11	9. 4728	20. 3600
	0. 13	9. 2575	20. 1250
	0. 15	8. 9853	19. 5333
1000	0. 05	9. 4428	20. 5278
	0. 07	11. 1128	21. 7897
	0. 09	10. 1600	21. 1667
	0. 11	9. 4522	20. 1111
	0. 13	9. 2120	19. 6000
	0. 15	8. 9275	19. 4076

3.2. Construction of BPNN
The hidden layer of the BPNN is designed to contain five neurons, and each node utilizes the hyperbolic tangent function. The output layer is configured with a linear activation function. The Levenberg-Marquardt gradient descent algorithm is selected as the training method. The maximum iteration number is set to 1000. The error threshold is set to 1×10^{-6}. The network conducts back propagation training to approximate the functional correspondence between input and output variables using the training set. The network performance is measured by minimizing the Mean Squared Error (MSE) during the training [8].

3.3. Steps of the PSO-BPNN algorithm
The flow of optimizing the BP neural network by the PSO algorithm is shown in Figure 2.

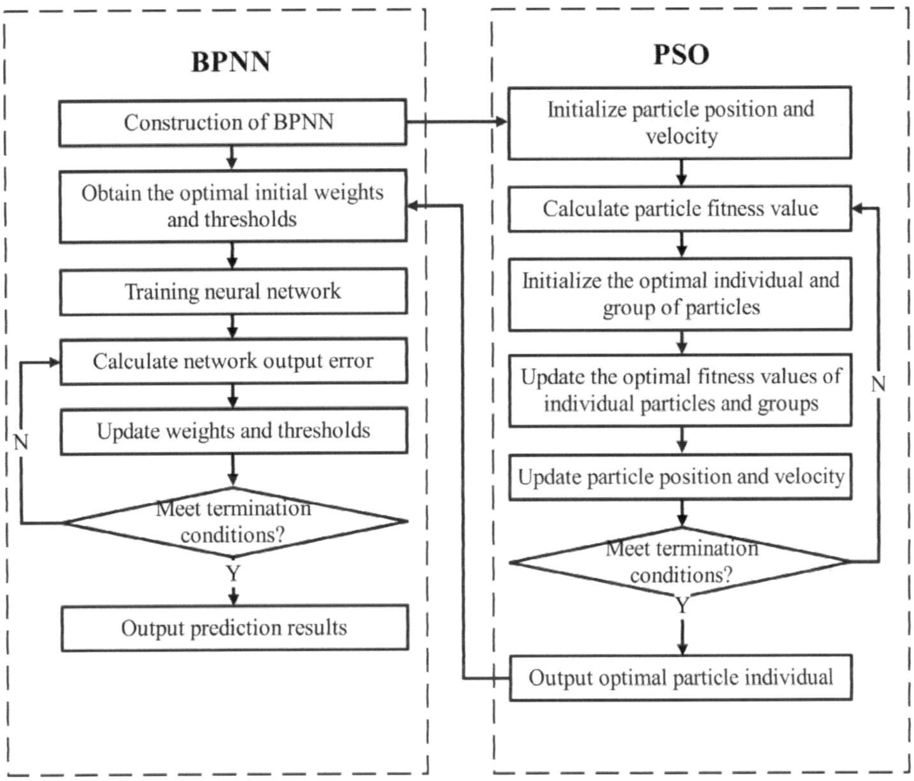

Figure 2. The graph of the PSO-BPNN algorithm

The experimental procedure is implemented through the following sequential steps:

Step (1): Construct the BPNN and initialize the group size, particle position and velocity, fitness function, and iteration numbers of the PSO algorithm.

Step (2): Calculate each particle's fitness value using its position as the network's weights and threshold values.

Step (3): Compare each particle's fitness value with its fitness value and select the higher individual fitness value as optimal.

Step (4) : Implement the iterative updating for particle positions and velocity according to equations (1) ~ (2).

Step (5): If the PSO algorithm meets the termination condition (meets the error precision requirement or reaches the maximum number of iterations), output the results; otherwise, return to step (2).

Step (6): The BPNN obtains the individual particles optimized by the PSO algorithm, uses them as the initial weights and threshold values to complete the model training, and outputs the prediction results

Based on the PSO-BPNN prediction model constructed above, the experimental temperature and molybdenum catalyst concentration are normalized and input into the model for learning and training. The output results are the corresponding mullite whiskers' length and aspect ratio.

4. Experiment results

4.1. *Evaluation indicators*

4.1.1. Coefficient of Determination (R^2). R^2 is a comprehensive indicator that measures a model's degree of data fitting, with a value range of $R^2 \in [0,1]$. Higher R^2 values indicate superior goodness-of-fit for prediction models. The calculation method is as follows:

$$R^2 = 1 - \frac{\sum_1^n (y_i - \hat{y}_i)^2}{\sum_1^n (y_i - \bar{y})^2} \tag{3}$$

Where n represents sample size, y_i represents the actual observed value, $\hat{y}_i$ signifies predicted value, and $\bar{y}$ corresponds to the average of the actual values.

4.1.2. Mean Absolute Error (MAE). MAE quantifies the mean absolute discrepancy between predicted values and actual values. It is a nonnegative value, and a more petite MAE indicates a better model. The calculation method is as follows:

$$MAE = \frac{1}{n}\sum_1^n |y_i - \hat{y}_i| \tag{4}$$

4.1.3. Mean Squared Error (MSE). MSE quantifies prediction accuracy by computing the square of the difference between each sample's predicted values and actual values. MSE $\in [0,+\infty]$, and a smaller MSE indicates a better model and stronger predictive ability. The calculation method is as follows:

$$MSE = \frac{1}{n}\sum_1^n (y_i - \hat{y}_i)^2 \tag{5}$$

4.1.4. Root Mean Squared Error (RMSE). RMSE is the square root of MSE. The calculation method is as follows:

$$RMSE = \sqrt{\frac{1}{n}\sum_1^n (y_i - \hat{y}_i)^2} \tag{6}$$

4.2. *Results analysis*

The Matlab R2023b platform was used to predict the mullite whiskers' length and aspect ratio on the test dataset. The experimental results are shown in Figures 3 and 4.

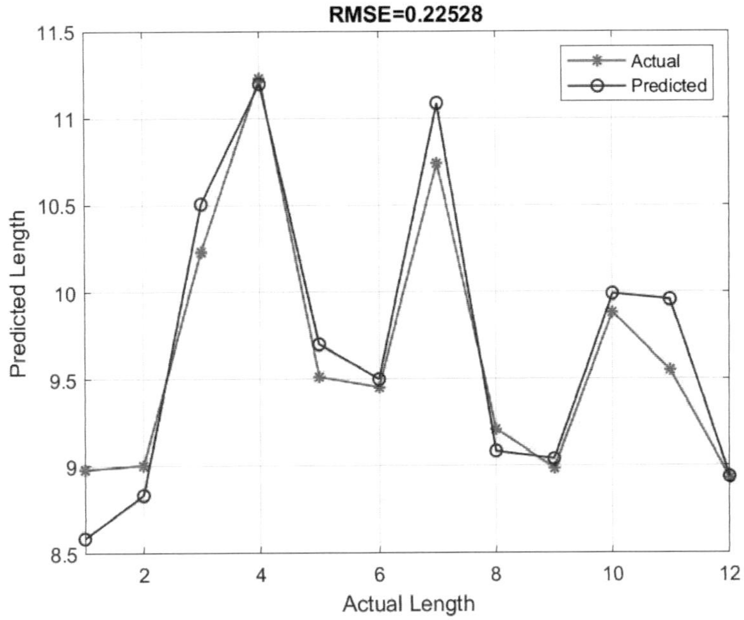

Figure 3. Predicted length by the PSO-BPNN

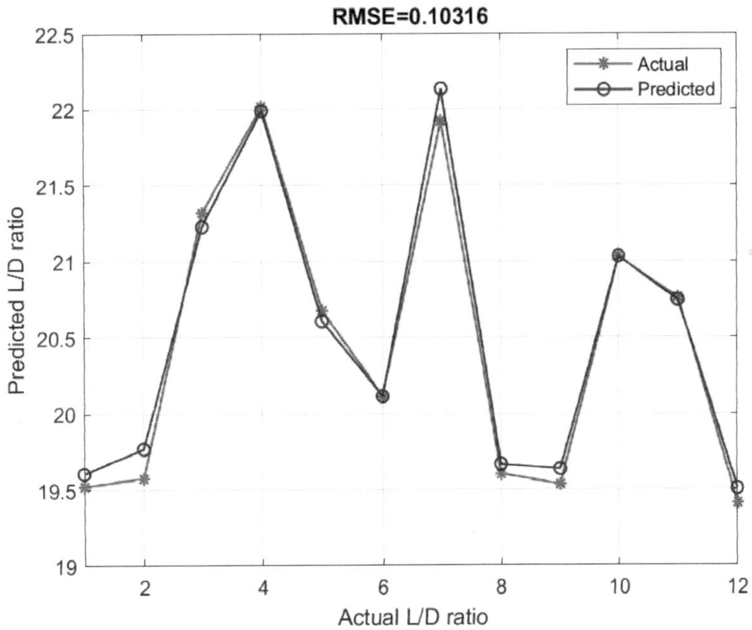

Figure 4. Predicted aspect ratio by the PSO-BPNN

Through comparison, the model's predicted values are very close to the actual sample values. The RMSE are 0.22528 and 0.10316, respectively, indicating that the constructed PSO-BPNN has a good fitting effect. In addition, the model's R-value (correlation coefficient) on the training dataset reaches 0.97465, and on the test dataset, it is 0.96396. This proves that the PSO-BPNN can map the input and output relationship well. The results are shown in Figures 5 and 6.

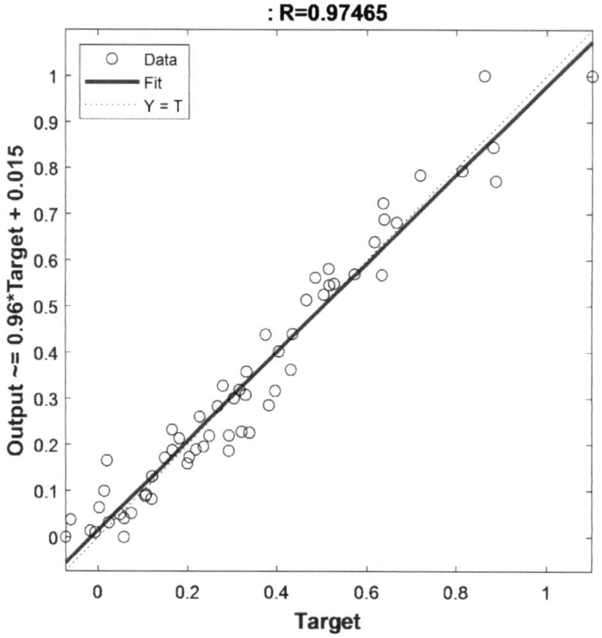

Figure 5. The Illustration of the correlation coefficient on the training dataset

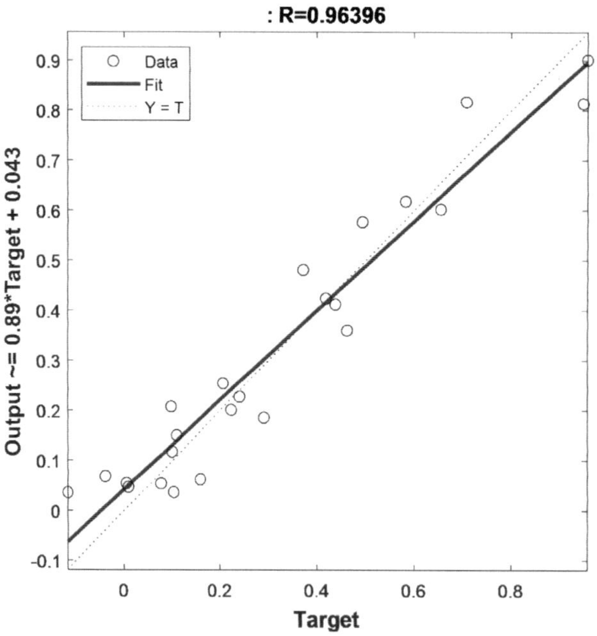

Figure 6. The Illustration of the correlation coefficient on the test dataset

A comparative analysis was conducted on the prediction results of four methods to compare their prediction accuracy: multiple linear regression (MLR), Random Forest (RF) [9], Radial Basis Function Neural Network (RBFNN) [10], and BPNN. The results are shown in Table 2.

Table 2. Experimental results of the different models

Length / Aspect Ratio	Method				
	MLR	RF	RBFNN	BPNN	PSO-BPNN
R^2	0. 7590 / 0. 9377	0. 7713 / 0. 8236	0. 7535 / 0. 9171	0. 8424 / 0. 9515	0. 9012 / 0. 9875
MAE	0. 3055 / 0. 1780	0. 2061 / 0. 1972	0. 3675 / 0. 2203	0. 2542 / 0. 1777	0. 1792 / 0. 0800
MSE	0. 1477 / 0. 0489	0.0766 / 0.0710	0. 2238 / 0. 0710	0. 0874 / 0. 0430	0. 0508 / 0. 0106
RMSE	0. 3843 / 0. 2211	0.2786 / 0.2664	0. 4731 / 0. 2665	0. 2956 / 0. 2073	0. 2253 / 0. 1032

As Table 2 shows, the PSO-BPNN performs outstandingly in various indicators. Its R^2 values reach 0.9012 and 0.9875, respectively, indicating that this model can explain the data well. There is a high linear correlation between the predicted values and actual values.

The values of this model's MAE, MSE, and RMSE are significantly lower than those of other models, suggesting that it has a high prediction accuracy, can effectively reduce prediction bias, and can provide relatively reliable prediction results in practical applications.

The PSO-BPNN has apparent advantages in prediction tasks and is the preferred model for handling such data.

5. Conclusions

The PSO-BPNN proposed in this paper establishes a growth prediction model for mullite whiskers under small samples. This model has significant advantages regarding R^2, MAE, and MSE. The PSO-BPNN can accurately predict the growth size of whiskers under a given temperature and molybdenum oxide catalyst concentration.

Although the PSO algorithm is employed to optimize the initial parameters of the BPNN, it remains susceptible to being trapped in the local optimum. In future work, we will improve the PSO algorithm or use other intelligent optimization algorithms, such as genetic algorithms.

Limited by experimental conditions, we use only forty-two groups of experimental data for model training and testing, and the number of samples is relatively small. Later, we will expand the data set to cover more data on mullite whisker preparation under different conditions so that the model can learn more comprehensive rules and enhance its generalization ability.

The BPNN is currently designed with five neurons in its hidden layer. Future research directions will involve systematically optimizing hidden layer neuron quantity through established neural architecture search methodologies, concurrently exploring network depth expansion through incorporating additional hidden layers, mining more complex features and relationships in data, and improving the model's expression ability.

In subsequent investigations, we intend to systematically examine the impact of additional experimental parameters, including but not limited to reaction duration and precursor particle dimensions, on the crystallization mechanisms and morphological evolution of mullite whiskers and construct a more complex and comprehensive multi-factor coupling prediction model to more accurately describe the growth process of mullite whiskers.

We will explore the application of the PSO-BPNN in predicting the preparation process of other materials, such as ceramic matrix composites, metal whiskers, and other fields. Migrate the experience and methods accumulated in the research of mullite whisker preparation and provide new ideas and technologies for the intelligent preparation of other materials.

Acknowledgments

This research was funded by the Research Foundation Ability Enhancement Project for Young and Middle-aged Teachers in Guangxi Universities under Grant (2023KY0816), University Engineering Research Center of Green Upgrade Key Technology for Energy Industry, Guangxi (202410-60). All authors have read and agreed to the published version of the manuscript.

References

[1] ZHANG L, HUANG J Y, XIAO Z H, et al. Progress in fabrication and growth mechanisms of mullite whiskers[J]. Journal of Ceramics, 2020, 41(6):880-893(in Chinese).

[2] Zhang T S, Kong L B, Du Z H, et al. Tailoring the microstructure of mechanoactivated Al2O3 and SiO2 mixtures with TiO2 addition [J]. Journal of Alloys and Compounds, 2010, 506(2):777-783.

[3] Zhang H H, Duan X K, Yu F L, et al. Effect of fluorinated catalyst on anisotropic growth of mullite whiskers[J]. Acta intraocular lens, 2019, 48(7):1326-1330.

[4] Abdullayev A, Klimm D, Kamutzki F, et al. AlF3-Assisted Flux Growth of Mullite Whiskers and their Application in Fabrication of Porous Mullite-alumina Monoliths[J]. Open Ceramics, 2021, 7(1):100145.

[5] Gao J M, Wang B, Du Z Y, et al. Molten salt synthesis of mullite whiskers entirely derived from fly ash for electronic packaging toughening ceramic applications[J]. Journal of Materials Research and Technology, 2022, 21(1):3719-3731.

[6] Abdullayev A, Zemke F, Gurlo A, et al. Low-temperature fluoride-assisted synthesis of mullite whiskers[J]. RSC Advances, 2020, 10(52):31180-31186.

[7] ZHOU Meng, LÜZhigang, DI Rouhai, et al. BP neural network modeling based on small sample data[J]. Science Technology and Engineering, 2022, 22(7):2754-2760.

[8] WANG P, FAN E. Neural network optimization method and its application in information processing[J]. Mathematical Problems in Engineering, 2021, 1(10):1-10.

[9] Thanh Noi P, Kappas M. Comparison of random forest, k-nearest neighbor, and support vector machine classifiers for land cover classification using Sentinel-2 imagery[J]. Sensors, 2017, 18(1):18.

[10] Ng K W, Huang Y F, Koo C H, et al. A review of hybrid deep learning applications for streamflow forecasting[J]. Journal of Hydrology, 2023:130141.

Interpretable Knowledge Transfer with a Novel Structure Discrepancy Metrics for Unsupervised Bearing Fault Diagnosis from Simulation to Reality

Kaitong Jia[1], Xin Wen[1], Gang Chen[1]*

[1] Shien-Ming Wu School of Intelligent Engineering, South China University of Technology, Guangzhou, Guangdong Province, 511442, P.R. China

E-mail: gangchen@scut.edu.cn

Abstract. This paper introduces a novel interpretable transfer learning framework, based on Wavelet Temporal Logic Network (WTLN) encoder, which leverages temporal logic for interpretable knowledge transfer in rolling element bearings fault diagnosis. Unlike traditional approaches, WTLN integrates Weighted Signal Temporal Logic (wSTL) with dynamic learning capabilities to robustly encode and interpret signal behaviors under varying conditions. The model uniquely combines a Temporal Logic Discrepancy Metric (TLDM) for enhanced loss function definition to address the challenges of transfer learning and knowledge structure alignment in fault diagnosis. Extensive experiments conducted with both simulated and real-world datasets show that the model is highly effective in diagnosing bearing faults under various operational conditions. By combining the WTLN with TLDM, the model achieves not only high diagnostic accuracy but also provides interpretability. This interpretability is vital for real-world applications, where it is crucial to understand the decision-making process.

1. Introduction

Given that failures in rolling element bearings (REB) can precipitate severe breakdowns or performance decline in machinery [1], there is a pressing need for intelligent fault diagnosis systems to enhance operational reliability, minimize maintenance costs, and prevent catastrophic failures. Unsupervised domain adaptation represents a crucial strategy within transfer learning, aimed at addressing the aforementioned challenges. In recent years, significant scholarly efforts have been directed towards investigating unsupervised cross-domain fault diagnosis in bearings. In 2021, Wang *et al.* [2] introduced a synthetic-to-real framework for unsupervised fault diagnosis that combines expert knowledge with domain adaptation (EKDA) to effectively bridge the gap between synthetic and real fault data, validated on laboratory and real-world datasets. In 2022, Tian *et al.* [3] proposed a new multi-source subdomain adaptation (MSSD) transfer learning method to enhance cross-domain fault diagnosis in industrial equipment by using a multi-branch network to match features across domains and conducting extensive testing on rotating machinery datasets. In 2023, Ding *et al.* [4] proposed Deep Imbalanced Domain Adaptation (DIDA) framework for fault diagnosis in bearings addresses both feature and label shifts across diverse working conditions by integrating cost-sensitive learning and categorical alignment to manage class-imbalanced label shifts. Although these methods have achieved considerable success in bearing fault diagnosis, they face two significant limitations. Firstly, these methods often function as black boxes and lack interpretability, meaning they do not provide insights into *how and why* decisions are made or reveal the underlying fault mechanisms.

To improve the interpretability of the transfer learning process, Liu *et al.* [5] introduced an explainable fault diagnosis method for oil-gas treatment stations using transfer learning, employing the two-dimensional class activation map algorithm and multi-dimensional dynamic time warping theory to elucidate the deep residual network's diagnostic process. Brito *et al.* [6] proposed FaultD-XAI, a new generic and interpretable method using transfer learning from augmented synthetic data for fault classification in rotating machinery, employing Grad-CAM and 1D CNN for post hoc interpretation to enhance diagnostic confidence. Logical inference-based methods have also been proposed to explain the decision process for bearing fault diagnosis, which involves deriving logical expressions that describe system properties from data [7]. It is often implemented in a formal language in the ante-hoc fault diagnosis methods. The fault diagnosis construction procedure can be formulated as a language generation process and the formal languages can be seen as interpretable classifiers, which provide interpretability for the fault diagnosis procedure. Recently, there has been applying temporal-logic-based formal language to diagnose faults and obtain good performance. Chen et al. [8] introduced a temporal logic neural network (TLNN) that can be interpreted as a weighted signal temporal logic. This approach preserves the beneficial properties of traditional neural networks while also providing a formal logical interpretation. Their experimental results, using real-world datasets, demonstrate that the formal language embedded within the network can offer insights into the decision-making process, enabling interpretable fault diagnosis. In a similar vein, Tian et al. [9] utilized weighted signal temporal logic (wSTL) as a formal language and developed a temporal logic network (TLN) to enable interpretable fault diagnosis of rolling element bearings. Additionally, Chen et al. [10] proposed a novel formal language for fault diagnosis, called signal spectral logic (SSL), which is inspired by signal temporal logic and defined over the spectral kurtosis of signals.

The methods mentioned previously have demonstrated robust transfer learning capabilities in various unsupervised cross-domain fault diagnosis tasks for bearings, offering a degree of interpretability. Nevertheless, there remain significant challenges that need addressing to enhance diagnostic performance:

1) Existing interpretative diagnostic approaches like Grad-CAM [6] provide insights into model decisions by emphasizing spatial or feature importance. However, they fail to adequately capture the temporal dynamics crucial for analyzing cause and effect. This limitation hampers their effectiveness in scenarios where understanding the timing of events is essential for accurate diagnosis.
2) While many studies focus on aligning the marginal distribution between source and target domains, they often overlook the structural integrity of the transferred knowledge. There is a pressing need to develop new metrics and loss functions that capture the similarities in knowledge structure across domains.

To tackle these issues, this paper introduces a novel structural discrepancy metric, termed Temporal Logic Discrepancy Metric (TLDM), to assess the similarity in knowledge structures. Additionally, we propose a transfer learning framework that utilizes formal languages for knowledge transfer and fault diagnosis, thereby elucidating the decision-making process from input to output. Our contributions are significant improvements over existing methods and include:

1) A novel neural network encoder, the WTLN encoder, which uses wSTL formulas defined over wavelet distance to encode transferred knowledge and invariant properties between domains. This makes the WTLN encoder both interpretable and comprehensible to humans.

2) An enhanced loss function incorporating TLDM has been developed to ensure simultaneous alignment of both marginal distribution and knowledge structure in an unsupervised setting. The effectiveness of the proposed framework is validated using real experimental datasets collected under various shaft speed conditions and fault magnitudes.

2. Wavelet Temporal Logic Network Encoder

When we use the wSTL to describe the behaviors of signals, learning a wSTL formula to represent the knowledge is a computationally costly process. To address this issue, the WTLN is proposed in this section to learn wSTL formulas and are robust to noise. The framework of proposed WTLN is shown in Fig.1. The details of each layer are introduced as follows.

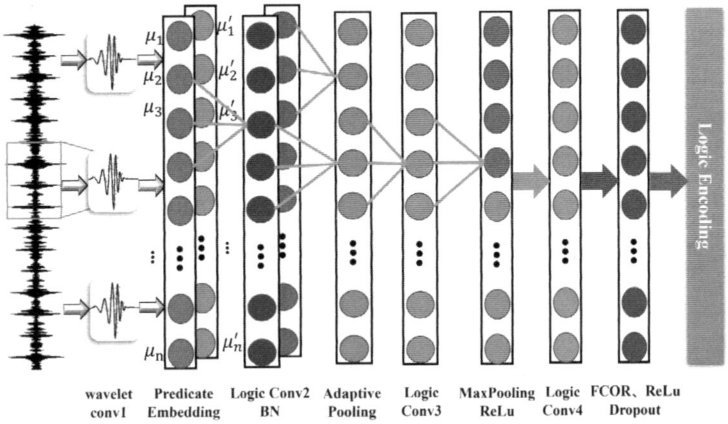

Figure 1. Structure of WTLN encoder for interpretable knowledge transfer from simulation to reality.

Input layer: The input of the network is a signal $\mathbf{x} = x(1), \cdots, x(n)$, and the input layer has n neurons. The output of $i-th$ neuron is o_i, denoted as:

$$o_i^1 = x(i) \tag{1}$$

Wavelet convolution layer: The wavelet convolution layer uses wavelet kernel as a basis, and maps the inputs from Layer 1 to the distance between the inputs and the basis. Assume the input of Layer 1 is x and the wavelet basis function is $\psi_{u,s}(t)$ as defined in [9], the output of $i-th$ neuron of Layer 2 is

$$o_i^2(t) = d(x(t), \psi_{u_i,s_i}) \tag{2}$$

where u_i denotes the translational parameter, t denotes the time,and s_i denotes a scale parameter that is inversely proportional to frequency. In this paper, the scale parameters are chosen from a set S sequentially, and $s_i = S(i)$, where $i = 1, \cdots, M$.

Predicate embedding layer: This layer constructs the basic predicates for STL formula construction, the output of the predicate embedding layer is the robustness of the predicates for each time t defined as:

$$\mu = \delta(o_i^2(t) - \lambda_i), \mu' = \delta(\lambda_i - o_i^2(t)) \tag{3}$$

where δ is the sigmoid function and λ_i is a real value.

Logic convolution layer: This layer is a convolution layer, in which the convolutional operator is applied. Compared with traditional convolution layer, the activation function applied to logic convolution layer is the robustness degree function. In this layer, we use robustness degree functions $g^{\Box}(w,\rho)$ and $g^{\Diamond}(w,\rho)$, respectively. Based on the robustness degree functions used, two types of wSTL formulas can be obtain in layer *Logic Conv2*, i.e., $\Box_{[m,h+m]}\mu$ and $\Diamond_{[m,h+m]}\mu$, respectively, where m is the time shift of the convolution process and h is the width of the filter for the convolution layer. Similarly, two types of wSTL formula can be obtain in layer *Logic Conv3*, i.e., $\Diamond_{[m,k+m]}\Box_{[m,h+m]}\mu$ and $\Box_{[m,k+m]}\Diamond_{[m,h+m]}\mu$, respectively, where m is the time shift of the convolution process and k is the width of the filter for the convolution layer. Moreover, the pooling layer after the convolution layer is the same as traditional pooling layer, in which maxpooling is used.

Fully Connected AND (FCAND) layer: This layer is a fully connected layer, but the activation function is the robustness degree function for "$\wedge$" operator, i.e., $g^{\wedge}(w,\rho)$. The formula obtained for each node can be described as $\varphi = \wedge^{w}_{i=1:N}\varphi_i$, where φ_i is the formula obtained from the previous layer.

Fully Connected OR (FCOR) layer: This layer is a fully connected layer, but the activation function is the robustness degree function for "$\vee$" operator, i.e., $g^{\vee}(w,\rho)$. The formula obtained for each node can be described as $\varphi = \vee^{w}_{i=1:N}\varphi_i$, where φ_i is the formula obtained from the previous layer.

3. Proposed Method

3.1. Source Domain Construction Based on Simulation

To generate a large set of fault samples with detailed labeled fault information and reduce reliance on real signals, a bearing system simulation model is developed. This model employs the lumped-mass method to simulate the vibration responses of bearings under varying conditions, which is used to create the source-domain dataset. In this model, the bearing system is represented as a nonlinear contact spring, with the inner race tightly fitted to the shaft, while the outer race is securely mounted in a rigid support, such as a bearing seat. The stiffness and damping of the primary components are treated as fixed values. Additionally, the Hertzian contact deformation theory is applied to calculate the contact force. The resulting motion of the bearing system is described by four differential equations [11].

$$\begin{cases} m_{in}\ddot{x}_{in} + c_{in}\dot{x}_{in} + k_{in}x_{in} = -F_x + e_{in}\omega_r^2\cos(\omega_r t) \\ m_{in}\ddot{y}_{in} + c_{in}\dot{y}_{in} + k_{in}y_{in} = m_{in}g - F_y + e_{in}\omega_r^2\sin(\omega_r t) \\ m_{out}\ddot{x}_{out} + c_{out}\dot{x}_{out} + k_{out}x_{out} = F_x \\ m_{out}\ddot{y}_{out} + c_{out}\dot{y}_{out} + k_{out}y_{out} = F_y + m_{out}g \end{cases}$$

where F_x and F_y are the support reaction forces of bearing, m_{in} are m_{out} are the equivalent mass of inner race and outer race, k_{in} and k_{out} are the contact stiffness between rolling ball and inner race, ball and outer race, c_{in} and c_{out} are the damping between ball and inner race, ball and outer race, e is the elastic modulus of ball and races materials, ω_c and ω_r are the cage speed and shaft speed, respectively. The cyclically varing contact forces F_x and F_y are the sum of the contact forces of the balls at different angles. The calculation process of F_x and F_y for inner ring defects, outer ring defects and rolling ball defects can be found in [12]. The simulation model adopts JUF6206RS bearing, and its main parameters are the same in [14].

3.2. Temporal Logic-based Discrepancy Metric

Most existing unsupervised domain adaptation methods use the MMD to calculate the discrepancy between the source domain and target domain, which does not consider the label of the data, thus they ignored the conditional distribution alignment and may misclassify fuzzy samples near the class boundary of the target domain. To address this issue, the joint MMD (JMMD) was proposed in [13] to take advantage of the labels in source domain and predicated labels in target domain. In fact, these methods use a "black box" deep learning framework, they ignores the transferred knowledge structure and focus only the feature vector itself. However, structure knowledge is very important in transfer learning since data from different sources usually follow the same physical mechanism and have similar structure knowledge. To capture the structure knowledge among the data, this article adopts the learned temporal logic to improve the loss function used in transfer learning framework.

Let $\mathcal{D}^s = \{x_i^s, y_i^s\}_{i=1}^{n_s}$ be the source domain with n_s labeled samples, and $\mathcal{D}^t = \{x_i^t\}_{i=1}^{n_t}$ be the target domain with n_t unlabeled samples. Since the learned wSTL formula of each node depends on the input formulas from the previous layer and the weight matrix associated, the structure information of each layer is embedded in matrix $\mathcal{H}_i = W_{i-1} \otimes \mathcal{O}_{i-1}$, where $\mathcal{O}_{i-1}$ is the output vector from layer $i-1$, W_{i-1} is the weight matrix from layer $i-1$ to layer i, $\otimes$ denotes element-wise multiple between a matrix and a vector. Let $\hat{f}_i^s = G_f(x_i^s)$, $\hat{f}_i^t = G_f(x_i^t)$, $\hat{y}_i^s = G_y(G_f(x_i^s))$, $\hat{y}_i^t = G_y(G_f(x_i^t))$, $h_{2_i}^s$ is the ith row of $\mathcal{H}_2^s$, $h_{2_i}^t$ is the ith row of $\mathcal{H}_2^t$, $h_{3_i}^s$ is the ith row of $\mathcal{H}_3^s$, and $h_{3_i}^t$ is the ith row of $\mathcal{H}_3^t$. Then the TLDM can be defined as

$$D_{\mathcal{K}} = \hat{D}(\theta_f, \theta_y) + \alpha D_{\mathcal{K}}^2(\mathcal{H}_2^t, \mathcal{H}_2^s) + \beta D_{\mathcal{K}}^3(\mathcal{H}_3^t, \mathcal{H}_3^s) \tag{4}$$

where α, β are the trade-off parameter, $\hat{D}(\theta_f, \theta_y)$ is the JMMD defined in [13] and $D_{\mathcal{K}}^m(\mathcal{H}_m^t, \mathcal{H}_m^s)$, $(m = 2, 3)$, are the structure discrepancy defined as

$$D_{\mathcal{K}}^m(h_m^s, h_m^t) = \frac{1}{n_s^2}\sum_{i=1}^{n_s}\sum_{j=1}^{n_s} k(h_{m_i}^s, h_{m_j}^s) + \frac{1}{n_t^2}\sum_{i=1}^{n_t}\sum_{j=1}^{n_t} k(h_{m_i}^t, h_{m_j}^t) - \frac{1}{n_s n_t}\sum_{i=1}^{n_s}\sum_{j=1}^{n_t} k(h_{m_i}^s, h_{m_j}^t) \tag{5}$$

where $k(\cdot)$ is the Gaussian kernel function, and we choose Gaussian kernel function with badwidth set to median pairwise square distances on the training data [14]. After taht, the proposed improved loss function embedded with TLDM is

$$L(\theta_f, \theta_y, \theta_d) = L_y(\theta_f, \theta_y) - \lambda L_d(\theta_f, \theta_d) + \gamma D_{\mathcal{K}}(\theta_f, \theta_y) \tag{6}$$

with

$$\begin{aligned}
L_y(\theta_f, \theta_y) &= -\frac{1}{n_s}\sum_{i=1}^{n_s}\sum_{k=0}^{K_s-1} I[y_i^s = k] \log\left(G_y^k(G_f(x_i^s))\right) \\
L_d(\theta_f, \theta_d) &= -\frac{1}{n_s}\sum_{i=1}^{n_s} \log\left(G_d(G_f(x_i^s))\right) - \frac{1}{n_t}\sum_{j=1}^{n_t} \log\left(1 - G_d(G_f(x_j^t))\right) \\
&\lambda, \gamma \sim 2M/(1 + e^{-Kt}) - M
\end{aligned} \tag{7}$$

where λ and γ are regularization coefficients, and the growth trajectory of regularization coefficients is given in (7), in which t is the current iterative training time, M is the upper limit of regularization coefficient, and K controls the rate of change.

3.3. Model Architecture and Optimization

The architectural framework of the proposed model, illustrated in Fig. 2, integrates three core components: a feature extractor (G_f), a classifier (G_y), and a domain discriminator (G_d). The feature extractor G_f incorporates a logic encoder followed by two 1-D logic convolutional layers. A max pooling layer and two fully-connected (FC) logic layers further refine feature representation. All convolutional components are supplemented with 1-D batch normalization layers and ReLU activation functions, while fully-connected layers employ ReLU activations combined with dropout regularization.

In contrast, the classifier G_y is structured as a single fully-connected layer, whereas the domain discriminator G_d comprises a convolutional layer followed by three FC layers for domain-specific differentiation. During training, the model is optimized through stochastic gradient descent (SGD), with parameters θ_f, θ_d, and θ_y being iteratively updated using the methodology detailed in [14].

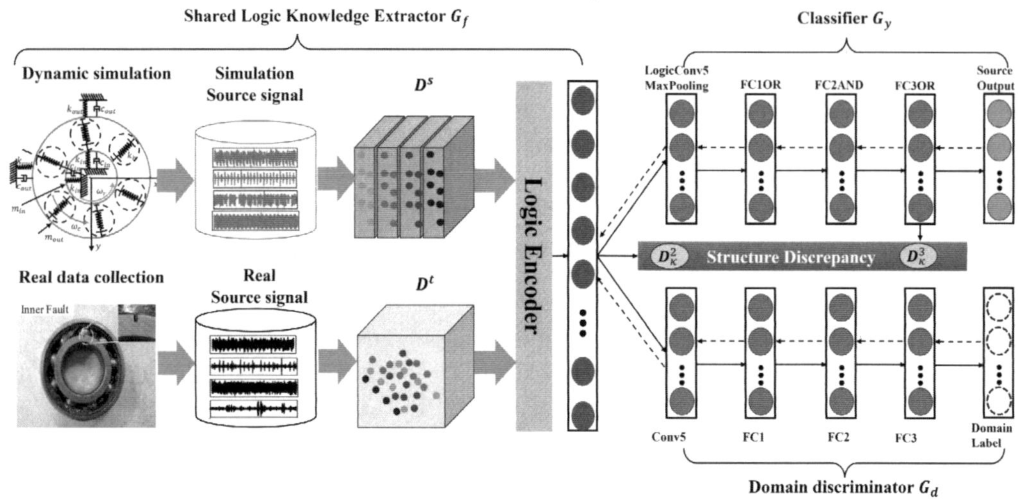

Figure 2. Model architecture of the proposed method.

4. Experiment and Result Analysis

4.1. Introduction of Target-Domain Datasets

To evaluate the diagnostic efficacy of the proposed methodology, two distinct cross-domain bearing fault diagnosis scenarios were designed under unsupervised conditions. Experiment 1 employs vibration data from the South China University of Technology (SCUT) bearing test platform, constituting Target Domain 1 with a sampling frequency of 16 kHz. Experiment 2 utilizes the widely-adopted Case Western Reserve University (CWRU) benchmark dataset, forming Target Domain 2 with 12 kHz sampling resolution. Both experimental configurations share three critical design parameters: 1) Fixed at 1024 data points per observation window;2) Source domain samples maintain parity with target domain characteristics;3) Implementation of domain-invariant feature learning without label supervision

This dual-validation framework examines the model's generalizability across different acquisition systems and operational environments. The SCUT dataset represents industrial-grade bearing degradation patterns, while the CWRU data provides standardized fault simulations under controlled laboratory conditions.

Table 1. Details of Target-Domain Datasets

Datasets	Bearing types	Health states	Number of training /test samples	Motor speeds
Target domain 1 (SCUT)	JUF6206RS	Normal Inner race fault Ball fault Outer race fault	300/80	1210 rpm 1240 rpm 1280 rpm 1180 rpm
Target domain 2 (CWRU)	NU205EM	Normal Inner race fault Ball fault Outer race fault	300/80	1797 rpm 1772 rpm 1750 rpm 1730 rpm

4.2. Comparison Experiments

In our effort to validate the superior performance of our proposed method, it is critically compared against seven other established techniques:

1) **Initial Benchmarking:** The proposed method, hereafter referred to as Method 1, is first evaluated against a conventional convolutional neural network, designated as Method 2. This step is crucial to demonstrate the utility of incorporating a transfer diagnosis strategy in our method.
2) **Cross-Domain Comparisons:** Next, Method 1 is assessed against renowned unsupervised cross-domain fault diagnosis methods, specifically the EKAD and the MSSD, which are denoted as Method 3 and Method 4 respectively. This comparison highlights the advantages of our method in handling cross-domain issues.
3) **Loss Function Evaluation:** Further, to demonstrate the efficacy of our proposed similarity function, we compare it against WTLN that operates without TLMD and use MMD, referred to as Method 5.
4) **Interpretable Diagnosis Strategy:** Lastly, the necessity and efficiency of the transfer diagnosis strategy for interpretable fault diagnosis are validated by comparing our method against well-known interpretable fault diagnosis methods, including WaveletKernelNet [15], time-frequency network [16], and frequency temporal logic based method [17], sequentially listed as Methods 6 to 8.

Each method is tested across ten trials to minimize the impact of randomness and ensure robust comparisons. Moreover, leveraging a greedy algorithmic approach, we strategically optimize one hyperparameter at a time while other variables are fixed, aiming to find the most effective combination of hyperparameters for enhanced performance.

4.3. Results Interpretation for Proposed Method

Table 2. WTLN associated STL formulas pieces for four conditions.

Fault Type	Interpretation
Inner Race	$\varphi_I = (\square_{[0,0.0031)}\Diamond_{[0,0.0125)}(d(x,\psi_{0.6,45}) < 0.43) \wedge \square_{[0,0.0031)}\Diamond_{[0.0169,0.0294)}(d(x,\psi_{5,34}) < 0.15)))\vee$ $(\quad \square_{[0,0.0031)}\Diamond_{[0.0262,0.0387]}(d(x,\psi_{2.3,6.4}) \geq 0.45) \vee \Diamond_{[0,0.0031)}\square_{[0.0031,0.0156)}(d(x,\psi_{0.3,44}) < 0.65))\wedge$ $(\quad \Diamond_{[0,0.0031)}\square_{[0.0350,0.0475)}(d(x,\psi_{0.2,4.4}) \geq 0.55) \vee \Diamond_{[0,0.0031)}\square_{[0.0100,0.0225)}(d(x,\psi_{0.2,3.4}) \geq 0.25))$
Outer Race	$\varphi_O = (\Diamond_{[0,0.0031)}\square_{[0.0156,0.0250)}(d(x,\psi_{0.8,25}) \geq 0.12) \wedge \square_{[0,0.0031)}\Diamond_{[0.0494,0.0588)}(d(x,\psi_{3.2,5.4}) < 0.48))\wedge$ $(\quad \square_{[0,0.0031)}\Diamond_{[0.0069,0.0194)}(d(x,\psi_{0.5,6.4}) \geq 0.45) \vee \Diamond_{[0,0.0031)}\square_{[0.0287,0.0412)}(d(x,\psi_{0.3,6.9}) < 0.66))\vee$ $(\quad \Diamond_{[0,0.0031)}\square_{[0.0094,0.0219)}(d(x,\psi_{0.3,34}) \geq 0.62) \vee \Diamond_{[0,0.0031)}\square_{[0.0269,0.0394)}(d(x,\psi_{0.1,2.41}) < 0.37))$
Rolling Element	$\varphi_R = (\Diamond_{[0,0.0031)}\square_{[0.0325,0.0419)}(d(x,\psi_{0.2,45}) < 0.28) \vee \Diamond_{[0.0006,0.0037)}\square_{[0,0.0094)}(d(x,\psi_{3.6,7.8}) < 0.38))\vee$ $(\quad \Diamond_{[0,0.0031)}\square_{[0.0150,0.0275)}(d(x,\psi_{0.8,65}) \geq 0.35) \vee \Diamond_{[0,0.0031)}\square_{[0.0013,0.0138)}(d(x,\psi_{0.9,5.6}) < 0.43))\wedge$ $(\quad \Diamond_{[0,0.0031)}\square_{[0.0362,0.0488)}(d(x,\psi_{0.7,54}) \geq 0.45) \wedge \Diamond_{[0,0.0031)}\square_{[0.0206,0.0331)}(d(x,\psi_{0.3,2.51}) < 0.48))$
Normal	$\varphi_N = (\Diamond_{[0,0.0031)}\square_{[0.0338,0.0431)}(d(x,\psi_{0.65,53}) < 0.45) \wedge \square_{[0,0.0031)}\Diamond_{[0.0494,0.0587)}(d(x,\psi_{0.3,2.4}) < 0.52))\wedge$ $(\quad \square_{[0,0.0031)}\Diamond_{[0.0094,0.0219)}(d(x,\psi_{0.06,4.4}) \geq 0.25) \vee \square_{[0,0.0031)}\Diamond_{[0.0250,0.0375)}(d(x,\psi_{0.2,0.24}) < 0.41))$

Table 2 presents selective diagnostic signatures derived from the Weighted Temporal Logic Network (WTLN), focusing on temporally-activated feature correlations rather than connection weights. The network's first two logical convolution operations establish critical temporal boundaries (t_1-t_8), with each interval's duration ($\Delta t = t_{2n} - t_{2n-1}$) being intrinsically governed by the convolutional filters' kernel dimensions. These generated temporal logic expressions provide causal explanations for specific fault patterns. For example, for the inner race fault, the first two terms of formula φ_I can be read "always within $[0, 0.0031]$ seconds, eventually within

[0, 0125] seconds, the signals' distance to wavelet $\psi_{0.6,45}$ should be smaller than 0.43, and always within [0, 0.0031] seconds, eventually within [0.0169, 0.0294] seconds, the signals' distance to wavelet $\psi_{5,34}$ should be smaller than 0.15." The other terms of these formulas in Table 2 can be interpreted in the same way and are omitted here.

The wSTL formulas identify signal components that exceed specific thresholds over designated time intervals, indicating significant discrepancies between the signals and wavelets, suggestive of rapid changes in the fault signals. Additionally, these formulas highlight differences in fault patterns: for instance, inner race faults display two regions of low similarity, whereas outer race faults show only one, suggesting a lower frequency of occurrence in outer race faults compared to inner race faults. This differentiation underscores the wSTL formulas' effectiveness in revealing insights about the fault mechanisms, enabling maintenance personnel to implement strategies, such as enhanced lubrication, to mitigate system failure risks.

4.4. Comparison Results Analysis

Table 3. Comparisons of The Eight Methods In Two Experiments.

#	Experiment 1 (%)	Experiment 2 (%)	#	Experiment 1 (%)	Experiment 2 (%)
1	**85.1250 ± 3.0573**	**91.2812± 4.6567**	5	82.5312 ± 2.8793	80.5313 ± 4.5843
2	82.6563 ± 5.3496	79.5938 ± 4.9096	6	70.7500± 3.2384	78.8125 ± 5.1126
3	73.1875± 7.3649	88.0938± 2.9531	7	68.7500 ± 3.8580	74.8125 ± 3.9744
4	80.7500 ± 2.3500	89.0000 ± 6.9119	8	81.6875 ± 5.3035	87.2188 ± 6.1275

In this study, the average accuracy is employed to comprehensively evaluate the diagnostic performance of each method on the target domain's test set. The results are presented in Table 3.

The comparative analysis in Table 3 demonstrates the superior performance of Method 1 in unsupervised cross-domain fault diagnosis. It outperforms seven well-established methods across two experiments. The key findings from this comparison are as follows.

1) Method 1 achieved the highest scores in both experiments (85.1250% ± 3.0573 and 91.2812% ± 4.6567), demonstrating its robustness and adaptability. The integration of a transfer diagnosis strategy significantly enhances its diagnostic accuracy over traditional models like the conventional convolutional neural network (Method 2).
2) Cross-Domain Techniques (Methods 3 and 4) also showed improvement in Experiment 2, yet none matched the consistency of Method 1. This highlights the superiority of Method 1's approach to handling domain variability.
3) Comparative Analysis of methods focusing on interpretable fault diagnosis (Methods 6 to 8) revealed that while they are effective to some degree, they do not consistently surpass Method 1, affirming the efficiency of Method 1's transfer diagnosis strategy.

The results affirm the design of Method 1, particularly its transfer diagnosis strategy and similarity function, as crucial for enhanced diagnostic precision. It establishes a new benchmark and broadens the potential for applications across varied fault conditions.

5. Conclusions

This study introduced a novel interpretable transfer learning framework for unsupervised bearing fault diagnosis, leveraging the WTLN and TLDM. Our approach enhances the accuracy and interpretability of fault diagnosis across varied operational conditions. The integration of WTLN with TLDM aligns knowledge structure and marginal distribution effectively, leading to superior diagnostic accuracy. Experimental validations demonstrate our method's robust performance compared to existing techniques. This research advances fault diagnosis technologies and suggests broad applicability in industrial settings, emphasizing the importance of interpretability in safety-critical applications. We anticipate its adoption will improve the reliability and efficiency of diagnostic processes across industries.

Acknowledgments

This work is partially supported by the Basic and Applied Basic Research Foundation of Guangzhou (grant number 2023A04J1582), the National Training Program of Innovation and Entrepreneurship for Undergraduates (202410561171)

References

[1] Hoang D T and Kang H J 2019 *Cognitive Systems Research* **53** 42–50
[2] Wang Q, Taal C and Fink O 2021 *IEEE Transactions on Instrumentation and Measurement* **71** 1–12
[3] Tian J, Han D, Li M and Shi P 2022 *Knowledge-Based Systems* **243** 108466
[4] Ding Y, Jia M, Zhuang J, Cao Y, Zhao X and Lee C G 2023 *Reliability Engineering & System Safety* **230** 108890
[5] Liu J, Hou L, Zhang R, Sun X, Yu Q, Yang K and Zhang X 2023 *Energy* **262** 125258
[6] Brito L C, Susto G A, Brito J N and Duarte M A V 2023 *Expert Systems with Applications* **232** 120860
[7] Kong Z, Jones A and Belta C 2016 *IEEE Transactions on Automatic Control* **62** 1210–1222
[8] Chen G, Lu Y and Su R 2022 *Automatica* **142** 110350
[9] Tian R, Cui M and Chen G 2024 *IEEE Transactions on Instrumentation and Measurement*
[10] Chen G, Wei P, Jiang H and Liu M 2020 *IEEE Transactions on Industrial Informatics* **18** 119–129
[11] Chen G 2008 *J Vib Eng* **21** 577–587
[12] Zhang Y, Ji J, Ren Z, Ni Q, Gu F, Feng K, Yu K, Ge J, Lei Z and Liu Z 2023 *Reliability Engineering & System Safety* **234** 109186
[13] Long M, Zhu H, Wang J and Jordan M I 2017 Deep transfer learning with joint adaptation networks *International Conference on Machine Learning* (PMLR) pp 2208–2217
[14] Xiao Y, Shao H, Han S, Huo Z and Wan J 2022 *IEEE/ASME Transactions on Mechatronics* **27** 5254–5263
[15] Li T, Zhao Z, Sun C, Cheng L, Chen X, Yan R and Gao R X 2021 *IEEE Transactions on Systems, Man, and Cybernetics: Systems* **52** 2302–2312
[16] Chen Q, Dong X, Tu G, Wang D, Cheng C, Zhao B and Peng Z 2024 *Mechanical Systems and Signal Processing* **207** 110952
[17] Chen G, Liu M and Chen J 2020 *Mechanical Systems and Signal Processing* **145** 106951

Research on the Dynamics of the Active Heave Compensation System for Shipborne Tethered Balloons

Liangpei Huang[1, 2*], **Zhuojun Li**[1]

[1] School of Mechanical and Electrical Engineering, Hunan University of Science and Technology, Xiangtan, Hunan 411201, China

[2] Hunan Provincial Key Laboratory of Mechanical Equipment Monitoring and Maintenance, Hunan University of Science and Technology, Xiangtan, Hunan 411201, China

*Corresponding author's e-mail: huanglp413@163.com

Abstract. Waves and gusts have a significant impact on the offshore operations of shipborne tethered aerostats. The radar, electronic reconnaissance equipment, and other systems carried by tethered aerostats experience vertical oscillations due to balloon motion, which reduces the operational accuracy of electronic devices. This paper establishes a virtual prototype dynamics model of an active heave compensation system for shipborne tethered aerostats and analyzes its dynamic characteristics. First, the measured irregular waves under sea state 3 conditions were fitted to obtain irregular wave data as external excitation. Second, aerodynamic simulations of a tethered aerostat were conducted to determine the lift coefficient under gust conditions, which was then converted into lift force using relevant formulas. Finally, a bushing force model was employed to simulate the flexible tethered cable, accurately reflecting the mechanical properties of real cables while integrating the entire mechanical system. Simulation analysis of the constructed virtual prototype demonstrates that, compared to systems without active heave compensation, the operational deployment of this system reduces both the vertical displacement fluctuation range of the aerostat and the tension between the aerostat and cable. This enhancement not only improves the operational precision of onboard electronic equipment but also effectively mitigates the risk of cable breakage caused by excessive tension. However, as lift force increases, fluctuations in the aerostat's displacement, velocity, and acceleration intensify, thereby diminishing the compensation effectiveness of the active heave compensation system.

1. Introduction

As a new - concept aerial platform that cannot be replaced by aircraft and satellites, the ship - borne tethered balloon can fill the gaps of aircraft and satellites in terms of time and space [1]. It is highly suitable for carrying various types of electronic mission equipment and has broad application prospects in both military and civilian fields [2]. However, when the tethered balloon operates in a marine environment, offshore operations will be affected by both waves and gusts [3]. Waves and gusts influence offshore operations by causing the tethered balloon and the engineering ship to exhibit six - degree - of - freedom motion [4]. Among them, horizontal movements (i.e., roll, surge, and yaw) can be addressed by the dynamic positioning system. Nevertheless, the remaining vertical movements, especially heave motion, will inevitably cause fluctuations in the dynamic load of the tether cable, resulting in changes in the position of the tethered balloon at the upper end of the tether cable and affecting the aerial stability and performance of the tethered balloon [5-6]. Usually, the compensation system solves this problem by decoupling the motion of the tethered balloon caused by gusts from the motion of the ship caused by waves.

This paper comprehensively considers the impacts of gusts and waves on tethered balloons and engineering ships. Combining with the compensation mechanism of the active heave compensation system of the marine winch, it conducts the virtual prototype modeling and simulation analysis of the active heave compensation mechanical system for ship - borne tethered balloons.

2. Obtaining and Processing of Measured Ship Heave Data

Under normal circumstances, the significant wave height is the decisive factor for the ship's heave displacement. The period of the ship's heave motion is between 5 seconds and 15 seconds, and the ratio of the ship's heave motion displacement to the significant wave height of the wave is generally less than 0.5. Assuming that the wave motion is sinusoidal wave excitation, the mathematical expression of the ship's heave displacement can be expressed as [7]:

$$y_0(t) = \frac{\mu H}{2}\sin(\frac{2\pi t}{T}) \tag{1}$$

In the formula, H represents the significant wave height of the wave, T represents the wave period, and μ represents the ratio of the ship's heave motion displacement to the significant wave height of the wave.

According to the wave - surface equation of irregular waves, simply regarding wave motion as sinusoidal or cosine excitation cannot accurately reflect the complexity of the actual marine environment. Considering that the wave amplitudes of sea conditions of level 2 and below are relatively small, and sea conditions of level 4, 5 and above are rather severe, with the ship fluctuating violently, which is not suitable for the operation of tethered balloons. Therefore, the measured ship heave data under level - 3 sea conditions are selected as the source of simulation excitation. Figure 1 shows the measured ship heaving data under level - 3 sea conditions, and the amplitude of the measured ship heave data under level - 3 sea conditions varies within $-0.8\text{m} \sim +0.7\text{m}$.

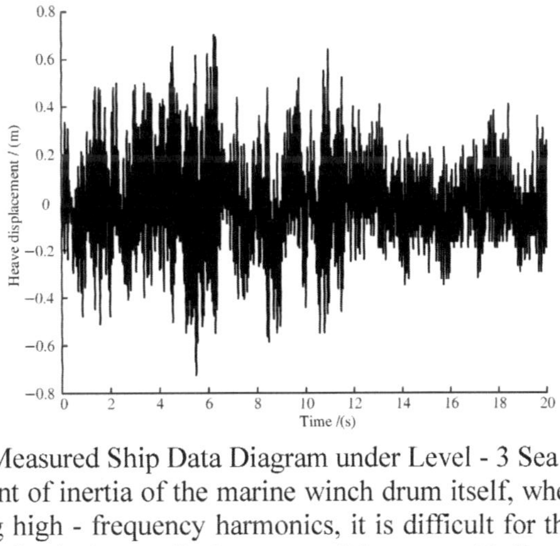

Figure 1. Measured Ship Data Diagram under Level - 3 Sea Conditions

Due to the large moment of inertia of the marine winch drum itself, when excited by the measured ship heave data containing high - frequency harmonics, it is difficult for the marine winch to quickly and accurately compensate for the heave displacement of the moored balloon. Therefore, it is necessary to perform specific processing on the measured ship heave data. That is, first perform a Fourier transform, and then perform an inverse Fourier transform to obtain the time - domain function after removing the high - frequency harmonics. Only the data processed in this way can be directly used for the compensation control of the winch. Figure 2 shows Ship data fitting results and filtering results under three-level sea conditions. After the Fourier transform and inverse Fourier transform, the obtained curve is highly consistent with the original data, and the high - frequency harmonic components in the data are successfully removed, making the processed data more accurate.

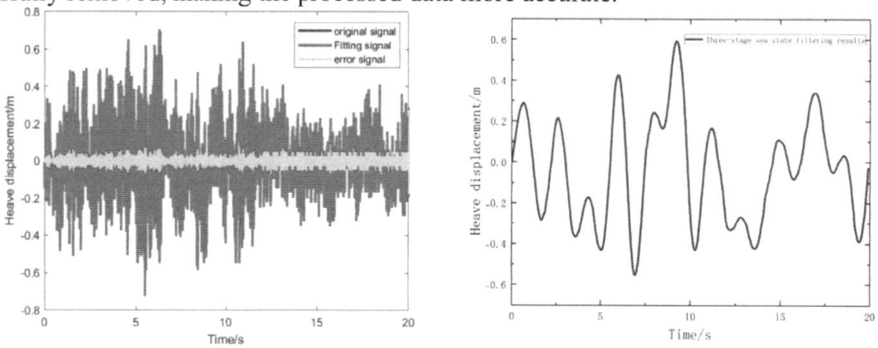

Figure 2. Ship data fitting results and filtering results under three-level sea conditions

3. Simulation Model of Tethered Balloon

In this paper, the Zhiyuan - 1 balloon developed by Shanghai Jiao Tong University [8] is selected as the simulation model of the tethered balloon. A three - dimensional model is built using SolidWorks and imported into Fluent to simulate its aerodynamic characteristics under wind - tunnel experiments,and the lift data is obtained using the lift calculation formula.

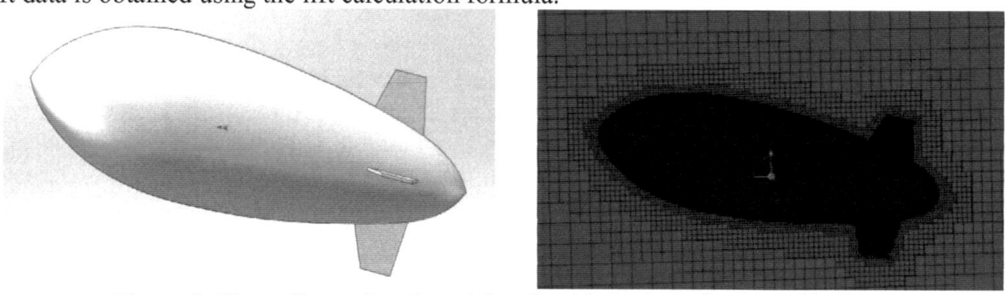

Figure 3. Three-dimensional model and grid planing of tethered balloon

As the tethered balloon ascends to a greater height, the atmospheric flow in the air is more likely to form rapidly - changing airflow disturbances, resulting in a more significant gust phenomenon. The influence of gusts on the tethered balloon cannot be ignored. The gust model used in this simulation is the extreme operating gust described in the International Electrotechnical Commission standard [9], and its formula is as follows:

$$V(t)=\begin{cases} V_m-0.37V_g\sin\left(\dfrac{3\pi t}{T}\right)\left(1-\cos\left(\dfrac{2\pi t}{T}\right)\right): & 0\le t\le T \\ V_m: & \text{otherwise} \end{cases} \tag{2}$$

In the formula, V represents the average wind speed, Vg represents the gust amplitude, and T is the period of the gust.

In this paper, the average wind speeds used in the simulation are respectively the average wind speeds corresponding to sea state 3, that is, 3.4 m/s, 5.4 m/s, and 7.9 m/s, which are taken as the simulation data. The angle of attack of the tethered balloon is set at 6°, and the amplitude is 10.7 m/s.

Table 1. Detailed working conditions for the simulation of the lift coefficient of the tethered balloon

Sea state level	Average wind speed magnitude（m/s）	Amplitude magnitude (m/s)	Angle of attack (°)
Sea state level 3	3.4，5.4，7.9	10.7	6

Calculate the lift force value from the simulated lift coefficient according to formula (3).

$$C_L = \frac{L}{0.5\rho v^2 A} \tag{3}$$

In the formula, D, L, and M represent the drag force, lift force, and moment respectively. ρ is the air density, v is the flow velocity of the airflow relative to the object, A is the reference area, and L is the reference length.

The simulation results of the lifting force of the tethered balloon are shown in Figure 4. Figure 4. (a) is the simulation result diagram where the average wind speed is 5.4 m/s, the gust amplitude is 10.7 m/s, and the angle of attack is 6°. Figure 4. (b) shows the simulation results of the lifting force received by the tethered balloon when the average wind speeds are 3.4 m/s, 5.4 m/s, and 7.9 m/s respectively, the gust amplitude is 10.7 m/s, and the angle of attack is 6°.

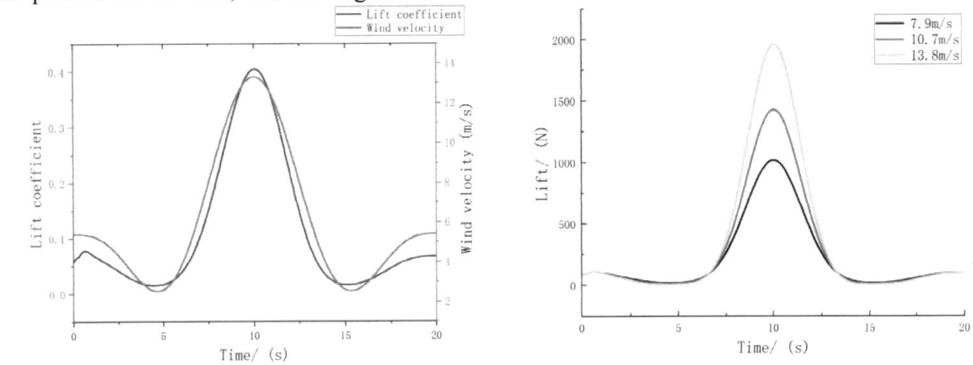

(a) (b)

Figure 4. Simulation results of the lift force of the tethered balloon

4. Virtual prototype modeling of active heave compensation system for shipborne tethered balloon

4.1. Modeling of the tethering cable

The tethering cable is usually composed of a tensile layer, signal - transmission cables, power - supply cables, etc. [10]. When an external tensile force is applied, there is often friction between the surfaces of the tethering cable. These characteristics bring difficulties to the simulation of the mechanical properties of the tethering cable. In this paper, the bushing force modeling method is used. By decomposing the continuous cable into a finite number of small rigid cylinders and connecting them with dampers. This method applies to a flexible force between two components by defining six components of force and torque. Connecting adjacent small rigid cylinders with bushing forces can better simulate the various properties of flexible ropes and more realistically reflect the mechanical properties of actual ropes. However, this method is relatively complex when establishing the model, and it also has relatively strict requirements for the length of the simulation time.

In this paper, the macro - commands in Adams are used to model the tethering cable. The modeling process of the tethering cable using macro - commands is as follows:

(1) First, determine the coordinate positions of the starting end of the tethering cable and the winch drum. According to the radius of the winch drum, write a loop statement to generate a spatial point system around the first - turn rope groove of the winch drum.

(2) Determine the coordinates of the key nodes where the cable passes. Use the loop statement to generate a series of subsequent spatial point systems based on the determined key nodes.

(3) Use the created spatial point system. Create small rigid cylinders with a radius of 16 mm and the length equal to the distance between adjacent two points and add bushing forces between adjacent cylinders.

The modeling result of the tethering cable using macro - commands is shown in Figure5.

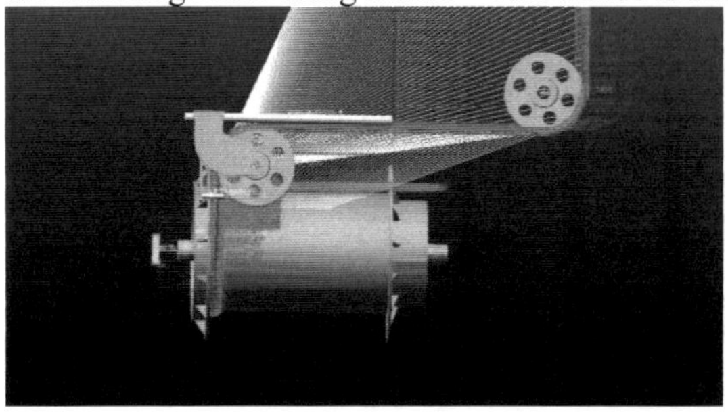

Figure 5. Modeling result of the tethering cable

4.2. Application of kinematic pairs and drives

When the tethering cable is created, the construction of the virtual prototype of the ship - borne tethered balloon can be completed by simply adding the correct constraints and corresponding drives. The main constraints are shown in Table 2.

Table 2. Main Kinematic Pairs

Object one	Object two	Constraint type
Vessel	Ground	Prismatic pair
Vessel	Guide wheel	Revolute pair
Vessel	Drum	Revolute pair
Vessel	Head sheave	Revolute pair
Tethered balloon	Tail section of the tethering cable	Hinge pair
Drum	Head section of the tethering cable	Hinge pair

The drive of the virtual prototype of the ship - borne tethered balloon is added using the SPLINE interpolation method. The measured heave motion of the ship is imported into the model as a drive through SPLINE interpolation. The lift force generated by the gust on the tethered balloon is imported into the model as an external force using SPLINE interpolation. The Archimedes' buoyancy formula is used to calculate the net buoyancy of the tethered balloon, and the calculation result is loaded into the virtual prototype model as an external force.

Finally, since the ship heave data drive is applied in the vertical motion direction of the ship, in order to achieve the active heave compensation function, the corresponding ship heave motion can be converted into the rotation of the winch drum. Meanwhile, a variable compensation amount is added, which constitutes the compensation drive on the winch drum, and then the heave displacement compensation of the tethered balloon is realized.

5. Dynamic simulation analysis of the active heave compensation system for ship - borne tethered balloons under sea state 3

The compensation effect of the shipborne tethered balloon active heave compensation system under sea state 3 is shown in Figure 6. The simulation under sea state 3 was conducted under the following conditions: an average wind speed of 3.4 m/s, gust amplitude of 10.7 m/s, and an angle of attack of 6° for the average lift force. Without active heave compensation, the heave displacement of the tethered balloon fluctuated between -0.6 m and +0.6 m. With the active heave compensation system, the heave displacement was reduced to a range of 0.00 m to +0.22 m. The motion of the tethered balloon closely matched the ship's movement, exhibiting minimal heave displacement fluctuations, demonstrating the excellent performance of the shipborne active heave compensation system under sea state 3 conditions.

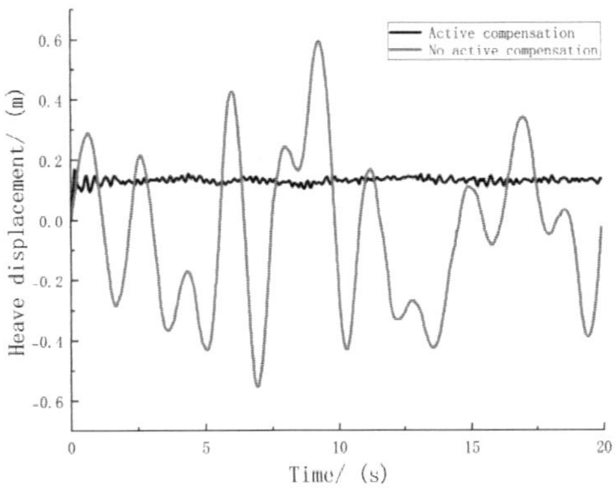

Figure 6. The compensation effect of the active heave compensation system under sea state 3

The lifting height of the tethered balloon is set at 300 meters, the amplitude of the gust is set at 10.7 m/s, and the angle of attack is 6°. The dynamic simulation of the active heave compensation system is carried out when the average wind speeds acting on the tethered balloon are 3.4 m/s, 5.4 m/s, and 7.9 m/s. Analyze the simulation results of the active heave compensation system of the shipborne tethered balloon under these three variable lifting forces.

As shown in Figure 7. (a), when the average wind speed acting on the tethered balloon is 3.4 m/s, the heave displacement of the tethered balloon after the active heave compensation effect varies within the range of -0.000 m to +0.221 m. At the same time, the average amplitude of its heave displacement is approximately 0.1213 m; when the average wind speed of the tethered balloon is 5.4 m/s, the heave displacement of the tethered balloon after the active heave compensation effect varies within the range of 0.000 m to +0.148 m, and the average value of its heave displacement is approximately 0.1327 m; when the average wind speed acting on the tethered balloon is 7.9 m/s, the heave displacement of the tethered balloon after the active heave compensation effect varies within the range of 0.000 m to +0.224 m, and the average amplitude of its heave displacement is approximately 0.145 m. By comparing the curves of the heave displacement of the tethered balloon changing with time under the three situations, it can be seen that as the lifting force caused by the increase in the average wind speed of the balloon increases, the average amplitude of the heave motion of the tethered balloon also increases.

The simulation results of the heave velocity of the tethered balloon changing with time are shown in Figure 7.(b). When the average wind speed acting on the tethered balloon is 3.4 m/s, the heave velocity of the tethered balloon varies within the range of -1.06 m/s to +1.46 m/s after the active heave compensation; when the average wind speed of the tethered balloon is 5.4 m/s, the heave velocity of the tethered balloon varies within the range of -1.22 m/s to +1.63 m/s after the active heave compensation; when the average wind speed acting on the tethered balloon is 7.9 m/s, the heave velocity of the tethered balloon varies within the range of -1.51 m/s to +2.21 m/s after the active heave compensation. As the lifting force acting on the balloon increases, the average amplitude of the heave motion of the tethered balloon also increases.

The simulation results of the heave acceleration of the tethered balloon changing with time are shown in Figure 7.(c). When the average wind speed acting on the tethered balloon is 3.4 m/s, the heave acceleration of the tethered balloon varies within the range of -14.6 m/s^2 to +15.2 m/s^2 after the active heave compensation; when the average wind speed of the tethered balloon is 5.4 m/s, the heave acceleration of the tethered balloon varies within the range of -16.2 m/s^2 to +15.6 m/s^2 after the active heave compensation; when the average wind speed acting on the tethered balloon is 7.9 m/s, the heave acceleration of the tethered balloon varies within the range of -19.4 m/s^2 to +19.2 m/s^2 after the active heave compensation.

The simulation results of the tension of the tethered cable changing with time are shown in Figure 7. (d). When the average wind speed acting on the tethered balloon is 3.4 m/s, the tension of the tethered cable varies within the range of 2.2 kN to 14.5 kN; when the average wind speed of the tethered balloon is 5.4 m/s, the tension of the tethered cable varies within the range of 2.2 kN to 15.6 kN; when the average wind speed acting on the tethered balloon is 7.9 m/s, the tension of the tethered cable varies within the range of 2.8 kN to 17.1 kN. The change in tension is consistent with the change in the lifting force acting on the tethered balloon.

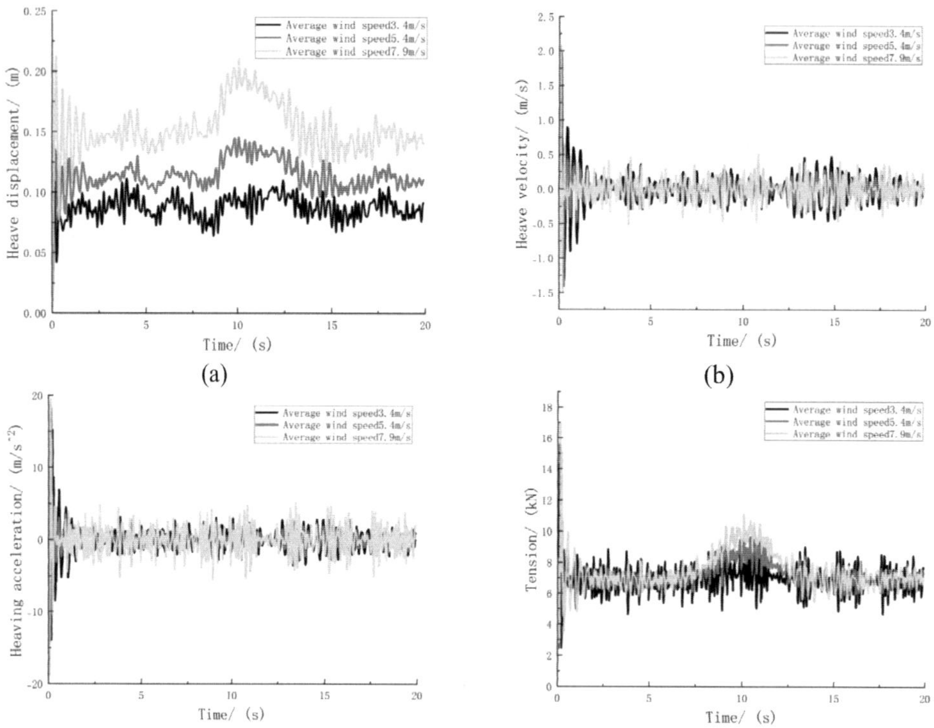

(a) (b)

(c) (d)

Figure 7. Simulation results under variable lifting force values

6. Conclusion

This paper conducts a dynamic analysis of the active heave compensation system for shipborne tethered balloons under the dual effects of irregular waves and unsteady gusts. The heave displacement data of the ship are obtained by filtering and fitting the measured signals under sea state 3. The lifting force data are obtained through the aerodynamic load simulation analysis of a certain tethered balloon under unsteady gusts. The flexible tethered cable modeled by the bushing force is used to connect the rigid model of the shipborne tethered balloon established in SolidWorks. The corresponding heave motion of the ship is converted into the rotation of the winch drum, and a variable compensation amount is added. At this time, the dynamic model of the virtual prototype of the heave compensation system for the shipborne tethered balloon is completed. Through the simulation analysis of the heave compensation dynamics of the shipborne tethered balloon, the following conclusions are obtained:

(1) It is very necessary to build the heave compensation system for the shipborne tethered balloon. In the simulation model with the active heave compensation system, the fluctuation range of the heave displacement of the tethered balloon will be reduced, and the tension between the tethered balloon and the cable will also decrease. This not only improves the accuracy of the operation of the electronic equipment carried by the tethered balloon during operation, but also effectively prevents the risk of the tethered cable breaking due to excessive tension.

(2) As the lifting force increases, the fluctuations in the displacement, velocity, and acceleration of the tethered balloon will all increase, which will reduce the compensation effect of the active heave compensation system.

Fund Project
Supported by the National Natural Science Foundation of China (Grant No. 52075163).

Author information
Huang Liangpei (born in 1971), male, professor, doctor, and doctoral supervisor.

References
[1] Wu Hongxun, Huang Wanning, Miao Jinggang, et al. Design and implementation of AIS data acquisition system for ship - based tethered balloon [J]. Computer Measurement & Control, 2023, 31(04): 167 - 173.
[2] Sheng Wanxing, Duan Qing, Sha Guanglin, et al. Application of tethered balloon platform in emergency communication support [J]. Electric Power Information and Communication Technology, 2022, 20(08): 22 - 28.
[3] Tiancai X, Liangpei H, Jingwei X, et al. Dynamic Analysis of Active Heave Compensation System for Marine Winch under the Impact of Irregular Waves [J]. Journal of Marine Science and Engineering, 2023, 11(2): 240 - 240.
[4] Chen Zhimei, Lu Yingbin, Shao Xuejuan, et al. Sliding mode predictive control of heave compensation for deep - sea crane [J]. Journal of Taiyuan University of Science and Technology, 2024, 45(01): 13 - 18.
[5] Hu Xiaojin, Shu Xin, Lai Zhenghua, et al. Dynamic modeling method for the complex system of near - ground tethered balloon and cable [J]. Journal of Dynamics and Control, 2024, 22(05): 48 - 55.
[6] Jeans T, Fagley C, Siegel S, et al. Irregular deep ocean wave energy attenuation using a cycloidal wave energy converter [J]. International Journal of Marine Energy, 2013, 116 - 32.
[7] Jing Biao, Zhu Keqiang, Zhang Tianyu. Dynamic simulation analysis of lifting system under regular waves [J]. Journal of Waterway and Harbor, 2015, 36(05): 398 - 403.
[8] Liu P, Fu G, Zhu L, et al. Aerodynamic characteristics of airship Zhiyuan - 1 [J]. Journal of Shanghai Jiaotong University (Science), 2013, 18(6): 679 - 687.
[9] IEC 61400 - 2:2013, International Standard, Wind turbines - Part 2: Small wind turbines, 2006.
[10] Kong Lei, Wang Lianchao. Strength failure analysis of Vectran fiber tethering cable [J]. China Synthetic Fiber Industry, 2023, 46(02): 84 - 88.

Journal of Physics: Conference Series **3024** (2025) 012005 doi:10.1088/1742-6596/3024/1/012005

Application of Mullite Nanowhisker-reinforced Polyurethane Composites in 3D Printing

Canyan Long[1], Xiaohua Gu[1,3,4*], Wenkai Wen[2], Jiahao Xu[5], Shaochun Sun[1], Zhong Liu[2], Zhonggang Xiong[2], Yuxuan Kang[1], Qingyong Su[1]

[1.] School of Energy and Building Environment, Guilin University of Aerospace Technology, Guilin 541004, P.R. China

[2.] School of Artificial Intelligence, Guilin University of Aerospace Technology, Guilin, 541004, P.R. China

[3.] Guangxi Key Laboratory of Special Engineering Equipment and Control, Guilin University of Aerospace Technology, Guilin 541004, P.R. China

[4.] University Engineering Research Center of Green Upgrade Key Technology for Energy Industry, Guilin 541004, P.R. China

[5.] Guangxi Key Laboratory of Green Building Materials and Construction Industrialization, Guilin University of Technology, Guilin 541004, P.R. China

[*]Corresponding author's e-mail: 2022049@guat.edu.cn.

Abstract: This study discussed the application of mullite nanowhisker-reinforced polyurethane composites in 3D printing. The composite materials were fabricated via in-situ polymerization technique, with their thermomechanical properties, microstructural characteristics, and thermal stability systematically investigated. The 3D printing process parameters were optimized, and the performance of the printed products was evaluated. Experimental results demonstrated that the incorporation of mullite nanowhiskers induced significant enhancements in the composite materials' mechanical properties and thermal stability. Compared with pure polyurethane, the composites' tensile properties, bending properties, and fracture toughness were increased by 60%, 62%, and 87%, respectively. At the same time, the coefficient of thermal expansion is reduced by about 20 units. Optimized 3D printing can produce products with good dimensional accuracy and surface quality. This study proposes a novel methodology for engineering advanced 3D printing materials, demonstrating significant potential for widespread application in industrial domains, including but not limited to aerospace engineering and automotive manufacturing sectors.

1. Introduction

The advancement of additive manufacturing technology has precipitated a substantial surge in the demand for specialized high-performance printing materials. As a kind of polymer, polyurethane has the advantages of low thermal conductivity, wear resistance, high hardness, and low water absorption [1-5]. Mullite nanowhisker is a new reinforcement material with excellent mechanical properties and thermal stability [6-7]. Introducing mullite nanowhisker into a polyurethane matrix can dramatically enhance the overall performance characteristics of the composite material. This study aims to explore the application potential of mullite nanowhisker-reinforced polyurethane composites in 3D printing.

By optimizing the composite formula and printing process parameters, it is expected to realize the rapid manufacturing of high-performance, complex structural parts and meet the requirements of high-end fields such as aerospace and automotive manufacturing for material properties.

2. Preparation of polyurethane composites reinforced with mullite nanowhiskers

2.1. Preparation of mullite nanowhiskers
Mullite nanowhiskers were prepared using the sol-gel method [8]. Tetraethoxysilane, aluminium nitrate, and other metal salts or alkoxides were used as precursors, dissolved in organic solvents, and formed a stable sol through hydrolysis and polycondensation. With the evaporation and reaction of water, the sol is transformed into a gel and then dried and heat treated to obtain mullite nanowhiskers, as shown in Figure 1. Scanning electron microscope of the mullite whisker, the process steps include raw material selection and ratio, sol preparation, gelation, drying, and heat treatment.

Figure 1. Scanning SEM of mullite nanowhiskers

2.2. Preparation of composites
Mullite nanowhisker-reinforced polyurethane composites were prepared by in-situ polymerization. Firstly, the surface of mullite nanowhiskers was treated to improve the interface compatibility between mullite nanowhiskers and the polyurethane matrix. Then, the nanowhiskers were uniformly dispersed in polyurethane prepolymer by ultrasonic dispersion. Finally, the treated nanowhiskers were mixed with polyurethane monomer to achieve in-situ formation and uniform dispersion of nanowhiskers in the polymerization process. After optimising process parameters, the composites with excellent mechanical properties and thermal stability were obtained.

3. Performance analysis of composite materials

3.1. Mechanical property analysis
A universal material testing machine tested the tensile properties, bending properties, and fracture toughness of the composites. The results show that the composites' tensile strength, bending strength, and fracture toughness significantly improve with increased mullite nanowhisker content. When the mullite nanowhisker content was 3wt %, the tensile strength of the composite reached the maximum value, about 45 MPa, which was about 60 % higher than that of pure polyurethane (about 28 Mpa). The bending strength reached the maximum value, about 68 mpa, about 62 % higher than pure polyurethane (about 42 MPa). The fracture toughness reaches the maximum value, about 2.8 MPa·m^1/2, about 87 % higher than pure polyurethane (about 1.5 MPa·m^1/2). When the content of mullite nanowhisker exceeds 3wt %, the tensile strength begins to decline, mainly because the high content of mullite nanowhisker leads to uneven dispersion and agglomeration, which reduces the material's overall performance, as is shown in Figure 2.

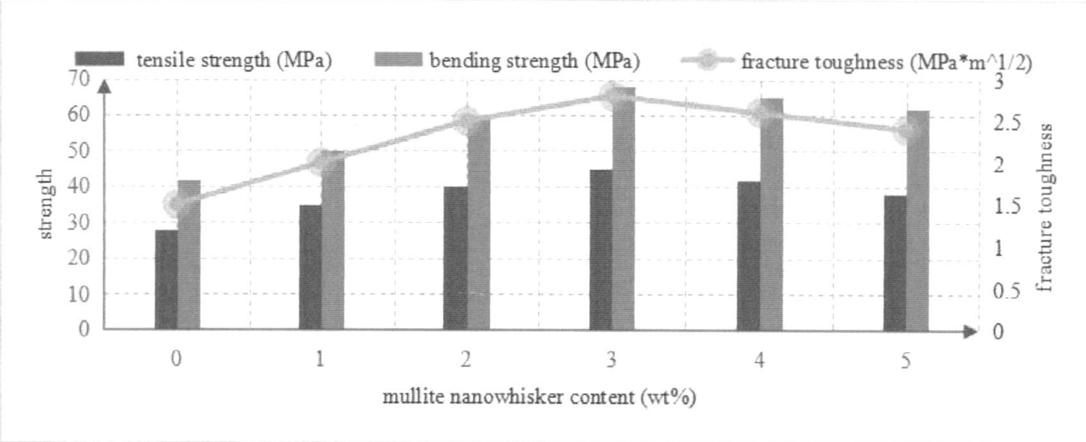

Figure 2. Mechanical property analysis

3.2. *Thermal performance analysis*

In the characterization of the thermal properties of mullite nanowhisker-reinforced polyurethane composites, determining the coefficient of thermal expansion (CTE) is essential to evaluate the dimensional stability of the materials when the temperature changes. Especially in 3D printing, it directly affects the printed parts' accuracy and long-term stability, providing a key basis for material design optimization. A thermal mechanical analyzer (TMA) was used to heat the sample from room temperature to high temperature at a constant heating rate, record the length change, and calculate the coefficient of thermal expansion by analyzing the length temperature curve. The results show that the thermal expansion coefficient of the composite is about 20 units lower than that of pure polyurethane, indicating that it has better dimensional stability, as is shown in Figure 3.

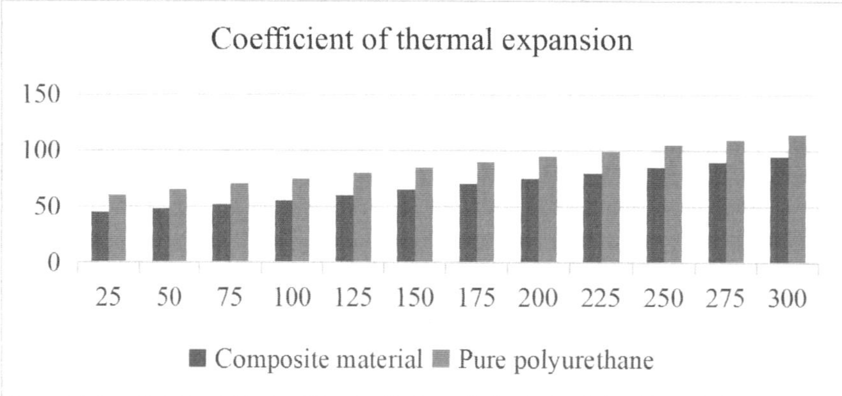

Figure 3. Thermal performance analysis

Thermogravimetric analysis revealed that the composite material exhibited markedly elevated residual content compared to pure polyurethane across varying temperature ranges, demonstrating superior thermal stability characteristics, as is shown in Table 1.

Table 1 Thermal stability comparison

Temperature range (°C)	Mass loss of pure polyurethane (%)	Mass loss of composites (%)
25 - 200	< 5 %	< 3 %
200 - 300	5 % - 20 %	3 % - 10 %
300 - 400	20 % - 60 %	10 % - 40 %
400 - 500	60 % - 90 %	40 % - 70 %
> 500	> 90 %	70 % - 80 %

3.3. *Micro morphology analysis*

The composite material's microstructural characteristics were examined using scanning electron microscopy (SEM). It was found that mullite nanowhiskers were evenly dispersed in the polyurethane matrix and formed a good interface with it. These observed microstructural characteristics, as shown in Figure 4, elucidate the underlying enhancement mechanisms responsible for the composite materials' concurrent mechanical performance and thermal stability improvement.

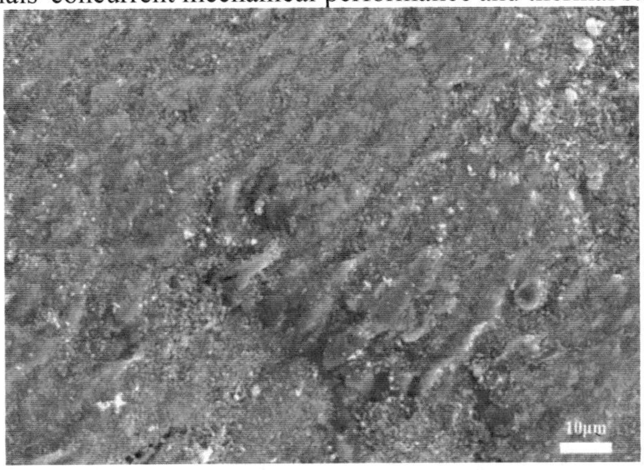

Figure 4. SEM of mullite nanowhisker/polyurethane nanocomposites

4. Experiment results

4.1. *Research on 3D printing technology*

Based on the rheological properties of composites, FDM was selected as the 3D printing process. The printing parameters, including printing temperature, printing speed, layer thickness, and filling rate, were systematically optimized through single-factor experimental analysis combined with response surface methodology (RSM). The experimental results demonstrate that 3D-printed specimens exhibit superior moulding quality and optimal mechanical properties when the printing temperature is 220 °C, the printing speed is 50 mm/s, the layer thickness is 0.2 mm, and the filling rate is 80 %.

X-ray tomography technology was used to analyze the internal defects of printed products to control the forming quality. The results show that the optimized process parameters can effectively reduce defects such as porosity and poor interlayer bonding. By adjusting the printing path and adding the surface treatment process, the product's surface roughness Ra value was reduced from 8.5 μm to 4.2 μm, and the surface quality was significantly improved.

4.2. *Performance evaluation of 3D printed products*

The mechanical performance of 3D-printed products were evaluated, and the thermal performance test showed that the printed products maintained good thermal stability, as shown in Table 2.

Table 2 Performance comparison of 3D printed products

Performance index	Pure polyurethane	Composite material	Change trend
Tensile strength (MPa)	25	38	improved
Elastic modulus (GPa)	0. 5	1. 2	improved
Bending strength (MPa)	30	52	improved
Initial decomposition temperature (°C)	250	300	improved
Coefficient of thermal expansion (μm / m·°C)	115	95	lower

The microstructure analysis shows that mullite nanowhiskers still maintain a good dispersion state in the printed products, but there is a certain degree of orientation. Investigation of interfacial bonding mechanisms reveals that the thermal processing history during the additive manufacturing process significantly influences the interfacial adhesion characteristics between nanowhiskers and the matrix material. The chemical corrosion resistance test showed that the printing products showed good stability in acid, alkali and organic solvents.

The printing accuracy evaluation results show that the product's dimensional error is controlled within ± 0.2 mm, which meets the accuracy requirements of general engineering applications. The surface roughness test showed that the Ra value of the product surface after post-treatment could be reduced to 3.5 μm, which was close to the surface quality level of injection moulding products.

5. Conclusions

In this study, mullite nanowhisker-reinforced polyurethane composites were successfully prepared, and their application in 3D printing was realized. The experimental results demonstrate that the incorporation of mullite nanowhiskers significantly enhances the mechanical properties and thermal stability of the composite materials. Compared with pure polyurethane, the composites' tensile properties, bending properties, and fracture toughness were increased by 60 %, 62 %, and 87 %, respectively. At the same time, the coefficient of thermal expansion is reduced by about 20 units.

By optimizing 3D printing process parameters, printing products with good comprehensive performance can be prepared. This study proposes an innovative conceptual framework for the design and fabrication of advanced 3D printing materials while demonstrating significant potential for multidisciplinary implementation in aerospace engineering, the automotive industry, the biomedical sectors, and related disciplines.

Acknowledgments

This research was funded by Scientific Research and Technology Development Program of Guilin (20230120-9). Natural Science Foundation of Guangxi Province (2025GXNSFAA069202).

References

[1] Acua P, Zhang J, Yin G Z, et al. Bio-based rigid polyurethane foam from castor oil with excellent flame retardancy and high insulation capacity via cooperation with carbon-based materials[J]. Journal of Materials Science, 2021,56(3): 2684-2701.

[2] Panda S S, Panda BP, Nayak S K, et al. A review on waterborne thermosetting polyurethane coatings based on castor oil: synthesis, characterization, and application[J]. Polymer-Plastics Technology and Engineering, 2018,57(6): 500-522.

[3] QU Q Q, WANG H, HE J, et al. Analysis of the microphase structure and performance of self-healing polyurethanes containing dynamic disulfide bonds[J]. Soft Matter, 2020,16(39): 9128-9139.

[4] BEHERA P K, RAUT S K, MONDAL P, et al. Self-healable polyurethane elastomer based on dual dynamic covalent chemistry using diels-alder "click" and disulfide metathesis reactions[J]. ACS Applied Polymer Materials, 2021,3(2): 847-856.

[5] Wendels S, Balahura R, Dinescu S, et al. Influence of the Macromolecular architecture on the properties of biobased polyurethane tissue adhesives[J]. European Polymer Journal, 2022,164: 110968.

[6] Abdullayev A, Klimm D, Kamutzki F,et al. AlF3-Assisted Flux Growth of Mullite Whiskers and their Application in Fabrication of Porous Mullite-alumina Monoliths[J]. Open Ceramics, 2021,7(1):100145.

[7] Gao J M, Wang B, Du Z Y,et al. Molten salt synthesis of mullite whiskers entirely derived from fly ash for electronic packaging toughening ceramic applications[J]. Journal of Materials Research and Technology, 2022,21(1):3719-3731.

[8] Abdullayev A, Zemke F, Gurlo A,et al. Low-temperature fluoride-assisted synthesis of mullite whiskers[J]. RSC Advances, 2020,10(52):31180-31186.

Life Prediction of Railway Relay Based on an Attention Mechanism TCN-BiGRU Model

Xianfeng LV[1,2*][0009-0001-5544-2338], Shuxin Liu[1,2], Xinzhi Qi[1,2], Xin Ming[1,2], Bingze Liu[1,2]

[1] Key Laboratory of Special Electric Machines and High Voltage Apparatus in the Ministry of Education, Shenyang University of Technology, Shenyang, Liaoning, 110870, China

[2] Shenyang University of Technology, Shenyang, Liaoning

*Corresponding author's e-mail: 2015564679@qq.com

Abstract. To address the problem of accurate and efficient prediction of railway relay lifespan, this study proposes an electrical lifespan prediction method for railway relays based on the integration of attention mechanisms, Temporal Convolutional Network and Bidirectional Gated Recurrent Unit. Considering the potential variability in the importance of temporal features, we introduce an attention mechanism to reallocate feature weights dynamically. The TCN module is employed to extract local temporal features from relay operation data, while the BiGRU component is utilized to capture long-term dependencies and obtain global temporal information. Experimental validation was conducted using characteristic parameters extracted from a railway DC relay electrical lifespan testing system platform. The results indicate that the proposed model outperforms in terms of prediction accuracy (with an average of 96.85% precision), minimal error rate, and enhanced stability compared with existing models, thereby confirming the practical feasibility of this prediction framework.

1. Introduction

Relays serve as critical control components in railway signaling systems, exerting substantial influence on system safety and reliability[1]. With the rapid expansion of high-speed rail networks and increasing demands for railway transportation, signaling systems face escalating requirements for operational complexity and precision. As core elements of these systems, relays must adapt to increasingly stringent operating environments and elevated technical standards. Beyond their traditional switching functions, modern relays are now required to provide stable and reliable support within complex automated and intelligent control architectures. Consequently, relay performance and quality directly impact railway safety, operational efficiency, and system stability, making accurate lifespan assessment and prediction crucial for ensuring high-speed rail operational safety. This positions relay lifespan prediction as a research area with significant theoretical value and broad practical applications.

Recent years have witnessed growing attention to predictive maintenance across various domains including photovoltaics[2-3], wind power, contactors, and circuit breakers. In photovoltaic applications, literature[4] proposed an innovative framework integrating improved LSTM networks with GAN for power generation forecasting. Another study[5] developed a hybrid model combining CNN and LSTM, demonstrating superior performance in photovoltaic power prediction through enhanced accuracy that correlates with input sequence length.

Journal of Physics: Conference Series **3024** (2025) 012006 doi:10.1088/1742-6596/3024/1/012006

To address the challenges of modeling long-term dependencies in sequential data, this study introduces an attention mechanism to construct a temporal attention module. By adaptively assigning distinct weights to individual temporal nodes within the BiGRU hidden units, the model achieves weighted processing of temporal features based on their contextual relevance to prediction outcomes. The proposed BiGRU-attention hybrid architecture not only enhances prediction accuracy but also effectively addresses feature interdependency issues caused by extended input sequences. Experimental validation demonstrates that this methodology significantly improves the precision of electrical lifespan prediction for railway DC relays, thereby advancing the operational reliability and safety assurance of railway systems.

2. Fundamental principle

2.1. TCN

Temporal Convolutional Network (TCN), an advanced architecture derived from convolutional neural networks (CNNs), integrates causal convolution with dilated convolution mechanisms for temporal sequence modeling. This framework demonstrates significant advantages in computational efficiency while maintaining extensive temporal receptive fields. The architecture exhibits enhanced parallel processing capabilities and flexible sequence memory control through adjustable receptive field dimensions, thereby enhancing model adaptability and performance across diverse temporal modeling tasks [6]. Dilated Convolution Architecture and Residual Block are as follows.

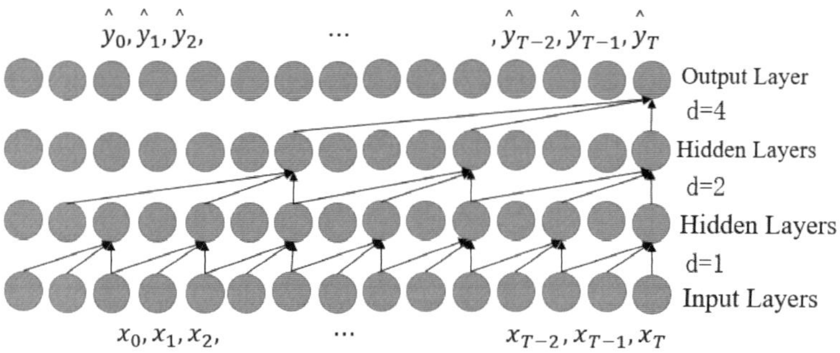

Figure 1. Dilated Convolution Architecture in TCN

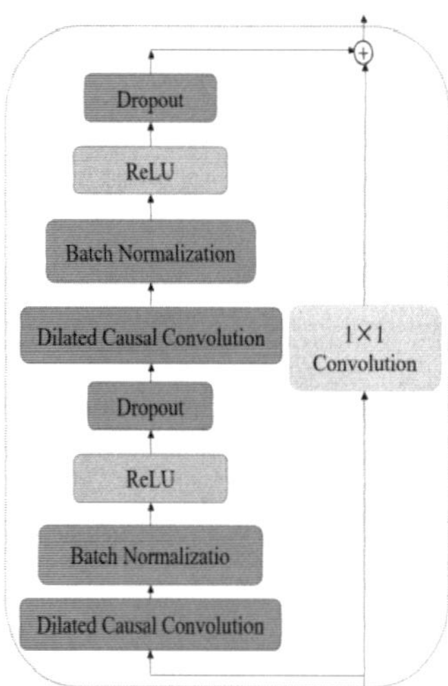

Figure 2. Residual Block in TCN

2.2. *BiGRU*

Compared to the recurrent neural network (RNN) framework, temporal convolutional networks (TCNs) may not demonstrate comparable efficacy in capturing long-term dependency patterns. In contrast, RNN architectures inherently possess the capability to model sequential dependencies of arbitrary lengths through their cyclic computational structure [7]. Both GRU [8] and (LSTM) networks [9], categorized under the RNN paradigm, introduce gated mechanisms to regulate information flow. The architectural distinction lies in their gating components: LSTM employs three distinct gates (input, forget, and output gates), whereas GRU, as a streamlined variant of LSTM, retains only two gates (reset and update gates). Relative to LSTM, GRU exhibits enhanced parametric efficiency through its simplified topology, as illustrated in Figure 3.

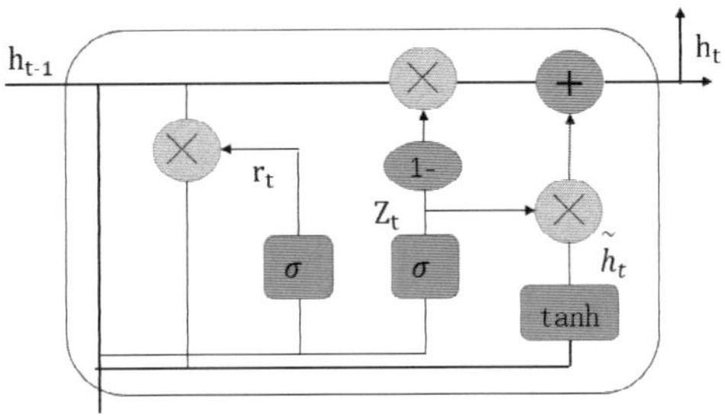

Figure 3. Architectural Schematic of GRU Model

The formulas for GRU are as follows :

$$r_t = \sigma\left(W_r \cdot [h_{t-1}, x_t]\right) \tag{1}$$

$$z_t = \sigma\left(W_z \cdot [h_{t-1}, x_t]\right) \tag{2}$$

$$h_t = \tanh\left(W_h \cdot [r_t \cdot h_{t-1}, x_t]\right) \tag{3}$$

$$h_t = (1 - z_t) \cdot h_{t-1} + z_t \cdot h_t \tag{4}$$

In the equation: r_t represents the reset gate, represents the update gate, h_t denotes the candidate output. W is the weight matrix that linearly transforms the input data. σ denote the logistic Sigmoid activation function. The indices t and t-1 correspond to the current and preceding time steps in the sequence. The four equations delineate critical computational operations within the gated recurrent unit [10]. The design enables simultaneous learning from past and future temporal contexts using GRU cells. The schematic representation of the BiGRU architectural configuration is provided in Figure 4.

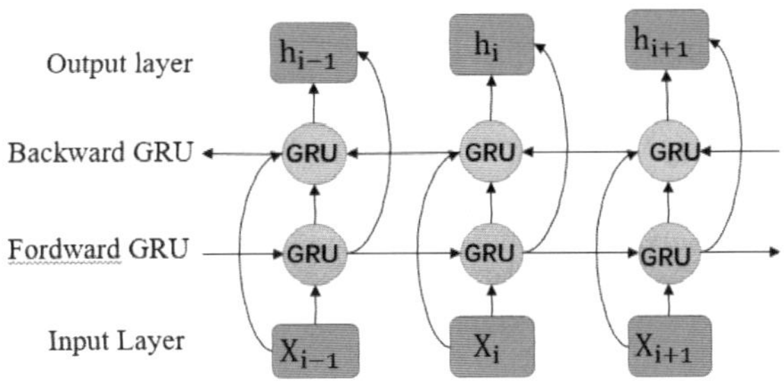

Figure 4. Architectural Schematic of Bidirectional Gated Recurrent Unit (BiGRU) Model

2.3. *Attention mechanism*

Through the incorporation of attention mechanisms [11], the model dynamically recalibrates temporal attention weights across different timesteps, enabling it to selectively focus on discriminative features critical to operational patterns.

$$e_t = u_a \tanh\left(w_a h_t + b_a\right) \tag{5}$$

$$a_t = \frac{\exp\left(e_t\right)}{\sum_{j=1}^{t} e_j} \tag{6}$$

where h_t is the state vector of the hidden layer of the neural network at time t, a_t represents the weight assigned to the feature by the attention module, w is the weight matrix, b is bias, and e_t is the probability distribution.

2.4. Comparative analysis of evaluation metrics and prediction results

Based on the prediction results, calculate four parameters *RMSE*、*MAE*、R^2 and Average accuracy to assess the model's performance. The formulas for *MSE*、*MAE*、$R^2$are as follows:

$$RMSE = \sqrt{\frac{1}{n}\sum_{t=1}^{n}(\hat{y}_t - y_t)^2} \tag{7}$$

$$MAE = \frac{1}{n}\sum_{t=1}^{n}\left|\hat{y}_t - y_t\right| \tag{8}$$

$$R^2 = \frac{\sum_{t=1}^{n}(\hat{y}_t - \bar{y})^2}{\sum_{t=1}^{n}(y_t - \bar{y})^2} = 1 - \frac{\sum_{t=1}^{n}(y_t - \hat{y}_t)^2}{\sum_{t=1}^{n}(y_t - \bar{y})^2} \tag{9}$$

$$\text{Average accuracy} = 100\% - \left(\frac{1}{n}\sum_{t=1}^{n}\frac{y_t - y_t}{y_t}\right) \tag{10}$$

In the equations, n is used to denote the total number of samples within the dataset, y_t represents the factual value of the remaining interruption counts, $\bar{y}$ is the mean of the factual values, and $\hat{y}_t$ corresponds to the final output of the predictive model. RMSE and MAE are inversely proportional to prediction accuracy, whereas the coefficient of determination (R2) is directly proportional to prediction performance.

3. Feature parameter extraction and data preprocessing

3.1. Feature parameter extraction.

The extraction process for these parameters is described as follows:

1)Extraction of Contact Resistance

$$R_n = \frac{\sqrt{\frac{1}{N}\sum_{n=1}^{N}u_n^2}}{\sqrt{\frac{1}{N}\sum_{n=1}^{N}i_n^2}} = \sqrt{\frac{\sum_{n=1}^{N}u_n^2}{\sum_{n=1}^{N}i_n^2}} \tag{11}$$

In the equation: N denotes the number of sampled data points during the steady-state closed contact condition; u_n, i_n represents the contact voltage and contact current, respectively.

2) Extraction of Pull-in Time

$$t_x = t_2 - t_1 \tag{12}$$

In the equation: t_1 denotes the coil energization time, t_2 represents the first contact time between the moving and stationary contacts.

3) Extraction of Overtravel Time:

$$t_c = t_4 - t_2 \tag{13}$$

In the equation, t_4 denotes the time of stable closure.

4) Extraction of Contact Bounce Time:

$$t_t = t_3 - t_2 \tag{14}$$

In the equation, t_3 denotes the end time of significant voltage and current fluctuations.

5) Extraction of Release Time:

$$t_s = t_6 - t_5 \tag{15}$$

In the equation, t_5 denotes the coil de-energization time, t_6 denotes the arc initiation time.

6) Extraction of Arc Energy:

$$E = \sum_{n=t_6}^{t_7} u_n \cdot i_n \cdot \Delta t = \frac{\sum_{n=t_6}^{t_7} u_n \cdot i_n}{f_s} \tag{16}$$

In the equation, t_7 denotes the arc extinction time, Δt denotes the Sampling Interval, f_s denotes the Sampling Rate.

3.2. Data preprocessing

Since the characteristic parameters in the optimal feature subset of the DC relay have different dimensions, it is necessary to perform normalization preprocessing before model training to obtain dimensionless sample parameters. The normalization formula is defined as Equation (19):

$$x_t^* = \frac{x_t - x_{\min}}{x_{\max} - x_{\min}} \tag{17}$$

where x_t represents the raw characteristic parameter data, $x_{\max}$, $x_{\min}$ denotes the maximum and minimum values of the raw characteristic parameters. x_t^* denotes normalized standard data.

4. Model construction and predictive results

4.1. Model construction

This study proposes a hybrid prediction framework integrating an attention-enhanced TCN with a BiGRU. The specific framework composition is presented in Figure 5.

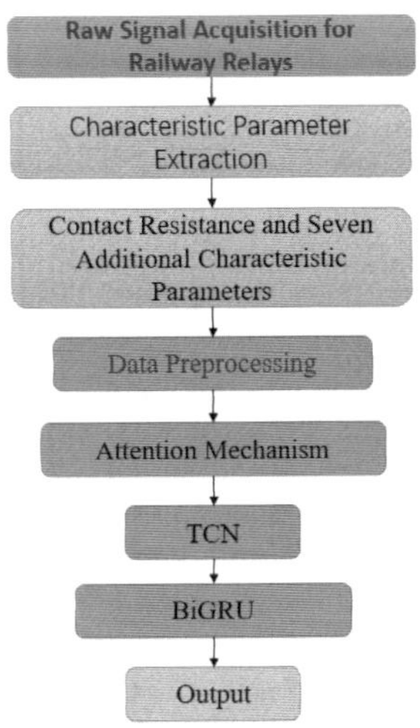

Figure 5. Process Flow Diagram of Relay Lifetime Prediction

4.2. Analysis of predictive results

Under identical experimental conditions, four models TCN, BiGRU, TCN-BiGRU, and TCN-BiGRU integrated with an attention mechanism were employed for railway relay life prediction, and their performance was systematically compared. The life prediction curves of the four models are illustrated in Figures 6, while their corresponding error curves are presented in Figures 7.

This comparative analysis highlights the predictive accuracy and robustness of each model under consistent testing scenarios. From Figures 6, by comparing the fitting degree between prediction results and actual lifespan curves, we can conduct the following qualitative analysis: When analyzing long-term time series data using the TCN, BiGRU, and TCN-BiGRU models, although these models can capture data trends to some extent, their predicted values often exhibit significant fluctuations and generally deviate substantially from the actual lifespan curves.

Comparative analysis reveals that the proposed model demonstrates significant performance advantages in predictive tasks. By leveraging a temporal attention mechanism, the architecture dynamically prioritizes critical temporal features and key time points within sequential data, thereby achieving enhanced prediction accuracy through context-aware feature refinement.

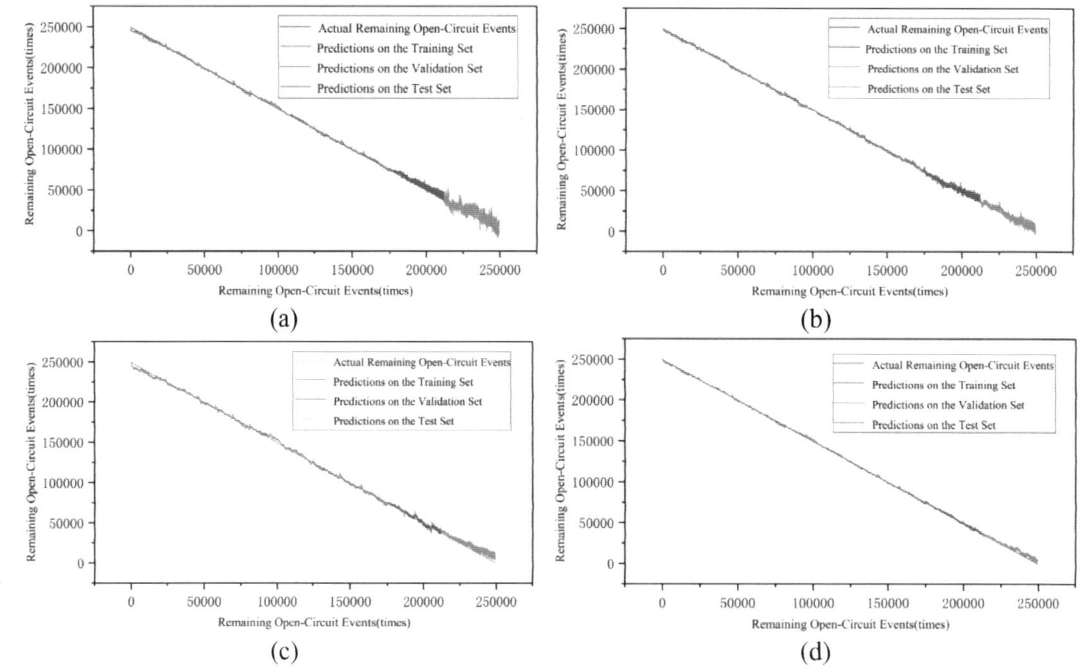

Figure 6. Comparative Visualization of Prediction Performance

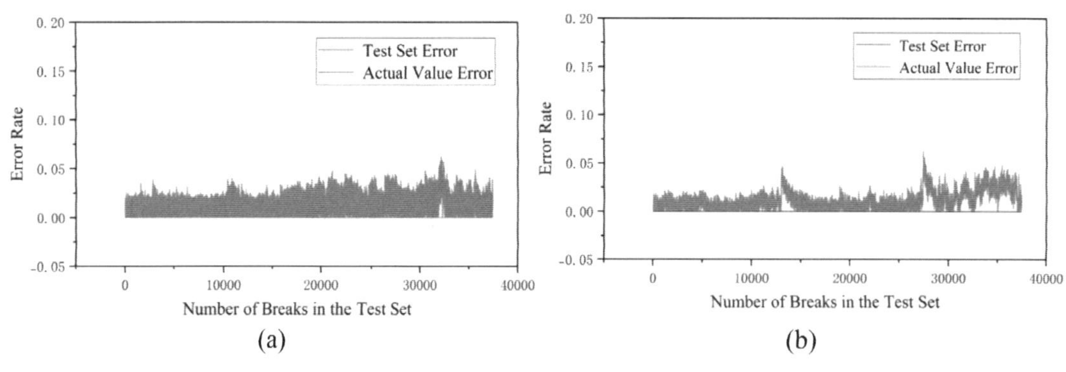

Journal of Physics: Conference Series **3024** (2025) 012006 doi:10.1088/1742-6596/3024/1/012006

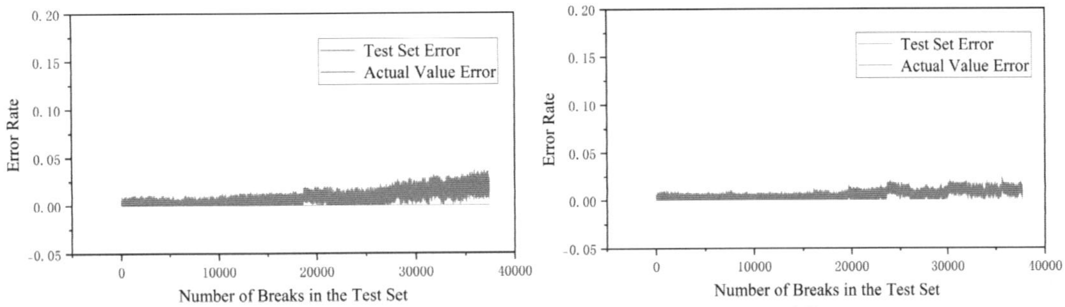

(c) (d)

Figure 7. Predictive Error

Through the analysis of the prediction error results in Figure 7, experimental results demonstrate that the proposed model has a significant advantage in predicting railway relay life. In terms of prediction error, the model presented in this paper shows a lower error rate compared to other comparison models.

From Table 1, the proposed model achieves significant reductions in RMSE, MAE, and maximum error rate alongside notably higher effective accuracy (96.85%) and R^2 values compared to the three benchmark models. These metrics collectively demonstrate the model's superior accuracy, enhanced prediction stability, and pronounced overall performance advantages in electrical lifespan prediction tasks.

Table 1. Comparison of Prediction Model Results

Prediction Model	TCN	BiGRU	TCN-BiGRU	Attention -TCN-BiGRU
RMSE	5475.70	4266.30	1566.30	1135.23
MAE	2809.12	2265.35	1249.57	401.29
R2	0.896	0.923	0.952	0.987
Average accuracy	85.33%	90.51%	94.02%	96.85%

5. Conclusions

This paper proposes a railway relay lifespan prediction method based on an attention-enhanced TCN-BiGRU framework. The TCN component extracts local temporal features from relay operation data, while the BiGRU captures long-term dependencies to obtain global temporal information. An attention mechanism is introduced to reallocate feature weights, further enhancing the model's capability.

The proposed prediction model outperforms the TCN, BiGRU, and TCN-BiGRU models in terms of accuracy, achieving an average accuracy of 96.85% with lower error rates and superior stability, thereby validating its feasibility for practical applications. This work provides a viable solution for addressing the lifespan prediction challenges of DC relays.

Acknowledgments

This work was financially supported by the Major Science and Technology Special Project of Liaoning Province 2020JH1/10100012, and by Project of Education Department of Liaoning Province LJGD2020001, and by Young and Middle-aged Science and Technology Innovation Talent Program of Shenyang RC210354.

References

[1] Wang X., Zhao X.H., Yao W., &Sun. H.D., 2023 Railway Relay Life Test System Based on PIC Microcontroller. Ailway Communication and Signal.

[2] Shi M., Xu K., Wang J., Yi R.,&Zhang P., 2021 Short-term Photovoltaic Power Forecasting Based on Grey Relational Analysis and GeoMAN Model. Transactions of China Electrotechnical Society, 36(11), pp. 2298-2305.

[3] Wu C.H., Dong A.L, Li Z.H.,& Wang F. 2022 Photovoltaic Power Forecasting Based on Graph

Similarity and PSO-XGBoost. High Voltage Engineering, 48(08), pp. 3250-3259.

[4] Yin H., Zhang Z., Ding W.F. , Chen J.M.& Chen S., 2021 Short-term Photovoltaic Power Forecasting with Few Samples Based on Generative Adversarial Network and LSTM-CSO.High Voltage Engineering, 48(11), pp. 1-10.

[5] Wang K, Qi X, Liu H., 2019 A Comparison of Day-ahead Photovoltaic Power Forecasting Models Based on Deep Learning Neural Network. Applied Energy. pp. 113-135.

[6] Hewage, P., Behera, A., Trovati, M., Pereira, E., Ghahremani, M., Palmieri, F., & Liu, Y. 2020, Temporal convolutional neural (TCN) network for an effective weather forecasting using time-series data from the local weather station. Soft Computing, 24, 16453-16482.

[7] Dudukcu, H. V., Taskiran, M., Taskiran, Z. G. C., & Yildirim, T. 2023, Temporal Convolutional Networks with RNN approach for chaotic time series prediction. Applied soft computing, 133, 109945.

[8] Mohsen, S. 2023, Recognition of human activity using GRU deep learning algorithm. Multimedia Tools and Applications, 82(30), 47733-47749.

[9] Wen, X., & Li, W. 2023. Time series prediction based on LSTM-attention-LSTM model. IEEE access, 11, 48322-48331. 187–193.

[10] Zhang Y.C., Wang H.B., 2021 Chinese Text Sentiment Analysis Based on Att-BiGRU-CRF Model. Journal of Tianjin University of Technology,37(06), pp. 31-35.

[11] Kang T.Y, Qin Z.J., 2022 Ultra-short-term Wind Power Forecasting Based on Hyperparameter Optimization and Dual-Attention Mechanism. Southern Power Grid Technology, 16(05), pp. 44-49.

Multi-scene trajectory planning of robot arm based on DQN algorithm

Jiaqiang Zhang, Huiqing Chen*, Yixiong Wang, Hongzhen Xu

Key Laboratory of Advanced Manufacturing and Automation Technology (Guilin University of Technology), Education Department of Guangxi Zhuang Autonomous Region, Guilin, 541006, China

College of Mechanical and Control Engineering, Guilin University of Technology, Guilin, 541006, China

E-mail: chenhuiqing-kmust@163.com

Abstract. To overcome the challenges of low accuracy and poor adaptability to high-dimensional spaces in deep reinforcement learning algorithms, presents an improved algorithm that integrates the APF method with the Q-network. The aim is to improve the obstacle avoidance efficiency of robotic arms in complex environments. By introducing a novel reward , avoid obstacles is greatly enhanced. Furthermore, the incorporation of a competitive network, mitigates the overestimation issue of the value function. This results in faster convergence, reduced training time, and guarantees the optimality of the trajectory. Experimental shows the improved algorithm significantly change the navigation capabilities of robotic arms in various environments, offering a new approach for trajectory planning in complex settings.

1. Introduction

In today's rapidly advancing era of automation and intelligence, the progress of robotic technology is particularly significant.One of the fundamental challenges in robotics is path planning, which directly impacts the robot's work efficiency, safety, and the scope of its applications.Therefore, in-depth research and continuous optimization of path planning algorithms are crucial for advancing the field of robotics and have a profound influence on the modernization and upgrading of related industries and technologies.

Traditional path planning algorithms, such as the heuristic-based A* algorithm, graph search-based Rapidly-exploring Random Tree (RRT) algorithm, Artificial Potential Field method, and Simulated Annealing method, have been widely studied and applied [1-3].These methods, while effective in certain environments, often face limitations related to real-time performance, scalability, and the ability to handle dynamic obstacles.Furthermore, these algorithms may struggle with complex or high-dimensional planning spaces.In contrast, swarm intelligence algorithms, such as Ant Colony Optimization, Particle Swarm Optimization, and Genetic Algorithms, have emerged as promising alternatives, offering more flexibility and robustness in various conditions, but still suffer from issues like local minima and convergence speed[4-6].

Recently, advancements in machine learning have led to the adoption of Deep Q-Networks (DQN) for path planning. Li introduced a method that utilizes DQN, significantly enhancing both learning efficiency and path planning accuracy by employing a densely connected network architecture[7]. While DQN provides notable improvements over traditional methods, it still faces challenges in dealing with the exploration-exploitation tradeoff and ensuring real-time performance in complex environments.Yi improved the DQN algorithm by decoupling the two steps of action selection and Q-value estimation within the neural network, effectively addressing the overestimation problem of Q-values [8].This approach, which enhances both learning efficiency and path planning accuracy, stands out by creating an experience evaluation network and a parallel exploration structure. These features adapt to various learning stages, improving the depth and scope of experience.Unlike traditional methods, Yi's improvement offers a more refined approach to exploration, making it more suitable for dynamic and unpredictable environments.

The contributions of this paper lie in further enhancing these machine learning-based methods. By building on Yi's improvements, this work aims to bridge the gap between traditional algorithms and deep learning approaches, ensuring a balance between accuracy, efficiency, and real-time responsiveness.This will not only improve the robot's path planning capabilities in complex environments but also offer a more scalable solution for real-world applications.

2. Related work

2.1. Kinematic model

The improved ABB IRB460 robotic arm is used as the research object in this paper. The improved three-axis 6-DOF robotic arm system consists of a fixed base, a rotating base, the A-axis, B-axis, C-axis, and a gripper. The fixed base supports the main body of the robotic arm; the A-axis drives the rotating base, with a rotation range of 0° to 70°; the B-axis directly drives the BC linkage to rotate via the B-axis servo, with a rotation range of 0° to 69°; the C-axis is connected to the gripper at its end to enable object grasping, with a rotation range of 30° to 180°. The 3D model of the robotic arm is shown in Figure 1.

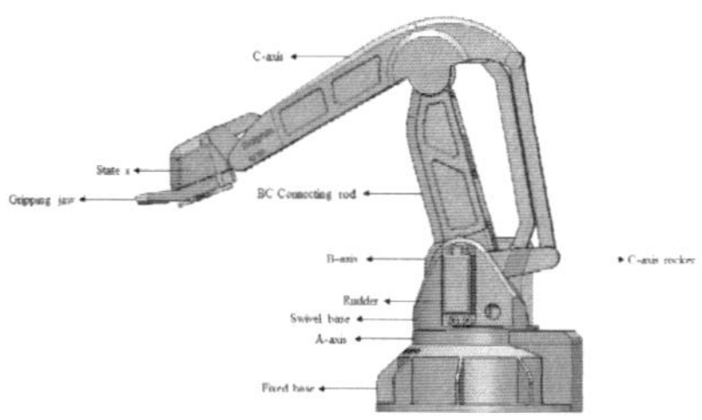

Figure 1. 3D model of robot arm.

2.2. Mechanical arm collision detection

In this paper, cylinders are used to represent the robotic arm's links, simplifying the collision detection problem to the detection of collisions between cylinders, spheres, and cuboids.As depicted in Figure 2, let the spherical obstacle be (x, y, z). Here, the cylinder is considered as a simplified model of the robotic arm, with A_i and B_i representing the two ends of the link i, and the radius being ω_i. For collision detection with a spherical bounding box, the sphere is treated as an obstacle in the configuration space,for a sphere with radius r and the separated from the connecting rod by being d_i. If $d_i > r + \omega_i$, It signifies that the robotic arm will avoid colliding with the obstacle; otherwise, a collision occurs. Additionally, for collision detection with an axial bounding box, if the line segment A_iB_i lies outside the obstacle's axial bounding box, no collision will occur. Otherwise, there is a risk of collision.

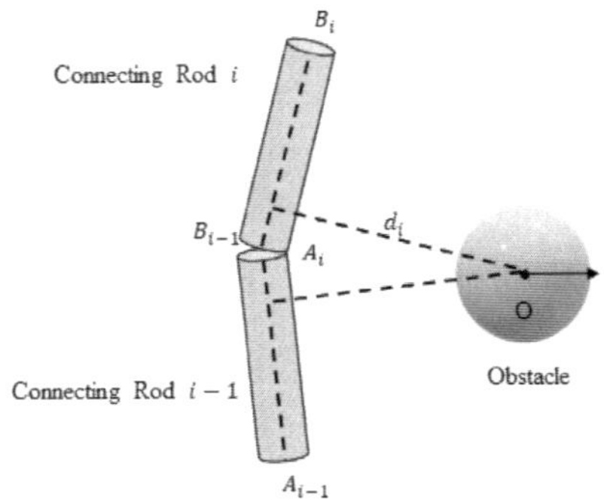

Figure 2. 3D model of robot arm.

3. Improved DQN algorithm

3.1. DQN algorithm

The Q-learning algorithm relies on a Q-table to store and update the Q-values corresponding to each state-action pair. In high-dimensional environments, when the number of states or actions is very large, the efficiency of Q-learning decreases significantly, leading to a sharp increase in storage and computation costs. DQN algorithm can manage high dimensional state space, as shown in Figure 3:

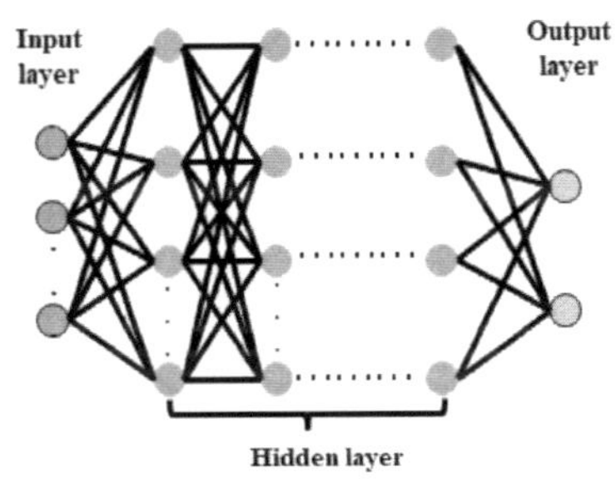

Figure 3. Competitive network architecture.

Formula (1) is the approximate q value:

$$Q(s_t, a_t) \approx Q(s_t, a_t, w) \tag{1}$$

In this context, w represents the network parameters, Q-function $Q(s_t, a_t)$ indicates the expected reward . Once w is determined, the Q-function is also determined. The DQN framework uses both an online Q-network, which selects actions, and a target Q-network, which evaluates the actions. This separation helps reduce bias in Q-value estimation. The online network selects actions as described in equation (2):

$$a_t = \arg\max_a Q(s_t, a; w) \tag{2}$$

The target Q-value is computed through the target network, which evaluates the actions selected by the online network as shown in equation (3):

$$y_t = r_t + \gamma \left(Q_{\text{online}}(s_{t+1}, \arg\max_a Q_{\text{online}}(s_t, a; w); w^-) \right) \tag{3}$$

Journal of Physics: Conference Series **3024** (2025) 012007 doi:10.1088/1742-6596/3024/1/012007

Where, r_t represents the immediate reward, s_{t+1} is the next state, and γ is the discount factor. The parameters of the target network, w^- are periodically updated from the online network.

3.2. Competitive network architecture

Initially, during the exploration and policy update stages of the robotic arm, an enhanced DQN algorithm employs two distinct networks. Next, by incorporating the APF method, the robotic arm is directed toward the target using an attractive field, while obstacles are avoided through a repulsive field. Additionally, a competitive network is introduced to change the model's learning performance and stability.

The Dueling Network Architecture is an improved deep Q-network structure.This decomposition allows the dueling network architecture to more effectively learn the value of each state of the robotic arm and the relative advantages of different actions. The dueling architecture splits the final convolutional layer of the existing network into two halves, each connected to a fully connected layer.Figure 4 shows the structure of the dueling network.

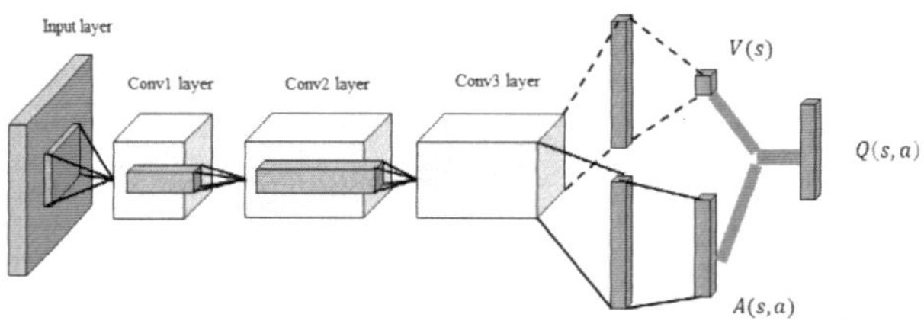

Figure 4. the structure of the dueling network.

Finally, the two estimated results are integrated into the function $Q(s, a)$ as equation (4):

$$Q(s,a) = V(s) + \left(A(s,a) - \frac{1}{|A|} \sum_{a'} A(s,a') \right) \tag{4}$$

Where, $|A|$ is the action space, a' refers to all possible actions.

In Dueling DQN, the variance of the advantage function is commonly used to measure the differences between actions. The variance can be calculated using the formula in equation (5):

$$V(A(s)) = \frac{1}{|A|} \sum_{a'} \left(A(s,a') - \mu \right)^2 \tag{5}$$

Where, μ is the mean of the advantage function.

3.3. APF reward function

In the process of reinforcement learning, the action taken by the robotic arm in each state interacts with the environment to receive a reward, which is then used to update the network parameters and find the optimal policy. Therefore, the reward function is a crucial component of reinforcement learning trajectory planning, and its design directly impacts both the trajectory planning performance of the algorithm and the convergence of the network. Under the reward function setup of the DQN algorithm, there are certain limitations, including the risk of getting stuck in local optima and the possibility of negative feedback.In view of these shortcomings, this paper proposes an artificial potential field to complete the trajectory planning of the robot arm. The design of the reward function is achieved by combining attractive and repulsive forces, enabling the robotic arm to effectively avoid obstacles and move toward the target.

Initially, specify the robotic arm's present location as $s = (x, y, z)$, the target position as $g = (g_x, g_y, g_z)$, and the set of obstacle positions as $O = \{(o_{x1}, o_{y1}, o_{z1}), (o_{x2}, o_{y2}, o_{z2}), \ldots, (o_{xn}, o_{yn}, o_{zn})\}$. Based on the above definitions, the reward r_t is as:

$$r_t = -|s-g| + \sum_{(o_x,o_y,o_z)\in O} R_{\text{obstacle}}(s, (o_x, o_y, o_z)) + r_{\text{goal}} \tag{6}$$

The term $-|s-g|$ indicates the distance , where negative values imply that the reward increases the target, the higher the reward. $R_{\text{obstacle}}(s, (o_x, o_y, o_z))$ represents the repulsive force , defined as:

$$R_{\text{obstacle}}(s, (o_x, o_y, o_z)) = \begin{cases} -\frac{1}{|s-g|}, & \text{if } |s - (o_x, o_y, o_z)| < 2 \\ 0, & \text{otherwise} \end{cases} \tag{7}$$

A negative reward is given when the robotic arm's distance to the obstacle is less than 2. The penalty increases as the arm gets closer to the obstacle, encouraging it to avoid collisions. r_{goal} is the reward for reaching the target position, and it is defined as:

$$r_{\text{goal}} = \begin{cases} 100, & \text{if } s = g \\ 0, & \text{otherwise} \end{cases} \tag{8}$$

Table 1. Algorithm performance comparison in environment.

Algorithm	Mean track length (mm)	Average time (s)	Success rate
DQN	142.68	150.28	80
Textual algorithm	112.56	87.54	97

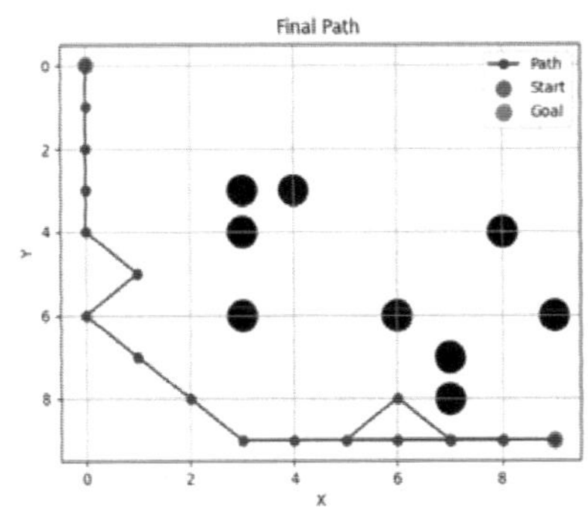

Figure 5. DQN algorithm test results in environment 1.

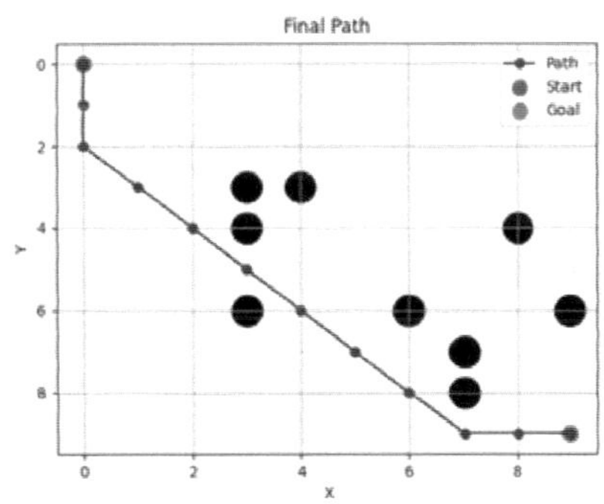

Figure 6. Textual algorithm test results in environment 1.

The enhanced reward function considers different reward and penalty mechanisms while integrating the Artificial Potential Field approach. Through this integrated design, the robotic arm can efficiently learn in the environment, avoid obstacles, and quickly achieve the target.

Journal of Physics: Conference Series **3024** (2025) 012007 doi:10.1088/1742-6596/3024/1/012007

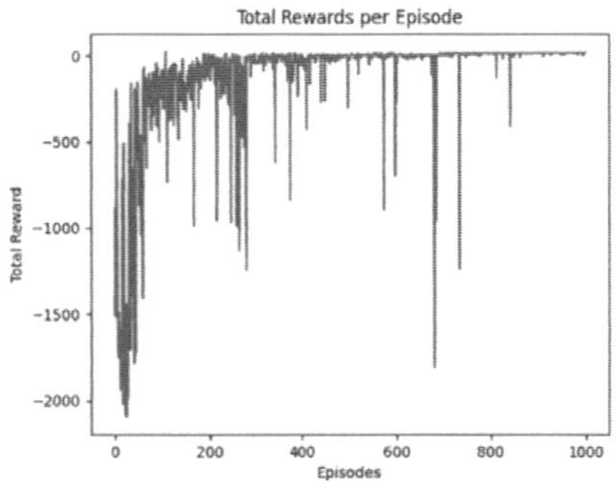

Figure 7. Convergence curve of DQN algorithm in environment 1.

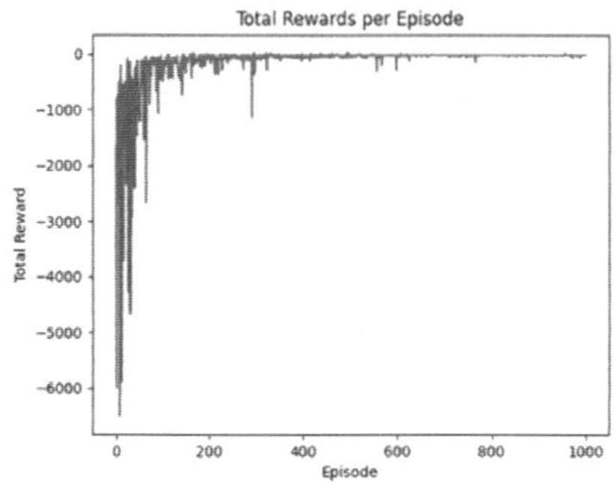

Figure 8. Convergence curve of textual algorithm in environment 1.

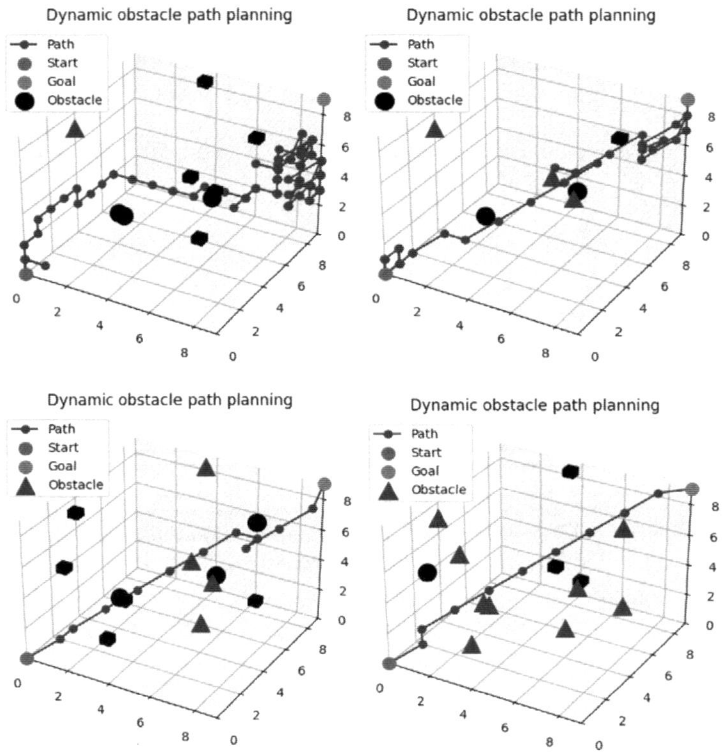

Figure 9. Three-dimensional dynamic obstacle avoidance trajectory.

4. Simulation and experimental analysis

We performed simulation and experimental analysis, comparing it with two different algorithms in various environments.

4.1. Two-dimensional simulation

To validate the algorithm's effectiveness in Python, a 10x10 environment is created . Obstacles, represented by black circles with a radius of 1, are randomly placed. The maximum iteration count is 1000. The path planning effect of DQN algorithm and textual algorithm are shown

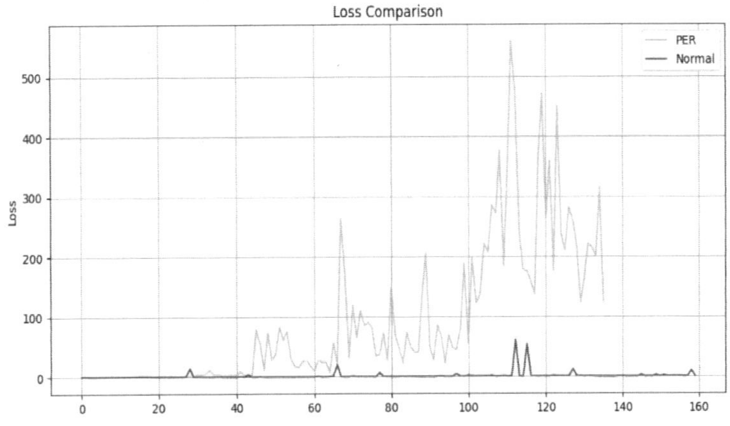

Figure 10. Loss comparison.

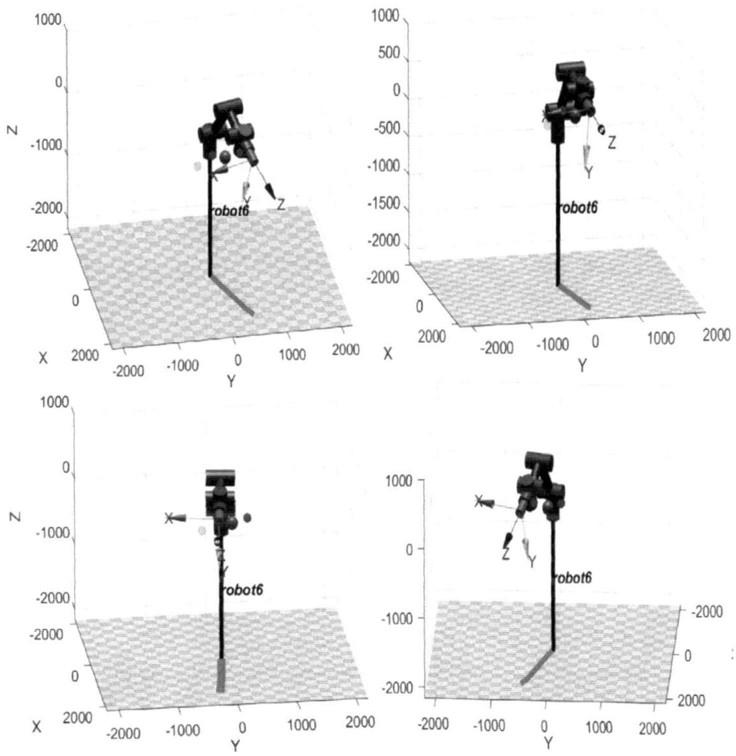

Figure 11. Result of virtual environment planning.

in the in Figure 5 and 6, the blue section represents the planned final trajectory curve, with the starting point and target point indicated in green and red, respectively. By comparing the trajectory lengths, number of turning points, and exploration time of each algorithm, the performance of the textual algorithm can be validated.

After 100 sets of repeated experiments, the experimental results are shown in Table 1.The comparison results indicate that the textual algorithm has a shorter planning time compared to the DQN algorithm.Figures 7 and 8 presents the reward value convergence curves of the two algorithms in Environment 1. Compared to the DQN algorithm, textual algorithm demonstrates better convergence, higher operational efficiency, and a higher obstacle avoidance success rate.

4.2. Three-dimensional simulation
Figure 9 shows the result, where the first two figures illustrate the algorithm's exploration process as the obstacles change, and the last two figures display the final trajectory generated after the algorithm completes. The experimental results show that, despite dynamic changes in obstacles, the algorithm can consistently explore and plan collision-free paths, effectively avoiding obstacles. This confirms that the proposed algorithm is well-suited for complex dynamic obstacle environments.

Figure 10 is used to compare the loss effects of preferred experience playback (PER) versus normal experience playback. It can be seen that the preferred experience playback method can reduce the value of the loss function more effectively during the training process, and the PER can converge faster and keep the loss low over multiple training cycles compared to the ordinary experience playback.

Figure 11 shows the planning results of the algorithm in the virtual environment. It can be seen that the algorithm successfully completed the obstacle avoidance path planning from the starting point to the end point

5. Conclusion
This paper presents an enhanced DQN algorithm for robotic arm trajectory planning across various scenarios. The algorithm was evaluated through two-dimensional and three-dimensional simulations in both Python and MATLAB. The results highlighted the improved DQN algorithm's effectiveness in terms of computational efficiency, convergence rate, and obstacle avoidance. The trajectory planning shows that the proposed method has better adaptability. Additionally, it achieves faster convergence and higher operational efficiency during the planning process. When the obstacle environment changes, the algorithm can quickly adapt and effectively plan the robotic arm's obstacle-free trajectory. This highlights the improved performance and suitability of the algorithm for dynamic and complex environments.

6. Acknowledgments
This work was supported by the Guilin University of Technology Research Start-up Fund (Project No. GUTQDJJ6615028).

7. References
[1] Li C, Huang X, Ding J, Song K, Lu SQ. Global path planning based on a bidirectional alternating search A* algorithm for mobile robots. Comput Ind Eng. 2022;168:108123.
[2] Liu Z, Li M, Zhang R, Zhang G, Han S, Chen K. Preoperative path planning of craniotomy surgical robot based on improved MDP-LQR-RRT* algorithm. Biomed Signal Process Control. 2025.
[3] Cheng J, Wang B. Flocking Control of Mobile Robots with Obstacle Avoidance Based on Simulated Annealing Algorithm. Math Problems Eng. 2020;2020:1–9.
[4] Mallidi SK, Ramisetty RR. Embedded-filter ACO using clustering based mutual information for feature selection. J Comb Optim. 2025;49:27.
[5] Tian S, Li Y, Kang Y, Xia J. Multi-robot path planning in wireless sensor networks based on jump mechanism PSO and safety gap obstacle avoidance. Futur Gener Comput Syst. 2021;118:37–47.
[6] Chen D. Fuzzy obstacle avoidance optimization of soccer robot based on an improved genetic algorithm. J Ambient Intell Humaniz Comput. 2019;11:6187–6198.
[7] Li Z, Luo X. Autonomous underwater vehicles (AUVs) path planning based on Deep Reinforcement Learning. 2022 9th Int Conf Digital Home (ICDH). 2022:257–262.
[8] Yi C, Qi M. Research on virtual path planning based on improved dqn. IEEE. 2020.

Journal of Physics: Conference Series **3024** (2025) 012008 doi:10.1088/1742-6596/3024/1/012008

Study on Deformablity and Material Flowing Behavior of Tailor-Welded Blanks of Dissimilar High-strength Steel

Po Liu[1], Shutong Li[2,3], Zhenhui Sun[2,3], Runzheng Wang[2,3], Pengbiao Han[2,3*]

[1] Guilin Aerospace Industry College Guangxi Key Laboratory of Special Engineering Equipment and Control, Guilin, Guangxi Zhuang Autonomous Region, 541004, PR China

[2] Hebei University of Science and Technology, College of Materials Science and Engineering, Shijiazhuang , Hebei Province, 050018, PR China

[3] Hebei Key Laboratory of Near-Net Forming Technology of Materials, Shijiazhuang Hebei Province, 050018, PR China

[*]Corresponding author's e-mail: hpb68@163.com

Abstract. In this study, tensile and cupping tests were carried out aimed at dissimilar high-strength steel laser tailor-welded blanks(TWBs) of DP590-TRIP590 and DP590-HC420, in order to study deformation ability and material flow rule. The influence of welding position on cupping performance was also analyzed. Results indicate that the welding seam and its location have a significant effect on the deformation capacity of the TWBs. The tensile properties of TWBs are primarily governed by the weaker base metal, and TWBs tensile strength is close to the weaker base metal. The DP590-TRIP590 TWBs(x=0) exhibit the highest ductility compared to those with other welding positions. The fracture occurs randomly on each side of TWBs(x=0 and ±8), whereas all fractures happen on the base metal with a larger proportion for TWBs(x=±16). For DP590-HC420 TWBs, the overall ductility is governed by the proportion of HC420. Fracture always occurs on the HC420 base metal, in which side material flow is more pronounced.

1. Introduction

Nowadays, energy crisis and environmental pollution are becoming key global issues. For the automotive industry, energy conservation and emission reduction can be achieved by the laser tailor welded blanks(TWBs) under the premise of maintaining safety and good performance[1,2]. This is because laser tailor welding has the advantage of high energy density, small heat-affected zone, high joint strength, and less fracture. TWBs parts can not only meet the requirements of thickness and performance differentiation of assembly parts, but it also realize structural optimization[3-7]. Therefore, they are widely used in automotive parts, such as door inner panels, front longitudinal beams, shock absorber support, beam, and bumpers[8,9].

However, due to the uneven performance of the welding line and base metal, it is easy to induce complex mechanical behaviors such as welding line shift and homogeneous deformation in the stamping stage or in operating conditions for TWBs. Compared with single base metal, it is more complex and difficult to predict the forming behavior of TWBs, which increases the probability of rupture on the weak side or thin side of TWBs parts. Besides, the overall forming performance is greatly reduced, which turns out to be a problem demanding prompt solution. Xue et al.[3] carried out uni-axial tensile test on DP600-H340LAD+Z TWBs, and explored the forming performance of TWBs with different ratios of thickness and strength. It was found that the strength of TWBs was between those base metals. The greater the proportion of weak base metal was, the better the plastic-forming performance of TWBs was.

During the plastic deformation process, it is quite difficult to visually observe material flow. At present, material flow and plastic strain distribution can be visualized by means of grid measurement, laser measurement, digital image correlation(DIC), and high-speed photography[10]. The grid measurement technology is usually used in sheet-forming limit test, in which a forming limit curve(FLC) is drawn[11]. In addition, Spena et al.[12] measured the strain of TWIP-TRIP tailor-welded blanks in uniaxial tension testing using DIC method, and pointed out that uneven plastic deformation happened in the weld fusion zone. The author ascribed the higher elongation to the uniform deformation of the heat-affected zone. However, the DIC method requires not only high-resolution cameras and advanced image acquisition systems, but also needs complex pre-processing on samples, which increases the measurement cost[13,14]. On the basis of DIC method, Yi et al.[15] proposed a strain calculation method based on circular grid labeling and artificial intelligence (AI) model, in which the circular grid was etched by laser and the strain of the sample was measured by AI model based on image data training. The results show that the measurement results are very close to that of traditional DIC. It thus provides a helpful solution to improve the automation degree of strain measurement.

In general, a consistent conclusion has been reached for TWBs with a large ratio of strength or thickness. However, when the ratio of thickness or strength is small, the flow stress of each base metal is relatively close. Moreover, the deformation of each base material is governed by its own hardening rule. In other words, TWBs deformation behavior should be affected, which means that more complicated material flowing and weld movement will be caused under certain stress states. Finally, the TWBs forming ability is weakened. In this paper, three kinds of equal-thickness steel plates with little difference in properties are selected. Based on DP590-TRIP590 and DP590-HC420 TWBs, uniaxial tensile tests, and cupping tests are carried out to study forming behavior and material flow rule of TWBs. Furthermore, the forming ability of TWBs is also analyzed. It thus can provide some reference for the design and forming of TWBs.

2. Experimental material and method

2.1. Experimental material and preparation of TWBs

The experimental materials are selected as 1.0 mm-thickness DP590, TRIP590, and HC420 steel plates. Their chemical composition and content are shown in Table 1.

Table 1. Chemical composition of base metal (mass fraction, %)

Material	C	Si	Mn	P	S	Als
DP590	0.15	0.6	2.5	0.04	0.015	0.01
TRIP590	0.16	1.08	1.46	0.01	0.005	0.03
HC420	0.06	0.27	1.07	0.024	0.003	0.05

Mechanical properties of base metal were measured by tensile tests, as shown in Table 2. The DP590-TRIP590 and DP590-HC420 TWBs were obtained by a laser welding machine, with the welding direction perpendicular to the rolling direction of base metal. Welding speed was set as 30 mm/s and the power was 3200 W. Argon is used as shielding gas during welding.

Table 2. Mechanical properties of base metal

	σ_s/MPa	σ_b /MPa	σ_s/σ_s	n	δ/%
DP590	415	684	0.607	0.186	23.4%
TRIP590	403	673	0.599	0.189	24.8%
HC420	426	642	0.841	0.154	25.7%

Journal of Physics: Conference Series **3024** (2025) 012008 doi:10.1088/1742-6596/3024/1/012008

2.2. *Uniaxial tensile test*

According to the ASTM E8/E8M metal material tensile test method, the samples were cut off the TWBs by a wire-cutting machine with the tensile direction perpendicular to the welding line. The weld seams are 0mm, ±8mm, and ±16mm away from the sample centre, respectively. Here, a positive value denotes that weld line moves toward DP590, and a negative value means the weld line is inclined to TRIP590 or HC420. Thus, the TWBs can be expressed in the form of x=0, ±8, ±16, respectively. Dimension of the testing sample is shown in Fig. 1. In order to facilitate analysis on material flow in the tensile test, grids with a diameter of 3 mm were carved in the main deformation surface of samples by LS-FL-30W laser engraving machine. Uniaxial tensile tests were conducted on a WDW-100 kN electronic universal testing machine with an extensometer gauge length of 50 mm, as shown in Fig. 2. Thus, the weld layout step size of 0, 8mm and 16mm can ensure a representative proportion of 50%, 34% and 18% for each metal material within the gauge region, respectively. Tensile speed was set at 3 mm/min. Tensile tests for samples with each weld position were repeated three times to avoid random error. The grid elongation was measured by a stereomicroscope with an accuracy of 1×10^{-5} mm.

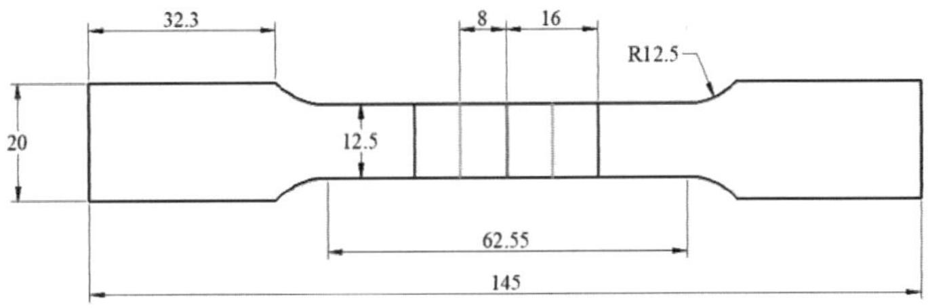

Figure 1. Schematic diagram of uniaxial tensile test specimen (mm)

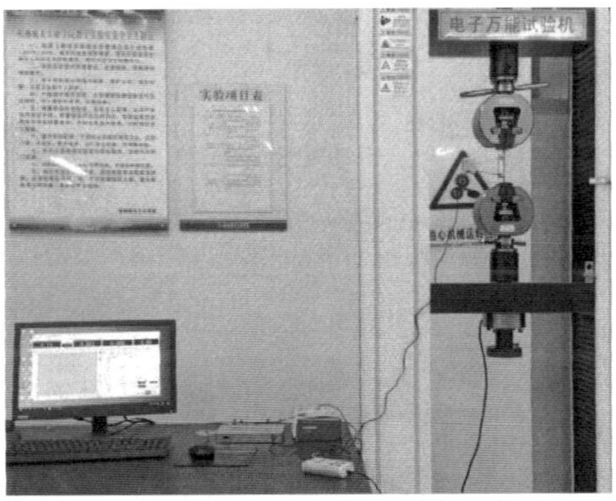

Figure 2. WDW-100kN electronic universal testing machine

2.3. *Microstructure observation*

Samples were cut along longitudinal section off the tensile tested specimens for microstructure observation. After mechanical thinning and ion thinning, the microstructure was observed by FEI Tecnai G2 F20 electron transmission microscope with an accelerating voltage of 200 KV.

3. Results and analysis

3.1. *Tensile deformation behavior of TWBs*

Fig. 3 shows the engineering stress-strain curves of DP590-TRIP590 TWBs. It is seen that the stress-strain curves of all TWBs are quite close. The TWBs tensile strength(σ_b) is near to that of TRIP590 base material, and the maximum deviation is 9 MPa. This is because the yielding stress(σ_s)

and σ_b of DP590 are quite close to that of TRIP590, which causes homogeneous performance for the DP590-TRIP590 TWBs. During uniaxial tension, tiny plastic deformation occurs first on the TRIP590 base material. With TWBs extending, the stress gradually increased. Once it reaches σ_s of DP590, plastic deformation also occurs in this base material. Subsequently, the whole TWBs undergoes coordinated deformation. It can be seen in Fig. 3 that the largest ductility is obtained with a welding line in the middle for all the TWBs. Nevertheless, its ductility is still between the two base metals. The elongation of TWBs (x=±8, ±16) is lower than that of base metals, and it decreases with an increase in the proportion of DP590. This is due to that TWBs(x=0) can bear uniform deformation and thus stress concentration is relieved. So the bigger plastic deformation is allowed before fracture. Moreover, the residual austenites in TRIP590 transform into martensite during deformation, as shown in the white region in Fig. 4. Due to phase transformation, energy is absorbed and plastic deformability is promoted[21]. As for TWBs(x=±8 or x=±16), the strain of each base metal is closely related to its corresponding parallel length according to Hooke law and hardening rule. The different parallel length of each part within the gauge region will lead to a different strain of each base material, as shown in Fig. 1. Therefore, each base material reaches the yielding point successively according to its own hardening rule, causing the asynchronous material flow behavior. As a result, the coordination deformation between two base metals is affected during the entire deformation process and the whole deformation capacity of the TWBs(x=±8 or x=±16) is reduced.

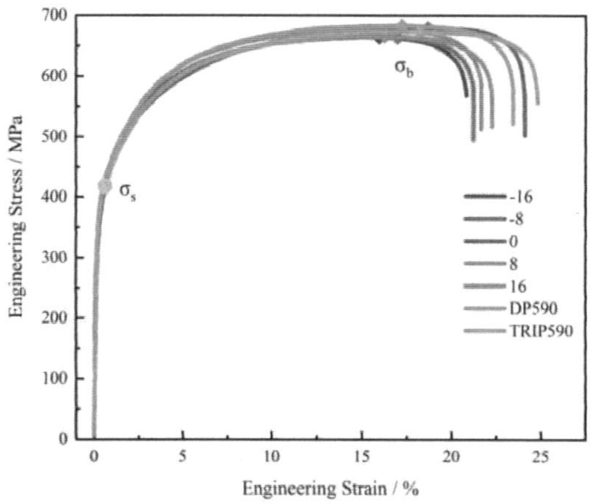

Figure 3. Engineering stress-strain curves of DP590-TRIP590 TWBs

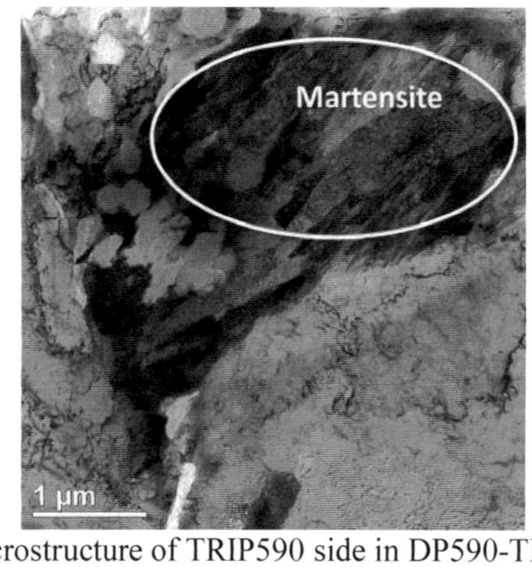

Figure 4. TEM microstructure of TRIP590 side in DP590-TRIP590 TWBs(x=0)

Fig. 5 shows engineering stress-strain curves of DP590-HC420 TWBs and tensile test results are obtained in Table 3. Obviously, all the tensile strength of DP590-HC420 TWBs are between that of two base metals, and yet they are quite close to that of HC420. This is because σ_s and elongation of HC420 are higher than those of DP590 (seen in Table 2). With specimen extending, plastic deformation first occurs on the DP590. With the applied load increasing, the internal stress reaches σ_s of HC420. At this point, plastic deformation is triggered on both the base metals. Due to that DP590 possesses the higher strain hardening exponent n and the lower yield-to-tensile strength ratio, it can be inferred that a greater stress increment is required for DP590 to achieve the same strain increment according to the Levy-Mises incremental theory[17]. Therefore, it is more difficult to induce deformation on this side. When TWBs stress reaches σ_s of HC420 base material, deformation mode on this side transitions from uniform deformation to local deformation, resulting in early necking. At this moment, the stress of DP590 base material has not yet reached its tensile strength. Hence, fracture always occurs on the HC420 side. This indicates that when a large strength difference exists between two base metals, the tensile properties of TWBs are mainly governed by the weaker base metal. In addition, the DP590-HC420 TWBs elongation decreases with an increase in the proportion of DP590 and it is lower than that of both base metals. The best ductility is achieved for the TWBs(x=+16), and it is approximately twice that of TWBs(x=-16). From the perspective of welding position, as the proportion of weak base metal increases, the overall TWB ductility will be enhanced.

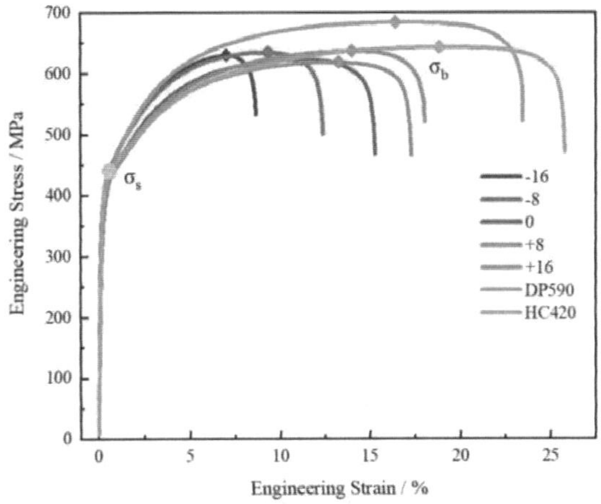

Figure 5. Engineering stress-strain curves of DP590-HC420 TWBs

Table 3. The DP590-HC420 TWBs uniaxial tensile test results

	x/mm	σ_s/MPa	σ_b/MPa	σ_s/σ_b	n	δ/%
DP590-HC420	-16	417	629	0.663	0.169	9.0
	-8	421	634	0.664	0.168	12.4
	0	410	623	0.658	0.171	15.3
	+8	411	618	0.665	0.159	17.3
	+16	423	636	0.665	0.160	18.0

3.2. Material flow behavior of TWBs

Fig. 6 shows the tensile tested specimens of DP590-TRIP590 TWBs. It can be seen that fracture appears randomly on DP590 and TRIP590 base metal. Nevertheless, it is found that fracture failure position is always located in base metal with a larger proportion of gauge in DP590-TRIP590 TWBs(x=±16), which is irrelative to the base material itself.

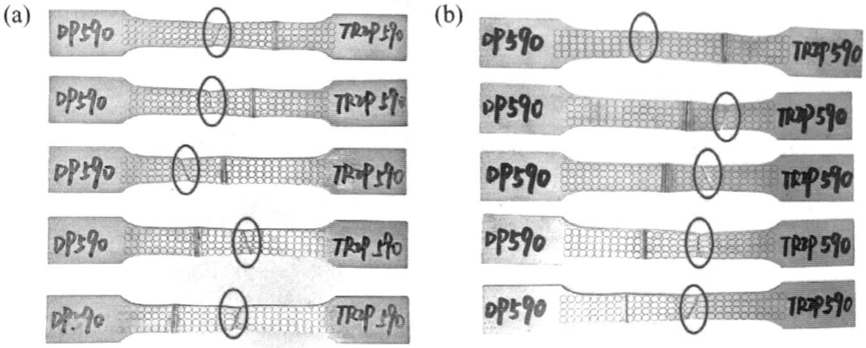

(a) Result 1 (b) Result 2

Figure 6. The tensile-tested DP590-TRIP590 TWBs specimens

It is assumed that tensile deformation of the TWBs only occurs within parallel sections. Moreover, width of the welding seam and the heat-affected zone are quite small compared to the specimen size, and the welding joint is characterized as having high strength and low plasticity[18,19]. The welding seam and the heat-affected zone are thus treated as rigid and no deformation occurs on it. The uniaxial tensile deformation on TWBs is always in a dynamic equilibrium state, and it obeys the Holloman hardening criterion as followed by Eq. (1):

$$\sigma = K\varepsilon^{n} \tag{1}$$

where σ denotes the engineering stress, ε the engineering strain, k is the material coefficient, n is the strain hardening exponent.

At the beginning, each base metal of TWBs undergoes slight elastic deformation under the uniaxial tensile load. When internal stress reaches the yielding strength of the weak base metal, plastic deformation occurs in this base material. At this stage, the difference between the cross-sectional area of the two base materials can be considered as equal. Assuming the welding seam is a rigid body, gauge elongation in the TWBs can be considered as the sum of the elongation of each base DP590 and TRIP590 material. Meanwhile, the plastic deformation of each base metal is also governed by the Holloman criterion. Here, a gauge of the DP590 part in TWBs is defined as A and the TRIP590 part is B. According to mechanical equilibrium condition and Eq. (1), it can be obtained as follows:

$$K_A\varepsilon_A^{n_A} = K_B\varepsilon_B^{n_B} \tag{2}$$

Here, ε_A and ε_B are engineering strains of A and B, respectively. K_A and K_B denote the material coefficients of A and B, respectively. n_A and n_B are strain hardening exponents of A and B, respectively.

From the discussion above, it can be seen that strain-hardening exponents(n_A, n_B) and the strength of each base material are quite close, meaning that the hardening characteristic and uniform plastic deformability are almost the same. As for TWBs(x=0), the individual gauge lengths of DP590 and TRIP590 sides are equal. Thus, the elongation on each part is nearly identical according to Eq. (2) during the tensile process. Since σ_s of TRIP590 is slightly lower than that of DP590 base material, plastic deformation happens first in the TRIP590 part. As elongation proceeds, the TWBs(x=0) stress reaches σ_s of DP590 base material, and plastic deformation occurs on this part. Because the elongation and tensile strength of the two original materials are relatively close, the fracture is likely to happen on the DP590 part or the TRIP590 part. It is known that the elongation rate increases with decreasing specimen gauge length according to existing literature[20-23]. As for TWBs(x=±8 and ±16), this means that the ductility of the longer gauge side is lower compared to that in TWBs(x=0), while the elongation of the shorter gauge is a little larger. Therefore, fracture easily occurs on the longer gauge section as for TWBs(x=+8 and ±16). However, an unusual phenomenon is that fractures occur on each base metal side in terms of TWBs(x=-8). It is due to that the plasticity of the TRIP590 part is increased because of the TRIP effect in tensile deformation.

As for DP590-HC420 TWBs, all the fractures appear in HC420 base material, as shown in Fig. 7. Although HC420 has a higher yielding strength, its tensile strength is much lower than that of DP590. When the internal stress of the TWBs reaches σ_b of HC420, necking is induced on the HC420 part. This necking area is subjected to concentrated load and main elongation. Thus, TWBs stress rarely increases and it is much lower than σ_b of DP590 base metal. As a result, the DP590 side only endures slight uniform deformation. In addition, the gauge region on the HC420 side undergoes greater deformation under the same stress because of its lower hardening exponent, which leads to a larger thinning rate and higher true stress in this region. As a result, deformation in this HC420 region is fastened, as well as the occurrence of fracture.

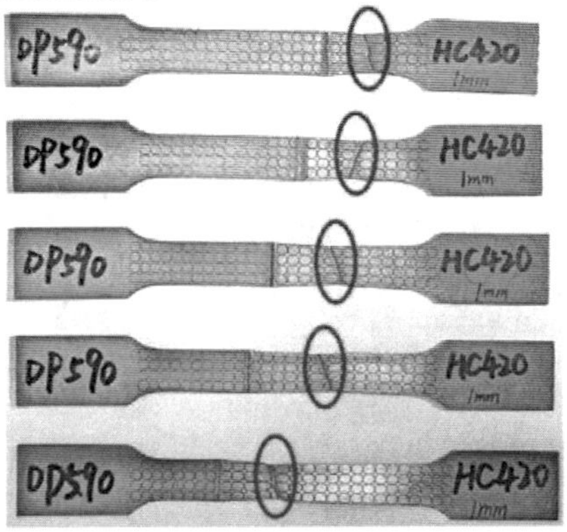

Figure 7. Specimens of DP590-HC420 TWBs after uniaxial tensile test

In order to validate the analysis as mentioned above, finite element analysis was conducted on the uniaxial tension of DP590-HC420 TWBs using ABAQUS software. According to the geometric size of the TWBs and the weld position, the geometric model was created. The weld zone, the base metal, and the clamping end area were partitioned respectively. Since the weld zone has high strength and hardness, it is thus regarded as a rigid strip with a width of 1mm, and the clamping end is set as a rigid body. Since fracture behavior is not the main research issue in this paper, the damage model is not considered. The established model is shown in Fig.8.

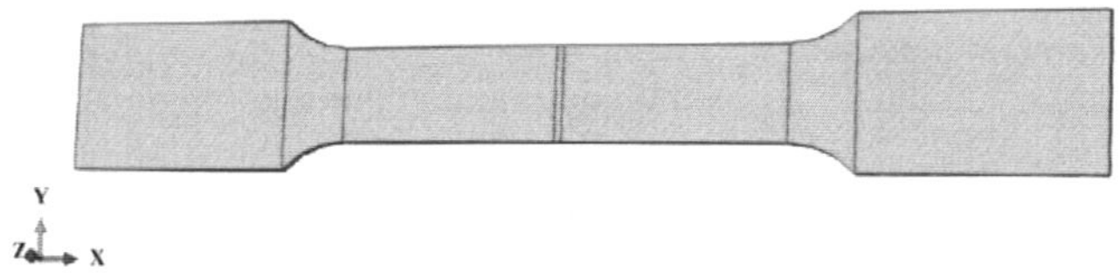

Figure 8. Finite element simulation model of DP590-HC420 TWBs(x=0)

The standard module was used for nonlinear deformation analysis, and the sheet was defined as a deformable body. The model was meshed with an eight-node linear reduced integral element, and the mesh size was 0.5 mm. One side of the clamping end was completely constrained. A speed of 0.05 mm/s along the X axis was applied on the other side, which is consistent with the experimental test. The distribution of equivalent stress and strain of DP590-HC420 TWBs(x=0) at different tensile displacements are shown in Fig. 9. It is clearly seen that the larger strain and stress are caused in the base metal of HC420 at each tensile stage, compared to the side of DP590 material.

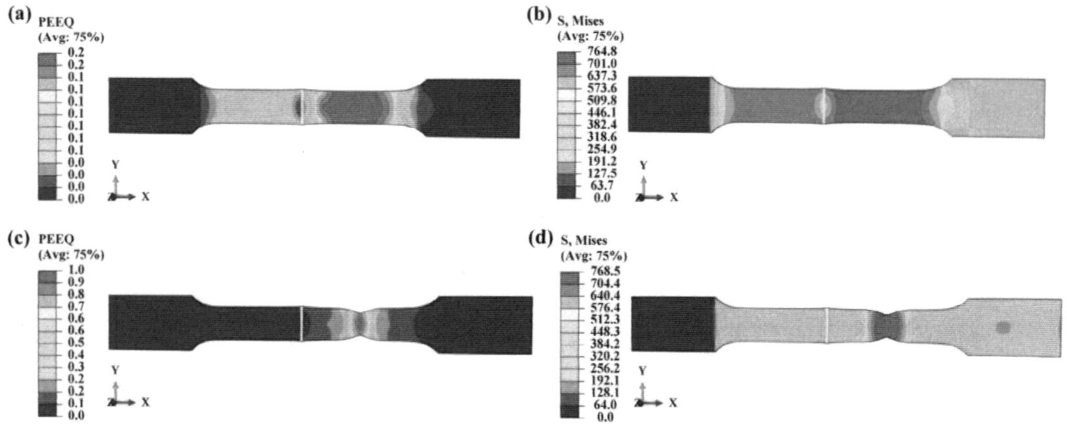

Distribution of strain (a) and stress (b) at displacement of 7.7mm and distribution of strain (c) and stress (d) at displacement of 12.2mm

Figure 9. Equivalent stress and strain contours of DP590-HC420(x=0) at different tensile displacements

The predicted and experimental true stress-strain curve of TWBs(x=0) was obtained, as shown in Fig. 10. It shows the FEM predicted result agrees well with the experimental one, demonstrating finite element simulation result is reliable. The equivalent strain of nodes symmetrical about the welding line was collected, and its variation curve with time is shown in Fig. 11. Slight plastic deformation, with a maximum equivalent strain of 0.058, is seen in the DP590 part. Meanwhile, the equivalent strain of the HC420 side is higher than that of the DP590 side, especially when the step time exceeds 130s. At this moment, the strain of the DP590 part tends to be steady. Meanwhile, deformation is mainly focused on the necking region of the HC420 side, which indirectly confirms the analysis as mentioned above. This finding differs from the tensile behavior of TWBs with a large difference in strength[2-3]. The reason lies that both base metals participate in TWBs tensile elongation when strengths (σ_s and σ_b) of base metals are close or overlapping. Therefore, each of the base materials should be considered as a deformation body, and not as a rigid body.

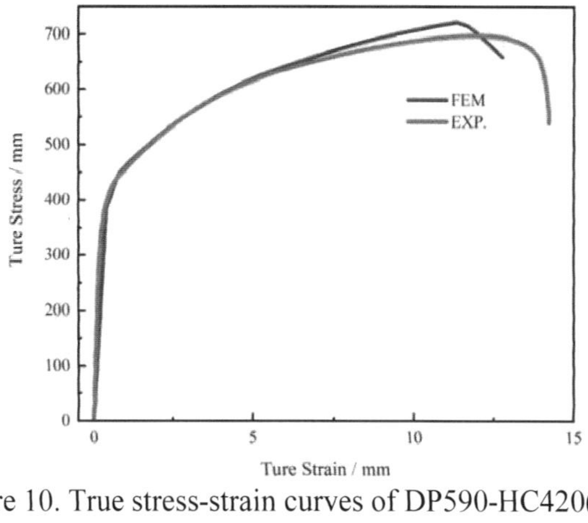

Figure 10. True stress-strain curves of DP590-HC420(x=0)

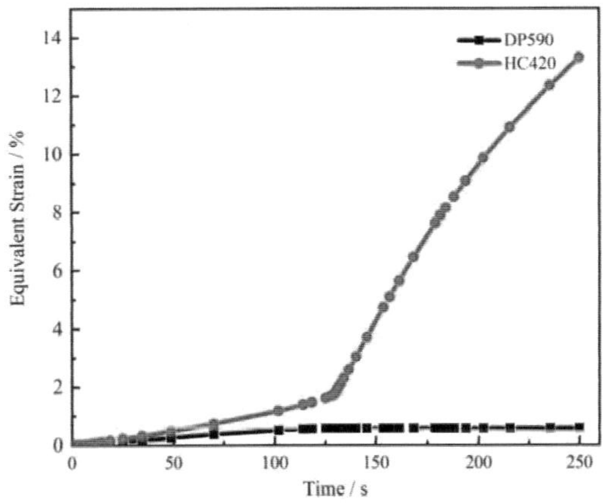

Figure 11. The change curve of equivalent strain with time of DP590-HC420(x=0)

The grid long-axis length of the tested TWBs is acquired by using a stereomicroscope. The welding seam of each TWBs is considered as the centreline. The DP590 base metal is positioned at the left side of the welding seam, and TRIP590 or HC420 can be placed at the right side. Thus, a grid long-axis length versus position is plotted. The negative x-axis coordinate denotes the position of the grid center of DP590 base metal, and the positive coordinate denotes the centre position of the others, as shown in Fig. 12. Since the weld zone has higher strength, elongation of the grid near to this welding zone is quite small. Due to the tiny strength difference for the DP590-TRIP590 TWBs, the variation of the grid long-axis length of two base metals is roughly symmetric about the welding line.

As for DP590-HC420 TWBs, obvious elongation is observed on a grid of the HC420 part, and a very tiny uniform deformation occurs on DP590. The grid length on the DP590 side is quite close to the original grid diameter. The grid near the necking area is obviously stretched, and some grids are even badly distorted. Except this, the remaining grid of the HC420 part is less elongated. In this HC420 side, the long-axis length presents a decreasing tendency with the distance away from the welding line.

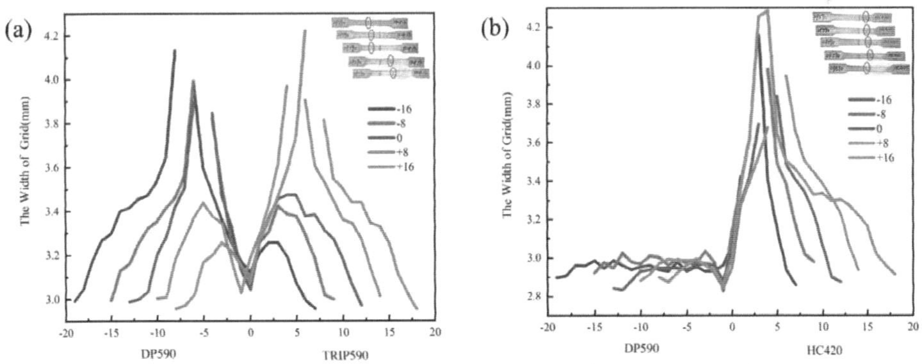

(a) DP590-TRIP590 (b) DP590-HC420

Figure 12. Curves of grid long-axis length versus grid center position of DP590-TRIP590 and DP590-HC420TWBs

4. Conclusion

In this paper, tensile tests were conducted on both equal-thickness DP590-TRIP590 and DP590-HC420 TWBs. Deformation behavior and material flow law of TWB are mainly studied through grid measurement and the FEM method. The main conclusions are drawn as follows:

(1) All the fractures happen in base metal for both DP590-TRIP590 and DP590-HC420 TWBs. However, the crack position is inconsistent for DP590-TRIP590 TWBs. When x=±16, the fracture

appears in the base material with the larger proportion. For the TWBs(x=0 and ±8), necking happens randomly on ether base metal side. The area adjacent to the necking area has a larger elongation, while the other side has a smaller one. All the fractures occur in the HC420 base metal in all DP590-HC420 TWBs, and the DP590 base metal endures small uniform deformation.

(2) The tensile behavior and material flow of the DP590-HC420 TWBs mainly depend on the weaker base metals. The tensile behavior of the DP590-TRIP590TWBs is governed by the interaction of the two base material. Each base metal within the gauge section is elongated according to its unique Holloman hardening criterion.

(3) The uniaxial tensile strengths of DP590-TRIP590 TWBs are close to that of the original TRIP590 material. The largest elongation can be obtained for the two kinds of TWBs(x=0), with a elongation of 24.1% for DP590-TRIP590 TWBs(x=0) and 17.1% for DP590-HC420 TWBs(x=0). The coordinated deformation is weakened for the TWBs(x=±8 and x=±16). DP590-TRIP590 TWBs elongation decreases with an increase in proportion of DP590. The average elongation of DP590-TRIP590 TWBs is decreased by 8.1% compared with DP590 original material, and decreased by 15.3 % compared with TRIP590 material. The ductility of DP590-HC420 TWB is dependent on the proportion of HC420. The highest ductility of 18% is obtained for TWBs(x=+16), which is nearly twice that of the TWBs(x=−16).

Acknowledgments
This work was funded by Natural Science Foundation of Guangxi Zhuang Autonomous Region (No.2020JJA160034) and the Guangxi Key Laboratory of Special Engineering Equipment and Control, Guilin University of Aerospace Technology (SEEC24ZR03).

References

[1] Bandyopadhyay K, Panda S K and Saha P 2014 Investigations into the Influence of Weld Zone on Formability of Fiber Laser-Welded Advanced High Strength Steel vol 23, ed Journal of Materials Engineering and Performance p 1465-1479
[2] Abbasi M, Ketabchi M and Shakeri H R 2012 Formability Enhancement of Galvanized If-Steel TWB By Modification of Forming Parameters vol 21, ed Journal of Materials Engineering and Performance p 564-571
[3] Xue S, Zhou J and He Y 2011 Formability Tensile Test of Different Gauge Tailor-Welded Blanks vol 36, ed Forging Stamping Technology p 30-33
[4] Zheng B, Tian X and Wang Z, et al. 2021 Research on Numerical Simulation Method for Dynamic Tensile Performance of Laser Tailor-Welded Joints of Different Thickness Vol 42, ed Iron Steel Vanadium Titanium p 62-68
[5] Wang J, Yao Y and Li C, et al. 2022 Influence of Hot Forming on Microstructure and Mechanical Properties of Laser Tailor-Welded Dissimilar Ultra-High-Strength Steels vol 5, ed Advanced Composites and Hybrid Materials p 1450-1459
[6] Gao X 2025 Laser Welding Technology Vol 15, ed Metals p 86
[7] Basak S, Katiyar Sb and Orozco-Gonzalez P, et al. 2019 Microstructure, Forming Limit Diagram, and Strain Distribution of Pre-Strained Dp-If Steel Tailor–Welded Blank for Auto Body Application vol 104, ed The International Journal of Advanced Manufacturing Technology
[8] Zhou T, Liu Y and Du A, et al. 2024 Effect of Laser Power and Defocusing Amount on Microstructure and Properties of Laser Welding Joints in 2000 Mpa Grade Hot-Formed Steel vol 16, ed Journal of Netshape Forming Engineering p 39-47
[9] Xian L, Wu K and Li L 2022 Analysis of Stamping Cracking of Laser Tailor-Welded If Steel Door Inner Plate vol 19, ed Journal of Physics: Conference Series
[10] Zhu X, Ma Y and Wu H, et al. 2024 In-Situ Tensile Testing of Fracture and Strain in a Selective Laser Melted Alsi10Mg Alloy vol 10, ed Heliyon
[11] Bardelcik A and Butler Q 2024 Strain Partitioning Characterization of Advanced High Strength

Steels Using in-Situ Tensile Tests With Micro Digital Image Correlation–Methodology and Analysis vol 218, ed Materials Characterization

[12] Spena Rp, Cortese L and Nalli F, et al. 2019 Local Formability and Strength of Twip-Trip Weldments for Stamping Tailor Welded Blanks (TWBs) vol 101, ed The International Journal of Advanced Manufacturing Technology p 757-771

[13] Youn S and Mathur K 2023 A Review of 3D Digital Garment Simulation Strategies for Enhanced Wearables and Medical-Grade Applications vol 16, ed Journal of Fiber Bioengineering and Informatics p 193-211

[14] Wen Y, Wang J and Zheng L, et al. 2024 Method of Generating Speckle Patterns for Digital Image Correlation Based on Modified Conway's Game of Life vol 32, ed Optics Express

[15] Yi S, Hyun D and Hong S 2024 Tensile Specimen Circular Grid Pattern and Ai-Based Strain Calculation Method vol 14, ed Applied Sciences p 7330-7343

[16] Nouri N, Hillenmeyer E and Thavarajan T, et al. 2025 Transformation Induced Plasticity (Trip) Due to Precipitation During Continuous Tempering of Quenched and Tempered Steels vol 20, ed Journal of Materials Research and Technology p 730-737

[17] Hou G and Niu X 2010 Perfect Elastoplastic Solution of Axisymmetric Cylindrical Cavity Based on Levy-Mises Constitutive Relation and Hoek-Brown Failure Criterion vol 29, ed Chinese Journal of Rock Mechanics and Engineering p 765-777

[18] Zheng Y, Yang B and Wang W, et al. 2024 High Strength Low Alloy Steel Joint Fabricated by Laser Welding with Real-Time High-Frequency Resistance Heating vol 35, ed Journal of Materials Processing Technology

[19] Miloš M, Martin F and Marek B 2024 Softening Effect in The Heat-Affected Zone of Laser-Welded Joints of High-Strength Low-Alloyed Steels vol 68, ed Welding In The World p 1497-1514

[20] Zhu L, Wang G and Jia D, et al. 2023 Determination of The Size Effect on the Tensile Properties of Miniaturized Specimens vol 65, ed Materials Testing p 524-535

[21] Zhu D and Ou Y 2016 Effects of Gauge Length and Strain Rate on Tensile Mechanical Properties of Kevlar49 Single Yarn vol 33, ed Acta Materiae Compositae Sinica p 225-233

[22] Motra B H, Hildebrand J and Dimmig-Osburg A 2014 Influence of Specimen Dimensions and Orientation on the Tensile Properties of Structural Steel vol 56, ed Materials Testing p 929-936

[23] Ezio C, Matteo D and Daniele F, et al. 2023 Influence of the Gauge Length on Dynamic Direct Tensile Test vol 6, ed Procedia Structural Integrity

Optimization and Efficiency Analysis of Lunar Water Vapor Condensation Collection Structure Devices

Yinchao Wang [1,2], Zihao Yin [1,2], Junnan Han [1,2], Weiwei Zhang [3,*], Lin Zu [4], Guanghong Tao [1,2], Suyang Yu [1,2], Chunying Jiang [1,2], Changlong Ye [2,5]

[1] College of Mechanical and Electrical Engineering, Shenyang Aerospace University, Shenyang 110136, China

[2] Key Laboratory of Rapid Development & Manufacturing Technology for Aircraft (Shenyang Aerospace University), Ministry of Education, Shenyang, 110136, China

[3] State Key Laboratory of Robotics and System, Harbin Institute of Technology, Harbin 150001, China

[4] Deep Space Exploration Laboratory, Hefei 230026, China

[5] College of Artificial Intelligence, Shenyang Aerospace University, Shenyang 110136, China

[*]Corresponding author's e-mail: zweier@hit.edu.cn (W.Z.)

Abstract. Aimed at the technical challenge of low water vapor collection efficiency in the in-situ thermal extraction of lunar polar water ice, this study, based on the heat and mass transfer characteristics of frost layer porous media, proposes a solution to enhance the gas-solid phase transition rate by optimizing the internal surface area of the water vapor collection unit. Through the establishment of a water vapor condensation dynamics model under vacuum and low-temperature conditions, the coupling effects of frost layer porosity and temperature field on the mass transfer process are revealed. A verification platform simulating the extreme lunar surface environment was constructed, and comparative experiments on different aluminum bead filling structures were conducted. The results indicate that the larger the internal surface area of the device, the higher the water vapor collection efficiency; the use of small-diameter aluminum beads significantly increases the internal surface area, but excessively small diameters may lead to pore blockage. The initial stage of water vapor collection is one of the key stages of mass transfer, where water vapor first comes into contact with the inner surface of the collection structure and undergoes a gas-solid phase change that affects the continued performance of the collection device. This research provides theoretical and experimental foundations for the design of water vapor collection units in in-situ lunar water ice extraction devices.

1. Introduction

The Moon is the sole natural satellite of the Earth that can monitor the Earth in a long-term, stable, large-scale, real-time manner. It functions as an observatory for outer space and a practice ground for deep space exploration technology. Research findings indicate that the temperature in the permanently shadowed regions of the Moon's polar areas is maintained at a remarkably low level, facilitating the capture and preservation of volatile water in the form of "cold traps." This process suggests the existence of substantial water ice resources[1]. Water ice resources on Earth's satellite are of critical importance as strategic space resources. Discovery of water resources and the subsequent construction of an environmentally controlled life-support system (ECLSS) within a spacecraft is imperative to the resolution of the problem of life-support in space exploration in the future, as well as being a fundamental element in the realization of a long-term on-orbit operational capacity of a manned space station and the establishment of extraterrestrial settlement.

Under the environmental conditions of the lunar polar regions, the operational efficiency of lunar equipment is constrained by limited energy resources. Additionally, the mechanical strength of ice-bearing lunar regolith is exceptionally high [2], while the thermal conductivity of regolith particles is extremely low. The porous structure of the regolith further increases gas flow resistance, significantly complicating the efficient collection of volatile water vapor in a vacuum environment. These factors collectively render the extraction of lunar water resources highly challenging.

In the preliminary experiments on the design of the cold trap capture device, Zancy, Luczek, et al. first utilized a glass jar as the shell to visualize the frost phenomenon [3]. The tin foil was then placed into the glass jar by forming a spherical shape with it. However, this method proved to be ineffective. In the subsequent optimization experiment, an aluminum can was selected as the shell, and the tin foil was folded and placed within the aluminum can to augment the heat transfer area. This modification yielded a superior outcome in comparison to the initial verification experiment, as evidenced by the cold trap capture function verification device depicted in Figure 1.

Figure 1. Cold trap capture function verification device

To address these challenges, Harbin Institute of Technology is developing an in-situ thermal extraction device for lunar regolith water ice, as illustrated in Figure 2. This device is designed to meet the demand for efficient extraction of surface frost layers and profile-enriched ice layers within the permanently shadowed regions of the Moon. The device comprises three primary components: a drilling unit, a water ice heating unit, and a water vapor collection unit. The drilling unit employs a spiral bit to penetrate the ice-bearing lunar regolith. Following drilling, the heating unit activates to warm the drill rods. In the Moon’s low-temperature vacuum environment, the water ice undergoes volatilization, transforming into water vapor. This vapor is then channeled into the water vapor collection unit, where it adheres to the internal surfaces, completing the gas-solid phase transition process. The development of this device not only provides a viable solution for China’s future in-situ utilization of water resources in the lunar polar regions but also enables the effective extraction, transformation, and utilization of lunar water ice resources. Furthermore, it offers critical technical support for China’s subsequent lunar exploration missions.

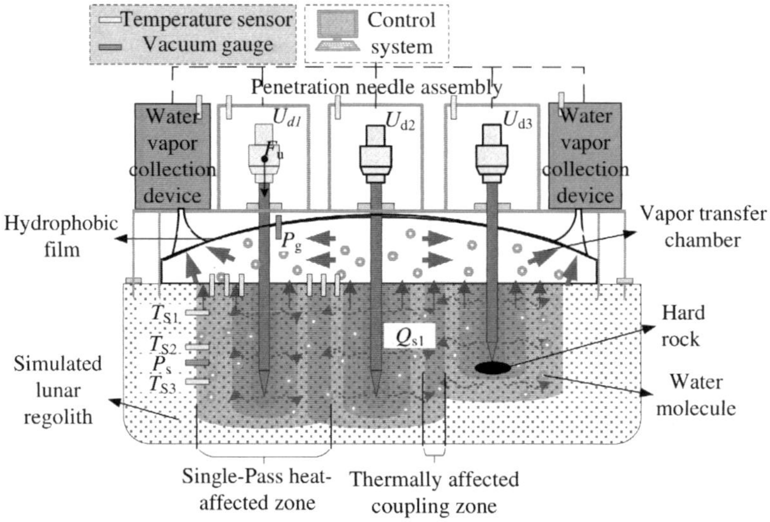

Figure 2. Schematic diagram of the in situ thermal extraction device for lunar soil water ice

The extreme and harsh environmental conditions within the permanently shadowed regions of the Moon render the extraction of water resources exceptionally challenging. Research in this field in China commenced relatively late, and significant breakthroughs are still required in both fundamental studies and technological development. There is an urgent need for low-energy-consumption, high-efficiency water extraction technologies, among which the water vapor collection unit is the most critical component [4-5]. During the development of the aforementioned lunar water ice thermal extraction device, it was observed that the Moon's vacuum environment classifies the water vapor generated from heating ice-bearing lunar regolith as a rarefied gas. To achieve rapid and efficient collection of this water vapor in a vacuum environment while minimizing its dispersion into the lunar atmosphere, it is essential to ensure the water vapor collection unit's high-efficiency heat and mass transfer capabilities. This necessitates maximizing the gas-solid phase transition rate of water vapor within the collection unit. Therefore, dedicated research on the water vapor collection unit is imperative to further enhance the efficiency of lunar water resource extraction.

2. Theoretical analysis of water vapor condensation collection on the lunar surface

The process of water vapor capture fundamentally entails the retention of water vapor as a frost layer on the inner surface of the collection device, achieved through sufficient heat exchange with the water vapor collection unit. However, existing research on water vapor collection predominantly employs experimental methods to analyze the mass and heat transfer processes during the initial stages of frost layer growth under atmospheric pressure and low-temperature conditions. These studies have yielded inconsistent conclusions, highlighting the need for further systematic investigation.

Studies conducted by Boxe [6] and Maye [7], among others, have demonstrated that ice particles within the frost layer exhibit smaller dimensions and enhanced adsorption capabilities under low-temperature vacuum conditions. As the ice layer grows, the pores within the layer gradually close, leading to a reduction in the adsorption-specific surface area of the crystals and a corresponding decrease in their ability to capture water vapor. Consequently, under vacuum and low-temperature conditions, the frost layer becomes thin and dense, with minimal variations in density and porosity, and the absence of macroscopic diffusion channels within the structure [8]. As demonstrated in the preceding analysis, the growth of the frost layer under the vacuum low-temperature environment is comprised of a sequence of distinct layers. In the initial stage, a loose frost layer forms. This is followed by the development of a dense frost layer, which subsequently transitions into a loose frost layer on the inner surface in the final stage. Based on the above analysis, it is evident that the frost layer growth process is significantly influenced by environmental factors such as pressure and temperature. To gain a more comprehensive understanding of this process, this study builds upon the work presented in [9], treating the frost layer as a porous medium composed of ice and water vapor. Furthermore, a frost layer growth model is established based on the Fourier heat transfer equation and Fick's law of diffusion.

$$\begin{cases} \rho_e C_e \dfrac{\partial T}{\partial t} = \lambda_e \nabla^2 T + Q_C \\ \dfrac{\partial \theta_g}{\partial t} = D_g \nabla^2 \theta_g - \dfrac{\dot{m}_v}{\rho_g} \end{cases} \tag{1}$$

Qc denotes the latent heat of phase transition (sublimation) released during water vapor capture, and $\dot{m}_v$ represents the water vapor capture rate. Since only ice and water vapor are involved in the capture process, the thermal parameters of the frost layer can be expressed as:

$$\begin{cases} \lambda_e = \lambda_g + \left(1-\theta_g\right)\lambda_i \\ \rho_e = \rho_g \theta_g + \left(1-\theta_g\right)\rho_i \\ C_e = \dfrac{\left(\rho C\right)_g \theta_g + \left(\rho C\right)_i \left(1-\theta_g\right)}{\rho_e} \end{cases} \tag{2}$$

D_g is the diffusion coefficient of water vapor in the frost layer, the value of which is mainly determined by the pressure and temperature data in the test, based on the above analysis and the empirical formula for the diffusion coefficient of water vapor in the frost layer proposed by Eckert and Drake [10], and the calculation formula was derived:

$$D_g = \begin{cases} \varepsilon \cdot \tau \cdot 2.302 \times 10^{-5} \cdot \left(\dfrac{0.98 \times 10^5}{P}\right)\left(\dfrac{T_0}{256}\right)^{1.81} & ,\varepsilon > \varepsilon_0 \\ 0 & ,\varepsilon < \varepsilon_0 \end{cases} \tag{3}$$

τ denotes the diffusion tortuosity; T_0 represents the cold wall temperature; P is the saturation pressure corresponding to T_0 ; and ε is the porosity of the frost layer. Since only ice and water vapor are present during the sublimation process, the pores can be considered fully occupied by water vapor, i.e., $\varepsilon = \theta_g$. Specifically, when the compactness of the frost layer exceeds a critical porosity ε_0 , the layer becomes sufficiently dense to preclude the existence of water vapor diffusion channels. This critical value ε_0 is dependent on the cold wall temperature T_0 , and it can be derived through theoretical formulation as follows:

$$\varepsilon_0 = 1 - 0.71 \cdot e^{0.229(T_0 - 273.15)} \tag{4}$$

Existing research has revealed that the closer to the cold source, the higher the ice content, the denser the frost layer, and the greater the thermal conductivity, which aligns with the results derived from the formula. These findings are consistent with the literature review. The inner surface of the collection device exhibits optimal capture efficiency. However, the frost layer formed by the gas-solid phase transition gradually increases the thermal resistance between the water vapor and the internal structure of the collection device, thereby reducing the heat transfer efficiency and impacting the overall water vapor collection performance. As the frost layer thickness increases, the collection efficiency of the device declines rapidly. To achieve better collection efficiency, the internal structure of the water vapor collection device should be designed to maximize the collision area between the water vapor and the cold trap wall on the inner surface, thereby enhancing the gas-solid phase transition rate of the water vapor.

Journal of Physics: Conference Series **3024** (2025) 012009 doi:10.1088/1742-6596/3024/1/012009

3. Construction of the efficacy verification platform for the water vapor condensation and collection device

To effectively verify the collection efficiency of the water vapor collection device, a performance verification platform for lunar water vapor collection has been established. The platform's parameters are designed to closely simulate the actual working environment on the lunar surface. As a ground-based verification system, this platform enables the evaluation of various collection structures under controlled conventional conditions. Compared to in-situ thermal extraction experiments using simulated lunar regolith as a sample, this approach offers a more efficient method for independently verifying the collection efficiency of the water vapor collection device.

The schematic diagram of the water vapor collection device efficiency verification platform is illustrated in Figure 3. The platform is primarily composed of four functional components: the thermal volatilization unit, the pipeline transportation unit, the condition monitoring unit, and the condensation collection unit.

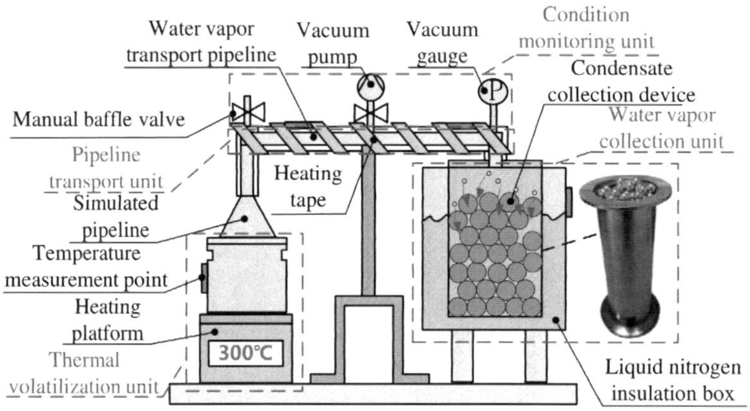

Figure 3. Schematic diagram of the efficacy verification platform

The thermal volatilization unit comprises a heating component and a glass fiber heating tape wrapped around the pipeline, enabling precise temperature regulation. The pipeline transportation unit consists of multiple sections of vacuum pipelines and vacuum valves, designed to transport water vapor generated by the thermal volatilization unit. The condition monitoring unit includes a vacuum gauge and a temperature monitoring module. The vacuum gauge is utilized to monitor pressure variations within the experimental platform, while the temperature monitoring module tracks temperature changes at key locations of the device. The condensation collection unit is composed of a condensation vessel and a condenser. The condensation vessel is filled with liquid nitrogen, and the water vapor collection device is placed inside the vessel to facilitate low-temperature collection. A photograph of the efficacy verification platform is shown in Figure 4.

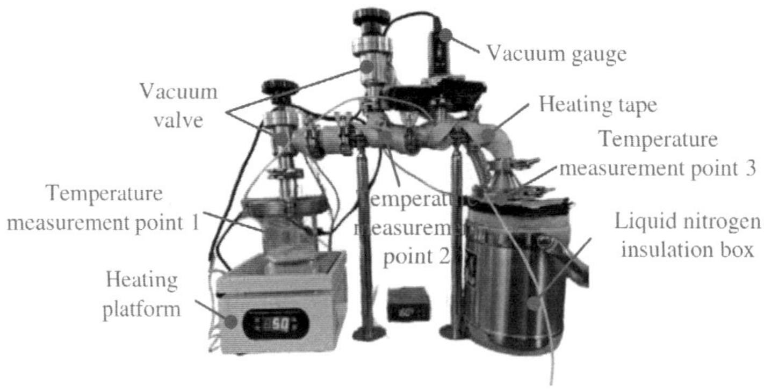

Figure 4. Photograph of the efficacy verification platform

4. Verification experiment of the water vapor condensation and collection structure

4.1. Functional verification experiment of mixed particle size aluminum beads for condensation

For the internal collection structure design of the water vapor collection device, aluminum beads are employed as the collection medium. When the total volume of aluminum beads remains constant, smaller-diameter beads result in reduced gaps between them and a denser arrangement. Since the design objective is to maximize the collision area between water vapor and the inner surface of the cold trap wall, the diameter of the aluminum beads should ideally be minimized. However, it is crucial to avoid pathway blockage caused by excessively small particle sizes and overly dense arrangements during the collection process. Based on these considerations, the functional verification test adopts a mixed particle size arrangement, as illustrated in Figure 5. The three types of aluminum beads are stratified into upper, middle, and lower layers and distributed within the collection device.

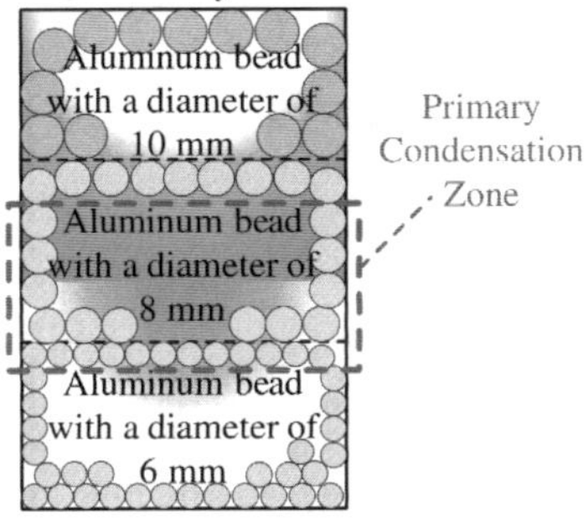

Figure 5. Arrangement of the collection structure for the functional verification experiment

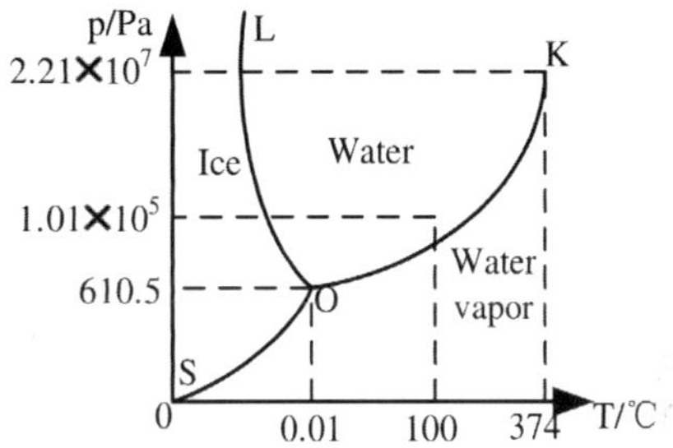

Figure 6. Phase diagram of water

The functional verification experiment utilized a 20 ml water sample. At the start of the experiment, a vacuum pump was activated to evacuate the internal environment of the device. After a certain period, the water in the sample container automatically condensed into ice, and the sample was then heated to generate water vapor. As illustrated in the water phase diagram shown in Figure 6, the triple point of water occurs at 0.01°C and 610.75 Pa. Since the water vapor collection device is immersed in liquid nitrogen, it is essential to maintain the internal pressure of the device below 610.75 Pa to ensure that the gas undergoes a gas-solid phase transition on the inner walls of the collection device.

Due to the generation of water vapor, the pressure readings inside the device exhibit continuous variations, as shown in Figure 7. Each inflection point in the curve represents a key operational node during the experiment.

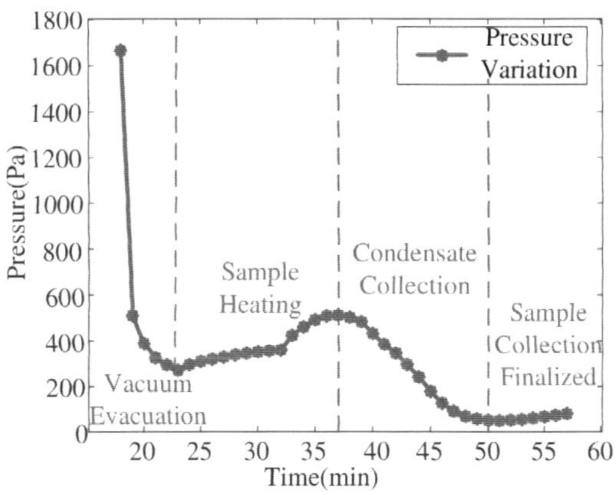

Figure 7. Pressure variation curve for the functional verification experiment

Journal of Physics: Conference Series **3024** (2025) 012009 doi:10.1088/1742-6596/3024/1/012009

After the experiment, the internal state of the water vapor collection device was analyzed, as shown in Figure 8. Figure 8(a) illustrates the state of all aluminum beads, Figure 8(b) shows the state after removing the top layer of loosely packed beads, and Figure 8(c) depicts the state of the remaining beads. Observations revealed that the gas-solid phase transition of water vapor primarily occurred in the region of 8 mm diameter aluminum beads. The bottom of the device contained loosely packed beads, while the middle section exhibited blockage. Significant phase transition did not occur in the lower half of the device, with the blockage located at the interface between the 8 mm and 6 mm diameter aluminum beads.

(a) (b) (c)

Figure 8. Results of the functional verification experiment

4.2. Efficacy verification experiment of water vapor collection

In the aforementioned experiment, the primary blockage occurred in the region of 6 mm diameter aluminum beads. To address this, the water vapor collection efficacy verification experiment employed aluminum beads with diameters larger than 6 mm, specifically 8–10 mm, as the collection structure for comparative validation, following the same experimental procedure. Since smaller-diameter aluminum beads result in denser packing and a larger specific surface area of the collection structure, the relationship between specific surface area and water vapor collection efficiency was analyzed by comparing the results after the experiment. The variation curve of internal pressure within the device over time is shown in Figure 9.

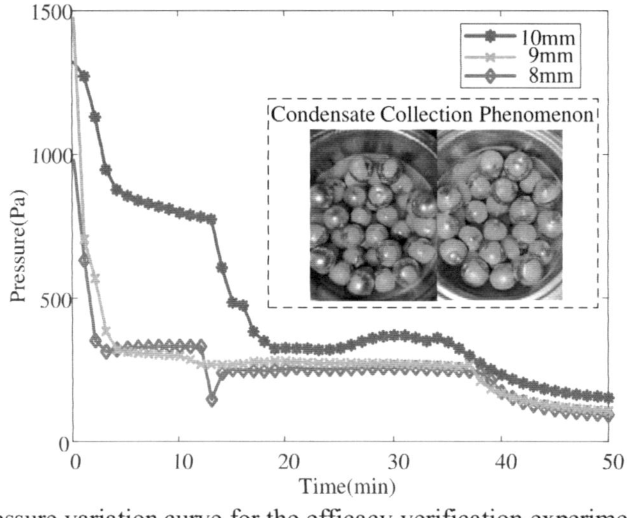

Figure 9. Pressure variation curve for the efficacy verification experiment (water)

Since the initial vacuum pumping process influences the state of water in the container, and to ensure the accuracy of the experimental results, copper(II) sulfate pentahydrate ($CuSO_4 \cdot 5H_2O$) was employed as a sample. This compound's state remains invariant under pressure changes and releases water vapor upon heating, making it suitable for this experiment. The pressure variation curve for this test is presented in Figure 10. To further investigate the relationship between specific surface area and collection efficiency, water vapor collection experiments were conducted using aluminum beads with diameters of 8 mm and 10 mm, respectively. These tests collectively validate the influence of specific surface area on collection performance.

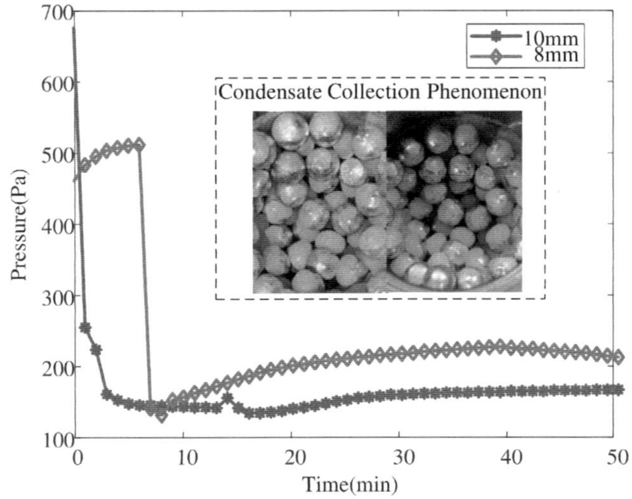

Figure 10. Pressure variation curve for the efficacy verification experiment ($CuSO_4 \cdot 5H_2O$)

By analyzing the comparative verification experiments conducted with water and copper(II) sulfate pentahydrate as samples, the relationship between water vapor collection efficiency and aluminum bead particle size was determined, as illustrated in Figure 11. The results indicate that smaller aluminum bead diameters correspond to higher water vapor collection efficiency.

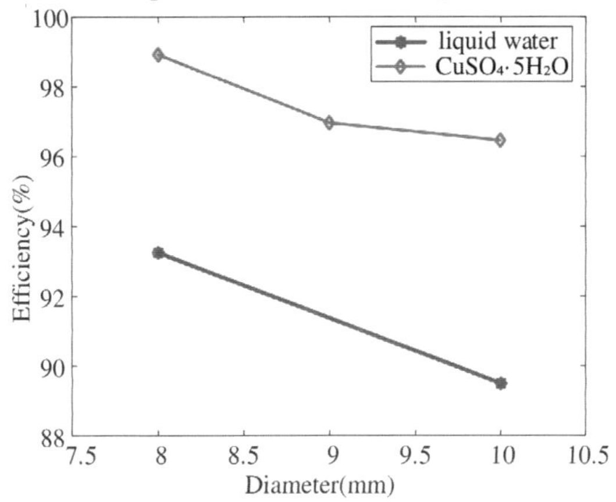

Figure 11. Relationship between collection efficiency and aluminum bead particle size

5. Conclusion

This study, through theoretical analysis and ground-based simulation experiments, reveals the primary factors influencing water vapor collection efficiency in the in-situ thermal extraction of lunar polar water ice.

1. Theoretical analysis demonstrates that changes in porosity during frost layer growth directly affect the permeability of water vapor diffusion channels, and the frost layer growth process is significantly influenced by environmental conditions (pressure and temperature).

2. Since the ice content increases closer to the cold source, the frost layer becomes denser and exhibits higher thermal conductivity. Consequently, the inner surface of the collection device achieves optimal capture efficiency. To enhance collection performance, the internal structure of the water vapor collection device should be designed to maximize the collision area between water vapor and the cold trap wall on the inner surface, thereby increasing the gas-solid phase transition rate of water vapor.

3. Experimental verification reveals that when aluminum beads are used as the collection structure in efficiency verification tests, smaller bead diameters result in a larger internal surface area of the device and more frequent collisions between water vapor molecules and the inner surface. However,

excessively small diameters lead to overly narrow pathways, which can cause internal blockages in the water vapor collection device.

4. The study confirms that optimizing the balance between the internal surface area and porosity distribution of the collection device is key to improving water vapor collection efficiency. Therefore, in the future, efforts will focus on the shape of the collection structure, pathway dimensions, and microstructures of the inner surface. By combining theoretical analysis, simulations, and experiments, the water vapor collection structure will be optimized, providing further insights for the design of in-situ thermal extraction devices for lunar water ice resources.

Acknowledgments

The author thanks the anonymous reviewers for their critical and constructive review of the manuscript. This study was supported by the National Major Engineering Key Technology R&D Program.

References

[1] Colaprete A, Schultz P, Heldmann J, et al. 2010. Detection of water in the LCROSS ejecta plume. Science. 330(6003): 463-468.
[2] Atkinson J, Zacny K. 2018. Mechanical properties of icy lunar regolith: Application to ISRU on the moon and mars. Earth and Space 2018: Engineering for Extreme Environments. Reston, VA: American Society of Civil Engineers, 109-120.
[3] Zacny K, Luczek K, Paz A and Hedlund M. 2016. Planetary Volatiles Extractor (pvex) for in Situ Resource Utilization (ISRU). Earth and Space.
[4] Zacny K, Vendiola V, Morrision P and Paz A. 2021. Planetary Volatiles Extractor (pvex) for Prospecting and in Situ Resource Utilization. EARTH AND SPACE 2021 SPACE EXPLORATION, UTILIZATION, ENGINEERING, AND CONSTRUCTION IN EXTREME ENVIRONMENTS. 713-723.
[5] Vendiola V, Zacny K, Morrison P, Wang A, Yaggi B, Hattori A and Paz A. 2018. Testing of the Planetary Volatiles Extractor (Pvex), EARTH AND SPACE 2018 ENGINEERING FOR EXTREME ENVIRONMENTS. 467-480.
[6] LI R, WU X and WANG W. 2005. Experimental Study of Frost Phenomena on Aluminum Plate. Journal of Thermal Science and Technology. 4.4: 315-318.
[7] Boxe C S , Bodsgard B R , Smythe W , et al. 2007. Grain sizes, surface areas, and porosities of vapor-deposited H2O ices used to simulate planetary icy surfaces. Journal of Colloid & Interface Science. 309(2):412-418.
[8] Mayer E , Pletzer R . 1986. Astrophysical implications of amorphous ice - A microporous solid. Nature. 319(6051):298-301.
[9] GUO Q, WU Y, SUN J, LI Q and WEI X. 2020. A Review of Research on Frosting Phenomenon. Vacuum. 57.3 : 67-72.
[10] Eckert E， Drake R. 1972. Analysis of Heat and Mass Transfer. McGraw-Hill. New York.

Journal of Physics: Conference Series **3024** (2025) 012010 doi:10.1088/1742-6596/3024/1/012010

A Novel Ultrasonic Longitudinal-Torsional Compound Vibration Lunar Soil Water Ice Drill

Yinchao Wang[1,2], Hang Jin[1,2], Zihao Yin[1,2], Weiwei Zhang[3*], Lin Zu[4], Guanghong Tao[1,2], Suyang Yu[1,2], Chunying Jiang[1,2], Changlong Ye[5]

[1] College of Mechanical and Electrical Engineering, Shenyang Aerospace University, Shenyang 110136, China;

[2] Key Laboratory of Rapid Development & Manufacturing Technology for Aircraft (Shenyang Aerospace University), Ministry of Education, Shenyang, 110136, China

[3] State Key Laboratory of Robotics and System, Harbin Institute of Technology, Harbin 150001, China; zweier@hit.edu.cn (W.Z.)

[4] Deep Space Exploration Laboratory, Hefei 230026, China; zulin@mail.ustc.edu.cn (L.Z.)

[5] College of Artificial Intelligence, Shenyang Aerospace University, Shenyang 110136, China;

[*] Correspondence: zweier@hit.edu.cn

Abstract. Under the extremely low-temperature environment of permanent shadow crater in the lunar polar region, the hardness, mechanical strength, and bonding degree of lunar soil water ice are high, and it is difficult to realize low-consumption and high-efficiency drill by traditional digging shovels and drilling. To solve the problem, an ultrasonic longitudinal-torsional composite drill is proposed. The longitudinal-torsional vibration coupling mechanism of the slant groove is analyzed, and the modal analysis, harmonic response analysis and transient analysis of the ultrasonic drill are carried out by the method of finite element. The experiment results show that the designed ultrasonic longitudinal-torsional composite drill can effectively generate longitudinal-torsional composite vibration, which provides a new technical means for the drill of lunar soil water ice.

1. Introduction

China's deep space exploration technology has constituted the core pillars of the country's emergence as a space power. At this stage, global space agencies and companies, including the United States, Europe, Russia, Japan and India, have all set their sights on the Moon, triggering a new race for deep space exploration. Since the launch of the Clementine exploration mission by the United States, scientists have aimed their attention at the polar regions of the Moon, and existing studies have found the existence of water ice on the Moon, various countries have adopted different means of detecting water ice on the lunar soil in the permanently shadowed region of the Moon and have made certain progress[1-2].

The distribution of resources, especially water resources, and the form of their existence in the lunar polar region is one of the important scientific objectives of lunar exploration, and the exploration of the existence of water is of great significance for establishing lunar scientific research stations. Due to the presence of extreme low temperature, low gravity, and electromagnetic strength in the lunar polar region, this makes the drill process extremely difficult[3-4]. The physical properties of lunar soil water ice in the polar region are extremely complex[5-8], and the macro-mechanical properties of lunar soil water ice, such as bearing capacity, compressibility and shear resistance, are closely related to the pore ratio, which gradually increases with depth, and the cohesion and friction angle gradually increase, which means that the degree of adhesion of the lunar soil water ice will also gradually increase. At present, the drilling sampling of lunar soil water ice in various countries is to carry the driller on patrol and use the spiral drill to carry out drilling sampling[9]. If we want to drill and sample the lunar soil water ice in the subsurface polar region, it also puts forward high requirements on the chip removal ability of the drilling apparatus. In recent years, ultrasonic drill based on the piezoelectric ceramic drive has achieved good results in asteroid detection, which can overcome the factors of low gravity, low temperature, and low power consumption[10-11], and the sonic drill apparatus has the outstanding advantages of simple structure, low required drill pressure, low power consumption, a wide range of temperature resistance (-200℃-500℃) and no need for lubrication. Ultrasonic drills can be categorized into percussion ultrasonic drills and rotary percussion ultrasonic drills[12]. According to the working principle. Impact ultrasonic drills have better drill results but poorer chip removal capability. There is a special longitudinal-torsional composite ultrasonic drill in the rotary percussion drill, longitudinal-torsional composite vibration ultrasonic drill generally realizes the longitudinal-torsional composite motion in two ways, one is to use the longitudinally polarized crystal pile and tangentially polarized crystal pile to generate a longitudinal-torsional composite[13], and the other is to realize the longitudinal-torsional composite motion by the vibration mode shift[14-15], this paper intends to present a piezoelectric In this paper, we propose a longitudinal-torsional composite drill with double rows of slanting grooves at the end of the transducer horn, which can increase the longitudinal-torsional transducer ratio as well as ensure the simplicity of the structure.

2. Structural design and theoretical analysis of longitudinal and torsional composite drill

2.1:Overall structural design.

The longitudinal-torsional composite vibratory ultrasonic drill is composed of four parts: piezoelectric stack, longitudinal-torsional composite kinematic unit, drill tool and support housing, as shown in Figure. 1.

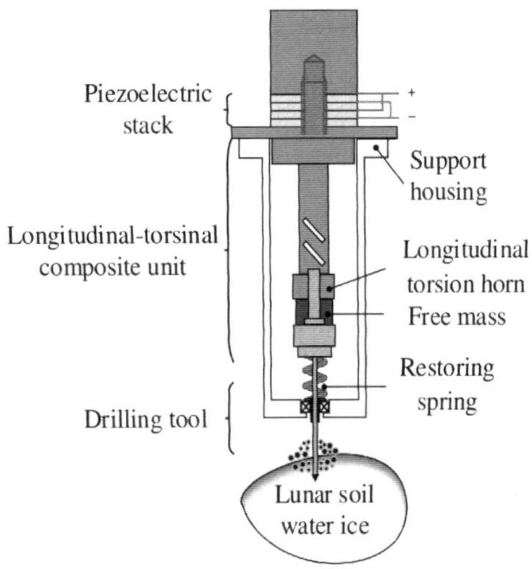

Figure 1. Ultrasonic longitudinal and torsional composite drill system components

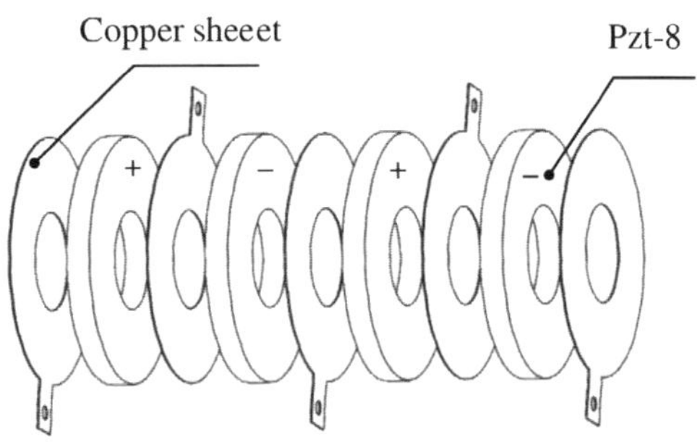

Figure 2. Piezoelectric stack

The piezoelectric ceramic stack consists of four piezoelectric ceramic wafers with alternating "+" and "-" poles and five copper wafers are shown in Figure. 2. The polarization direction is along the thickness direction and the operating mode is d_{33} mode.

The longitudinal-torsional composite motion unit is located at the front end of the piezoelectric stack, which contains a longitudinal-torsional amplitude-change rod, a return spring and a self-lubricating bushing. After the piezoelectric stack is subjected to high-frequency AC voltage, the piezoelectric ceramics will generate high-frequency mechanical vibration along the polarization direction, which puts the horn in resonance, and longitudinal-torsional coupled vibration will occur at the end of the horn, and the longitudinal vibration will be amplified by its mass. The existence of oblique grooves on the horn converts part of the axial motion into circumferential motion, eventually forming an elliptical motion at the end. The return spring provides the preload force, which provides the friction required for the rotation of the contact between the end of the horn and the free mass. The friction force drives the free mass to torsion, and the free mass and the drill tool are mated by the spline, and the self-lubricating sleeve is an interference fit with the bearing.

2.2. Piezoelectric transducer design

The horn is made of aluminum alloy with low density and high strength, the rear cover plate should be made of 304 stainless steel with high density, and the piezoelectric ceramic sheet should be made of pzt-8. The physical properties of each material are shown in Table 1.

Table 1. Material properties

Material	Density /(kg/m^{-3})	Young's modulus /(GPa)	Velocity /(m•s^{-1})
304 stainless steel	7930	209	5130
aluminum	2700	70	5253
PZT-8	7450		3170

The size of each part of the piezoelectric transducer is theoretically calculated using the piezoelectric vibration theory. The theoretical operating frequency of the transducer is set to 20 kHz, and the dimensional parameters of the piezoelectric transducer are obtained by deriving the piezoelectric vibration equation as shown in Table 2.

Table 2. Piezoelectric transducer structural parameters

Parameters	Unit (of measure)	Numerical value
Rear end cap cross-sectional area S_1	mm^2	843.46
Rear end cap length L_1	mm	27
Cross-sectional area of piezoelectric ceramic sheet S_2	mm^2	761.1
Thickness of piezoelectric ceramic sheet L_2	mm	12
Cross-sectional area S_3 of the third part of the horn	mm^2	1023.72
Length of the third part of the horn L_3	mm	7
Cross-sectional area S_4 of horn part 4	mm^2	94.98
Length of horn part 4 L_4	mm	62
Cross-sectional area S_5 of horn part 5	mm^2	267.62
Length of horn part 5 L_5	mm	10

2.3. Analysis of longitudinal and torsional coupling mechanism.

After the structural dimensional parameters of the piezoelectric transducer are determined, a slanted groove is to be opened at the end of the amplitude-variable rod. Due to the existence of the oblique groove, after the longitudinal vibration wave is transmitted to the area of the oblique groove, the geometrical characteristics of the oblique groove will decompose the axial displacement into longitudinal and tangential components. At this time, the tangential component excites torsional vibration around the axis through asymmetric shear deformation.

The structural parameters of the diagonal groove will affect the effect of longitudinal torsional vibration, and the number, depth, angle and width of the diagonal groove will affect the generation of its torsional vibration, which is a coupled relationship, so the effect of the structural parameters of the diagonal groove on the torsional effect is to be analyzed. Firstly, the ratio of longitudinal vibration displacement to torsional displacement is defined as longitudinal torsional transmutation ratio ($\beta = A_t/A_l$), and the mathematical model between longitudinal torsional transmutation ratio and structural parameters of inclined grooves is established.

The coupled kinetic equations are established according to D' Alembert equation:

$$\begin{cases} \rho A_{eff} \dfrac{\partial^2 u}{\partial t^2} = \dfrac{\partial F_z}{\partial z} + F_\theta \\ \rho J_p^{eff} \dfrac{\partial^2 \phi}{\partial t^2} = \dfrac{\partial M_z}{\partial z} + F_\theta \cdot R \end{cases} \quad (1)$$

where ρ is the density of the material, kg/ m^2; A is the cross-sectional area, m^2; E is the modulus of elasticity, Pa; J_p is the cross-sectional polar moment of inertia, N/m; G is the shear modulus, Pa, and G=E/2(1+v); R is the radius of the horn, F_θ is the shear coupling force caused by the diagonal groove;

The shear coupling force is brought into the above equation to obtain the final coupling equation as:

$$\begin{cases} \rho A_{eff} \dfrac{\partial^2 u}{\partial t^2} = EA_{eff} \dfrac{\partial^2 u}{\partial z^2} + \dfrac{nGh^2}{4R} \sin(2\theta) \dfrac{\partial^2 u}{\partial z^2} \\ \rho J_p^{eff} \dfrac{\partial^2 u}{\partial t^2} = GJ_p^{eff} \dfrac{\partial^2 \phi}{\partial z^2} + \dfrac{nGh^2 R}{4} \sin(2\theta) \dfrac{\partial^2 u}{\partial z^2} \end{cases} \quad (2)$$

K is defined as the longitudinal-torsional coupling coefficient:

$$K = \frac{nGh^2 \sin(2\theta)}{4R} \cdot \frac{G}{E} \quad (3)$$

The solution to longitudinal and torsional vibration is obtained as:

$$\begin{cases} U(z) = A_l \sin\left(\dfrac{\omega}{c_l} z\right) & \phi(z) = A_t \sin\left(\dfrac{\omega}{c_t} z\right) \\ c_l = \sqrt{\dfrac{E(1+K)}{\rho}} & c_t = \sqrt{\dfrac{G}{\rho}} \end{cases} \quad (4)$$

Finally, the longitudinal torsional translational ratio is obtained by matching the boundary conditions:

$$\beta = \frac{A_t}{A_l} = \frac{K \cdot c_t^2}{c_l^2 - c_t^2} \quad (5)$$

Through the establishment of the longitudinal torsional coupling equation to obtain the mathematical model between the structural parameters of the diagonal groove and the longitudinal torsional conversion ratio, and analyze the influence of each parameter on the longitudinal torsional conversion ratio, it can be seen that the number of diagonal grooves, diagonal groove angle and the depth of the diagonal groove will increase the longitudinal torsional conversion ratio. This paper visualizes the multi-parameter coupling, and in the process of analysis, it is assumed that one of the diagonal groove width, diagonal groove angle, and number of diagonal grooves is a fixed value, and the specific parameters are shown in Table 3, and the influence of the other two variables on the longitudinal torsional conversion ratio is observed, as shown in Figures. 3.

In the parameter setting of the inclined groove, the depth, angle and number of inclined grooves should be considered in accordance with the priority of the inclined groove depth, according to the image, it can be seen that when the depth of the inclined groove is 2mm, and the angle of the inclined groove is 45° the longitudinal torsional conversion ratio is the largest, and the preliminary setting of the inclined groove structure parameters are n=8, h=2mm, θ= 45°, and the double-row layout.

Table 3. Parameters of Inclined Groove

Structural Parameters of Inclined Groove	numerical value
Depth of inclined groove/mm	1-5
Number of inclined grooves/pc	4-8
Groove angle/°	30-60

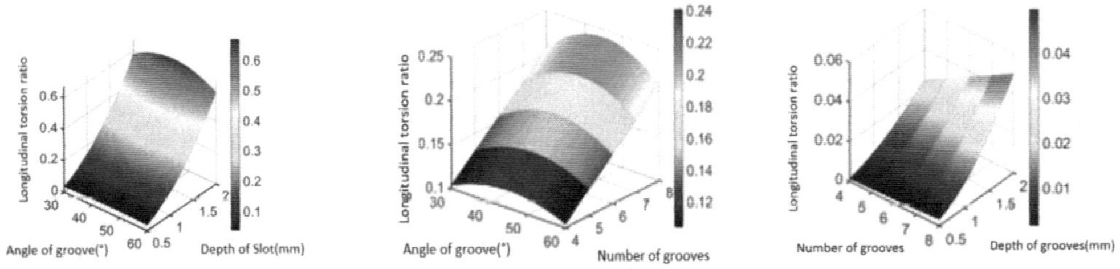

Figure 3. Relationship between grooves and longitudinal conversion ratio

3. Simulation verification of longitudinal-torsional composite motion

3.1 Harmonic response analysis of piezoelectric transducers

The modal analysis of the above model is carried out first to obtain the resonance frequency of the piezoelectric transducer, and then the harmonic response analysis is carried out by using the complete method, according to the resonance frequency obtained from the modal analysis is 13831 HZ, the frequency search range is set to be 13000 HZ-14000 HZ, and the harmonic response analysis is carried out by using the complete method for the piezoelectric transducer, after meshing the piezoelectric transducer, and then fixed constraints are added on the flange to the piezoelectric ceramic stack was applied with a sinusoidal excitation voltage of 300v, and a point was taken at the end of the horn to observe its vibration displacement in the X, Y and Z directions. As shown in Figure. 4, the vibrator has a resonant frequency of 13680 Hz. It indicates that both longitudinal vibration and torsional vibration reach the resonance state at this point, which verifies the reasonableness of the design.

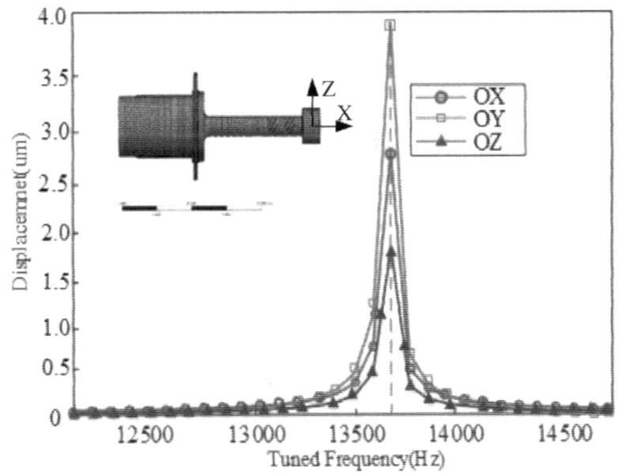

Figure 4. Piezoelectric transducer harmonic response analysis results

When the piezoelectric transducer reaches the resonance frequency, the displacement in the X direction is 4um, which is the longitudinal displacement, and the displacement in the Y direction and Z direction is 2.5um, and the longitudinal torsional transducer ratio is 0.6, which is basically the same as that obtained from the three-dimensional image of longitudinal torsional coupling analysis, and this also verifies that the mathematical model of the longitudinal torsional coupling analysis between the slanting groove parameter and the longitudinal torsional transducer ratio is Correctness.

3.2 Transient simulation

In order to further verify the correctness of the design of the longitudinal-torsional composite piezoelectric ultrasonic transducer, the ultrasonic longitudinal-torsional composite drill drive transducer is analyzed by transient analysis using the FULL method. According to the resonant frequency of 13680 Hz obtained from the harmonic response analysis to set up the appropriate step length and number of steps, a sinusoidal excitation voltage of 300v is applied to the piezoelectric ceramic sheet, and a point is taken at the output end face of the horn. The longitudinal vibration displacement change and torsional vibration displacement change of the point are extracted, and the all-directional displacement of the point over time is fitted into a curve, and it is learned that the motion of the point is an ellipse, as Figure. 5shown, which verifies the preliminary design concept.

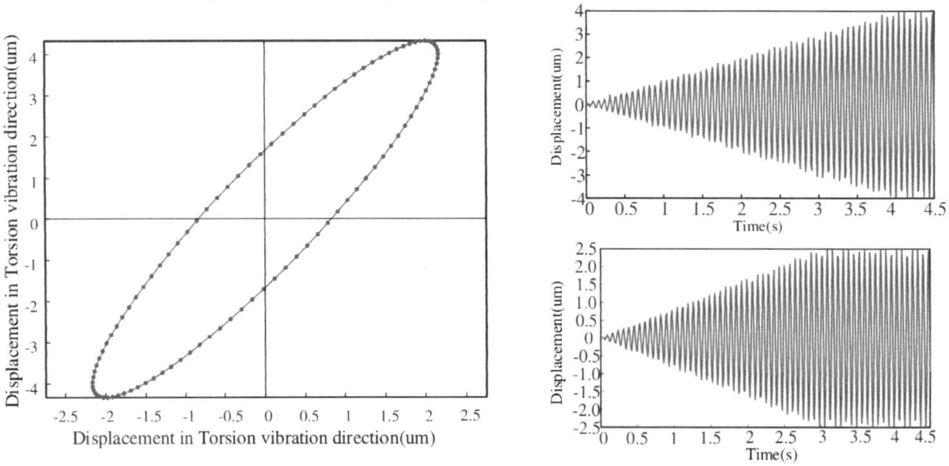

Figure 5. Transient analysis results

4. Experimental Result

4.1 Vibration Displacement Measurements

In order to compare the theoretical vibration displacements with the actual vibration displacements, a laser displacement sensor (LK-H020, KEYENCE) was used to measure the vibration displacements at the end of the horn. The measured nodes are the same as those in the harmonic response analysis, and the measurements were performed under the following conditions: excitation voltage 300 V and frequency 13.8 kHz, as shown in Figure.6. The vibration displacements of node in the OX and OZ directions are 3.5 μm and 1.8 μm, respectively. Since the vibration displacement is close to the measurement limit of the sensor, it is easy to be disturbed by external vibration, which is consistent with the results of the harmonic response simulation from the experimental results.

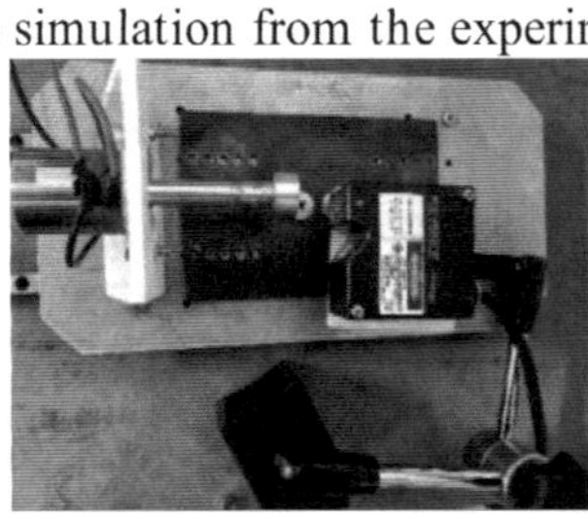

Figure 6. Measuring the end displacement of horn

4.2 Feasibility Drilling Experimental Verification

The driller piggybacked on the test rig was used to test the feasibility of the design as shown in Figure 7. The driller was mounted on a vertical sliding table using weight to provide the drilling load. All tests were conducted with the driller under its own weight load and an excitation voltage of 300 V and a response frequency of 13.8 kHz. Initial drilling tests were conducted on sandstone using an aluminum-gold drill bit.

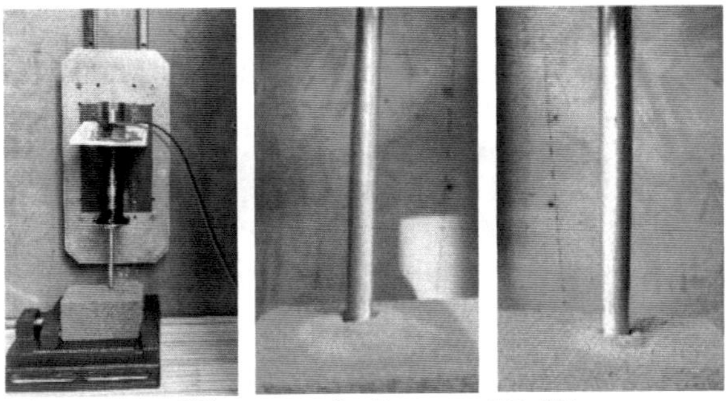

Figure 7. Drilling test platform and Drilling process

In order to verify the drilling function of the drill, as shown in Figure.8, a drilling experiment was set up in the vertical state. Figure shows the drilling process at the same time interval. The drilling depth has reached 20mm, so the feasibility of the drill is proved. The change of drilling depth with time is shown in the Figure.11.

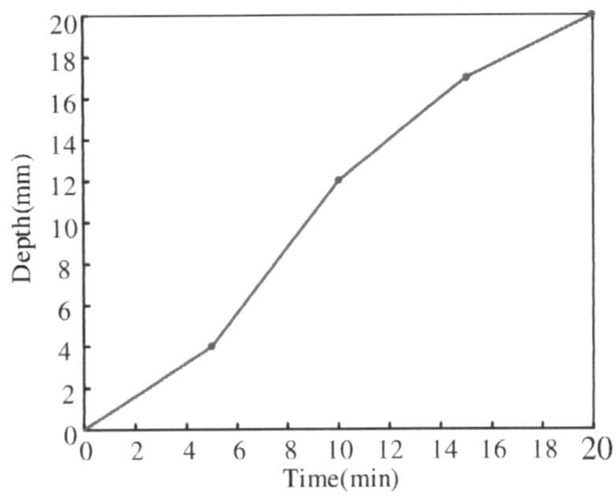

Figure 8. Drilling depth change

5. Conclusion

In this paper, we propose a novel ultrasonic drill designed to achieve longitudinal and torsional composite motion using a single excitation source. This is accomplished by creating a double row of diagonal grooves at the end of the horn. We established a mathematical model that relates the structural parameters of the inclined channel to the longitudinal-torsional translational ratio. Through visual analysis, we determined the optimal parameters for the inclined channel structure, and the accuracy of the mathematical model was validated through finite element simulations. Additionally, we conducted a transient analysis of the transducer using the finite element method to confirm the working principle of the ultrasonic drill. Based on the design theory, we fabricated a prototype of the drilling tool, and preliminary experiments confirmed the validity of the theoretical design.

Future drilling experiments should be conducted using lunar water ice simulants in a low-temperature vacuum environment to assess the sampling performance of the drilling apparatus. These experiments should also involve comparative analyses with conventional percussive ultrasonic drills.

Acknowledgments

The author thanks the anonymous reviewers for their critical and constructive review of the manuscript. This study was supported by the National Major Engineering Key Technology R&D Program.

Reference

[1] Hayne, P. O., Hendrix, A., Sefton-Nash, E., et al. 2015. Evidence for exposed water ice in the Moon's south polar regions from Lunar Reconnaissance Orbiter ultraviolet albedo and temperature measurements. Icarus, 255, 58-69.

[2] Pieters, C. M., Goswami, J. N., Clark, R. N., et al. 2009. Character and spatial distribution of OH/H2O on the surface of the Moon seen by M3 on Chandrayaan-1. Science, 326(5952), 568-572.

[3] Bart, G. D., Colaprete, A. 2009. Shadow depths and other characteristics of potential LCROSS impact sites. In Proceedings of Lunar and Planetary Institute Science Conference; [s.n.].

[4] Feldman, W., Barraclough, B., Maurice, S., et al. 1998. Major compositional units of the Moon: lunar prospector thermal and fast neutrons. *Science*, 281(5382), 1489-1493.

[5] Davies, J. G., Lovell, B., Pritchard, R. S., et al. 1966. Observations of the Russian Moon probe Luna 9. Nature, 209(5026), 848-850.

[6] Karachevtseva, I., Oberst, J., Scholten, F., et al. 2013. Cartography of the Lunokhod-1 landing site and traverse from LRO image and stereo-topographic data. Planetary and Space Science, 85, 175-187.

[7] Severny, A. B., Terez, E. I., Zvereva, A. M. 1975. The measurements of sky brightness on Lunokhod-2. The Moon, 14(1), 123-128.

[8] Bao, X., Bar-Cohen, Y., Chang, Z., et al. 2003. Modeling and computer simulation of ultrasonic/sonic driller/corer (USDC). *IEEE Transactions on Ultrasonics, Ferroelectrics, and Frequency Control*, 50(9), 1147-1160.

[9] Liu D., Zhang Q., Yang S., et al. 2020. Lunar Polar Regolith Sampling Techniques. Journal of Deep Space Exploration, 7(3): 278-289.

[10] Harkness, P., Lucas, M. 2012. A brief overview of space applications for ultrasonics. Ultrasonics, 52(8), 975-979.

[11] Berden, Y., Wang, Y., Li, H., et al. 2018. An ultrasonic driller with piezoelectric transducer directly driving drilling tools. Journal of Harbin Engineering University, 39(05), 806-812.

[12] Wang, Y., Quan, Q., Yu, H., et al. 2018. A rotary impact ultrasonic drill driven by a single piezoelectric stack. Journal of Beijing University of Aeronautics and Astronautics, 44(09), 1850-1859.

[13] Zhang, D., Xu, L., Zhang, H., et al. 2022. Design of a double excitation cross-shaped slotted longitudinal and torsional composite vibrational piezoelectric ultrasonic transducer. Journal of China University of Weights and Measures, 33(2), 197-203.

[14] Harkness, P., Lucas, M., Cardoni, A. 2012. Coupling and degenerating modes in longitudinal-torsional step horns. Ultrasonics, 52(8), 980-988.

[15] Tsujino, J., Ueoka, T. 1998. Vibration characteristics of one-dimensional longitudinal-torsional converter with multiple slitted parts. In: Proceedings of 1998 IEEE Ultrasonics Symposium, 723-728.

Three-Dimensional Modeling-Driven AMESim Virtual Prototype Development and Flow Characterization of Swashplate Axial Piston Pumps

Jiabao Li[1,2], Zhonggang Xiong[2,3*], Zhong Liu[1,2], Deqing Liu[1,2], Xintao Liu[2,4], Huixian Li[2,3]

[1] School of Mechanical and Electrical Engineering, Guilin University of Electronic Technology, Guilin Guangxi 541004,China

[2] Guangxi Key Laboratory of Special Engineering Equipment and Control, Guilin Guangxi 541004, China

[3] Guangxi University Special Engineering Equipment Design and Intelligent Drive Technology Laboratory Guilin University of Aerospace Technology, Guilin Guangxi 541004, China

[4]School of Mechanical and Electrical Engineering, Guangxi University, Nanning Guangxi 530004, China

[*]Corresponding author's e-mail: xzglsl2013@163.com

Abstract. The swashplate-type axial piston pump, a core component of hydraulic systems, encounters challenges in optimizing performance due to the high costs and lengthy cycles of traditional experimental methods.This paper proposes a virtual prototyping approach that integrates 3D modeling with AMESim simulation to overcome the limitations of conventional design.We employ a high-precision geometric model of the pump and derive piston motion equations and a transient flow mathematical model based on kinematic theory. This establishes a theoretical framework for analyzing the pump's dynamics and flow and pressure characteristics.Using the AMESim platform, we developed virtual prototypes for fixed-displacement and constant-pressure variable pumps to systematically study how key parameters—such as swashplate tilt angle and outlet displacement—affect output flow, pressure pulsation, and dynamic response.The results show that:Increasing the swashplate tilt angle significantly boosts flow but worsens pulsations;Outlet displacement requires a balance between response speed and pulsation suppression.By implementing a digital thread, this study seamlessly integrates geometric design with dynamic simulation, offering greater efficiency in multidisciplinary coupling analysis and performance optimization compared to traditional HCD library modeling methods.This approach provides effective technical support for low-pulsation design, parameter optimization, and engineering applications of hydraulic components.

1. Introduction

In hydraulic systems, the swashplate-type axial piston pump has attracted widespread attention due to its high efficiency and excellent dynamic response.However, flow fluctuations and parameter coupling effects under real operating conditions present significant challenges for its design and optimization.Traditional experimental testing methods are often limited by cost and long test cycles, whereas virtual prototyping for simulation analysis has emerged as an efficient research approach.Haitao Qi developed a novel multi-domain simulation model for an electric servo variable pump based on AMESim. The model's credibility was verified through simulations of its flow-pressure characteristics and variable adjustment properties, thereby providing simulation support for pump system optimization[1]. Junzhe Lin addressed the issue of source pressure pulsation in a seven-piston pump by developing a simplified AMESim model and constructing an experimental platform for validation. It was observed that under constant speed, an increase in load pressure resulted in a larger amplitude of source pressure pulsation while the pulsation rate decreased, with the fundamental frequency amplitude being most significantly affected. The simulation results closely matched the experimental data, thereby providing theoretical support for suppressing pump source pulsations and controlling hydraulic pipeline vibrations[2]. Xinjie Zhang developed a dynamic simulation model of a piston pump with multi-parameter coupling based on AMESim software, systematically analyzing the influence of outlet displacement on pulsation characteristics—a parameter that has received relatively little attention in traditional studies. Through theoretical derivation and simulation validation, the study revealed the differentiated effects of swashplate tilt angle, motor speed, and outlet displacement on pulsation amplitude and pulsation rate[3].

In recent years, advances in 3D modeling technology have introduced novel approaches for rapid pump body modeling and optimization, while advanced simulation platforms such as AMESim offer reliable tools for dynamic system simulation.This paper aims to develop a corresponding virtual prototype by integrating a detailed geometric model of a swashplate-type axial piston pump with the AMESim platform. It investigates the flow characteristics of the piston pump under various parameters and operating conditions, providing both theoretical and technical support for pump optimization.

2. Swashplate Axial Piston Pump Flow and Pressure Theory Analysis

Based on kinematic principles, this section theoretically models the core working mechanism of the swashplate-type axial piston pump. It focuses on analyzing piston movement and the resulting flow and pressure changes, providing a theoretical foundation for subsequent simulation and optimization design.

2.1. Kinematic Analysis of Axial Piston Pumps

The operating principle of swashplate axial piston pumps relies on the reciprocating motion of the plungers in the cylinder body, which is generated by the uniform rotation of the cylinder body in conjunction with the tilting of the swashplate. In order to ensure the overall output balance, each plunger is usually evenly distributed in the cylinder body, and its motion can be described by the following mathematical model[4].

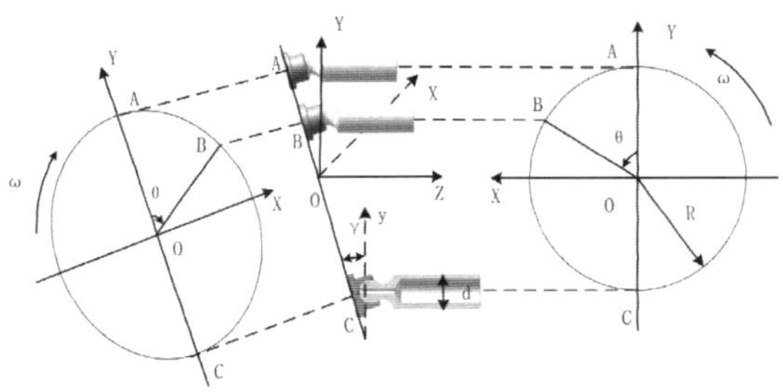

Figure 1. Dynamic Analysis of Swashplate Axial Piston Pump

Inclination angle of swashplate is γ;
Plunger distribution circle radius is R;
The angular speed of rotation of the cylinder is ω;
The instantaneous angle of rotation of the plunger in the cylinder is $\theta=\omega t$;
The motion of the piston in the swashplate projection direction is simple harmonic, with a maximum stroke of $2R\tan\gamma$. The axial displacement of the piston, denoted as $s(t)$, is derived from geometric relationships as follows:

$$s(t) = R\tan\gamma(1-\cos\theta) \tag{1}$$

Based on this displacement expression, the instantaneous velocity of the piston is obtained by differentiating with respect to time:

$$v(t) = \frac{ds}{dt} = R\tan\gamma\omega\sin\theta \tag{2}$$

and the piston acceleration equation is:

$$a(t) = \frac{d^2s}{dt^2} = R\tan\gamma\omega^2\cos\theta \tag{3}$$

The kinematic equations of the swashplate-type axial piston pump not only reflect the periodic variations in the piston's motion, but also provide the necessary kinematic foundation for subsequent calculations of flow and pressure.

2.2. *Theoretical Modeling of Single Plunger Flow Pressure*

The tilt angle of the swashplate determines the piston's stroke length, thereby affecting the pump's displacement. By varying the swashplate tilt angle (via mechanical or hydraulic control), the pump's output flow can be continuously adjusted to achieve variable functionality.

The reciprocating motion of an individual piston directly generates local flow; its instantaneous flow is the product of the piston's velocity and its effective cross-sectional area. Assuming a piston diameter of d, the effective cross-sectional area is given by:

$$A_p = \frac{\pi d^2}{4} \tag{4}$$

Thus, the instantaneous flow $Q_k(t)$ of the kth piston can be expressed as:

$$Q_k(t) = \frac{\pi d^2}{4}R\tan\gamma\cdot\omega\sin(\omega t - \frac{2\pi(k-1)}{Z}) \tag{5}$$

The total number of pistons is Z;
For a pump with Z pistons, the overall instantaneous flow is the sum of the instantaneous flows from each piston[5]:

$$Q_{total}(t) = \sum_{k=1}^{Z} Q_k(t) = \frac{\pi d^2}{4}R\tan\gamma\cdot\omega\sum_{k=1}^{Z}\sin\left(\omega t - \frac{2\pi(k-1)}{Z}\right) \tag{6}$$

Table 1 Comparison of Stroke Phases in Swashplate Piston Pump

Phase	Angular Range	Piston Motion	Flow Rate	Pressure
Discharge Stroke	$0<\theta<\pi$	Outward Motion	$Q_{out} = \frac{\pi d^2}{4}\omega R\tan\gamma\sin\theta$	High Pressure (Outlet)
Suction Stroke	$\pi<\theta<2\pi$	Inward Motion	$Q_{int} = -\frac{\pi d^2}{4}\omega R\tan\gamma\sin\theta$	Low Pressure (Inlet)

By varying the swashplate tilt angle γ, the pump's variable functionality can be achieved, thereby enabling continuously adjustable output flow.

3. 3D Modeling-Driven Modeling Approach for AMEsim Piston Pump Submodels

This paper investigates the Rexroth A10V series swashplate axial piston pump. It employs an established three-dimensional model of the pump, which is imported into AMESim using the CAD import function to generate the submodel for the swashplate axial piston pump. Initially, the imported three-dimensional model is simplified, and the fluid domains for the inlet and outlet are extracted, as shown in Figure 2.Subsequently, the "activate 1D modeling" function was used within the primitive association to select components and set parameters for the axial piston pump, as shown in Figure 3. After the adjustments, the "model generation" function was employed to create the submodel of the swashplate-type axial piston pump, as illustrated in Figure 4[6].

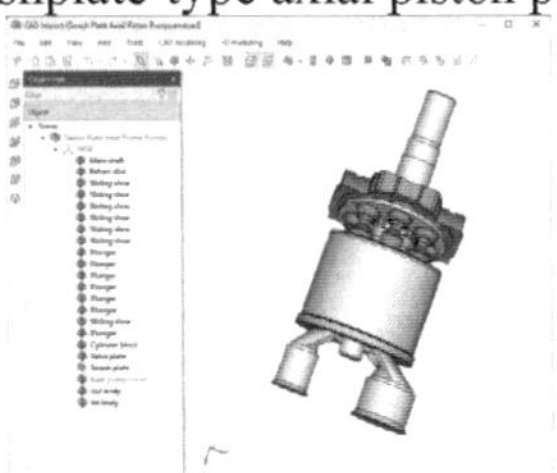

Figure 2. Extraction of Inlet and Outlet Fluid Domains

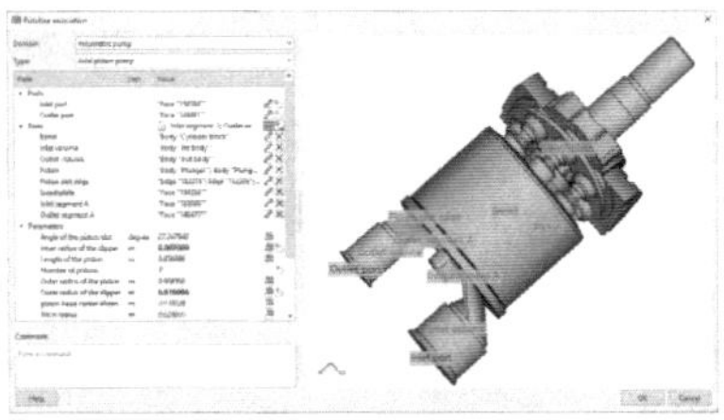

Figure 3. Selection of Components for the Axial Piston Pump

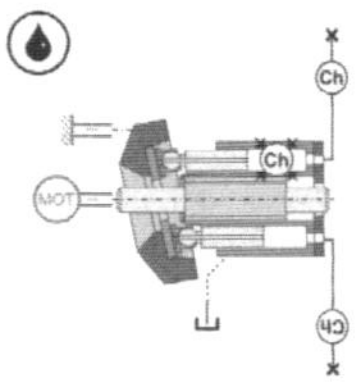

Figure 4. Axial Piston Pump Submodel

Figure 5 illustrates the classical piston pump modeling method: the swashplate-type axial piston pump model, constructed using standardized submodels from the HCD library, clearly demonstrates the multiphysics coupling mechanisms within the pump body and facilitates independent optimization of key components; however, it is not conducive to the rapid optimization of the pump's overall performance[7].

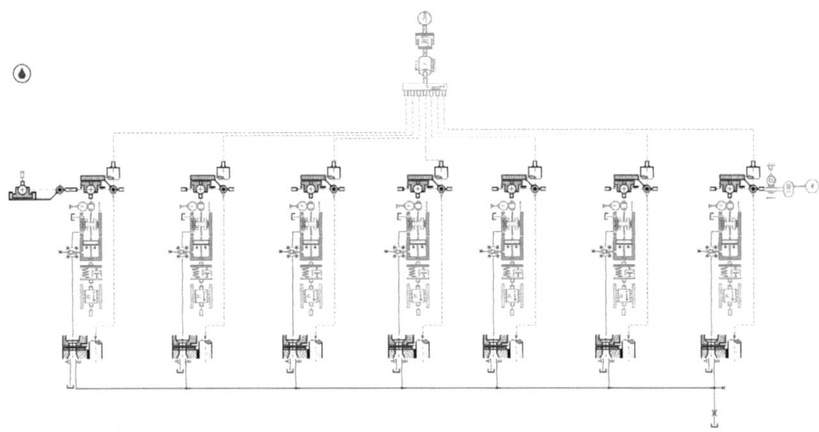

Figure 5. Classical Piston Pump Modeling Method

By directly generating the axial piston pump submodel from CAD 3D model files, a digital thread from geometric design to physical simulation is established, significantly enhancing accuracy, efficiency, and collaborative capabilities. This method is particularly suited for advanced piston pump designs with high complexity and multidisciplinary coupling, and represents a key technological pathway for constructing digital twins of hydraulic systems.

4. Simulation Analysis of the Virtual Prototype for Swashplate Axial Piston Pump

In the preceding analysis, we explored the kinematic characteristics of the Swashplate axial piston pump and the flow and pressure models for a single plunger.Next, we will build a fixed displacement pump model based on the generated plunger pump submodel to further investigate the key factors affecting the overall performance of the pump, including the influence of parameters such as Swashplate angle and outlet volume on flow and pressure pulsations[3].

4.1. *Modeling and Simulation Analysis of Fixed Displacement Pump*

As shown in Figure 6, the fixed-displacement pump model was constructed in AMESim. The structural parameters of the swashplate-type axial piston pump submodel were compared with those of the corresponding 3D model, and once their accuracy was confirmed, the simulation settings were initiated.

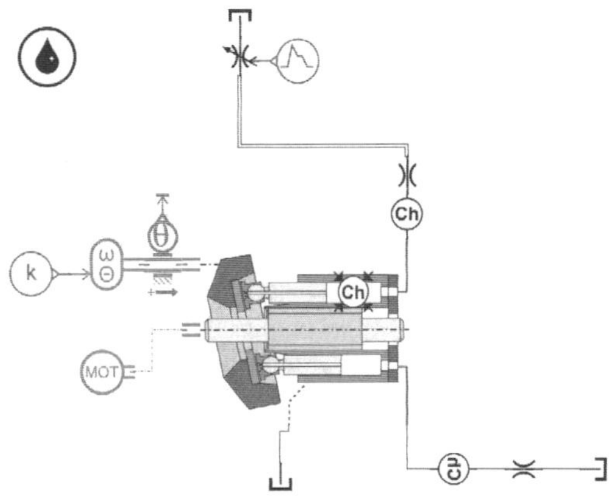

Figure 6. Fixed-Displacement Pump Model

The AMESim quantitative pump model was configured with variable Swashplate inclination angles: The motor speed was maintained at 3000 rpm with an outlet volume of 60 cm³. Batch processing was implemented by adjusting the K-value parameter to sequentially set the Swashplate angles of the piston pump to 10°, 12°, 14°, 16°, and 18°.Following simulation completion, the characteristic curves depicting the output pressure and flow rate of the axial piston pump under various Swashplate inclination angles are presented in Figs. 7 and 8.

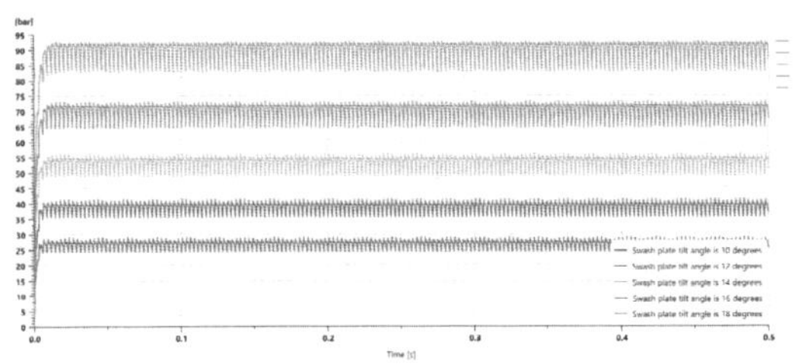

Figure 7. Pressure-Time Characteristic Curves under Variable Swashplate Inclination Angles

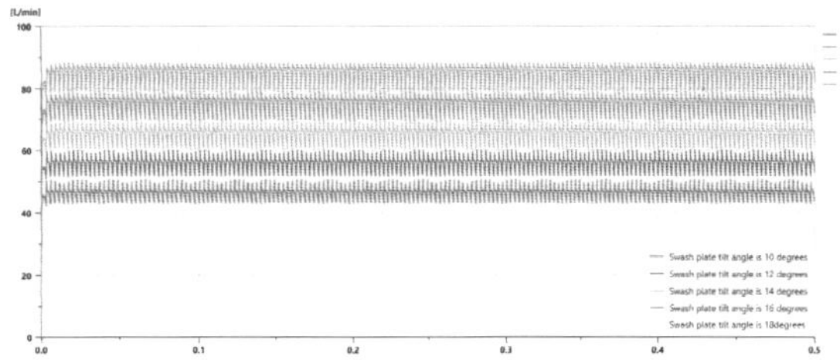

Figure 8. Flow Rate-Time Characteristic Curves under Variable Swashplate Inclination Angles

Analysis of simulation data with variable Swashplate inclination angles: As derived from the theoretical formulation, flow pulsation exhibits positive correlation with Swashplate inclination, which is corroborated by the graphical data.Increased Swashplate inclination extends the reciprocating stroke of pistons, thereby elevating displacement per revolution and consequently increasing theoretical flow rate. Conversely, reducing the Swashplate inclination diminishes displacement per revolution, resulting in decreased flow output.

Configuration of AMEsim fixed-displacement pump model: The system was configured with variable outlet volumes by adjusting parameter K. Key operational parameters included: Swashplate angle fixed at 18°, motor speed maintained at 3000 rev/min, and outlet volumes systematically varied across 20 cm³, 40 cm³, 60 cm³, 80 cm³, and 100 cm³ through batch processing.Post-simulation analysis:

Journal of Physics: Conference Series **3024** (2025) 012011 doi:10.1088/1742-6596/3024/1/012011

Characteristic curves depicting the axial piston pump's pressure and flow rate outputs under varying outlet volumes are presented in Figs. 9 and 10, demonstrating distinct operational patterns corresponding to volume modifications.

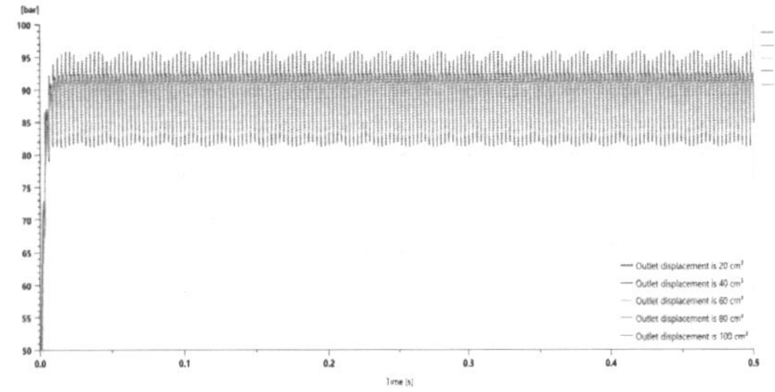

Figure 9. Variable-outlet-volume pressure versus time curve

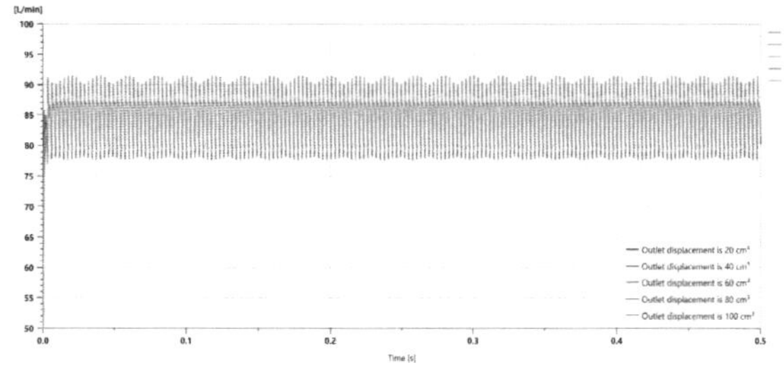

Figure 10. Variable-outlet-volume flow rate versus time curve

Variable-outlet-volume simulation analysis: The simulation data reveal a nonlinear correlation between outlet volume and pulsation suppression efficiency. Notably, the pulsation attenuation rate exhibits asymptotic stabilization when the volume exceeds 80 cm^3.While outlet volume variation minimally affects the pump's mean flow rate (determined by displacement), it significantly modulates both flow pulsation characteristics and pressure parameters, particularly influencing the pressure rise rate and stability metrics.Larger outlet volumes demonstrate enhanced pulsation buffering capacity and reduced pressure fluctuations, whereas smaller volumes facilitate faster pressure buildup at the expense of increased fluctuation amplitude and transient peak pressures.If rapid system response and sensitivity to transient pressure variations are prioritized, smaller outlet volumes are recommended. Conversely, operational scenarios emphasizing system stability and pulsation mitigation would benefit from larger volume configurations.These findings establish a methodological framework for advanced investigations into Swashplate axial piston pumps, particularly regarding dynamic response optimization under variable-volume conditions.

4.2. Development and Simulation Analysis of a Constant Pressure Variable Displacement Pump Model

To further predict the pump's performance, the axial piston pump submodel was connected to an external hydraulic system to establish a constant-pressure variable piston pump simulation model, as shown in Figure 11[8].

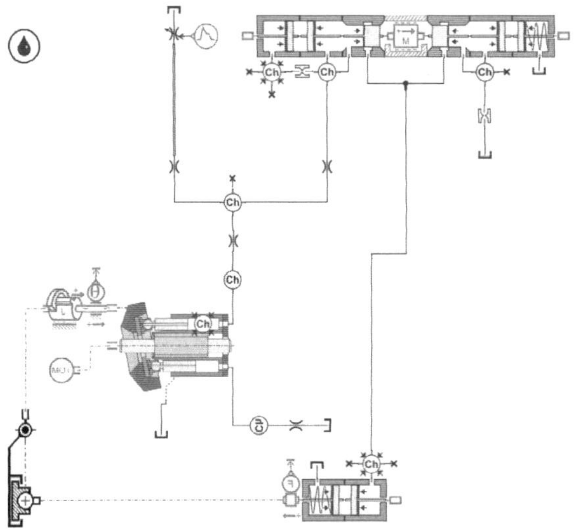

Figure 11. Constant-pressure variable-displacement pump simulation model

The AMESim constant-pressure variable pump model was configured with a motor speed of 3000 rpm and an initial swashplate angle of 18° for the piston pump.The pressure variation of the axial piston pump depends on the magnitude of the load; the load pressure was regulated by adjusting the variable throttle signal, with a smaller throttle opening corresponding to a larger load. The variable load signal was set to decrease from 0.5 to 0.1 within 4 seconds.

For the constant-pressure variable pump, the spring pre-load within the control mechanism is critical for establishing the operating pressure. By adjusting the spring pre-load, the pump's constant pressure is modified. Spring pre-loads of 900 N, 700 N, and 500 N were set, and batch processing was performed with a simulation duration of 6 seconds.After the simulations, the characteristic curves of output pressure and swashplate tilt angle of the axial piston pump under different pre-load conditions are shown in Figures 12 and 13.

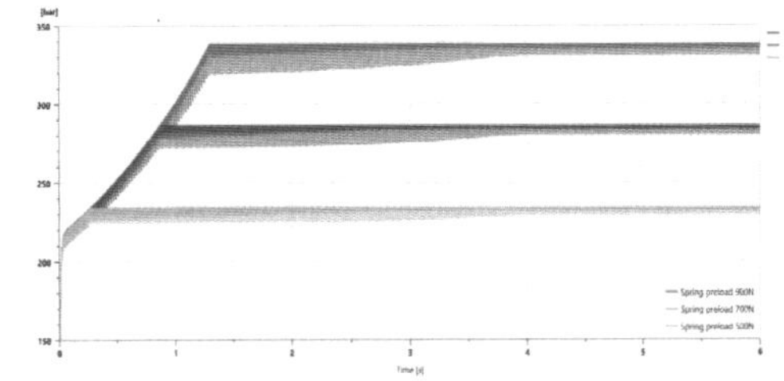

Figure 12. Constant-Pressure Variable Pump Pressure vs. Time Curve

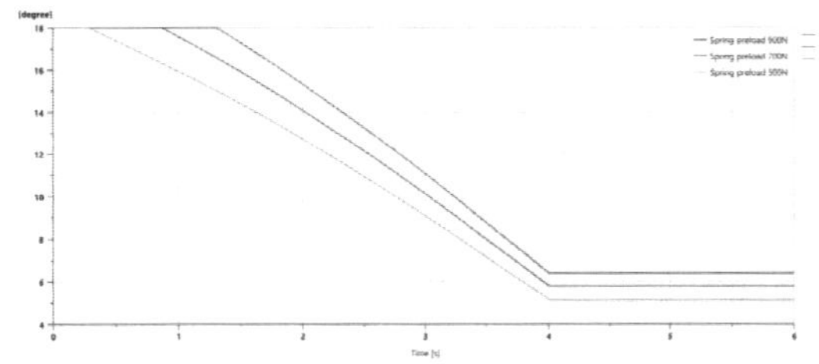

Figure 13. Constant-Pressure Variable Pump Swashplate Tilt Angle vs. Time Curve

Simulation Data Analysis for the Constant-Pressure Variable Pump: In the constant-pressure variable pump, the spring pre-load within the control mechanism adjusts the pump's response to maintain the specified pressure; increasing the spring pre-load raises the operating pressure setpoint. According to the simulation data, a 500 N spring pre-load reaches a set pressure of 23.4 MPa at 0.26 s, a 700 N pre-load attains 28.7 MPa at 0.86 s, and a 900 N pre-load achieves 33.8 MPa at 1.33 s. Under a constant variable load maintained at 4 s, the pressure amplitude decreases and tends to stabilize.

5. Conclusion

This paper constructs a geometric model of the Swashplate axial piston pump using three-dimensional modeling technology and develops a virtual prototype on the AMESim platform to conduct an in-depth simulation study of the pump's dynamic behavior and flow characteristics.Based on kinematic principles, mathematical models for plunger motion and flow are derived, providing a theoretical foundation for the simulation. Through simulation analysis of fixed displacement and constant pressure variable displacement pumps, the study systematically reveals the influence of key parameters—such as Swashplate angle and outlet volume—on pump performance, validating the effectiveness of the virtual prototype method in predicting performance under multiple operating conditions.The research proposes an integrated method based on three-dimensional modeling and AMESim simulation. Through the digital thread of CAD models and dynamic simulation, this approach overcomes the limitations of traditional HCD library standardized modeling, demonstrating higher precision and efficiency in overall performance optimization and multidisciplinary correlation analysis.The simulation results not only accurately characterize the pressure-flow characteristics of the Swashplate axial piston pump under different operating conditions but also reveal the coupling effects between parameters, providing technical support for the pump's optimized design that balances theoretical depth and engineering practicality, significantly reducing development costs and time.

Acknowledgments

Funding: National Natural Science Foundation of China(62241304); Guangxi Natural Science Foundation(2025GXNSFAA069293); Special Project of Guilin University of Aerospace Technology (TS2024111).

References

[1] QI HAITAO, FU YONGLING, LANG YAN. Modelling and Simulation of Electrical Servo Variable Displacement Piston Pump Based on AMESim[J]. Machine Tools and Hydraulics, 2015, 43(5): 115-118+31.

[2] JUNZHE LIN, YUANYUAN WANG, SHENGHAO ZHOU, et al. Simulation and Experimental Analysis of Pressure Pulsation Characteristics of Pump Source Fluid[J/OL]. Applied Sciences, 2021, 11(20): 9559.

[3] ZHANG XINJIE, CHEN YANHUI, ZHOU ZHIJIN. Influence of Axial Piston Pump Parameters on Flow and Pressure Pulsation[J].

[4] ZHAO XI, WU HUAICHAO, HOU GANG, et al. Analysis and Optimization of Plunger Cavity Force on the Swashplate Overturning Moment of a Miniature Axial Piston Pump[J].

[5] Junze Tan, Yueliang Lu, Tao Cao, et al. Optimization Design of Pressure Pulsation for a Certain Type of Aerospace Hydraulic Piston Pump[J]. Hydraulics and Pneumatics, 2022, 46(4): 156-164.

[6] Hunyeyang Hu, Hewu Yang, and Zhenyin Zhu. Simulation Study of a Swashplate-Type Axial Piston Pump Based on AMESim[J/OL]. Pearl River Water Transport, 2024(23): 11-14.

[7] JIANPING WANG, QUAN LI. Characteristics Simulation of Axial Pump Based on AMESim[C/OL]//2021 IEEE Asia-Pacific Conference on Image Processing, Electronics and Computers (IPEC). Dalian, China: IEEE, 2021: 191-195.

[8] Wei Luo, Dapeng Zou, Tibing Xiao, et al. Modeling and Simulation Analysis of a Constant-Pressure Variable Piston Pump Based on AMESim[J]. Machine Tools and Hydraulics, 2017, 45(5): 100-103.

Journal of Physics: Conference Series **3024** (2025) 012012 doi:10.1088/1742-6596/3024/1/012012

Development of Pulse Power Supply for EDM Dressing of Metal-bonded Diamond Grinding Wheel

Yong Zhang[1,3], Ping Sun[2*], Ping Pan[1,3]

[1] Guilin University of Aerospace Technology, Guilin, Guangxi, 541004, P.R. China

[2] Department of Mathematics and Computer Technology, Guilin Normal College, Guilin, Guangxi,541199, P.R. China

[3] Guangxi Key Laboratory of Special Engineering Equipment and Control, Guilin, Guangxi,541004, P.R. China

*Corresponding author's e-mail: sunping@mail.glnc.edu.cn

Abstract. To solve problems such as outdated human-machine interaction, easy damage of power tube, and unsatisfactory discharge waveform of pulse power supply used in EDM dressing of metal-bonded diamond grinding wheel, a new special pulse power supply was developed in this study. The experimental results show that the pulse power supply has good interoperability, stable and reliable performance, and high-quality machining waveform, which can fully meet the demand for EDM dressing pulse power supply.

1. Introduction

Metal-bonded diamond grinding wheels are widely used in precision machining of hard and brittle materials due to their excellent grinding performance. However, issues such as grit wear and clogging may occur during wheel usage, necessitating regular dressing to restore their grinding performance[1]. The dressing of diamond grinding wheels comprises two stages: truing and sharpening. Truing refers to material removal (including both diamond abrasives and bonding agents) from the wheel surface to achieve required geometric accuracy, while sharpening aims to expose protruding abrasive grains on the wheel surface to form sharp cutting edges and chip storage space.

EDM (Electrical Discharge Machining) dressing, as an efficient and precise method for grinding wheel conditioning, has garnered significant attention in recent years. The principle of the EDM dressing method for metal-bonded diamond grinding wheels involves generating pulsed spark discharges between the tool electrode and the rotating diamond grinding wheel[1-2]. This process utilizes the electro-erosion effect to remove the metal binder from the grinding wheel and expose the diamond abrasive grains on the wheel's surface.

The pulse power supply, as the core component of the electrical discharge dressing system for metal-bonded diamond grinding wheels, directly influences the dressing's effectiveness and efficiency[3-4]. However, existing pulse power supplies for electrical discharge dressing generally suffer from issues such as outdated human-machine interaction methods, susceptibility of power transistors to burnout, and suboptimal discharge waveforms, hindering the further development and application of electrical discharge dressing technology.

In response to the aforementioned issues, this paper conducted an in-depth study on the pulse power supply for the EDM dressing of metal-based diamond grinding wheels. A special pulse power supply was developed, and experimental tests were carried out to verify the machining performance of the designed pulse power supply.

2. Traditional Transistor-type Pulse Power Supply

Generally, a transistor-based pulse power supply consists of three main components: the oscillator stage, amplification stage, and power stage[5-6]. The oscillator stage essentially serves as a pulse generator, producing rectangular pulse signals with specific pulse widths and intervals. These signals are then amplified by the amplification stage, ultimately driving the final-stage power transistor to switch on or off, as illustrated in Figure 1. To meet the high-current requirements, multiple channels of power MOSFETs (8 or 16 channels) are typically connected in parallel. The discharge current magnitude is adjusted by controlling the number of conducting/switching MOSFETs. Under such circumstances, the current value flowing through each MOSFET is

$$I_m = \frac{U_{open} - U_{gap}}{R} \tag{1}$$

where U_{open} is the open voltage, U_{gap} is the gap voltage between the electrode and the workpiece, and R is the resistance value.

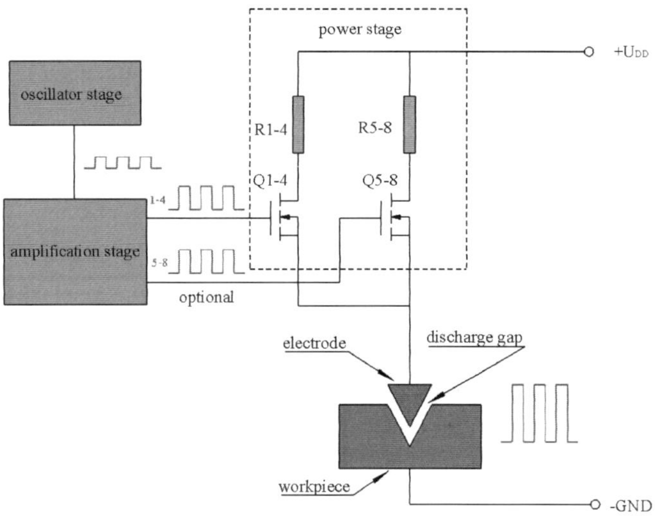

Figure 1. Transistor-type pulse power supply

Traditional transistor-type pulse power supplies generally adopt the 51 microcontroller as the pulse generation controller, combined with the independent buttons and digital LED display panel for human-machine interaction, As shown in Figure 2. Key parameters of pulse power supplies, including pulse width, pulse interval, and current magnitude, are all set through button switches on the operation panel. Under this human-machine interaction mode, the adjustment of discharge parameters requires manual operation, which proves unfavorable for continuous discharge processing in multi-process application scenarios of diamond grinding wheels.

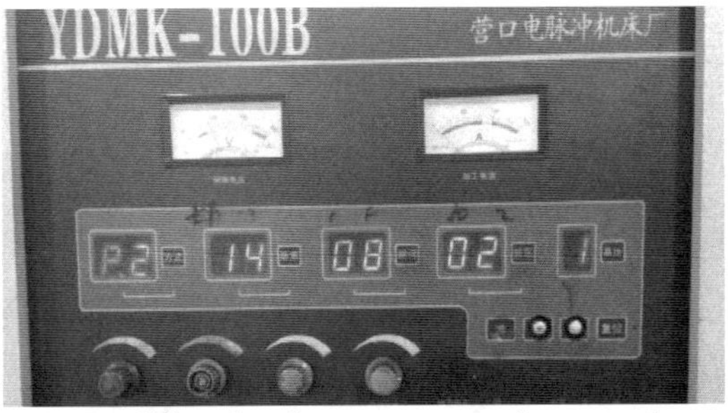

Figure 2. Traditional pulse power supply control panel

Additionally, since diamond grinding wheels use high discharge parameters during the initial stage of electrical discharge dressing (rough machining phase) and require long processing times (typically over 2 hours), the power MOSFET in traditional electrical discharge dressing machines is subjected to high processing currents. This often leads to frequent burnout of the power MOSFETs, which is detrimental to normal production and processing.

On the existing EDM dressing machines, the voltage waveforms between the electrode and the workpiece are measured via an oscilloscope and shown in Figure 3. The following issues exist:

(1) Due to the presence of energy storage capacitors in the discharge circuit, the machining voltage cannot drop instantaneously during discharge processing, resulting in a curved pulse falling edge.

(2) Parasitic inductance in the circuit prevents the power control transistor from shutting off immediately, leading to reverse voltage generation[7].

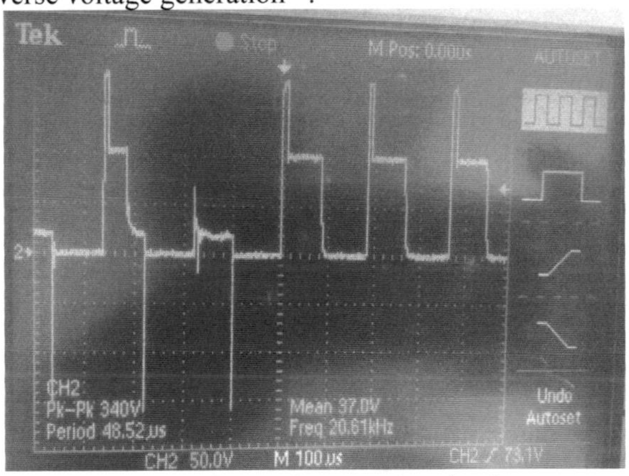

Figure 3. Discharge waveform of conventional pulse power supply

3. Development of Pulse Power Supply for EDM Dressing

Aiming at the problem of traditional pulse power supply used in EDM dressing of metal-based diamond grinding wheels, a special power supply was developed.

3.1. *The overall structure of the pulse power supply*

To enhance the interactivity between the pulse power supply and the machine tool control system for EDM dressing, an integrated design scheme is proposed as illustrated in Figure 4. This scheme adopts an STM32 microcontroller as the core controller of the pulse power supply, interfacing with the host computer via the RS485 serial bus. The pulse power supply implements Modbus RTU protocol-based communication with the host computer, enabling dynamic parameter configuration and real-time acquisition of discharge status data.

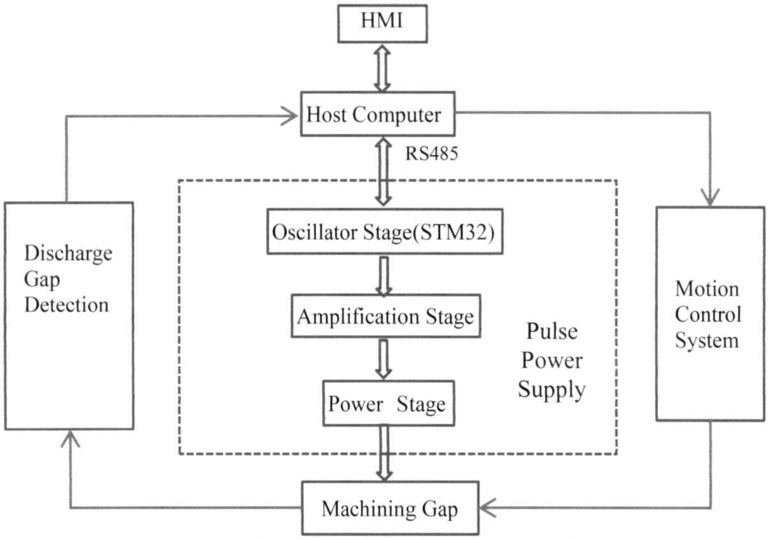

Figure 4. Structure Diagram of Pulse Power Supply and EDM System

Under the fieldbus working mode, it not only enables the development of a human-machine interface for the pulse power supply on the host computer to replace traditional control panels but more importantly facilitates the integration of the machine control system with the pulse power supply. This integration realizes fully automated discharge processing with flexible adjustment of multi-stage discharge parameters across multiple processes from rough shaping and fine shaping to sharpening. The multi-stage machining discharge parameter setting interface developed for the pulse power supply is shown in Figure 5.

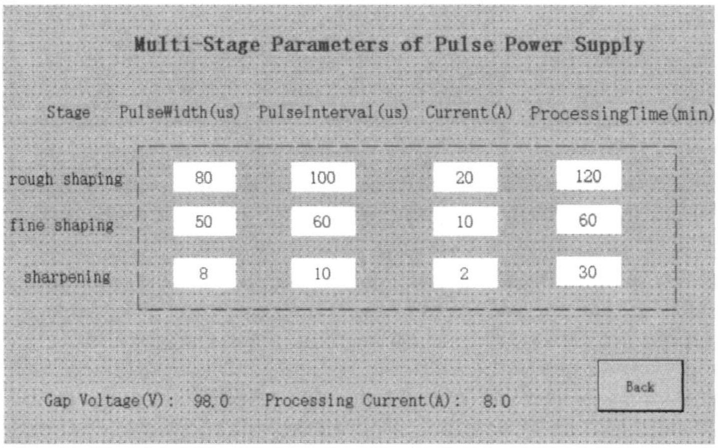

Figure 5. Multi-Stage Machining Discharge Parameters Setup Interface of Pulse Power Supply

3.2. High-low voltage pulse power supply

Because of the existence of abrasive particles in the process of metal-based wheel dressing, especially when the particle size is large, the discharge rate of EDM will be reduced[8]. Based on this, the high-low voltage composite scheme is adopted to increase the discharge gap and improve the discharge rate. Therefore, the developed power supply can be used to dress large-grain grinding wheels. Figure 6 shows the schematic diagram of the main circuit. A typical work cycle operates as follows: the high voltage circuit is turned on firstly by MOSFET Q0 to break down the gap. After the forming of the discharge channel, the high-voltage circuit is switched off, and the low-voltage circuit is switched on by MOSFET Q1~Q8 to maintain the current. After reaching the set pulse on time, the low-voltage circuit is turned off.

The design proposed in this study involves keeping all power MOSFETs working together during the discharge process while adjusting the discharge current magnitude through the switching of multiple relays K1-K8， as shown in Figure 6. In this configuration, the current flowing through each MOSFET becomes

$$I'_m = \frac{N_k(U_{open} - U_{gap})}{N_m R} \quad (2)$$

where N_k is the total number of relays that are enabled by parameters,U_{open} is the open voltage, U_{gap} is the gap voltage between the electrode and the workpiece, N_m is the number of the power MOSFETs,and R is the resistance value. Since $N_k < N_m$, then$I'_m < I_m$,this significantly reduces the workload on each MOSFET, thereby extending its operational lifespan.

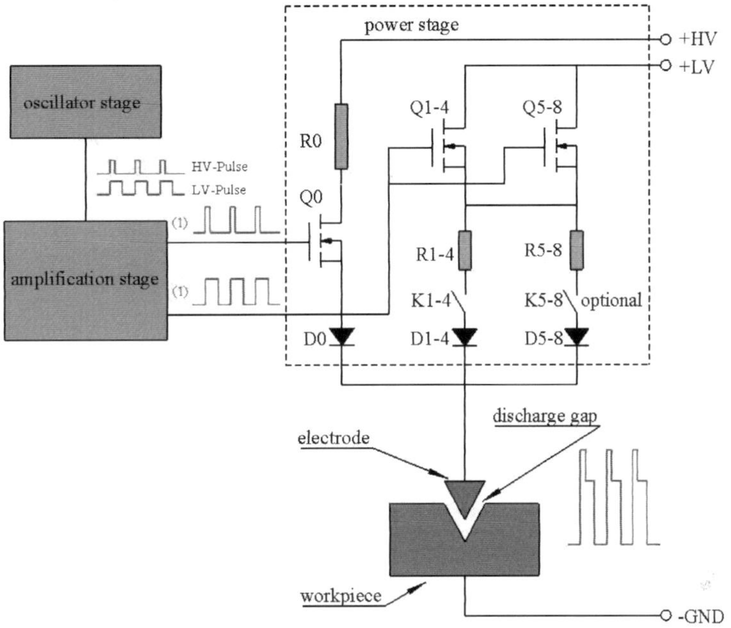

Figure 6. Schematic diagram of high-low voltage pulse power supply

3.3. Optimization of Discharge Waveform

The design methodology proposed in this paper involves connecting a low-power resistor in parallel between the electrodes during pulse intervals to dissipate energy stored in parasitic capacitances and inductances, thereby improving turn-off characteristics. As illustrated in Figure 7, the snubber circuit operates as follows: Upon the falling edge of the control pulse, the main MOSFET Q1-Q8 in the power circuit is turned off through the pulse signal in reverse phase, while simultaneously activating the snubber MOSFET Q9. This configuration enables rapid discharge of the inter-electrode voltage through the low-power resistor, resulting in abrupt voltage collapse during pulse termination without observable voltage tailing phenomena.

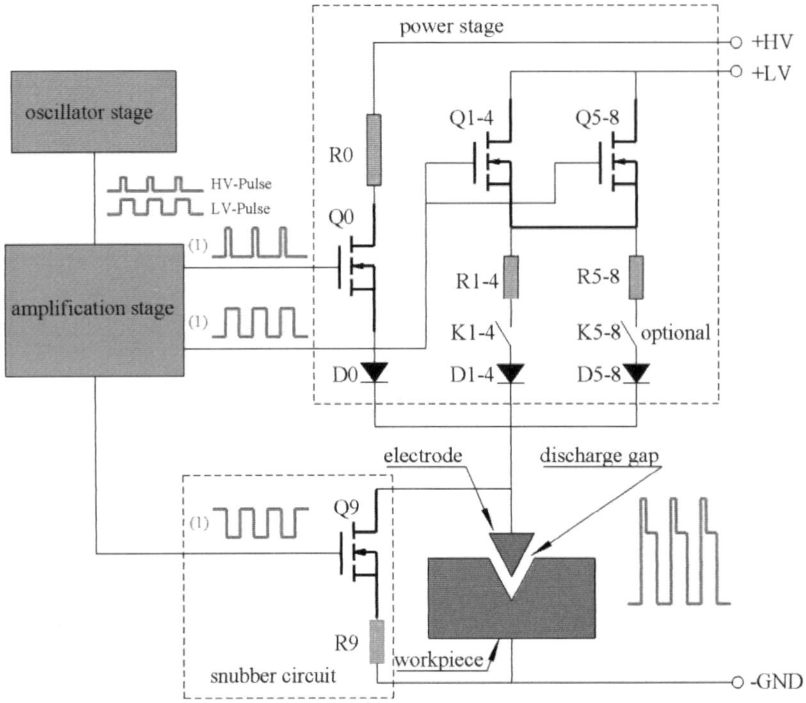

Figure 7. Schematic of the Snubber Circuit

4. Pulse power supply experimental testing

Based on the above design principles, a pulsed power supply circuit board has been developed. The physical prototype is shown in Figure 8, where (a) represents the control circuit board and (b) shows the power circuit board.

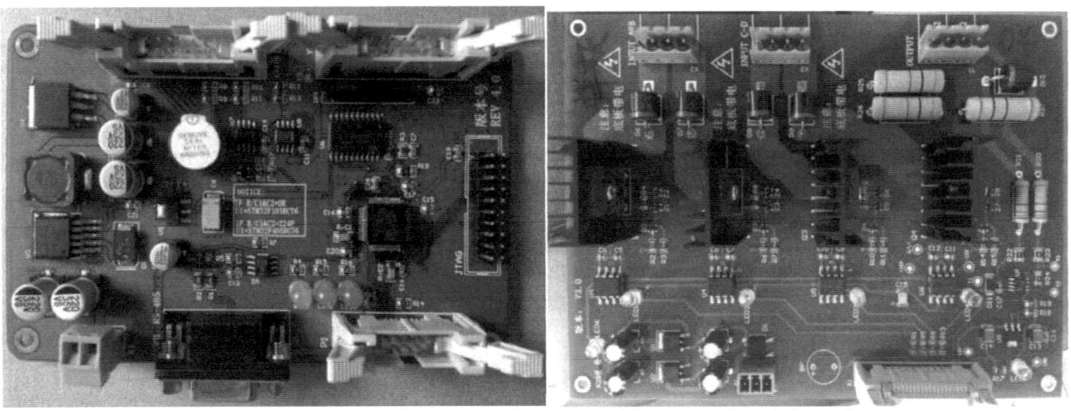
(a)Control circuit board (b)power circuit board
Figure 8. Pulse power supply circuit board

In the existing electrical discharge diamond grinding wheel dressing machine, a new type of pulse power supply was installed for processing experiments under the conditions listed in Table 1.

Table 1. Machining Test Conditions

	Parameters	Content/Value
	Machining Polarity	Negative Polarity
	Machining Medium	EDM Oil
Machining Voltage (V)	*High Voltage*	175
	Low Voltage	100
Pulse Width (μs)	*High Voltage*	10
	Low Voltage	80
	Pulse Interval (μs)	100
	Machining Current(A)	10

With a consistent external processing environment, the improved pulse power supply's discharge machining waveform was measured using an oscilloscope as shown in Figure 9. Compared with Figures 3 and 9, it can be observed that: the designed pulse power supply demonstrates cleaner discharge waveforms with minimal noise, higher discharge pulse frequency, and reduced reverse waveform duration(the peak-to-peak voltage has dropped from 340V to 180V) caused by parasitic inductance shutdown. These improvements thereby enhance discharge machining efficiency and elevate the dressing machine's discharge processing performance.

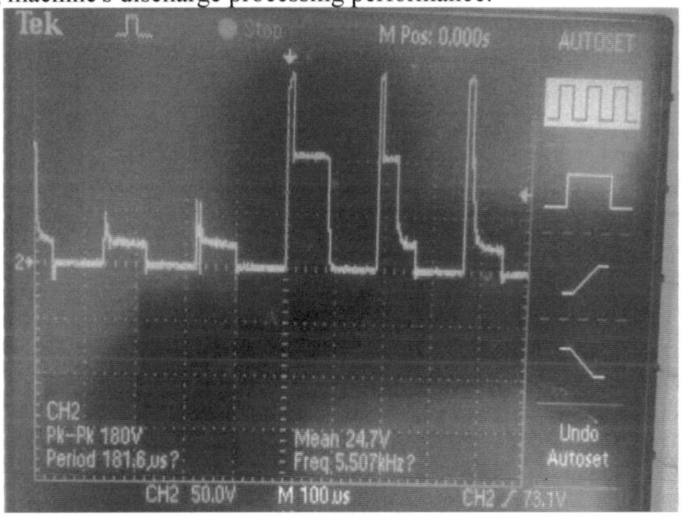

Figure 9. Discharge waveform of improved pulsed power supply

5. Conclusion

The pulse power supply designed in this study for EDM dressing not only optimizes the discharge waveform but also enhances reliability and intelligence, effectively meeting the specific requirements for dressing metal-bonded diamond grinding wheels. The experiment tests have verifed that the developed pulse power supply presents the following main innovative aspects in design and application:

1) High-Low Voltage Composite Power Supply Design, which significantly enhances the discharge efficiency of large-grain diamond grinding wheels;

2) Optimization of MOSFET Current Load, extending the lifespan of power tubes through relay-controlled parallel operation of multiple MOSFETs;

3) Enhanced Intelligence and Interactivity, achieved via STM32 microcontroller and RS485 bus communication for dynamic parameter configuration and real-time status monitoring with the upper computer, improving system intelligence and adaptability to multi-stage machining.

This pulse power supply provides an efficient and stable energy supply for the precision dressing of super-hard abrasive tools, driving the evolution of EDM dressing technology toward greater performance.

Acknowledgments

The authors are grateful to the financial support from the Fund Project:

1.Project for Enhancing Young and Middle-aged Teacher's Research Basis Ability in Colleges of Guangxi in 2022, "Research on Key Technology of High Efficiency and Energy Saving EDM" (2022KY0787), Leader: Yong Zhang.

2.Project for Enhancing Young and Middle-aged Teacher's Research Basis Ability in Colleges of Guangxi in 2023, "Research on Data Transmission and Receiving System of Internet of Things Based on CAN Bus" (2023KY0983), Leader: Ping Sun.

References

[1] JIAO F, LI CL, NIU Y 2021 Review and Prospect of Grinding Wheel Dressing Technique. China Mechanical Engineering, 32(20), 2435-2448.

[2] WANG X and ZHU K 2018 Influence of Discharge Parameters on Efficiency of EDM Dressing Diamond Grinding Wheel. Machine Tool & Hydraulics, 46 (23):99-101.

[3] LIU L and FANG M 2024 Research on the Grinding Performance of Grinding Wheel in the Initial Grinding Stage after Electric Spark Dressing of Grinding Wheel.The International Journal of Advanced Manufacturing Technology Vol 131 pp 3103-3111

[4] ZHANG L and WANG X 2024 Metal-bonded Diamond Wheel Dressing by Using Wire Electrical Discharge: Principle and Prototype Machine Design.The International Journal of Advanced Manufacturing Technology Vol 131 pp 2677-2690

[5] WANG L, LIAN H 2021 Development of Pulse Power Supply of Insulated Ceramics Based on FPGA. Machine Tool &Hydraulics,Vol 49 pp 88-93

[6] YIN H, WU X 2019 Development of Transistor-type Micro-EDM Pulse Power Supply Based on MOSFET. Mechanical & Electrical Engineering Technology,48(09):1-3.

[7] JI X, GU L and LU X 2023 Pulse Power Supply Design and Application of Diamond Grinding Wheel Trimming Machine[J]. Machine Tool & Hydraulics,51(11): 133-139.

[8] WANG Z 2020 Research on High and Low Voltage Compound Inverter Pulse Power Supply for Dressing of Metal-based Superhard Abrasive Wheel. Taiyuan University of Technology

Design and Analysis and Testing of a Hydraulic Loading System for Pull Force Testing of Grinding Mill

Sibo LIU[1,2], Hongwang ZHAO[2,3*], Jingyun CAI[1,2], Dandan WU[1,2], Zhong LIU[1,2], Hao LI[1,2]

[1] School of Mechanical and Electrical Engineering,Guilin University of Electronic Techonlogy,Guilin Guangxi 541004, China

[2] Guangxi Key Laboratory of Special Engineering Equipment and Control,Guilin Guangxi 541004, China

[3] Guangxi University Special Engineering Equipment Design and Intelligent Drive Technology Laboratory Guilin University of Aerospace Technology,Guilin Guangxi 541004, China

*Corresponding author's e-mail: zhaohongwang@guat.edu.cn

Abstract. Winch mill Winch mills are essential for power line cableway and tower hoisting operations, with their performance critically impacting construction efficiency and safety. The current challenge is the absence of practical tools for detecting and controlling work performance parameters in grinding mills. This paper introduces a hydraulic loading system for strand mill tension testing, complete with a mathematical model and simulation analysis. It addresses issues like wire rope inertial forces due to stress imbalances and wear from friction and fatigue. A prototype tension test bench and accompanying hydraulic system experiments are developed, demonstrating a piston displacement speed error of 0.99% to 2.29% between simulations and experiments, and a flow rate within an acceptable range for the rodless chamber. This innovation supports the advancement and quality assurance of strand mills.

1. Preface

As an important mechanical equipment for power line construction and lifting of goods in construction terminals, the application of the winch mill in tower erection and lifting and hauling is especially critical[1,2].

Nowadays, the winch mill lacks a device which can timely detect its tension and wire rope speed in the online variable load conditions during the winch mill's development and factory process. For the current hydraulic loading system of mechanical equipment, scholars at home and abroad have conducted some researches. Tang Xiaolong et al [3] established a simulation model of mining hydraulic cylinder dynamic load loading system based on AMESim to simulate and analyze the pressure characteristics of the rodless cavity of the hydraulic cylinder under dynamic load conditions. The simulation parameters obtained basically match with the experimental parameters, which verifies the correctness of the model. Bai Guobao et al [4] used a hydraulic winch load test device with a combination of a fixed pulley and a movable pulley to evaluate the winch loading capacity by monitoring the system pressure change when pulling the wire rope, which effectively reduces accidents and improves production efficiency. However, the system does not yet have the function of real-time adjustment of load pulling force, which has certain application limitations. Lv Shaoli et al [5] used a new type of hydraulic loading system to simulate the counterweight, realizing the active-passive cylinder conversion loading which ensures that the test cylinder piston rod throughout the constant load. Otherwise, it also improves the efficiency and reliability and meets the high standards of the experiment. However, the size of the loading force can only be adjusted by adding or subtracting weights and the weights are limited to vertical movement of the vertical structure, which has certain limitations. At present, there's a shortage of accessible performance testing for strand mill tension tests, which are costly and pose risks and limitations. This paper addresses this by designing a hydraulic loading system for strand mill tension testing, encompassing dynamic loading, pressure control, cooling, and fluid guidance modules. It offers innovative testing methods and devices, enhancing safety, and efficiency by correlating modeling, simulation, and test bench analysis.

2. Hydraulic loading system's principles and model

2.1. The process of testing the pulling force of the hydraulic cylinder of the grinding mill

In the winch mill tension test bench illustrated in Figure 1, wire ropes are guided through three fixed pulleys, optimizing the traction path via dynamic pulley steering to the lugs.

A load hydraulic cylinder is hinged at one end of the frame, with the other end connecting to the piston cylinder hinge ear through a dynamic pulley, facilitating unloading. During the load tension test, the winch mill reverses to initiate lifting, with wire ropes pulling the hydraulic rod linked to the movable pulley. The hydraulic cylinder simulates a heavy load, propelling the piston rod upwards. In descent, the winch mill rotates positively. As the hydraulic cylinder decreases, the piston rod, lifting ring, dynamic pulley, and hydraulic pressure control the descent speed, with system pressure opposing the piston. Friction between the piston and cylinder wall also contributes to the controlled descent speed.

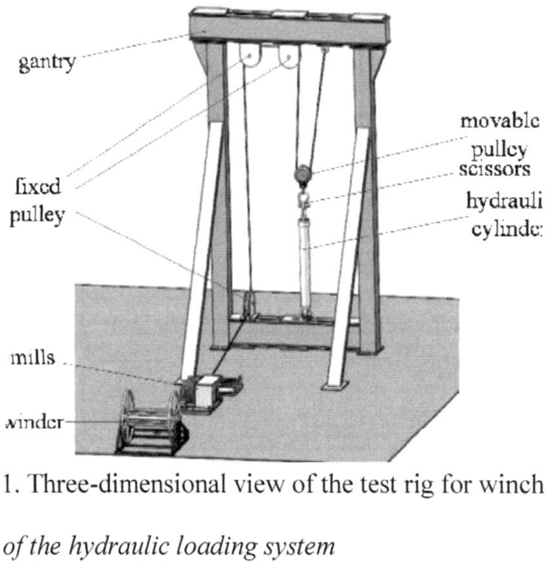

Figure 1. Three-dimensional view of the test rig for winch testing

2.2. Working principles of the hydraulic loading system

In Figure 2, the hydraulic cylinder (1) is in a passive pulling state for the winch mill's lifting.

The solenoid relief valve (8, 9) is open for system unloading due to power loss, while the solenoid reversing valve (11) is neutral, directing oil flow through a low-pressure bypass loop back to the tank for low-pressure unloading, and a high-pressure working circuit via the Y-type directional valve (6) for the passive lifting process. The check valve (14) limits flow paths, guiding oil to pilot-operated relief valves (15) for reflux control. The relief valve's set pressure indicates the load value.During the winch mill's lowering, the hydraulic cylinder (1) actively releases pressure, driving the piston rod down. The solenoid relief valve (8, 9) is closed, and the left solenoid reversing valve (8) is energized for the left working state. The rod cavity receives high-pressure fluid for driving force, while the rodless cavity is low-pressure, creating a pressure differential for smooth piston rod descent. To ensure a smooth and safe lowering, the descending speed and winch mill's lowering rate must be finely controlled to prevent rapid hydraulic system responses causing shock, which can damage components and decrease efficiency[6].

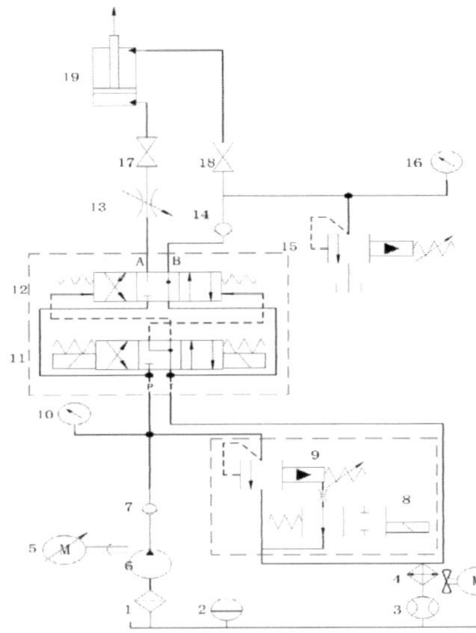

Figure 2. Schematic diagram of hydraulic loading system
1.Filter; 2.Thermometer; 3.Flow Sensor; 4.Air Cooler; 5.Frequency Conversion Motor; 6.Gear Pump; 7.Check Valve; 8.Two-position Four-way Solenoid Valve; 9.Pilot Operated Relief Valve; 10.Pressure Sensor; 11.Three-position Four-way Y-type Solenoid Valve; 12.Pilot Operated Directional Valve; 13.Speed Regulator Valve; 14.Check Valve; 15.Pilot Operated Relief Valve; 16.Pressure Sensor; 17~18.Cut-off Ball Valve; 19.Hydraulic cylinder.

3. Mathematical model of hydraulic loading system

3.1. Mathematical model of the piston rod lifting and lowering

The hydraulic system's control valve is pivotal in the winch mill's lifting and lowering operations, with the wire rope's external tension being a secondary factor.

Focusing on the tensile load test hydraulic loading system, this study disregards wire rope quality in extreme lifting or lowering scenarios influenced by inertial forces, friction, and rope wear. The stress distribution's unevenness in tensile tests introduces a measurable error, as referenced [7].

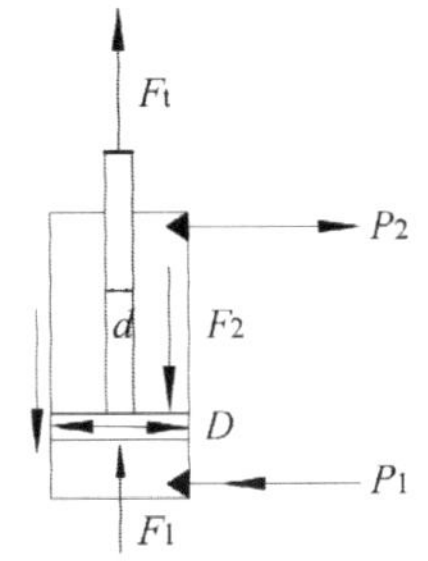

Figure 3. Passive lifting force analysis of piston rod

The model of the hydraulic cylinder is HSG125 single rod engineering hydraulic cylinder, shown in Figure 3. The oil lifts from the rodless cavity and unloads from the rod cavity when the hydraulic cylinder is lifted. Otherwise, setting the pressure value of the rodless cavity P_1, and then achieving the pressure value of the rod cavity P_2 according to the hydraulic cylinder to provide the load force formula. The driving force of the hydraulic cylinder is as followed[8]:

$$F_{h1} = (F_1 - F_2)\eta = (P_1 A_1 - P_2 A_2)\eta \quad (1)$$

$$F_{h1} = \left[(p_1 - p_2)(\pi D^2 / 4) + (p_2 \pi d^2 / 4)\right]\eta \quad (2)$$

Where Fh1 is the hydraulic cylinder rising driving force, kN; F_1 is the force which the rodless cavity acts on the piston , kN; F_2 is the force which the rod cavity acts on the piston , kN; P_1 is the piston rod pressure, MPa; P_2 is the piston rod cavity's pressure, MPa; A_1 is the cross-sectional area of the rodless cavity, m2; A_2 is the ross-sectional area of the rod cavity , m^2; D is the piston's diameter, mm; d is the piston rod's diameter mm; η is the hydraulic cylinder's loading rate.

Considering that the hydraulic cylinder will be subjected to friction and other resistance during the movement, the equilibrium equation for the force on the piston rod of the hydraulic cylinder under the action of the pulling force is as followed:

$$F_t = F_{h1} - F_f \quad (3)$$

As shown in Figure 4, the hydraulic cylinder descending speed depends on the the pressure oiginating in the piston rod, the lifting ring, the gravity of the dynamic pulley Fg, as well as of the hydraulic system in the lowering process. Meanwhile, the friction between the piston and the cylinder wall will slow down the descending speed.

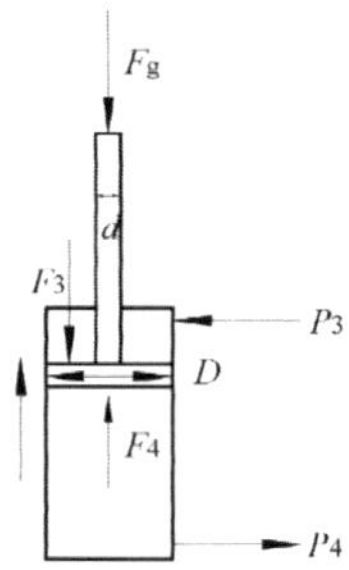

Figure 4. Piston rod drop force analysis

The oil comes from the rod cavity during the descent period, comeing back from the rodless cavity. Meanwhilr, the pressure value of the rod cavity is P3, the rod cavity's pressure value is P_4. The hydraulic cylinder driving force is:

$$F_{h2} = (P_3 A_2 - P_4 A_1)\eta \quad (4)$$

$$F_{h2} = [(P_3 - P_4)(\pi D^2 / 4) - (P_3 \pi d^2 / 4)]\eta \quad (5)$$

Where F_{h2} is driving force of the hydraulic cylinder descent ,kN; P_3 is the pressure of the piston rob's rod cavity ,MPa; P_4 is the pressure of piston rob's rodless cavity ,MPa;

According to Newton's second law, you can get the dynamic equations of the piston rod rising and falling process as followed:

$$F_h = m_1(\mathrm{d}^2 y / \mathrm{d}t^2) + B_1(\mathrm{d}y / \mathrm{d}t) + F_L \quad (6)$$

Where F_h is the driving force of the hydraulic cylinder ,kN; m_1 is the mass of the piston and other parts ,kg; y is the piston's displacement ,m; B_1 is the viscous damping coefficient of the hydraulic cylinder ,N-s/m; F_L is the external load force of the piston ,kN.

As the fluid is often mixed with insoluble gases, as different elastic properties exists in the cylinder, pipe, valve body, etc. Considering these factors and referring to the research results [9], the single-rod engineering hydraulic cylinder lifting and lowering process of the inlet and return oil flow equations with the equivalent volume modulus of elasticity is:

$$\begin{cases} Q_1 = A_1(\mathrm{d}y / \mathrm{d}t) + C_t(P_1 - P_2) + (V / \beta_e)(\mathrm{d}P_1 / \mathrm{d}t) \\ Q_2 = A_2(\mathrm{d}y / \mathrm{d}t) + C_t(P_1 - P_2) + (V_2 / \beta_e)(\mathrm{d}P_2 / \mathrm{d}t) \\ Q_3 = A_2(\mathrm{d}y / \mathrm{d}t) + C_t(P_3 - P_4) + (V_2 / \beta_e)(\mathrm{d}P_3 / \mathrm{d}t) \\ Q_4 = A_1(\mathrm{d}y / \ t) + C_t(P_3 - P_4) + (V_1 / \beta_e)(\mathrm{d}P_4 / \mathrm{d}t) \end{cases} \quad (7)$$

In the above formula, Q_1 is the inlet flow for lifting the rodless chamber, L/min; Q_2 is the outlet flow for lifting the rod chamber, L/min; Q_3 is the outlet flow for lowering of the rod chamber, L/min; Q_4 is the outlet flow for the lowering of the rodless chamber, L/min; C_t is the equivalent leakage coefficient of the hydraulic cylinder; β_e is the equivalent volumetric elasticity modulus of the oil fluid, MPa; V_1 is the effective volume of the rodless chamber piston, m^3; V_2 is the effective volume of rod chamber piston, m^3;

3.2. Mathematical Model of Electromagnetic Directional Valve

This paper utilizes an electro-hydraulic directional valve comprising a main valve and a soleno id pilot valve.

Upon receiving an external control signal, the solenoid in the pilot valve generates a magnetic force that actuates the spool.This movement alters the oil circuit's state within the pilot valve, permitting pressure oil to flow into or out of the main valve's control cavity, thereby modifying the main valve's fluid path and controlling the hydraulic system's fluid flow. The spool's shift, driven by oil pressure, facilitates this control mechanism[10].

Dynamic equations for the spool of a solenoid-directed valve which is energizing is as followed:

$$m_2(\mathrm{d}^2 x_1 / \mathrm{d}^2 x_1) = F_m - F_b - B_2(\mathrm{d}x_1 / \mathrm{d}t) - k(x_1 + x_2) \tag{8}$$

Where m^2 is the mass of the electromagnetic directional valve spool ,kg; x_1 is the opening degree of the electromagnetic directional valve spool ,m; F_m is the electromagnetic suction added to the spool, kN; F_b is the hydraulic pressure acting on the valve force, kN; B_2 is the viscous damping coefficient, N-s/m; k is the balance of the spool spring stiffness, N/m; x_2 is the initial compression of the spring,m.

Equation for the volume flow rate at the valve port after commutation of a solenoid-directe d valve is:

$$Q_V = C_d A_0 \sqrt{2(P_0 - P_V) / \rho} \tag{9}$$

Where C_d is the flow coefficient; A_0 is the valve port overflow area, m_2; P_0 is the valve port inlet pressure, MPa; P_V is the hydraulic cylinder inlet pressure, MPa; ρ is the density of the fluid, kg/m^3.

4. Simulation study of hydraulic loading system

4.1. Hydraulic loading system model and parameters

In order to obtain piston rod displacement and hydraulic cylinder velocity under different loads, a simulation model was constructed in AMEsim using the established mathematical model.

As shown in Figure 5, the system has a one-way gear pump and an actuator that drives a single rod hydraulic cylinder with a mass block. The tensile test uses a Y-shaped electromagnetic directional valve, but there is no signal in the simulation. When the displacement signal fails, the system relies on the force signal to simulate additional load force and adjust the signal source parameters to replicate experimental results under different loads.

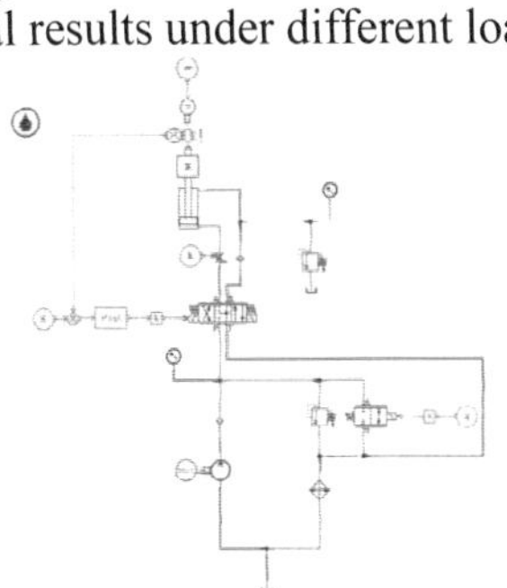

Figure 5. Lifting simulation model of hydraulic loading system

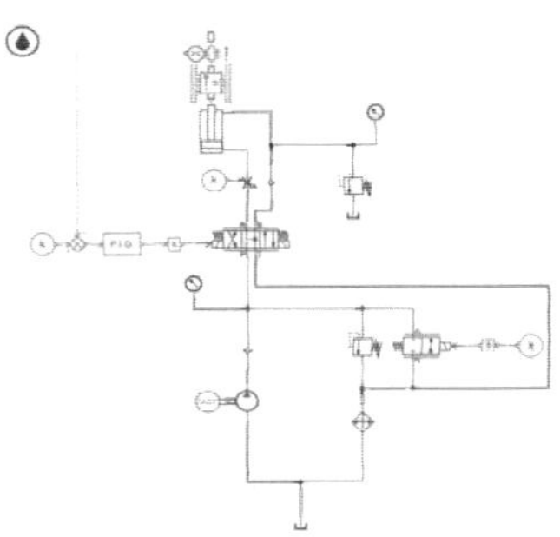

Figure 6. Simulation model of hydraulic loading system devolution

During the descent of the hydraulic rod, Figure 6 shows a simplified model characterized by a single rod hydraulic cylinder with mass blocks, external displacement signals, and limiting devices. During the simulation process, the electromagnetic relief valve was activated at 40mA. The electromagnetic directional valve was controlled by a PID controller with a proportional gain of 10. The electromagnetic directional valve operated in the left position, affecting the gain, and was set to -40. Adjust the displacement signal to reflect the movement of the piston rod during the lowering process, in order to obtain displacement and velocity curves.

Table 1. Simulation parameter values

Parameters	Parameter value
Piston diameter,D/mm	125
Piston rod diameter,d/mm	70
Gear pump displacement,L/min	80
Gear pump speed,v/rev·min^{-3}	2000
Hydraulic Fluid Index	68
Piston rod stroke,s/mm	1000
rated pressure,P/MPa	16
Mass Block Mass,m/Kg	250
Fluid density,ρ/k·m^{-3}	860
Inherent frequency of solenoid-directed valves,/Hz	50
Flow rate of solenoid-operated directional valves,Q/L·min^{-1}	80

The parameters in Table 1 were input into the corresponding hydraulic components of the simulation system for simulation and analysis, so the velocity and displacement curves of the hydraulic rod passively ascending under different load pulling forces as well as the velocity and displacement curves of the hydraulic rod during active decentralization were obtained, as shown in Figs.7-8.

Table 2. Results of some parameters of the experimental bench

Experimental tonnage/speed	Rodless cavity passive flow(L/min)	Rod cavity pressure(MPa) (efficiency1-0.8)
1 ton (16m/min)	98.17	2.5-3.2
3 tons (8m/min)	49.08	7.7-9.6
5 tons (6m/min)	36.8	12.8-16

Table 2 details the winch mill tension test bench's experimental parameters, revealing load tensions at 1 ton, 3 tons, and 5 tons, alongside rodless cavity passive flow and pressure. This study assumes a rod cavity pressure efficiency of 0.8. The bench simulates pulley set tension, emphasizing force conservation over distance, doubling the experimental pulling force and halving the speed. Simulations at various loads provide insights into the system's response.

From Figure 7(a), at tensions of 1.0×10^{5}N and 0.6×10^{5}N, piston rod vibrations intensify, contrasting with the stability at 0.2×10^5N. Increased tension raises hydraulic cylinder pressure and flow, potentially causing abrupt hydraulic component state changes, leading to shock and vibration. These are transmitted to the piston rod, causing oscillations in its velocity curve before stabilizing at zero.

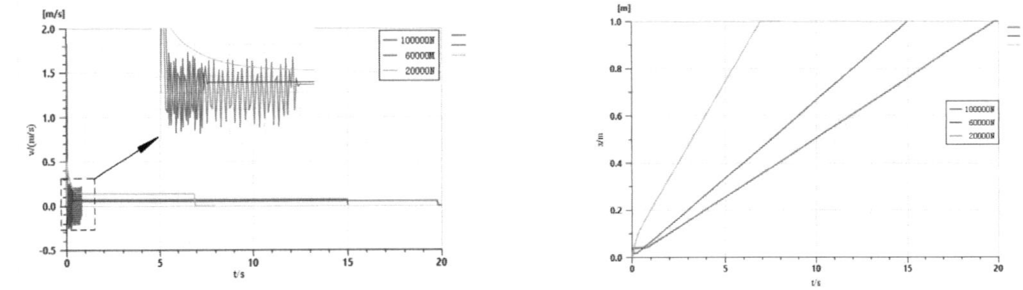

(a) Piston rod passive rising speed curve under each load pulling force

(b) Displacement curve of piston rod under each load pulling force

Figure.7 Passive rising speed and displacement curve of piston rod under each load pulling force.

When the simulation time is set to 8s, the piston rod is allowed to descend 1m by the displacement signal and no load is provided, the velocity and displacement curves of the piston rod actively lowered can be obtained, as shown in Figure 8.

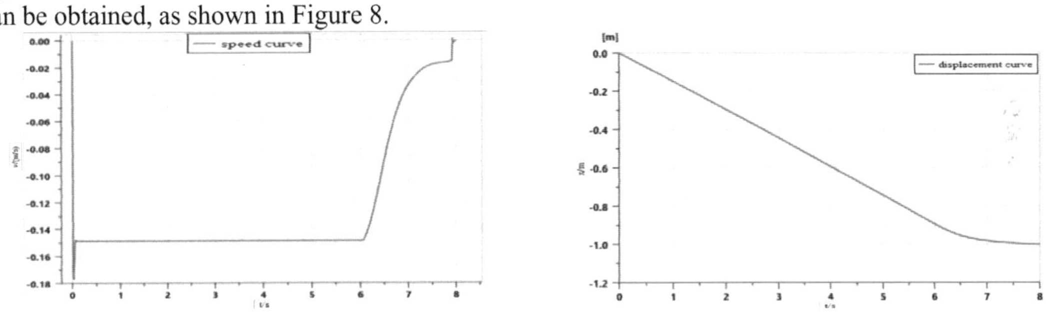

(a)Velocity profile of piston rod lowering process

(b) Displacement curve of piston rod lowering process

Figure 8. Lowering process velocity profile and displacement profile

4.2. *Simulation and experimental results' analysis of hydraulic loading system*

As shown in Fig. 9 for the hydraulic loading system test bench, the AMESim simulation model is used to compare simulation outcomes with experimental data from Table 2, as summarized in Tables 3 and 4, to validate the experiment's feasibility. The simulation adjusts the solenoid reversing valve's pressure drop to 0.001MPa at 2 tons load, 3MPa at 6 tons load, and 8MPa at 10 tons load, aligning simulation results with experimental values.

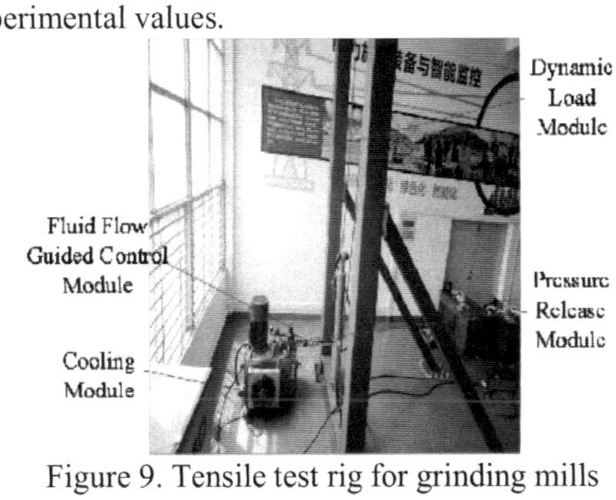

Figure 9. Tensile test rig for grinding mills

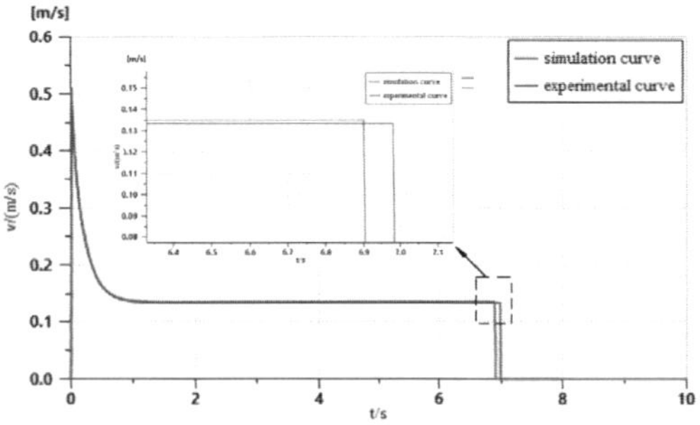

(a) Comparison of simulated and expected values of piston rod rising speed under 0.2×105N load

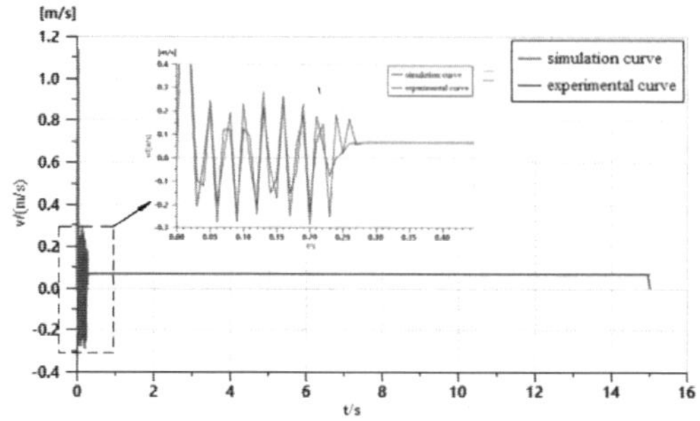

(b) Comparison of simulated and expected values of piston rod rising speed under 0.6×105N load

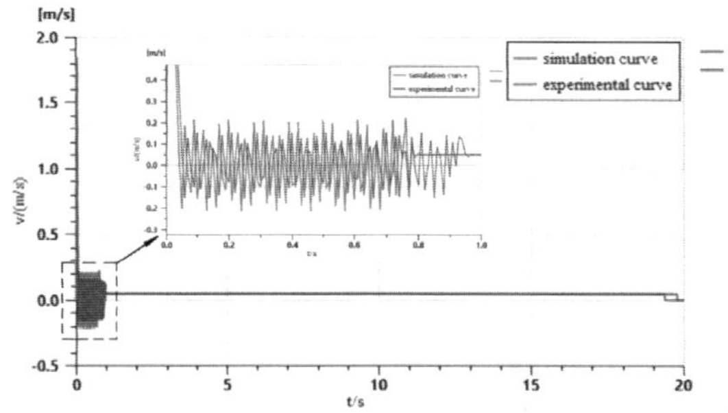

(c) Comparison of simulated and expected values of piston rod rising speed under 1.0×105N load

Figure 10. Comparison of expected and simulated values of piston rod velocity at each load

Table 3. Comparison of theoretical and simulated values

Simulated Tension(N)	Tempo(m/min)	Rodless cavity flow9L/min)	rod cavity pressure(MPa)	Simulation time (s)
	Experimental value/simulation value	Experimental value/simulation value	Experimental value/simulation value	
0.2×105	8.02/8.10	98.17/72.94	2.5/2.2	8
0.6×105	4.04/3.99	49.08/49.16	7.7/8.7	15
1.0×105	3.06/2.99	36.8/36.84	12.8/13.4	20

Table 4. comparison of errors

Simulated Tension(N)	speed Error(%)	Rodless Cavity Flow Error(%)	Rodded Cavity Pressure Error(%)
0.2×105	0.99	25	12
0.6×105	1.24	0.162	11.49
1.0×105	2.29	0.108	4.47

From Table 3 and Table 4, under simulated tensions of 0.6×10^5N and 1.0×10^5N, the speed error rates are 1.24% and 2.29%, respectively, while the rodless cavity flow errors are 0.162% and 0.108%. At a tension of 0.2×10^5N, the flow error is 25%, and the speed error is 0.99%. The flow rate error exceeds expectations due to low system resistance and incomplete valve opening under light load, causing some oil to return to the tank. The directional valve, in a half-open state, contributes to this return flow at low loads but fully opens under higher loads to ensure adequate oil supply. Safety features like relief valves are incorporated to prevent overloading by diverting fluid during light loads.

The hydraulic pump's characteristics are crucial, with high low-pressure leakage and easy backflow, and reduced high-pressure leakage and backflow. Overall, the simulation model in this study effectively mirrors real-world conditions, with errors that are acceptably low, indicating good model reliability.

5. Conclusion

In this paper, a hydraulic loading system for winch mill tension test is designed and the AMESim software is used to build the model.

The study develops a hydraulic loading test bench for winch mills, featuring a closed-loop control system with displacement, speed, and tension sensors for precise rod movement control. It includes a fault protection system for operational safety. The dynamic load module adapts to variable loads, offering an innovative method for winch mill development and inspection. The system's performance is confirmed through AMESim simulations and experimental validation, establishing it as a reliable tool for online testing and a reference for hydraulic loading systems.

Acknowledgments

Funding: National Natural Science Foundation of China(51765014); Guilin Science and Technology Bureaunder key project (2020010313)..

References

[1] XU Xiangjun, YANG Kang, JIA Laiqiang, et al. Development of intelligent electric winch for overhead transmission lines[J]. Electrotechnology,2020,(05):47-49.
[2] Cai Jinyun, Liu Zhong, Wang Gang, et al. Optimization design of auxiliary tail rope pulling device for strand mill based on response surface method[J]. Journal of Engineering Design, 2024,31(02):178-187.
[3] TANG Xiaolong, WANG Xiaodong. Simulation and experimental study on the dynamic overload characteristics of hydraulic cylinder for hydraulic support mine[J]. Hydraulic and Pneumatic, 2020,(03):124-130.
[4] BAI Guobao, XU Li. Research on loading test of coal mine winch[J]. Coal Mine Machinery, 2018,39(11):51-52.
[5] Lv Shaoli, Ma Yanping. A principle of hydraulic cylinder loading simulating counterweight mass force[J]. Hydraulic Pneumatic and Sealing,2021,41(09):62-64.
[6] ZHAO Min, YANG Bo, LI Wei. Research status and prospect of hydraulic shock suppression methods[J]. Chinese Journal of Agricultural Machinery Chemistry, 2023,44(09):123-130.
[7] MA Jun, GE Shirong, ZHANG Dekun. Research on three-dimensional contact model and inter-wire stress distribution of steel wire rope[J]. China Mechanical Engineering, 2012,23(07):864-868.
[8] Biao-Bing Li, Cheng-Wen Wang, Er-Hui Zhao. Coordinated control of flow pressure of pump-valve composite electro-hydraulic load simulator[J]. Machine Tools and Hydraulics, 2024,52(04):140-147.
[9] YU Zhuoping, SHI Biaofei, ZHUO Guirong, et al. Theoretical study on displacement-pressure characteristics of integrated electro-hydraulic braking system[J]. Journal of Mechanical Engineering, 2022,58(22):294-303.
[10] Sheng Xiaoming. Hydraulic and pneumatic transmission [M]. Beijing:Science Press,2014.6.

Journal of Physics: Conference Series **3024** (2025) 012014 doi:10.1088/1742-6596/3024/1/012014

Research on the Matching of Power System Parameters of Pure Electric Winch Mill

Zhong Liu[1,2,3], Jinyun Cai[1,2*], Hongwang Zhao[1,2,3], Han Wang[1], Zhonggang Xiong[1,2,3], Hongxiao Li[1,2]

[1] Guangxi Key Laboratory of Special Engineering Equipment and Control, (Guilin University of Aerospace Technology), Guilin, Guangxi, 541004, PR China

[2] School of Mechanical and Electrical Engineering, Guilin University of Electronic Technology, Guilin, Guangxi, 541004, PR China

[3] Key Laboratory of Special Engineering Equipment Design and Intelligent Driving Technology (Guilin University Of Aerospace Technology), Education Department of Guangxi Zhuang Autonomous Region, Guilin, Guangxi, 541004, PR China

*Corresponding author's e-mail: 1789991528@qq.com

Abstract. To promote the development of the energy transition of traditional power construction equipment, A design method for matching the power system parameters of a purely electric winch mill is proposed. Firstly, on the basis of studying the working mechanism and mechanical analysis of the winch mill, theoretical calculations are carried out on its driving motor, power battery and transmission system to determine the parameter values of the power system and the optimization feasible domain of the transmission ratio, in order to achieve the power and economic performance requirements of the winch mill for power transmission and substation engineering; secondly, the simulation model of the power system of this winch mill is established based on MATLAB/Simulink, and after the simulation The study shows that: the maximum traction force of the winch mill can reach 50.25kN, the maximum traction speed can reach 16.9m/min, and the endurance under the extreme working condition can reach 4.5h, which meets the demand of the winch mill in the construction of power transmission and transformation projects, and verifies the reasonableness of the results of the matching of the parameters of the power system and the validity of the design method; finally, the transmission ratio of the drive train is further analyzed and optimized, which improves the traction performance of the winch mill by 5.2%, and increases the traction performance by 5.2%. Finally, the transmission ratio of the drive system is further analyzed and optimized, and the traction performance of the winch mill is improved by 5.2%. This study provides theoretical and technical guidance for the power system design of similar pure electric construction equipment.

1. Introduction

In the power transmission project, the plateau complex mountainous area of the tower set up complex terrain, harsh environment, the power performance requirements of the winch mill is higher, in some large-scale equipment can not enter the area, can only rely on manual handling or cable transport[1], to set up the tower construction can be completed smoothly and efficiently, the winch mill requires a small size, lightweight, strong power, and sufficient range. A pure electric winch mill using storage

Journal of Physics: Conference Series **3024** (2025) 012014 doi:10.1088/1742-6596/3024/1/012014

batteries as energy for motor drive, to meet the essential function of its long-time lifting extensive weight tower material, there needs to be enough power for the motor and sufficient capacity of the battery, which means that the storage battery and the motor itself is a larger load, there is a strong power and range between these and the contradiction between the small size and lightweight, this is the power system design needs to face the problem[2,3]. To solve this problem, the matching and optimization of powertrain parameters of pure electric grinding mills is an important means to improve the power performance and economy of pure electric grinding mills[4].

At present, pure electric power system parameter matching and optimization in the field of pure electric winch mills and other group tower construction equipment is still very little research, mainly focusing on new energy vehicles in the field of power system parameter matching research.LIU[5] and other establishment of the mass house model of the electric vehicle as well as the joint simulation and control model carried out a number of groups of orthogonal tests, analysed the coupling law between the characteristics of the parameter and the vehicle performance indexes, and then according to the The sensitivity of the vehicle power and drive train parameters are optimised to improve the technical indicators of the vehicle; HU et al[3] studied from the aspects of power parameter matching and transmission ratio of the drive mechanism, and used multi-objective genetic algorithm to solve the optimisation problem, and got the final low-energy transmission scheme; Li Shengqin et al[6] proposed the control strategy of power distribution based on the sequence quadratic programming method, so as to ensure that the drive system works in the interval of optimal efficiency, and improve the efficiency of the drive system, and to ensure that the drive system works in the interval of optimal efficiency. The interval of optimal efficiency and improved energy utilization.

Taking the pure electric winch mill as the research object, firstly, the working principle and force situation of the winch mill are analyzed, and the main performance indexes of the pure electric winch mill are defined based on the electric power industry standards, and the parameter matching design and analysis of the power system are carried out; secondly, the simulation model of the power system of the winch mill is set up to carry out the simulation analysis to verify the reasonableness of the parameter matching design; finally, the transmission ratio of the drive system is analyzed and studied. Research to improve the dynamics of the winch mill.

2. Winch mill working principle and force analysis

As shown in Figure 1 is a three-dimensional schematic diagram of the traditional motorized winch mill, which is mainly composed of a petrol engine, a belt drive, a transmission, a pass-through box, and two drums.

On its basis, this paper proposes a dual motor-driven pure electric winch mill structural scheme, which can distribute power between the two motors according to the working conditions, optimize the use of energy, and improve the overall energy utilization rate. As shown in Fig. 2 and Fig. 3, the three-dimensional schematic diagram and system structure diagram of the pure electric winch mill are respectively shown, which mainly consists of the electric control system, power battery pack, dual electric motors, transmission system, and dual drum parts. The main difference between it and the traditional motorized winch mill is: that the mechanical connection part of the power source will be replaced with dual electric motors, directly connected to the transmission system, the design of a higher degree of integration of the transmission mechanism, and no longer use the transmission form of the transmission box + overpass box; the electrical part of the increase in battery packs for the electric motor to provide power, but also the use of the electronic control system as a winch grinding mill work as an intelligent centralized control center.

Figure 1. Three-dimensional diagram of a traditional motorized winch mil

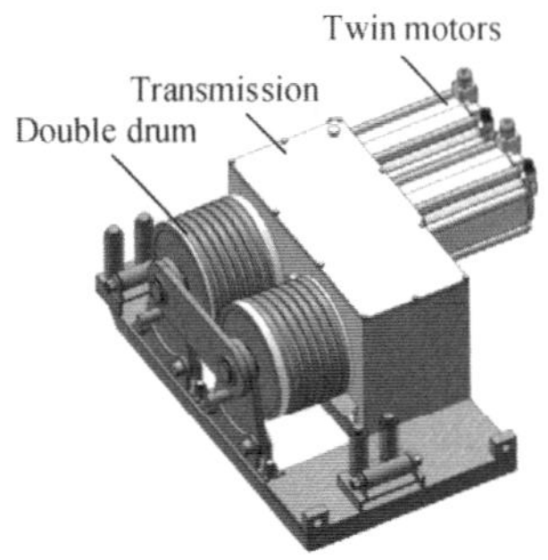

Figure 2. Three-dimensional diagram of pure electric winch mill

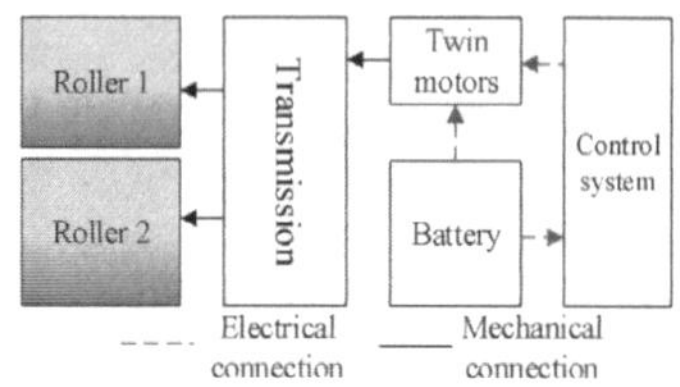

Figure 3. System structure diagram of pure electric winch mill

As shown in Figure 4 for the working schematic diagram of the winch mill, the winch mill in the execution of the work, relies on the traction rope wrapped around the bottom of the double drum groove through the holding bar to achieve the traction of the tower tower material. Drum as the main part of the implementation of the work, needs to establish a mechanical model of its working state of force analysis, to facilitate the matching of power system parameters and the construction of a power system simulation model.[7]. As shown in Figure 5, the winch drum for force analysis, the wire rope wrapped around the two drums, one end of the connection of the weight and lifting, the load F_1 that is the weight of the weight of the weight of the force exerted on it, the other end of the pulling the tail rope workers to pull tightly, the force exerted by the tail rope tension F_2, according to the Euler's formula for flexible bodies can be obtained

$$F_1 = e^{\mu\alpha} \cdot F_2 \tag{1}$$

Where: μ is the friction coefficient between the wire rope and the drum of the winch mill; α is the angle of wrapping of the wire rope on the drum of the winch mill.

The friction of the wire rope on the drum of the winch mill is:

$$f = \left(1 - 1/e^{\mu\alpha}\right) F_1 \tag{2}$$

Then the torque required to be output to the drum shaft by the power system of the winch is:

$$T = f \cdot R = \left(1 - 1/e^{\mu\alpha}\right) F_1 \cdot R \tag{3}$$

Where: R is the radius of the groove bottom of the winch drum, mm

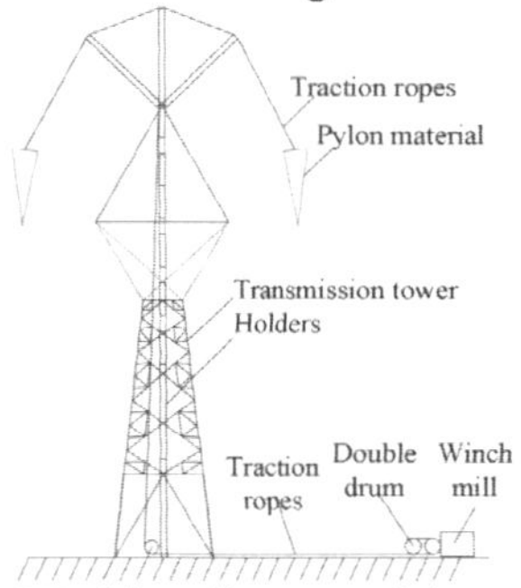

Figure 4. Working diagram of the winch mill

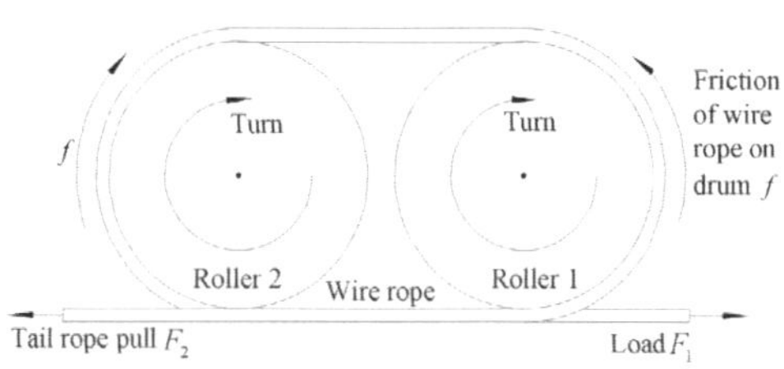

Figure 5. Sketch of the force on the drum

3. Powertrain parameter matching design

3.1. Pure electric winch system configuration

The power of a pure electric winch mill is mainly measured by the maximum traction force and lifting speed, and the economy is mainly measured by the range time and working energy consumption. Among them, the transmission system drive motor and storage battery pack parameters have a coupling relationship, so the determination of each parameter and the reasonable matching between parameters Therefore, the determination of each parameter and the reasonable matching between parameters are extremely important. According to the national grid power line group tower construction process and safety specification requirements, in the large high-voltage line project, the use of single drum winch mill construction equipment is not allowed [8], so the pure electric winch mill is developed based on the traditional double-drum motorized winch mill model, based on the national electric power industry standards in the DL_T 733-2022, ***Motorised Winch Mills for Transmission and Substation Engineering***, considering the actual working conditions. The main technical index parameters of the winch mill are shown in Table 1.

Table 1 Winch mill main performance index parameters

Parameter name	Parameter value	Parameter name	Parameter value
Rated traction force/kN	50	Common traction speed/(m/min)	0~16 step-less speed
Rated traction speed/(m/min)	0~8 step-less speed		
Common traction force/kN	25	Total duration/h	4

3.2. *Drive motor parameter matching*

3.2.1. *Drive motor power*

In the motor parameters matching, need to consider the rated power of the motor, speed, efficiency, and other parameters. The rated power of the motor needs to meet the rated traction force index of the winch mill and the maximum rated traction speed, in the pure electric winch mill, the rated traction force of the maximum rated traction speed condition, the motor demand power is:

$$P_{e1} = F_e v_{emax} / 60\eta_w \eta_e \tag{4}$$

Where: F_e for the rated traction force of the winch mill, kN; V_{emax} for the maximum rated traction speed of the winch mill, m/min; η_e for the total efficiency of the drive motor; η_w for the transmission efficiency of the entire transmission system.

When the winch mill is in the rated traction force of the accelerating condition, at this time the motor demand power of the winch mill, can be calculated by the following formula:

$$P_{e2} = F_e v_e / 60\eta_w \eta_e \cdot (1 + v_e / gt) \tag{5}$$

Where: V_e is the target speed of acceleration within the range of rated traction speed of the winch mill, rpm; g is the acceleration of gravity; t is the time for the winch mill to accelerate to the target speed, s.

3.2.2. *Motor speed*

The rotational speed of the motor is calculated according to the common traction speed index of the pure electric winch mill, to meet the maximum traction speed requirement of the winch mill, the input rotational speed of the motor is required:

$$n_e = i v_{s\max} / 2\pi R \tag{6}$$

Where: i is the transmission ratio of the winch mill drive system; V_{smax} for the maximum common traction speed of the winch mill, m/min.

3.2.3. *Motor torque*

The torque of an electric motor can be found in the equation relating torque to speed and power:

$$T_e = 9549 P_e / n_e \tag{7}$$

According to the above analysis and calculation, to simultaneously meet the above drive motor power, speed, and torque requirements, as well as a certain safety coefficient, according to the pure electric winch mill adopts the dual-motor drive scheme, resulting in dual-motor matching parameters as shown in Table 2.

Table 2 Dual Motor Matching Parameters

Parameter name	parameter value	Parameter name	parameter value
Rated power of single motor/kW	6.8	Total rated power/kW	13.6
Rated torque of single motor/(Nm)	22	Total rated torque/(Nm)	44
Total rated speed/rpm	3000		

3.3. Power Battery Parameter Matching

Select the number of battery packs need to meet the pure electric winch mill drive motor voltage, range indicators, and maximum power requirements[9]. According to the pure electric winch mill rated traction under the performance index requirements to determine the capacity of the battery pack requirements for:

$$C = W/U = 1000 P_{e1} t / \eta_{DoD} U \tag{8}$$

Where: W is the total energy of the battery pack, Wh; P_{e1} is the demanded power when the winch mill is running under the rated traction force, kW; t is the continuous working length of the winch mill, h; n_{DoD} is the depth of discharge of the battery; U is the rated voltage of the battery pack, V.

The rated voltage of the battery pack U needs to be determined according to the rated voltage of the selected drive motor, and is required to be greater than the rated voltage of the drive motor. When the rated voltage of a single battery does not meet the rated voltage of the drive motor, the required total voltage can be obtained by connecting the single batteries in series, at least the number of single batteries required to be connected in series is:

$$N_1 = U_e / U_0 \tag{9}$$

Where: U_e is the rated voltage of the winch drive motor, V; U_0 is the rated voltage of the single battery, V

Similarly, when the capacity of a a single cell does not meet the required capacity, the required capacity can be obtained by connecting the single cells in parallel, at least the number of single cells required to be connected in parallel is:

$$N_2 = C / C_0 \tag{10}$$

Where: C_0 is the single cell capacity, Ah

The maximum output power of the battery pack also needs to be based on the output power of the selected drive motor, at least the number of batteries required:

$$N_3 = P_{e1} / P_{amax} \eta_w = 4 R_a P_e / E^2 \tag{11}$$

Where: P_{amax} is the maximum power of a single cell, kW; R_a is the equivalent resistance; E is the rated voltage of a single cell, V.

According to the above analysis and calculation, in order to meet the working voltage of the drive motor and the demand for range capacity and output power, the power battery scheme used in the pure electric winch mill is shown in Table 3.

Table 3 Power Battery Pack Matching Parameters

Parameter Name	parameter value	Parameter Name	parameter value
Cell Voltage/v	3.2	Total Battery Voltage/V	76.8
Capacity of single battery/Ah	120	Total battery capacity/Ah	480
Number of batteries	4 Series 4 Parallel		

3.4. Preliminary calculation of transmission ratio of transmission mechanism

In the design of transmission ratio, large transmission ratio can enhance the output torque of the winch mill, that is, to improve the traction capacity; small transmission ratio can increase the output speed of the winch mill drum, that is, to improve the traction speed, but the corresponding output torque will be reduced, and the stability of traction will also be reduced[10]. Therefore, the reasonable choice of transmission ratio is related to the performance of the winch mill.

3.4.1. Upper limit of transmission ratio

To ensure that when the drive motor is at the maximum speed output, after the transmission system transmission, the drum speed can still meet the maximum common traction speed performance indicators, need to determine the upper limit of the transmission system ratio. Determined by the maximum lifting speed of the winch that is the maximum common traction speed:

$$i_{\max} \le 2\pi R n_{\max} / v_{s\max} \tag{12}$$

Where: $n_{\max}$ is the maximum output speed of the motor, rpm.

3.4.2. Lower limit of transmission ratio

To ensure that when the drive motor with the maximum torque output, after the transmission system transmission, the drum torque can still meet the rated traction force performance indicators, need to determine the lower limit of the transmission system ratio. Determined by the rated traction force of the winch:

$$i_{\min} \ge F_e R / T_{\max} \eta_e \cdot \left(1 - 1/e^{\mu\alpha}\right) \tag{13}$$

In summary, it can be calculated that the transmission ratio range of the pure electric winch mill is $133 \le i \le 141$, in the parameter matching process, the transmission system transmission ratio can be pre-set a basic value, and then verify the reasonableness of the set value after matching.

4. Pure electric winch mill machine power model establishment and simulation research

4.1. Mathematical modeling of drive motors and batteries

The mathematical model for building a servo motor is usually derived from the derivation of transformations based on the magnetic chain equation, torque equation, voltage equation, and mechanical equation[11]. The mathematical model of the drive motor is:

$$\begin{cases} \omega_m(s) = (T_e(s) - T_L(s)) / (J_L + B_m) \\ T_e(s) = K_t I_q(s) \\ I_q(s) = (U_q(s) - K_e \omega_e(s)) / (L_q(s) + R_s) \end{cases} \tag{14}$$

Where: T_e is the motor torque, N-m; TL is the motor load torque, N-m; J_L is the motor moment of inertia, kg • m^2; B_m is the motor damping coefficient; K_t is the motor torque coefficient; I_q is the two-phase rotating coordinate system under the current in the q-axis component, A; U_q is the two-phase rotating coordinate system under the voltage in the q-axis component, V; K_e is the coefficient of inverse electromotive force; ωe is the angular velocity of rotor, rad/ s; L_q is the stator inductance in the q-axis inductance in the two-phase rotating coordinate system, H; R_s is the stator resistance, Ω.

The charge value of the current state of the Li-ion battery, i.e., the state at the moment tK, can be expressed by the following equation[12]:

$$SOC_k = SOC_0 - \int_{t_0}^{t_k} I dt / C_e \tag{15}$$

Where: SOC_0 is the starting state of lithium battery discharge or charge; I is the output current of lithium battery; C_e is the maximum capacity of lithium battery.

4.2. Simulation modeling of the power system of the grinding machine

To verify the reasonableness of the above power transmission system parameter matching, based on the mathematical model of the drive motor and battery and the working principle and force analysis of the winch mill, MATLAB/Simulink is used to establish the power simulation model of the pure electric winch mill to simulate the dynamics and economy of the winch mill. The simulation model of the power system of the winch mill is built as shown in Fig 6 below:

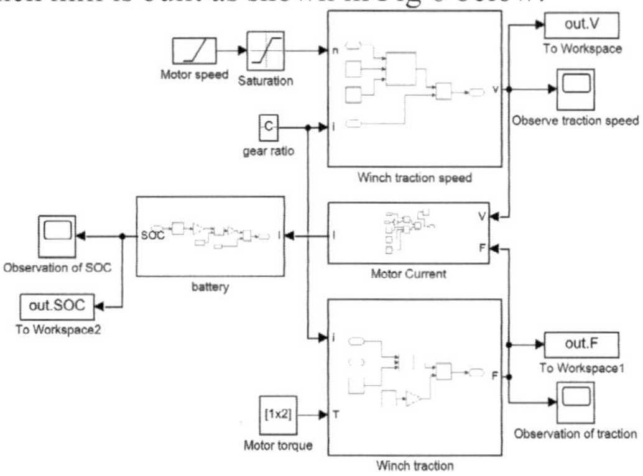

Figure 6. Simulation model of the power system of the winch mil

After completing the model construction, the parameters are adjusted and set within the model framework based on the matching results of the power system parameters mentioned above, aiming to ensure that the simulation can accurately reflect the actual power and economic performance of the grinding mill. Through the computational analysis of MATLAB/Simulink, the power and economic performance of the strand mill were comprehensively evaluated.

4.3. Analysis of the simulation results of the power system of the winch mill

4.3.1. Dynamic performance analysis

The power performance of the winch mill is the key factor in judging the advantages and disadvantages of its performance indexes. In the analysis of the power performance, the focus is on the key indexes such as the power output of the winch mill, the maximum traction force, the fastest traction speed, and so on. Figure 7 shows the traction speed curve of the strand mill, describing the time required for the strand mill to accelerate to a certain stable working speed from a standstill state, and it can be analyzed that the strand mill can reach a maximum traction speed of 16.9m/min within 10s of the response time set by the drive motor, which meets the requirements of the rated and commonly used traction speeds in the performance indexes.

The output characteristic curve of the strand mill is shown in Figure. 8, which demonstrates the relationship between the traction power, traction force, and traction speed of the strand mill under different operating conditions. Analyzing the mechanical characteristics of the winch mill under rated traction power and common traction power conditions, it can be seen that the winch mill can stably provide a maximum traction power of 50.25kN in the low-speed range of the continuous use area of the drive motor, and the winch mill can stably provide a traction power of more than 25kN in the high-speed range, which can satisfy the requirements of the rated traction power and the commonly used traction power in the performance indexes.

4.3.2. Dynamic performance analysis

The economic performance of a winch mill is closely related to its range time, which reflects the sustainability of the winch mill in practical engineering application scenarios. Shown in Figure 9 is the SOC curve of the battery of the winch mill, the SOC curve can visualize the whole process of the battery from full charge to depletion, and it can be seen through the analysis that the endurance time of the winch mill reaches 4.5h under the limiting working condition of the commonly used traction force, and 4.8h under the limiting working condition of the rated traction force, which both satisfy the requirement of the endurance time in the performance index.

According to the above simulation analysis results, the performance indexes shown in Table 4 below can be obtained as a comparison table with the simulation results, and the simulation results can meet all the performance indexes of the pure electric winch mill, which verifies the reasonableness of the matching of the power system parameters.

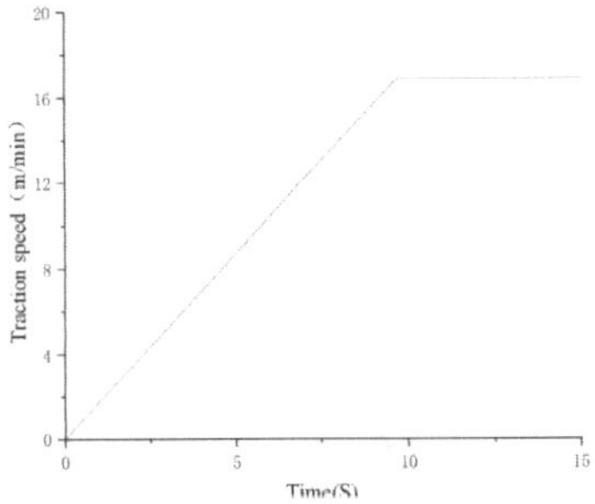

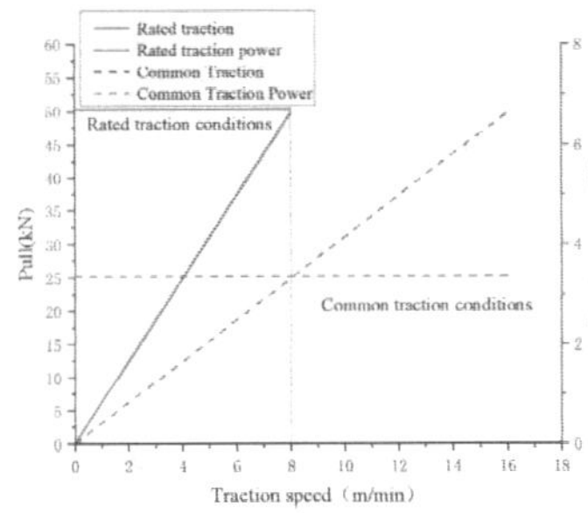

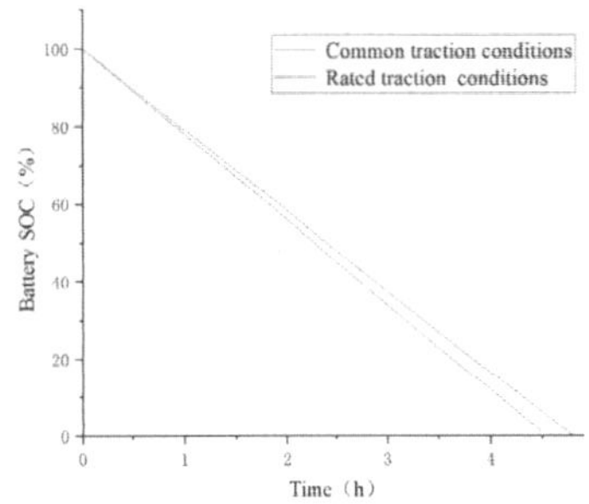

Figure 7. Traction speed curve Figure 8. Output characteristics Figure 9. Battery SOC curves

Table 4 Comparison table of performance indexes and simulation results

	Rated traction force/kN	Rated traction speed/(m/min)	Usual traction force/kN	Usual traction speed/(m/min)	Duration/h
Performance Indicators	50	8	25	16	4
Simulation results	50.25	>8	>25	16.9	≥4.5
Satisfaction	meet	meet	meet	meet	meet

4.4. Ratio optimization analysis

To further improve the performance of the winch mill, based on the feasible domain of transmission ratio optimization obtained in the above power parameter matching calculation, different transmission ratio schemes are formulated and sequentially substituted into the power system simulation model for solving. According to the above analysis, the performance index boundary parameters are rated traction force, common traction speed, and common traction force operating conditions of the duration, and finally obtained the following Figure 10 shows the performance parameters of the winch mill under different transmission ratios.

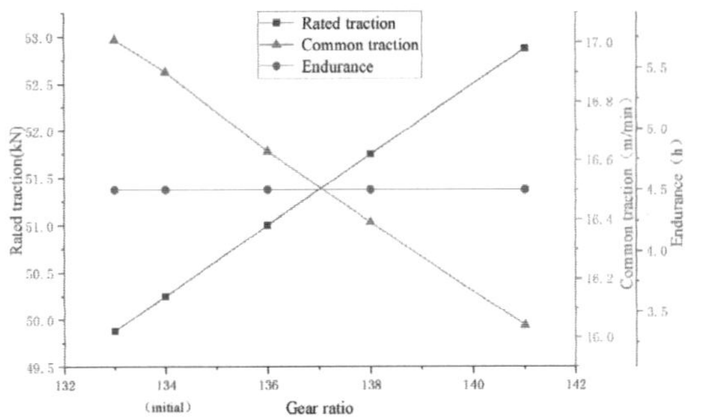

Figure 10. Performance parameters of the winch mill under different transmission ratios

Comparing the calculation results of different transmission ratio schemes, it can be seen that with the increase of transmission ratio, the traction performance of the winch mill is improved, but the traction speed is reduced. For the duration, the change in the transmission ratio does not have a significant effect as the operating power remains constant. For the selection of transmission ratio, considering the weight of safety in the process of electric tower erection[13], under the premise of satisfying the performance index of traction speed, the traction performance is regarded as a more important index, which determines whether the winch mill can stably and reliably complete the task in various working environments. Under the comprehensive analysis, the transmission system scheme with a transmission ratio of 141 is selected, which can increase the rated traction force of the winch mill by 5.2% compared with the initial design value, and has better traction performance.

5. Conclusion

(1) On the basis of the force analysis of the actuating parts of the winch mill, according to the requirements of the performance index of the winch mill, the power system of the winch mill is parameter-matched through theoretical calculations and the optimization feasible domain of the transmission ratio is derived.

(2) MATLAB/Simulink was used to establish the power system simulation model of the pure electric grinding mill, and the maximum traction force of the grinding mill was 50.25kN, the maximum traction speed was 16.9m/min, and the range was up to 4.5h, which meet the performance index requirements of the grinding mill, and verified the reasonableness of the matching results of the power system parameters.

(3) Calculated and analyzed the impact of the transmission ratio on the performance of the winch mill, considering the performance index weight factors, choose 141 large transmission ratio as the transmission system scheme to enhance the traction performance of the winch mill.

Acknowledgments

Funding: National Natural Science Foundation of China (51765014); XXX Engineering Science and Technology Research Project commissioned by CEC-Sichuan Electric Power Design and Consulting Co. (007-HTSEDC-2023); Guilin Science and Technology Bureau under key project (2020010313).

References

[1] Liu Z, Cai JY, Wang G, et al. Development and application of miniature steel pipe pile drilling rig in complex mountainous area [J]. Electromechanical Engineering Technology, 2023,52 (10): 6-10.

[2] Liu J, Ge S, Zhu H, et al. Mine Rescue Robot Power Matching Based on Multi-objective Particle Swarm Optimization [J]. Journal of Mechanical Engineering, 2015,51 (3): 18-28.

[3] Hu J, Cao W, Jiang F, et al. Study on Multi-Objective Optimization of Power System Parameters of Battery Electric Vehicles[J]. Sustainability, 2023, 15(10): 8219.

[4] Geng J, Chi Q. Parameter matching and simulation analysis of power system of pure electric vehicle[J]. Journal of Physics: Conference Series, 2021, 2076(1): 012091.

[5] Liu H, Lei Y, Fu Y, et al. Parameter matching and optimization for power system of range-extended electric vehicle based on requirements[J]. Proceedings of the Institution of Mechanical Engineers, Part D: Journal of Automobile Engineering, 2020, 234(14): 3316–3328.

[6] Li S, Ding X, Yu B. Parameter Matching Design and Control Strategy of Dual Drive Powertrain for Pure Electric Vehicle [J]. Journal of Chongqing Jiaotong University (Natural Science Edition), 2022,41 (11): 145-152.

[7] Hu JJ. Mechanical Research and Analysis on the Key Parts of a 5T Twin-Drum Winch [D]. Xiangtan University, 2016.

[8] Zhu H. Differential Constant Torch Ratio Transmission System Research of the 5 t Double Drum Winching [D]. Xiangtan University, 2017.

[9] Li Z, Lu RH, Yu QH, et al. Research and Analysis of Thermal Runaway Characteristics and Prevention and Control Technology of Power Battery [J]. Automotive Engineering, 2024,46 (1): 139-150.

[10] Niu J, Wang ZJ, Zhang PB, et al. Matching and Optimization of Two-speed Electric Vehicle Transmission System based on Cruise and Isight [J]. Mechanical transmission, 2021,45 (5): 133-138.

[11] Zhao J. Modeling and Simulation of SOC Estimation for Lithium-ionBattery Based on Second-order EKF [D]. Xi'an University of Science and Technology, 2018.

[12] Yang J. Construction and Simulation of Hybrid PowerSystem for Proton Exchange Membrane Fuel Cell of Forklift [D]. Northeast Forestry University, 2020.

[13] Zheng HQ, Xiong ZM, Tower Assembling and Erection Technologies for 1 000 kV UHV AC Transmission Line [J]. Power Grid Technology, 2008 (20): 15-19.

Research and Application of Multi-Source Sensor Fusion Localization Algorithm for Complex Scenarios

Qingshan Wang[0000-0003-4169-0661], Dengquan Zhang[0000-0002-2148-5006], Yongxiang Cai, Cong Cai, Jiaochen Wang, Peng Guo*[0000-0002-8321-6653]

CATARC (Tianjin) Automotive Engineering Research Institute Co., Ltd, No. 68 Xianfeng Road, Dongli District, Tianjin

*Corresponding author's e-mail: guopeng@catarc.ac.cn

Abstract. Due to remote locations such as mines and ports, GNSS signals have not achieved full coverage. They are weak, resulting in significant positioning errors for autonomous vehicles and making them unable to meet task requirements. In response to the problem of substantial positioning errors in autonomous driving vehicles relying solely on GNSS in complex scenarios, this paper first adopts the multi-sensor fusion of GNSS, LiDAR, and IMU. It proposes a loop detection based on the NDT algorithm SLAM architecture to construct a lightweight scene map. It provides high-precision and low-data volume scene information for autonomous driving positioning. Secondly, based on constructing a lightweight scene map, a multi-feature matching method combining point cloud features and point cloud strength information is adopted to achieve precise positioning of autonomous driving. Finally, the volumetric Kalman filter is used to reduce the influence of noise points on the filter by weighting the observed and predicted values of sensors, thereby achieving high-precision positioning of autonomous driving in complex scenes through multi-sensor positioning fusion. The experimental results show that the multi-feature matching algorithm is less than 36ms, the backend optimization of lightweight scene maps is 80ms, the compression ratio of high-precision point cloud maps is 92%, and the fusion positioning accuracy is less than 5cm.

1. Overview

With the continuous iteration of autonomous driving technology and the constant improvement of rules, L2-level autonomous driving and L4-level have the most significant potential for the landing of autonomous driving. Both foreign Tesla models and domestic car manufacturers such as Xiaopeng, GAC, and Weilai have added L2-level auto-drive systems to their models to help drivers better manage vehicle driving in a specific environment, to a certain extent, increasing the sense of technology and driver experience. L4 autonomous driving has been deployed in some closed/semi-closed environments in China, especially in promoting the construction of smart mines, smart ports, and other scenarios, where smart transportation and smart logistics play an important role. Not only does it solve the problem of difficulty in recruiting workers in harsh environments, but it also increases the operational efficiency of autonomous driving in mines and ports.

The perception module and positioning module are essential key technologies in autonomous driving, especially in L4-level autonomous driving, where high-precision maps are indispensable. High-precision maps provide positioning and path-planning services for autonomous vehicles. In the early days, based on the perception of the surrounding environment by vehicle-mounted LiDAR sensors, a point cloud feature matching method was used to construct maps. However, due to the single sensor and algorithm limitations, the constructed point cloud maps had a large amount of data and poor consistency, resulting in poor real-time positioning and positioning accuracy of autonomous driving. With the proposal and development of SLAM (Simultaneous Localization and Mapping) architecture [1-4], more and more scholars are using multi-sensor fusion methods such as LiDAR, GNSS, and IMU to construct high-precision maps, which can meet the positioning and navigation needs of autonomous vehicles [5-8].

An essential part of the perception module in autonomous driving is positioning. Without precise positioning results, subsequent decision-making and execution cannot be discussed. With the development of electronic maps in daily life, the positioning error has been reduced from over 10 meters to meter-level positioning error. This positioning error makes it difficult to meet the centimeter-level positioning of autonomous driving. Although centimeter-level positioning is achieved through autonomous driving equipped with GNSS, interference from complex environments such as underground parking lots, tunnels, indoors, remote mining areas, and ports makes it impossible to achieve centimeter-level positioning for autonomous driving solely with a single onboard sensor. Therefore, a multi-sensor fusion approach is proposed to achieve high-precision positioning in complex scenarios.

In response to the problems of large data volume and poor accuracy in building high-precision maps, this paper proposes to use the NDT algorithm based on SLAM architecture to achieve loop detection and complete the construction of point cloud maps in large-scale, lightweight, and high-precision complex environments. This paper proposes an algorithm based on the volume Kalman filter to achieve accurate positioning in complex scenes by fusing multiple sensors to address issues such as single-sensor positioning failure. This article uses "Tianjin" as a verification platform to ultimately achieve the construction of large-scale lightweight point cloud maps and high-precision positioning in complex scenes.

This article implements high-precision positioning based on multi-source sensor fusion in complex environments, solving the problems of significant positioning deviation and positioning failure caused by weak and unstable GNSS signals. Firstly, multi-sensor joint calibration is performed based on LiDAR, IMU, and GNSS, completing steps such as time synchronization, space synchronization, and data preprocessing, Secondly, high-precision maps are the foundation for providing fusion positioning, so the key factor is to construct rich high-precision point cloud maps based on unstructured roads in parks and irregular and remote mining areas; Finally, by combining the communication based GNSS absolute positioning method with the LiDAR based feature matching relative positioning method, high-precision positioning is achieved through fusion in complex environments. The overall architecture is shown in **Figure 1**.

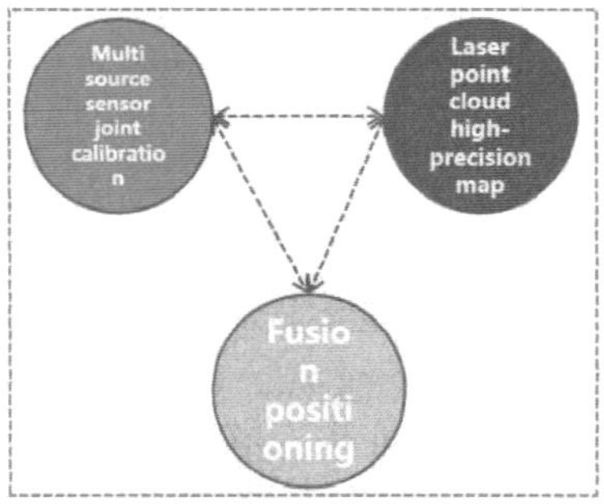

Figure 1. Overall Architecture Diagram

2. Construction of high-precision maps

This article uses GNSS, IMU, and LiDAR multi-sensors to construct high-precision point cloud maps, including data preprocessing, point cloud map construction, and factor map trajectory optimization. The data preprocessing completes the time and space synchronization of GNSS, IMU, and LiDAR, and completes the conversion of geodetic coordinates based on GNSS communication and the calculation of navigation confidence, Using IMU for distortion correction of LiDAR point cloud and completing multi-sensor data processing. The construction of high-precision point cloud maps adopts the NDT algorithm[9] , map optimization, and sliding maps to improve real-time mapping and lightweight data storage. The factor map trajectory optimization uses a point line feature matching algorithm to complete loop detection, correct the cumulative error of the front-end odometer, and incorporate GNSS factor map optimization to improve the accuracy of point cloud maps in large-scale scenarios, thereby completing the construction of point cloud maps, as shown in **Figure 2**.

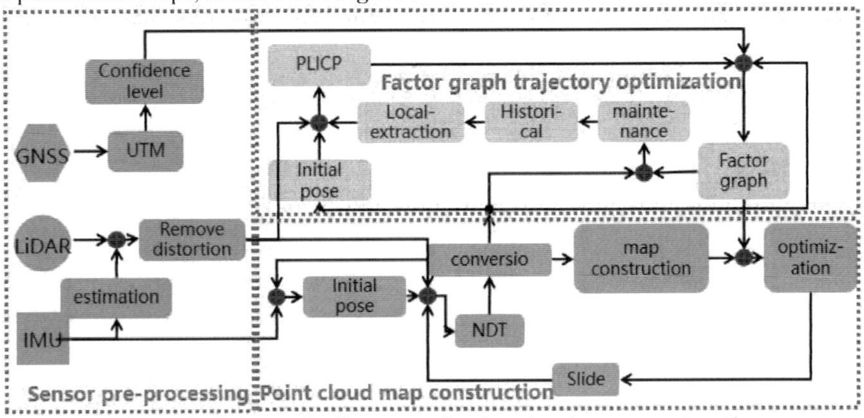

Figure 2. Schematic diagram of point cloud map construction

2.1 Data Preprocessing

The current latitude and longitude information is obtained through a vehicle-mounted satellite receiver, and the conversion of geodetic coordinates is completed using UTM projection. When the vehicle is in motion, many noise points are generated due to environmental influences. Therefore, Kalman filtering is used to filter out the noise points and obtain the absolute position of the vehicle based on the confidence of positioning. Due to the distortion of the laser point cloud caused by the vehicle's speed during autonomous driving, the laser point cloud at the current moment is distorted and has low accuracy. This article utilizes the high-frequency characteristics of IMU to compensate for and correct the distortion of laser point clouds during motion. The period for LiDAR to rotate once is 10Hz, and the period for IMU motion is 50HZ. The distortion phenomenon of the laser point cloud is shown in **Figure 3**.

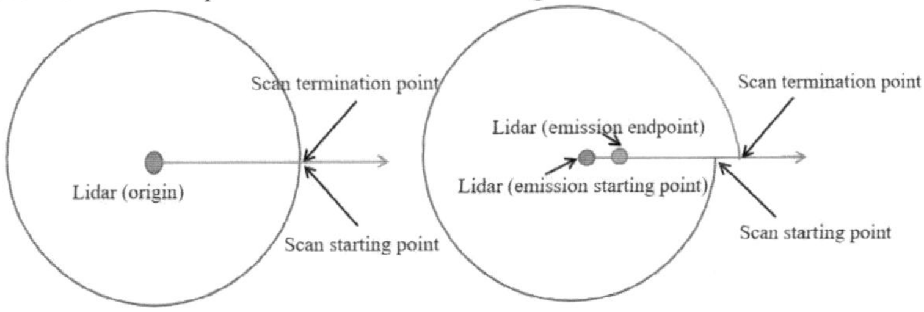

Figure 3. Schematic diagram of laser point cloud distortion phenomenon

IMU angle calculus:

$$\alpha_t = \frac{w_{t_0} + w_t}{2} \cdot dt \tag{1}$$

Where dt is the adjacent time difference, w is the angular velocity at the current time, and α is the angular velocity at the current time.

Due to the particularity of the storage method of laser point clouds, the pose transformation at the current moment is represented by quaternions.

$$q = \begin{bmatrix} \cos(\frac{\theta}{2}) \\ \sin(\frac{\theta}{2}) \cdot r \end{bmatrix} = \begin{bmatrix} W \\ X \\ Y \\ Z \end{bmatrix} \tag{2}$$

The transformation of the IMU angle $\frac{\theta}{2}$ for adjacent moments is close to 0. So there are:

$$q_t = q_{t_0} \cdot q_{dt} \qquad q_{dt} = (1, \frac{\alpha_x}{2}, \frac{\alpha_y}{2}, \frac{\alpha_z}{2}) \tag{3}$$

Among them, $\alpha_x\ \alpha_y\ \alpha_z$ represents the angle changes on three axes, and the attitude quaternion is represented by q.

The coordinate unification between IMU and LiDAR is achieved through the measured attitude angle changes mentioned above.

$$R_{lidar0}^{lidar1} \cdot R_{lidar}^{IMU} = R_{lidar}^{IMU} \cdot R_{IMU0}^{IMU1} \tag{4}$$

Among them, R_{lidar}^{IMU} is the rotation matrix from LiDAR to IMU, R_{lidar0}^{lidar1} is the change from the initial state to the final state of LiDAR, and R_{IMU0}^{IMU1} is the change from the initial state to the final state of IMU (attitude quaternion q).

Linear interpolation transformation is used to correct the distortion of LiDAR point clouds. Firstly, the angle of each point cloud is calculated as follows:

$$\begin{cases} \theta_i = \arccos(\frac{x_i}{\sqrt{x_i^2 + y_i^2}}), if \ \ y_i < 0 \\ \theta_i = 2\pi - \arccos(\frac{x_i}{\sqrt{x_i^2 + y_i^2}}), if \ \ y_i > 0 \end{cases} \tag{5}$$

Transform the current point cloud state to the terminal state to complete the linear difference, and then correct the point cloud distortion.

$$q_{slerp} = (q_1 q_0^{-1})^t q_0 \tag{6}$$

Among them, $t \in [0,1]$ and q_0 are quaternions in the starting state, and q_0 is a quaternion in the ending state.

Coordinate transformation is achieved by converting the rotation matrix into quaternions:

$$p_1 = \begin{bmatrix} X_1 \\ Y_1 \\ Z_1 \\ 1 \end{bmatrix} = T_{slerp} \cdot p_0 = \begin{bmatrix} R_{slerp} & 0 \\ 0 & 1 \end{bmatrix} \cdot \begin{bmatrix} X_0 \\ Y_0 \\ Z_0 \\ 1 \end{bmatrix} \tag{7}$$

Among them, p_0 is the original point cloud data, and p_1 is the corrected point cloud data.

2.2 High precision point cloud map construction

By using point cloud data collected by vehicle-mounted LiDAR for inter frame point cloud matching to estimate vehicle attitude, this paper adopts the NDT algorithm SLAM point cloud map construction architecture. It proposes map optimization and sliding window downsampling of point cloud maps to reduce data size and improve the computational efficiency of attitude estimation. As shown in **Figure 4**, a sliding window is a fixed window set on the timeline that slides over time. Only the variables within the window are optimized each time, and the remaining variables are marginalized. Since all variables will be re-linearized during the optimization iteration, the cumulative linearization error is small and the accuracy is guaranteed; At the same time, due to the fixed window size, the number of optimization variables remains unchanged, which can meet real-time requirements. Set a sliding window for the time series t, extract a local map $\{f_{t-k},\dots,f_{t-1}\}$ with a step size of k from the global map, and perform grid partitioning and calculate the average and covariance of each grid point cloud after obtaining the point cloud data of the $f_p^{'}$ keyframe.

$$\mu = \frac{1}{m}\sum_{k=1}^{m} y_k \tag{8}$$

$$\Sigma = \frac{1}{m-1}\sum_{k=1}^{m}(y_k - \mu_k)(y_k - \mu_k)^T \tag{9}$$

Among them, y_k is the keyframe of the laser point cloud.

Secondly, based on the keyframe $f_p^{'}$, map matching is performed with the local map $\{f_{t-k},\ldots,f_{t-1}\}$ through the transformation function $x_i^{'}=T(x_i,p)$, where $x_i^{'}$ is obtained from the transformation pose of x_i, and p is defined as:

$$p=[t_x,t_y,t_z,\phi_x,\phi_y,\phi_z]^T \tag{10}$$

Where (t_x,t_y,t_z) is the initial coordinate and (ϕ_x,ϕ_y,ϕ_z) is the initial pose.

$$T(p,x)=\begin{bmatrix} c_yc_z & -c_ys_z & c_y \\ c_xs_z+s_xs_yc_z & c_xc_z-s_xs_ys_z & -s_xc_y \\ s_xs_z+c_xs_ys_z & c_xs_ys_z+s_xc_z & c_xc_y \end{bmatrix} x+\begin{bmatrix} t_x \\ t_y \\ t_z \end{bmatrix} \tag{11}$$

Among them, $s_k=\sin\phi_k$, $c_k=\cos\phi_k$.

Calculate the probability density function of x_k based on the mean and covariance matrix:

$$p(x_k)=\frac{1}{(2\pi)^{\frac{3}{2}}\sqrt{|\Sigma|}}e^{-\frac{(x_k-\mu)^T\Sigma^{-1}(x_k-\mu)}{2}} \tag{12}$$

Sum up $s(p)$ based on the probability density function of each point, and then use Newton's optimization algorithm to optimize $-s(p)$, that is, find the optimal transformation matrix between keyframe $f_p^{'}$ and local map $\{f_{t-k},\ldots,f_{t-1}\}$.

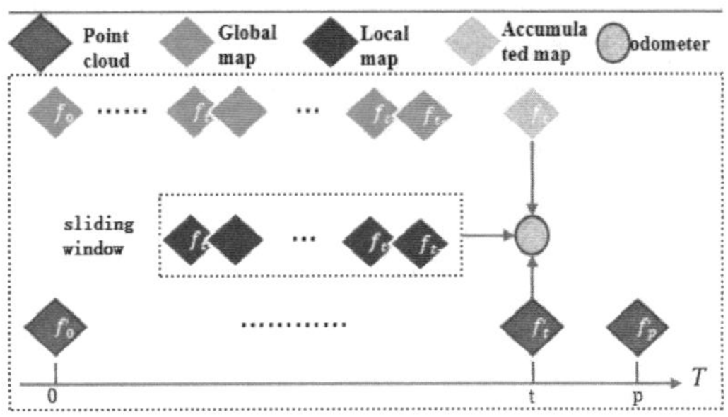

Figure 4. Schematic diagram of sliding window

Factor graph is a probability graph based on nonlinear least squares optimization, which can be used to fuse data from multiple input sources. When it is necessary to consider state variables, various factors can be added to the original graph. If the reliability of the measured values is low or unknown, the number of factors may be reduced because the factor map contains observation information from multiple sensors. Therefore, the graph can be described as a constraint problem mapping between the current pose and the historical pose during SLAM, that is, the constraint relationship between nodes and edges. This article introduces a factor graph optimization model and incorporates loop detection based on the PLICP (point-to-point ICP) algorithm [10-12] and GNSS absolute information factor to construct a large-scale point cloud map, as shown in **Figure 5**.

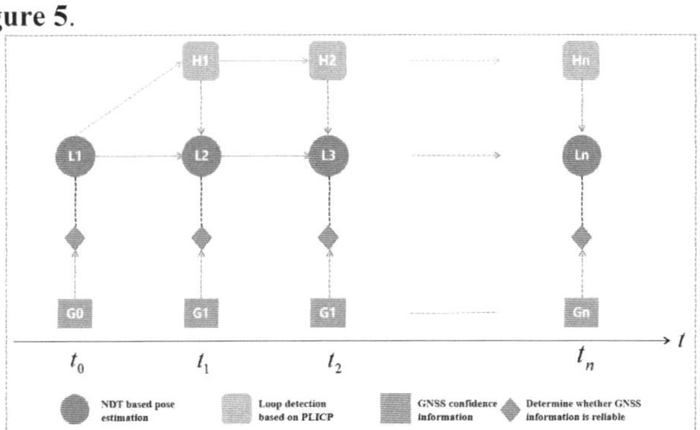

Figure 5. Schematic diagram of factor diagram optimization

3. Multi-sensor fusion positioning

This article is based on the multi-sensor fusion of GNSS, IMU, and LiDAR to achieve high-precision positioning in complex environments. Firstly, an online calibration model is proposed for joint calibration of GNSS and LiDAR, while completing the collection and updating of high-precision point cloud maps; Secondly, absolute positioning is achieved based on GNSS signals, and relative positioning is achieved through multi-feature point cloud matching based on high-precision point cloud maps; Finally, based on the multi-feature positioning information, the volume Kalman filter is used to filter out the positioning noise points, completing the fusion positioning and outputting stable navigation information, as shown in **Figure 6**.

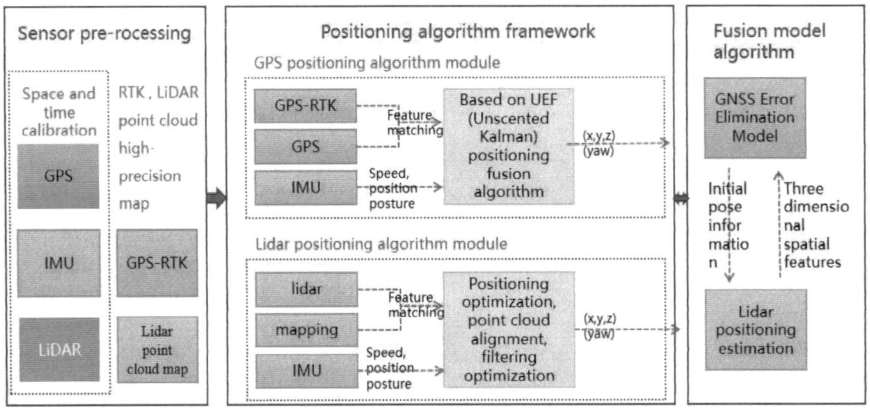

Figure 6. Schematic diagram of fusion positioning architecture

3.1 Joint calibration of sensors based on online models

The calibration of GNSS and LiDAR includes extrinsic calibration (x, y, z) and intrinsic calibration $(roll, pitch, yaw)$. The external parameters are calibrated based on the measurement of the installation position of the onboard equipment. In contrast, the internal parameters are adjusted by comparing the trajectory output from point cloud mapping with the GNSS positioning trajectory. Therefore, this article proposes an online positioning model based on the NDT algorithm, which adaptively adjusts the real-time output pose trajectory from point cloud mapping and the pose estimation output from GNSS positioning until the root mean square error of both is optimized, completing the calibration of the internal parameters.

3.2 Multi feature based point cloud matching algorithm

The positioning algorithms based on point cloud matching algorithms for pose estimation include mainstream algorithms such as the ICP algorithm and NDT algorithm[13-14]. Among them, the NDT algorithm can meet the real-time requirements of vehicle positioning, but it may lead to positioning failure when encountering a single scene with few feature points. Therefore, based on the NDT algorithm, this paper proposes a point cloud matching algorithm NDT based on point cloud strength similarity and standard distribution transformation for multi-feature inter-frame point cloud matching, achieving fast and high-precision LiDAR SLAM pose estimation.

The point cloud intensity feature information represents the degree of correlation between adjacent frame point cloud data. It determines the relationship between adjacent frames by calculating the covariance and correlation coefficient between adjacent frames. The intensity information between adjacent point clouds is as follows:

$$A = [a_1, a_2, ..., a_n] \tag{13}$$

$$B = [b_1, b_2, ..., b_n] \tag{14}$$

Calculate the covariance function based on point cloud intensity information as follows:

$$\text{cov\_}ba = F(dev\_B * dev\_A) \tag{15}$$

Among them, dev_B and dev_A are the deviation matrices, respectively. Covariance can simply reflect the correlation between two sets of statistical samples. If the covariance value is positive, it is a positive correlation; If the value is negative, it is negatively correlated. The larger the absolute value, the stronger the correlation.

By calculating the correlation coefficient to represent the degree of correlation between adjacent frame point clouds, the calculation is as follows:

$$K = \frac{\operatorname{cov}\_ba}{std\_b * std\_a} \tag{16}$$

Among them, cov_ba is the covariance of point cloud intensity information, std_a is the standard deviation of point cloud sample A intensity, std_b is the standard deviation of point cloud sample B intensity, K is the correlation coefficient, and a trend towards 1 indicates positive correlation.

The objective function of the NDT algorithm is:

$$s(p) = \sum_i \exp[\frac{(X_i^{'} - q_i)^T \sum_i^{-1} (X_i^{'} - q_i)}{2}] \tag{17}$$

The matrix transformation of $X_i^{'}$ adjacent frame mapping, q_i is the mean of the current point cloud frame, and the maximum value of $-s(p)$ is calculated according to the Newton optimization algorithm.

Finally, based on the correlation between point cloud strength and the optimal solution of the NDT algorithm, calculate the pose estimation for point cloud matching:

$$H = \alpha * s(p) + \beta * K \tag{18}$$

Among them, α and β are the coefficients of the correlation between NDT algorithm and point cloud strength, respectively.

3.3 Fusion Model Localization

The currently mainstream multi-sensor fusion algorithms include the Kalman filter (EK), the Extended Kalman filter (EKF), and the Unscented Kalman filter (UKF). EK is a standard random linear filter that requires a high level of mathematical modeling. EKF and UKF are random nonlinear filters, and an improved data model based on EK solves second-order nonlinear optimization problems. With the upgrading of the auto drive system and its application in complex scenarios, various sensors share the demand for autopilot tasks. Therefore, due to the fusion of multiple sensors and the impact of noise points generated during fusion on the filter, this paper adopts the Volume Kalman Filter (CKF) for multi-sensor fusion. At the same time, the noise points generated by the observation values of multiple sensors are processed online during the fusion process to ensure that the filter is optimal during fusion.

$$\begin{cases} X_k = F(X_{k-1}) + W_{k-1} \\ Z_k = H(X_k) + V_k \end{cases} \tag{19}$$

Among them, $Z_k = [Z_{1,k}{}^T, Z_{2,k}{}^T, ..., Z_{d,k}{}^T]^T$ is the fused observation value with dimension M, $H(X)$ is the fusion observation equation $F(X)$, which is the fusion prediction equation; $X = [x, y, z, v_x, v_y, v_z, \alpha_x^n, \alpha_y^n, \alpha_z^n, \theta, \gamma, \varphi]^T$ is the target state information, with a dimension of N. W_k is process noise, V_k is measurement noise.

4. Experimental results

As shown in **Figure 7**, the testing and verification platform of this article is the "Tianjin" solar vehicle, which is equipped with sensors such as laser radar, millimeter wave radar, camera, combined inertial navigation, and domain controller. The core autonomous driving technology is provided by the China Automotive Center Engineering Institute, and it has the functions of autonomous cruising, lane merging, overtaking and lane changing, side parking, adaptive cruise control (ACC), obstacle autonomous path planning, intelligent collision avoidance, and automatic parking. At the same time, it has the characteristics of green energy saving, real-time efficiency, safety, and comfort, and has reached the level of L4 autonomous driving.

Journal of Physics: Conference Series **3024** (2025) 012015 doi:10.1088/1742-6596/3024/1/012015

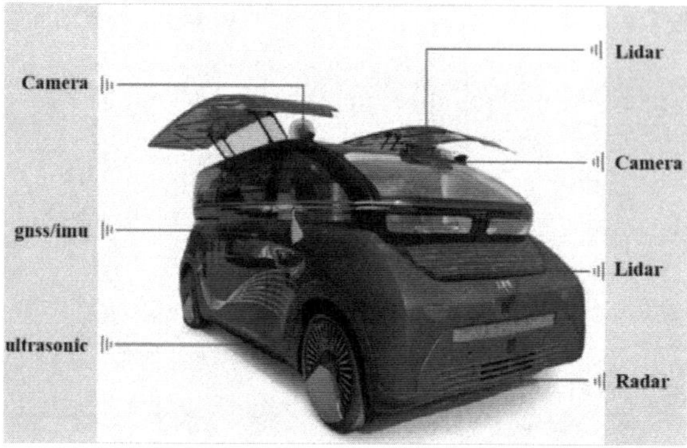

Figure 7. Schematic diagram of solar energy on the "Tianjin"

4.1 Results of mapping experiment

The testing environment is in the Zhongqi Central Park. The surrounding point cloud information is collected by a vehicle-mounted LiDAR (M1), and the NDT algorithm and the multi-feature electric fish matching algorithm proposed in this paper are used to construct high-precision point cloud maps. In terms of consistency of the point cloud map, the point cloud map constructed by the algorithm in this paper (**Figure 9**) is more complete than the consistency in **Figure 8**. Especially in this article, loop detection and GNSS factors were added to correct cumulative errors during the construction of point cloud maps, making pose estimation more accurate during the construction process.

Figure 8. NDT algorithm point cloud map construction

Figure 9. Algorithm based point cloud map construction in this article

In the construction process of point cloud maps, the stationary state of vehicles is determined based on the real-time output of vehicle heading angle transformation. Our algorithm is compared with the NDT algorithm to compare the heading angle changes. Our algorithm has a smaller heading angle change, as shown in **Figure 10**.

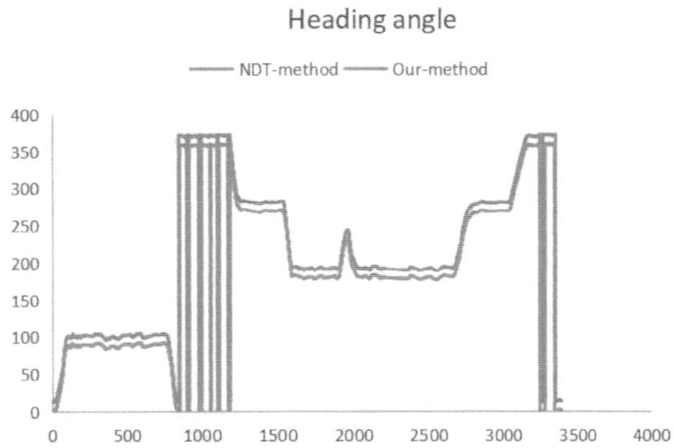

Figure 10. Schematic diagram of heading angle change

Table 1. Comparison of indicators between NDT algorithm and the algorithm proposed in this paper

Method	**NDT-method**	**Our-method**
Accuracy	0.84(dm)	0.57(dm)
Frame odometer	525.7ms	35.2ms
Frame mapping	110.4ms	80.5ms
Mapping storage	536M	15.3M

In Table 1, the algorithm proposed in this paper is compared with the NDT algorithm in terms of positioning accuracy, pose estimation time consumption of the front-end odometer and mapping time consumption. By comparing with GNSS true values, our algorithm has a root mean square error of 5.7cm in trajectory, while the NDT algorithm has a root mean square error of 8.4cm; The time consumption of the NDT algorithm is 525.7ms, and the pose estimation time consumption of the front-end odometer of this algorithm is 35.2ms. The NDT algorithm for point cloud mapping in the backend consumes 110.4ms, while the algorithm in this paper consumes 80.5ms, resulting in a 27% increase in time efficiency; The compression ratio of point cloud maps has increased by 98%.

4.2 Localization experiment results

This article preprocesses the noise in the observed values before multi-sensor fusion. By comparing the observed values, predicted values, and true values through data simulation, it can be seen that when the noise generated by the observed values is very large, the weighted average of the sensor observed values and predicted values are used to reduce the impact of noise on the filter. The average error between the observed values and true values is 0.57, and the average error between the predicted values and true values after data preprocessing is 0.02. This provides good predictive data for multi-sensors based on volume Kalman filtering, as shown in **Figure 11**.

Journal of Physics: Conference Series **3024** (2025) 012015 doi:10.1088/1742-6596/3024/1/012015

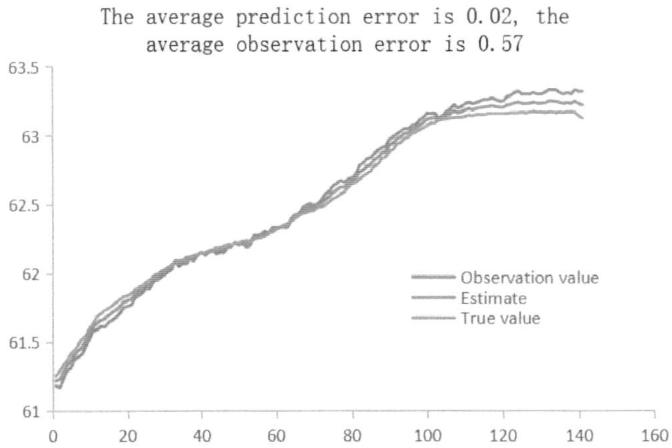

Figure 11. Simulation Fusion Data Processing

The test park is shown in **Figure 12**. Due to environmental and magnetic interference, the GNSS signal is unstable within the park (the yellow trajectory indicates the GNSS signal interference area). Therefore, to address issues such as positioning failure solely based on GNSS signals, the algorithm proposed in this paper constructs a large-scale, high-precision, and lightweight point cloud map for the park, as shown in **Figure 13**.

Figure 12. China Automotive Center Park

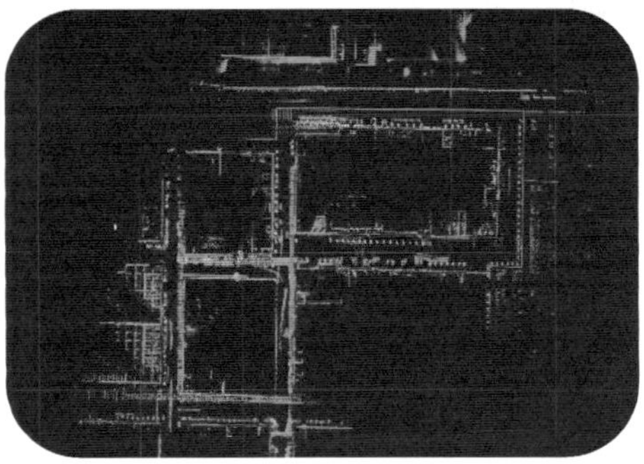

Figure 13. The point cloud map constructed by the algorithm in this article

This article is based on GNSS signals to achieve absolute positioning for autonomous driving, and based on LiDAR point cloud feature matching to achieve relative positioning for autonomous driving. As shown in **Figure 14**, due to the lack of GNSS signals in the local environment, absolute positioning fails. Therefore, this paper adopts a multi-sensor fusion positioning method. As shown in **Figure 15**, due to unstable positioning caused by noise points, high-precision positioning in complex environments has been achieved through the algorithm proposed in this paper, with an accuracy of less than 5cm.

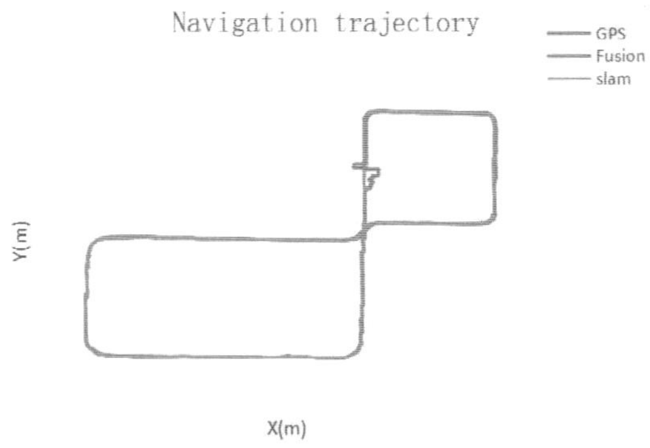

Figure 14. Positioning and Navigation Trajectory

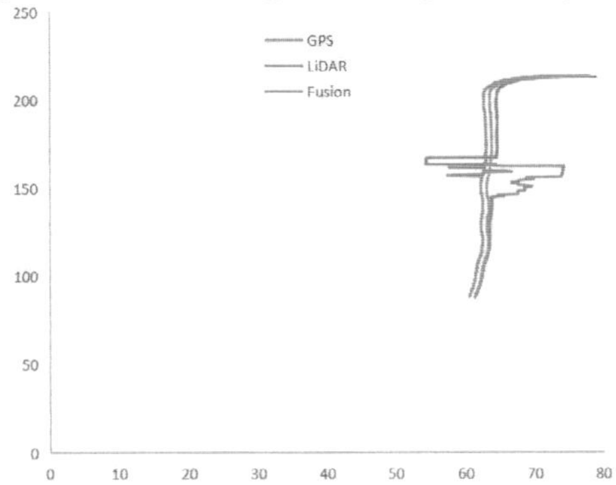

Figure 15. Local fusion trajectory

5. Summary

This article is based on the "Tianjin" solar-powered vehicle, which achieves multi-sensor fusion positioning in complex environments. Firstly, by studying high-precision point cloud maps, a loop detection based on the PLICP algorithm and graph optimization with GNSS factor were proposed, achieving high-precision and lightweight point cloud maps in large-scale scenarios. The inter-frame odometer only takes 35.2ms, and the backend mapping consumes 80.5ms, resulting in a 27% increase in time rate; The compression ratio of point cloud maps has increased by 98%, meeting the real-time requirements of autonomous driving positioning and the storage capacity of vehicle side maps. A method based on volume Kalman filter was studied in multi-sensor fusion positioning. This paper proposes a multi-feature point cloud matching algorithm to achieve high-precision positioning in a single scene. At the same time, it proposes to reduce the influence of noise points on the filter through weighted averaging of sensor observation values and predicted values in fusion, thereby completing multi-sensor positioning fusion. The positioning is less than 5cm, which meets the positioning requirements of autonomous driving.

Acknowledgments

The author(s) disclosed receipt of the following financial support for the research, authorship, and/or publication of this article: This work was financially supported by " High precision map and fusion positioning demonstration verification(2021YFB2501105)" and Research on the Integration Perception and Risk Warning Method of Intelligent Connected Vehicles under the Integration of Vehicle, Road and Cloud(24JCZXJC00490).

References

[1] Qiu D, May S, Nüchter A. GPU- accelerated nearest neighbor search for 3D registration[C]//Proceedings of International Conference on Computer Vision Systems. Berlin, Heidelberg: Springer, 2009:194-203.

Journal of Physics: Conference Series **3024** (2025) 012015 doi:10.1088/1742-6596/3024/1/012015

[2] Neumann D, Lugauer F, Bauer S, et al. Real-time RGB-D mapping and 3-D modeling on the GPU using the random ball cover data structure[C]//Proceedings of IEEE International Conference on Computer Vision Workshops, 2011: 1161-1167.
[3] Biber P, Strasser W. The normal distributions transform: a new approach to laser scan matching[C]//Proceedings of 2003 IEEE/RSJ International Conference on Intelligent Robots and Systems, 2003: 2743-2748.
[4] Magnusson M, Lilienthal A, Duckett T. Scan registration for autonomous mining vehicles using 3D-NDT: Research articles[M].[S.l.]: John Wiley and Sons Ltd, 2007:803-827.
[5] Zhang J, Singh S. LOAM: Lidar odometry and mapping in real- time[C]//Proceedings of Robotics: Science & Systems Conference, 2014.
[6] Zhang J, Singh S. Low-drift and real-time lidar odometry and mapping[J]. Autonomous Robots, 2017, 41(2):401-416.
[7] Dubé R, Dugas D, Stumm E, et al. SegMatch: Segment based place recognition in 3D point clouds[C]//Proceedings of 2017 IEEE International Conference on Robotics and Automation, 2017: 5266-5272.
[8] Shan T, Englot B. LeGO-LOAM: Light-weight and ground optimized lidar odometry and mapping on variable terrain[C]//Proceedings of IEEE/RSJ International Conference on Intelligent Robots and Systems, 2018: 4758-4765.
[9] Besl P J, Mckay N D. A Method for registration of 3-D shapes[J]. Proceedings of SPIE, 1992, 14(3):239-256.
[10] Rusinkiewicz S. Efficient variants of the ICP algorithm[C]//Proceedings of the Third International Conference on 3-D Digital Imaging and Modeling, 2001:145-152.
[11] Yang C, Medioni G. Object modelling by registration of multiple range images[J]. Image and Vision Computing, 1992, 10(3):145-155.
[12] Segal A, Haehnel D, Thrun S. Generalized- ICP[J]. Robotics: Science and Systems, 2009,2(4).
[13] Wang Qingshan et al. Point Cloud Registration Algorithm Based on Combination of NDT and ICP[J]. Computer Engineering and Applications, 2020,56(07):88-95.
[14] Wang Qingshan et al. High-Precision and Fast LiDAR Odometry and Mapping Algorithm[J]. jaciii, 2022, 26(2) : 206-216.

Journal of Physics: Conference Series **3024** (2025) 012016 doi:10.1088/1742-6596/3024/1/012016

ORB Feature Uniform Distribution Algorithm Integrating Quadtree and Adaptive Thresholding

Weiqing Wang[1,2], Zhonggang Xiong[1,2], Zhong Liu[1,2*]

[1] Guangxi Key Laboratory of Special Engineering Equipment and Control, Guilin University of Aerospace Technology, Guilin, Guangxi Province, 541004, P.R. China

[2] Key Laboratory of Special Engineering Equipment Design and Intelligent Driving Technology (Guilin University of Aerospace Technology), Education Department of Guangxi Zhuang Autonomous Region, Guilin University of Aerospace Technology, Guilin, Guangxi Province, 541004, P.R. China

[*]Corresponding author's e-mail: liuzhong678@163.com

Abstract. To enhance the uniform distribution characteristic of ORB extraction and leverage the global information of the image for improved positioning accuracy in visual SLAM, a novel ORB feature point extraction method combining quadtree and adaptive threshold was introduced. This method involves building a multi-resolution image pyramid and dividing the images into 30 pixels × 30 pixels grids to determine the number of feature points needed for each layer. The adaptive threshold for detecting feature points is derived from the image's gray mean and variance. Feature points and their descriptors are then extracted using this adaptive threshold, followed by refinement through quadtree management and non-maximum suppression. Experimental data proves that the proposed solution enhances both the balanced distribution of ORB and the reliability of their tracking.

1. Introduction

Visual simultaneous localization and mapping (VSLAM) is a key technology of robot navigation, which perceives the surrounding environment through the camera and completes its positioning and mapping[1] [2]. In feature-based VSLAM, the front end estimates the robot's motion through the camera, and its accuracy depends on the uniformity of feature point extraction. Because the feature points with high uniformity can provide better overall coverage[3]. Therefore, improving the uniformity of feature points has attracted the attention of researchers.

Although ORB feature points are widely used in VSLAM, the extracted feature points are easy to aggregate and unevenly distributed, which leads to low positioning accuracy due to the use of only local information of the image in feature point tracking[4]. Therefore, Mur-Artal[5] introduced the quadtree technique to manage the feature points and improve their uniformity. Although Mur-Artal's method can improve the uniformity of ORB feature point extraction, there are some problems, such as fixed meshing, quadtree division depth, and setting gray thresholds. To solve the problem of fixed meshing, Yang[6] et al. proposed a Qtree_ORB improvement algorithm for adaptive meshing, which uses adaptive meshing images based on Qtree_ORB to limit the division depth of the quadtree and retain the feature points. At the same time, some researchers consider the influence of ambient light brightness on the extraction of feature points. To improve the threshold of fixed grayscale, Li[7] et al.

proposed to dynamically adjust the adaptive gray threshold. Zong[8] et al. used a coefficient to adjust the gray threshold of the extracted feature points. All of the above methods have achieved some results, but some areas can be improved, such as meshing and the adjustability of gray thresholds.

Inspired by this, to improve the uniformity of ORB extraction, and then use the global information of the image to improve the positioning accuracy of visual SLAM, this paper proposes an ORB feature point extraction method that combines quadtree and adaptive threshold. In this method, the construction of multi-resolution image pyramids, adaptive threshold extraction of feature points, and quadtree management are adopted.

More concretely, the main contribution of the proposed method is that the ORB feature point extraction method combines quadtree and adaptive threshold, and the uniformity of ORB feature points and the robustness to light and shade are improved.

2. Method

In order to improve the uniformity of ORB feature point extraction, and then use the global information of the image to improve the positioning accuracy of visual SLAM, an ORB feature point extraction method combining quadtree and adaptive threshold is proposed. In this method, a multi-resolution image pyramid is constructed, and the images are meshed by using 30 pixels × 30 pixels to calculate the number of feature points required to extract each layer of images. When the required number of feature points is not met, the adaptive threshold is established by analyzing the image's gray mean and variance, the feature points and their descriptors are extracted by using the adaptive threshold, and the feature points are screened by quadtree management and non-maximum suppression. Finally, RANSAC was used to eliminate mismatches. The following describes the extraction of adaptive threshold ORB feature points, quadtree management feature points, and the proposed method.

2.1. ORB feature point extraction using adaptive thresholding

In the extraction of adaptive threshold ORB feature points, the oFAST feature points are detected according to the set threshold on the grid image. When the number of feature points is not satisfied, the adaptive threshold of the detected feature points is obtained through the gray mean and variance of the image, ORB and its descriptors are extracted by the adaptive threshold, and the feature points are filtered by quadtree management and non-maximum suppression.

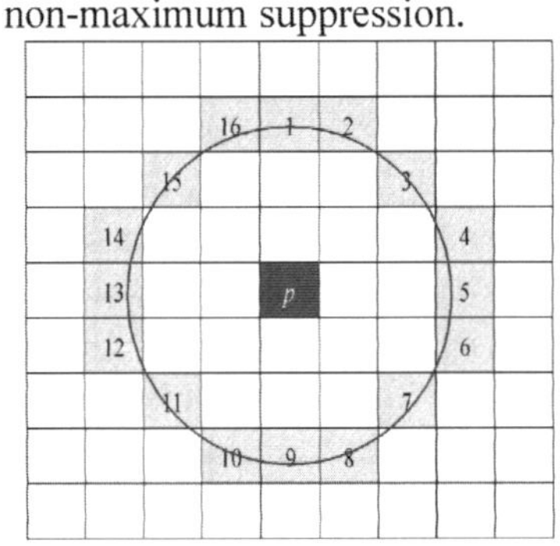

Figure 1. Schematic diagram of the neighborhood arc with a radius of 3 pixels at the center point

The algorithm of the FAST feature defines a corner point: if there are enough pixels in the target pixel neighborhood that are not similar to it, the pixel is considered to be a corner. The FAST algorithm typically selects a discretized Bresenham circle with a radius of 3 as the area of study for the target pixel. A total of 16 pixels pass through a circle with a radius of 3 pixels, centered on the candidate pixel p, and numbered 1 to 16 in a clockwise direction, as shown in Figure 1. If the gray value of the ith pixel is greater than the threshold σ, it will be denoted as 1. Otherwise, it is denoted as 0.

$$f(I_x, I_p) = \begin{cases} 1, & |I_p - I_i| \geq \sigma \\ 0, & |I_p - I_i| < \sigma \end{cases} \tag{1}$$

where I_P is the grayscale value of the center point p, I_x represents the gray value of the midpoint numbered from 1 to 16 around the center point p.

If the absolute value of the gray difference between N consecutive pixels and the center pixel p is greater than or equal to σ among the 16 pixels, the detection point is determined to be a feature point.

$$N = \sum_{x \in circle(P)} f(I_x, I_p) \tag{2}$$

N is usually 9 or 12. Non-maximum suppression can eliminate pseudo-corners within the neighbourhood of the corners, so define a score function to calculate the score value of each detected corner, and the calculation formula is as follows:

$$S_{\text{FAST}} = Max[\sum_{x \in S_{brighter}} |I_{p \to x} - I_p| - t, \sum_{x \in S_{darker}} |I_p - I_{p \to x}| - t] \tag{3}$$

where

$$\begin{cases} S_{brighter} = \{x \mid I_{p \to x} \geq I_p + t\} \\ S_{darker} == \{x \mid I_{p \to x} \leq I_p - t\} \end{cases} \tag{4}$$

After non-maximum suppression filtering, the maximum score obtained on the local window is the final corner.

In the proposed method, the adaptive grayscale threshold t used is calculated as follows:

$$\begin{cases} \mu = \frac{1}{n} \sum_{i}^{n} I(x_i) \\ t = \frac{\frac{1}{n} \sum_{i}^{n} (I(x_i) - \mu)^2}{\mu} \end{cases} \tag{5}$$

Where, $I(x_i)$ is the grayscale value of the pixel. When detecting feature points, the initial gray threshold is 20. If the number of detected feature points is not satisfied, the gray threshold is reduced to $t/4$ in the Equation (5).

2.2. Feature point organization using quadtree management

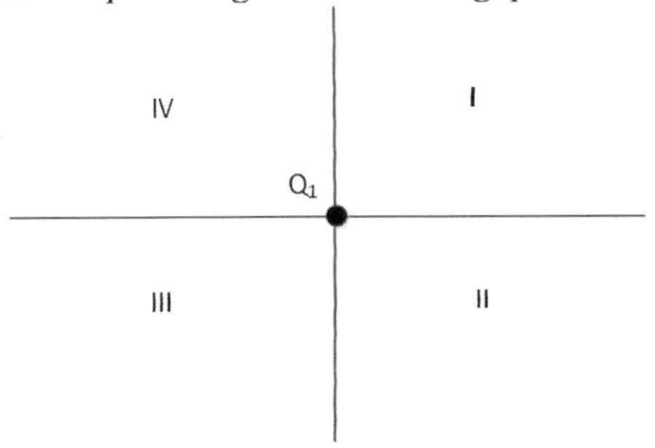

(a)Schematic diagram of the quadtree region

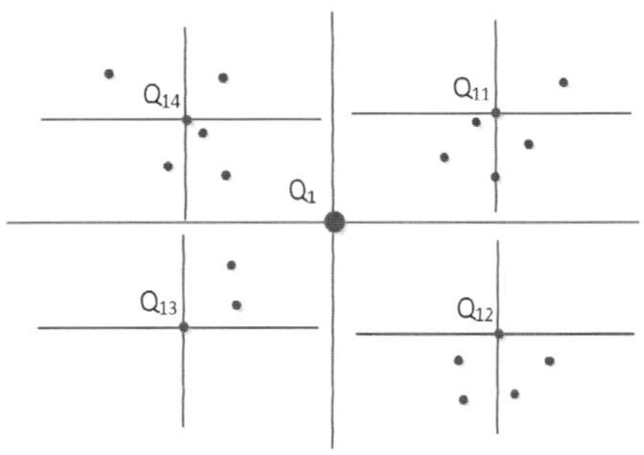

(b)Schematic diagram of the quadtree subregion

Figure 2. Schematic diagram of the process of dividing the region of the quadtree

The quadtree decomposition method is widely used in image processing, two-dimensional spatial data indexing, and sparse data storage. In the quadtree decomposition method, there are at most four sub-nodes under each node, and the two-dimensional spatialization is often divided into four regions, and the data information in the four regions is stored in the nodes at the same time. As shown in Figure 2(a), the area where node Q_1 is located is divided into four sub-regions, starting from the upper right region, which are numbered I, II, III, and IV clockwise. As shown in Figure 2(a), the quadtree starts from node Q_1 and stores the data information of the area where node Q_1 is located to the node. Node Q_1 can be divided into four sub-nodes, namely Q_{11}, Q_{12}, Q_{13}, and Q_{14}, as shown in Figure 2(b). Similarly, the data information of the area where the four child nodes are located is stored in these four child nodes.

2.3. *An ORB feature point extraction method based on quadtree and adaptive threshold*

To improve the uniformity of ORB feature point extraction, and then use the global information of the image to improve the positioning accuracy of visual SLAM, an ORB feature point extraction method integrating quadtree and adaptive threshold was proposed. The structure of the method is shown in Figure 3. In the first step of meshing the multi-resolution image pyramid, a multi-resolution image pyramid is constructed, each layer image is meshed by meshing, and the number of feature points required to extract each layer of the image is calculated. In the second step of the extraction of adaptive threshold ORB feature points, the oFAST feature points are detected on the grid image according to the set threshold. When the number of feature points is not satisfied, the adaptive threshold of the detected feature points is obtained through the gray mean and variance of the image, ORB and its descriptors are extracted by the adaptive threshold, and then the feature points are filtered by quadtree management and non-maximum suppression. In the third step, the initial feature point matching dataset is obtained by brute force matching, and the RANSAC is used to eliminate the mismatch.

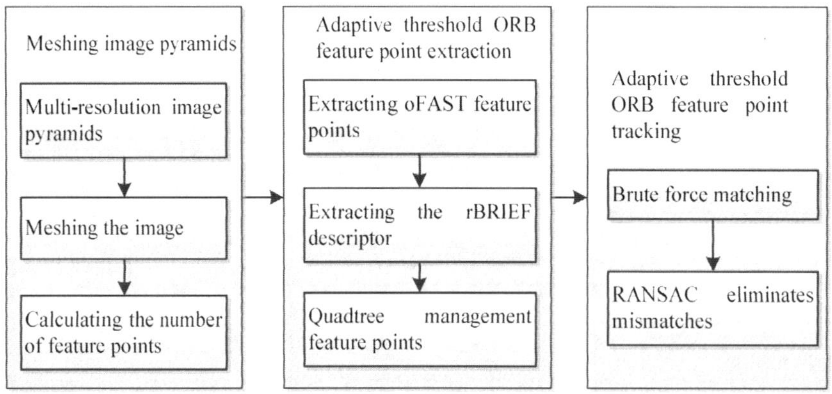

Figure 3. Schematic diagram of ORB feature point extraction method fusing quadtree and adaptive threshold

3. Experimental results

To measure the capability of the proposed ORB feature point extraction method combining quadtree and adaptive threshold, it was compared with the ORB, Qtree_ORB, and FDQtree_ORB[9] methods. In the experiment, the proposed method of fusing quadtree and adaptive threshold ORB feature point extraction is called QAT_ORB. The evaluation methods of the experiment include the uniformity of image feature point distribution, and the algorithm running time. The evaluation method of the uniformity of the distribution of image feature points divides the image into horizontal direction, vertical direction, upper left to lower right direction, lower left to upper right direction, and center and surrounding directions. After the image is segmented in the above directions, the number of feature points in these ten areas is counted, and then a description vector is generated

$[n_1, n_2, n_3, n_4, n_5, n_6, n_7, n_8, n_9, n_{10}]^T$. The standard deviation of the vector is defined as the uniformity S of the distribution of feature points on the image, and it is calculated as follows:

$$\begin{cases} \mu = \dfrac{1}{10}\sum_{i=1}^{10} n_i \\ S = \sqrt{\dfrac{1}{10}\sum_{i=1}^{10}(n_i - \mu)^2} \end{cases} \tag{6}$$

Where n_i is the number of feature points in each area, and μ is the average of the number of these ten feature points.

The test platform is a laptop computer with Intel(R) Core(TM) i5-7200U CPU @ 2.5 GHz, 16GB RAM, 64-bit Ubuntu 16.04 operating system, OpenCV 4.5.5. Five pairs of images were used in the experiment, which was in the open dataset, as shown in Figure 4 below. In Figure 4, Bikes, War, Venice, Port, and Ceiling are affected by blurring, angle and scale changes, and light and shade shifts, respectively.

(a)Bikes (b)War (c)Venice (d)Port (e)Ceiling

Figure 4. Images from the experiment dataset

Some of the parameters in the test are set as follows:

(1) Extract 1000 ORB feature points on the tested image.

(2) For the initial matching of feature points, the distance ratio between the nearest neighbor and the second nearest neighbor is less than or equal to 0.7.

(3) The maximum number of iterations of RANSAC is 2000, the confidence level is 0.995, and the threshold for judging the inner point is 3 pixels.

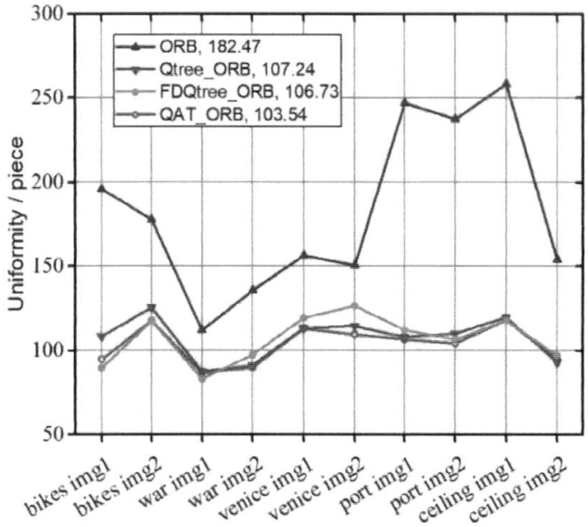

Figure 5. The performance of the four methods to extract the uniformity of the distribution of feature points

Figure 5 shows the uniformity of the distribution of ORB, Qtree_ORB, FDQtree_ORB, and QAT_ORB extracted feature points on 10 images. As can be seen from Figure 3.2, the average distribution uniformity of ORB, Qtree_ORB, FDQtree_ORB, and QAT_ORB extracted feature points was 182.47, 107.24, 106.73, and 103.54, respectively. It can be seen that the average distribution uniformity of the feature points extracted by the ORB method is the largest, indicating that the feature points extracted by the ORB method are very easy to gather. There is little difference between the Qtree_ORB, and FDQtree_ORB methods in the distribution uniformity of the extracted feature points, which indicates that the adaptive threshold in the FDQtree_ORB is not significantly improved compared with the threshold set in the Qtree_ORB. The average distribution uniformity of the extracted feature points is the smallest in QAT_ORB.

Table 1. The average consumption time of the four methods to extract 1000 feature points

Method	ORB	Qtree_ORB	FDQtree_ORB	QAT_ORB
time/ms	47.36	49.98	49.44	49.32

Table 1 shows the average time spent by ORB, Qtree_ORB, FDQtree_ORB, and QAT_ORB to extract 1000 feature points and their descriptors in the above dataset. As can be seen from Table 3.1, the average consumption time of ORB, Qtree_ORB, FDQtree_ORB, and QAT_ORB to extract 1000 feature points and their descriptors is 47.36 ms, 49.98 ms, 49.44 ms, and 49.32 ms, respectively. The ORB method takes the least average time to extract 1000 feature points and their descriptors. Due to the addition of meshing images, adjusting gray thresholds, and managing feature points in the quadtree tree in the feature point extraction, the average consumption time of Qtree_ORB, FDQtree_ORB, and QAT_ORB extraction of 1000 feature points and their descriptors is about 2ms more than that of the ORB method, which is about 5.53%. The average consumption time of FDQtree_ORB, and QAT_ORB extraction feature points and their descriptors are shorter than that consumed by Qtree_ORB, indicating that the adaptive threshold has certain advantages in real-time. QAT_ORB of the three improvement methods, the average time spent extracting feature points and their descriptors is the least. However, in practice, the consumption time of Qtree_ORB, FDQtree_ORB, and QAT_ORB extracting feature points and their descriptors varies with the different detection images, but there is little difference between them.

4. Conclusion

To solve the problem that visual positioning accuracy is affected by the uniformity of feature points, this paper proposes an ORB feature point extraction method that fuses quadtree and adaptive threshold. In this method, a multi-resolution image pyramid is constructed, the image grid is divided, the number of feature points required to be extracted from each layer of the image is calculated, and the feature points are detected by partitioning. When the number of feature points is insufficient, the adaptive threshold of the detected feature points is obtained through the grayscale information of the image, to increase the number of extracted features. On this basis, the quadtree is used to manage the feature points. The experimental results show that the ORB feature point extraction method by fusing quadtree and adaptive threshold can improve the uniformity of ORB feature point extraction, and then improve the accuracy of feature point tracking by using the global information of the image.

Acknowledgments

This work was funded by the Guangxi Key Laboratory of Special Engineering Equipment and Control, Guilin University of Aerospace Technology (SEEC24ZR06, 2024JJA160098). This research was supported by the Guangxi Natural Science Foundation under Grant No. (2025GXNSFAA069293, 2025GXNSFAA069469).

References

[1] Xu K, Hao Y, Yuan S, et al. AirSLAM: An Efficient and Illumination-Robust Point-Line Visual SLAM System[J]. IEEE Transactions on Robotics, 2025, (3):1-20.
[2] Adkins A, Chen T, Biswas J. ObVi-SLAM: Long-Term Object-Visual SLAM[J]. IEEE Robotics and automation letters, 2024, 9(3):2909-2916.
[3] Mianshu C, Lulu Y, Xiaoni L, et al. Loop detection based on uniform ORB[J]. Journal of Jilin University (Engineering and Technology Edition), 2023, 53(9):2666-2675.
[4] Xin X, Qiang T, Hongbin Z. Local Dynamic Threshold ORB Algorithm Based on Feature Enhancement[J]. Science Technology and Engineering, 2024, 24(33):14339-14346.
[5] Mur-Artal R, Montiel J. Orb-slam2: An open-source slam system for monocular, stereo, and rgb-d cameras[J]. IEEE transactions on robotics, 2017, 33(5):1255-1262.
[6] Shiqiang Y, Lele B, Cheng Z, et al. Improved Qtree_ORBAlgorithm[J]. Computer Engineering and Applications, 2021, 57(20):82-89.
[7] Guojun L, Yanhai X, Jiewen D, et al. ORB-SLAM method based on local adaptive threshold extraction feature points[J]. 2021, (9):32-37.
[8] Zehua Z, Haijun Z, Jinfeng Z. ORB feature extraction and tracking based on adaptive threshold[J]. 2022, 29(3):110-114.
[9] Yao J, Zhang P, Wang Y, et al. ORB feature uniform distribution algorithm based on improved quadtree[J]. Computer Engineering and Design, 2020, 41(6):1629-1634.

Journal of Physics: Conference Series **3024** (2025) 012017 doi:10.1088/1742-6596/3024/1/012017

Structure Design and Performance Analysis of Linear Actuator for Active Suspension

Chao Yang[1*], Shudi Jia[1], Mengfei Li[2], Wenju Feng[2]

[1] School of Automobile and Transportation, Henan Polytechnic, Henan Province, Zhengzhou, 450046, China;

[2] School of Intelligent Mechatronics Engineering, Zhongyuan University of Technology, Henan Zhengzhou 450007, China)

[*]Corresponding author's e-mail:yclovejsd@126.com

Abstract. In view of the high energy consumption and low power density of the pneumatic-hydraulic active suspension actuator,a type of permanent magnet linear actuator with high electromagnetic thrust and low ripple is designed.An analytical analysis is conducted on the radial magnetic field of air gap in actuator by using separated variable method based on the scalar quantity magnet in cylindrical coordinate,and its finite element model built verified.The characteristics of actuator are studied in terms of electromagnetic thrust, its fluctuation and its peak time. The results show that electromagnetic thrust has a close to linear relationship with input current, meanwhile, its fluctuation increases with the rise in electromagnetic thrust.

1. Introduction

Currently, active suspension systems have emerged as a pivotal new direction in suspension technology development. Current research in this domain predominantly focuses on suspension system modeling and control algorithm optimization[1-2]. Among the core components of vehicle active suspension systems, the actuator plays a critical role, significantly influencing not only the suspension system but also overall vehicle performance. Consequently, in-depth research into actuator design and performance is indispensable. Extensive literature reviews reveal that active suspension actuators for vehicles can be broadly categorized into several types, including pneumatic-hydraulic active suspension systems, rotary-to-linear active suspension systems, magnetic levitation active suspension systems, and electromagnetic linear active suspension systems[3].

To address the limitations of existing active suspension systems, reference[4] introduced a two-phase axially magnetized cylindrical permanent magnet linear DC actuator. Building on this, reference[5] developed a three-phase permanent magnet synchronous linear actuator featuring external magnetic poles and a slotless structure, which demonstrated excellent performance in experimental tests. Additionally, references[6] and [7] designed a new induction actuator based on electromagnetic induction principles. Compared to permanent magnet actuators, this design boasts a simpler structure and lower manufacturing costs. Furthermore, reference[8] proposed a reluctance electromagnetic linear actuator inspired by switched reluctance motor principles, characterized by its compact design and high reliability.

Despite these advancements, current active suspension systems still face challenges such as high energy consumption, low power density, and insufficient research into electromagnetic force fluctuations. In this study, a novel fractional-slot permanent magnet linear actuator is proposed, aiming to achieve high electromagnetic force with minimal fluctuations. The electromagnetic force and its fluctuation characteristics under varying excitation and currents are thoroughly analyzed, with the actuator performance evaluated.

2. The actuator structure

Figure 1 shows a schematic diagram of the installation structure of the designed actuator. In the active suspension system, the stator of the actuator is connected to the unsprung mass, which includes the wheel shaft, hub, motor, and tire. The rotor is connected to the sprung mass. When current is applied, the stator and rotor undergo relative motion, generating electromagnetic damping force.

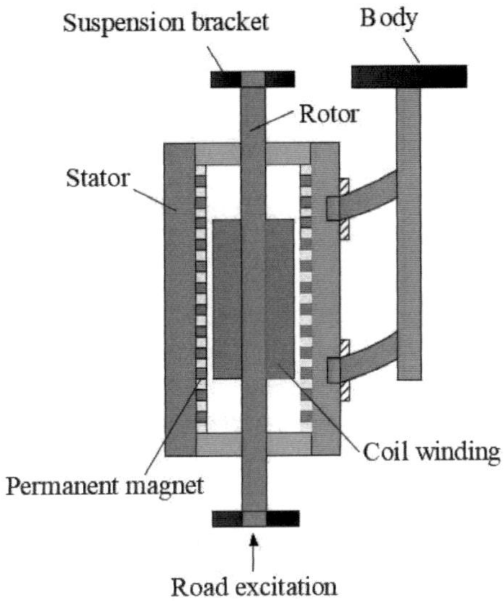

Figure 1. The schematic diagram of the installation structure of the designed actuator

To comply with the actuator's travel requirements and facilitate its installation in the test vehicle, external dimensions of the actuator are constrained as follows:

Stator outer diameter of the cylindrical actuator must not exceed 90 mm. The ratio of rotor outer diameter to stator outer diameter is approximately 0.5 considering the slot shape design and winding installation.The stator length must not exceed 200mm. The air gap length is set to 1 mm. A flat-bottom slot design is selected.

To avoid unbalanced radial magnetic tension and minimize thrust fluctuations, This paper recommends a fractional-slot actuator structure with 12 grooves and 14 poles. The specific dimensions are provided in Table 1. 3D model of actuator is illustrated in Figure 2

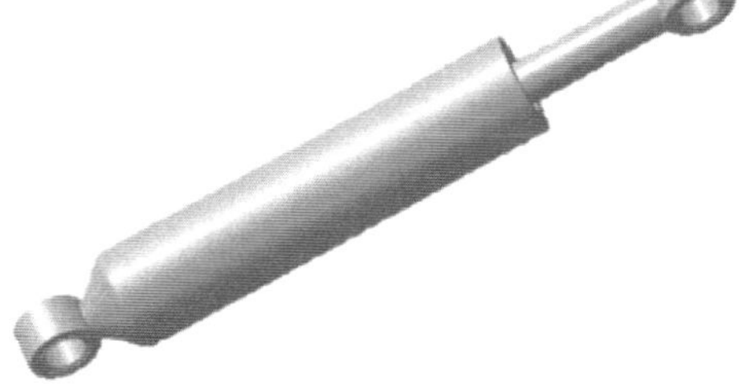

Figure 2. 3D model of actuator

Table 1. Structural parameters of actuator

Parameter	Numerical value	Parameter	Numerical value
Slot height	14/mm	Winding coil turns	100
Slot pitch	15/mm	Stator outside diameter	84/mm
Polar distance	13/mm	Rotor outer diameter	40/mm
Pole embrace	1	Stator length	182/mm
Slot width	12/mm	Radius of rotor axis	16/mm
Thickness of permanent magnet	4/mm	Air gap length	1/mm
Slot opening width	3/mm	Stator yoke height	7/mm

The actuator rotor and stator are constructed using DW465-50 cold-rolled silicon steel sheets. Permanent magnet is made of NdFe30, with a coercive force of 838kA/m and a residual magnetic flux density of 1.1T. Based on the structural parameters provided in Table 1, due to the cylindrical structure of the actuator, model exhibits rotational symmetry about the Z-axis, finite element model of the actuator is established using Ansoft Maxwell electromagnetic simulation software, as illustrated in Figure 3. The material properties of each component are defined as follows: Winding is made of copper wire with a diameter of 1 mm.Air gap space is set to vacuum.Model area is divided, and a balloon boundary condition is applied to the outer region.

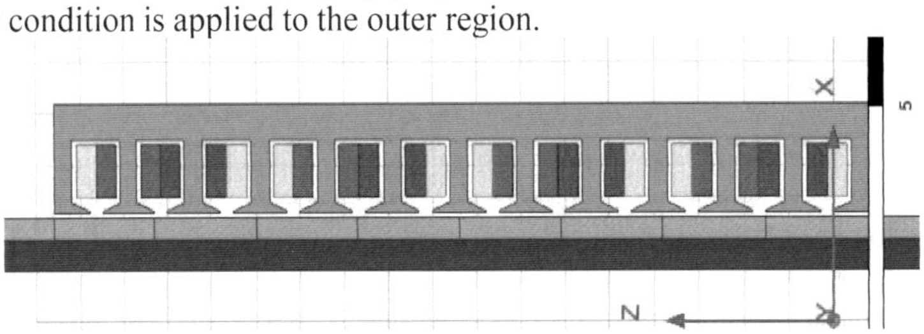

Figure 3. Finite element model of actuator

3. Actuator Magnetic Field Analytical Calculation

Based on the scalar magnetic potential separation method, magnetic field distribution in the air gap of the actuator is calculated.The structure of a 12-slot 14-pole radial magnetized cylindrical permanent magnet actuator is shown in Figure 4. Here, g is the air gap length, R_s is the outer radius of the coil,Rm is the outer radius of the magnetic pole, R_r is the shaft radius, h_w is the coil thickness, h_m is the thickness of permanent magnet, τ_m is the length of the permanent magnet, and b_0 is the slot opening width.

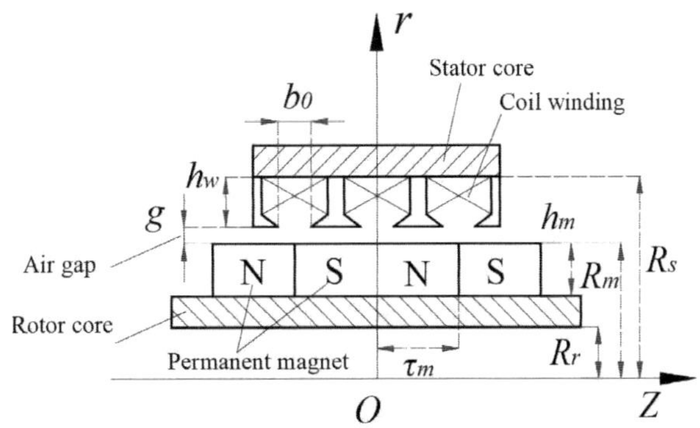

Figure 4. Schematic diagram of the magnetic field of a 12-Slot 14-Pole actuator

Assuming the permanent magnet surface is an equipotential magnetic field, setting the stator outer boundary magnetic potential is zero, the magnetomotive force F(z)distribution for a pair of magnetic poles is given by:

$$F(z)=\begin{cases}F_0, & -r_m \le z \le 0\\ -F_0, & 0 \le z \le r_m\end{cases} \tag{1}$$

Where z is the axial length. Fourier expansion is made for F(z) is given by:

$$F(z)=\sum_{n=1}^{\infty} F_m \sin(mz) \tag{2}$$

$$F_{\rm m}=\frac{4F_0 \sin(\frac{m\tau_m}{2})}{m\tau_p} \tag{3}$$

Where, τ_p is the pole pitch.The scalar magnetic potential in the air gap satisfies Laplace's equation:

$$\frac{\partial^2\varphi(r,z)}{\partial^2{}_r}+\frac{1}{r}\times\frac{\partial\varphi(r,z)}{\partial_r}+\frac{\partial\varphi(r,z)}{\partial^2{}_r}=0 \tag{4}$$

The boundary conditions for the solution are:

$$\begin{cases}r=R_{\rm m}, & \varphi(R_{\rm m}, z)=F_{\rm m}\\ r=R_{\rm s}, & \varphi(R_{\rm s}, z)=0\end{cases} \tag{5}$$

Using the separation of variables method, the solution is:

$$\varphi(r, z)=\frac{F_{\rm m}}{\gamma_1}\gamma_2(r)\sin(mz) \tag{6}$$

$$\gamma_1=I_0(mR_s)K_0(mR_s)-I_0(mR_m)K_0(mR_s) \tag{7}$$

$$\gamma_2=I_0(mR_s)K_0(mr)-I_0(mr)K_0(mR_s) \tag{8}$$

I0( ) and K0( )are the modified Bessel functions of the first and second kind of order zero, respectively. The radial magnetic flux density B(r,z)in the air gap can be calculated as:

$$B_r(r,z)=\sum_{n=1}^{\infty}\frac{\mu_0 mF_n}{\gamma_1}\gamma_3(r)\sin(mz) \tag{9}$$

$$\gamma_3(r)=I_0(mR_s)K_1(mr)+I_1(mr)K_0(mR_s) \tag{10}$$

I_1( )and K_1( )are the modified Bessel functions of the first and second kind of order one, respectively; μ_0 is the permeability of free space, μ_0 =4π×10-7N/A^2 .

The results of theoretical calculation and finite element simulation are as follows figure 5:

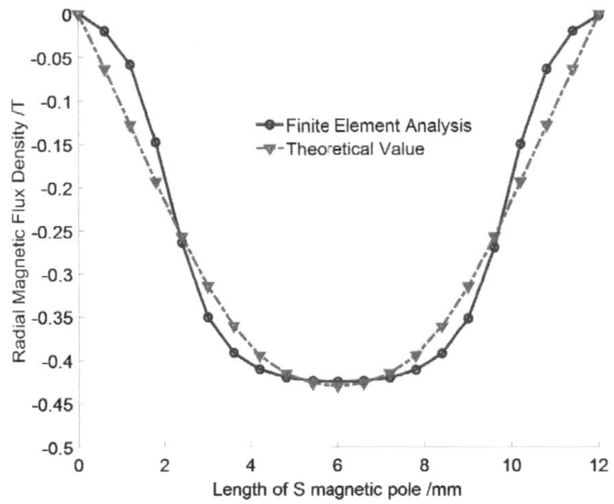

Figure 5. The radial magnetic flux density of S-pole

In figure 5, it can be seen that The results of theoretical calculation and finite element simulation are very similar, which verified the correctness of the finite element model constructed.

4. Performance Analysis of the Actuator

4.1. Analysis of Induced Electromotive Force(EMF)

The relative motion between the actuator's rotor and stator is determined by the velocity of the road excitation. Therefore, the actuator operates in a stochastic motion state. In simulation and experimental studies, this excitation is typically characterized as a sinusoidal signal with varying amplitude and frequency. Assuming the road excitation directly acts on the actuator, the displacement excitation can be expressed as：

$$z(t)=z_0+A_{\max}\sin\omega t \tag{11}$$

$$\omega=2\pi f \tag{12}$$

$$v(t)=2A_{\max}\pi f\cos 2\pi ft \tag{13}$$

The induced EMF of the actuator is related to the amplitude and frequency of the road excitation. In this study, sinusoidal excitations with an amplitude of 0.012 m (one pole pitch) and frequencies of 2, 4, 6, 8, and 10 Hz were employed as road excitations. Five different frequency excitations were applied sequentially to the actuator, each for one motion cycle. The analysis of the induced EMF is as follows in figure 6.

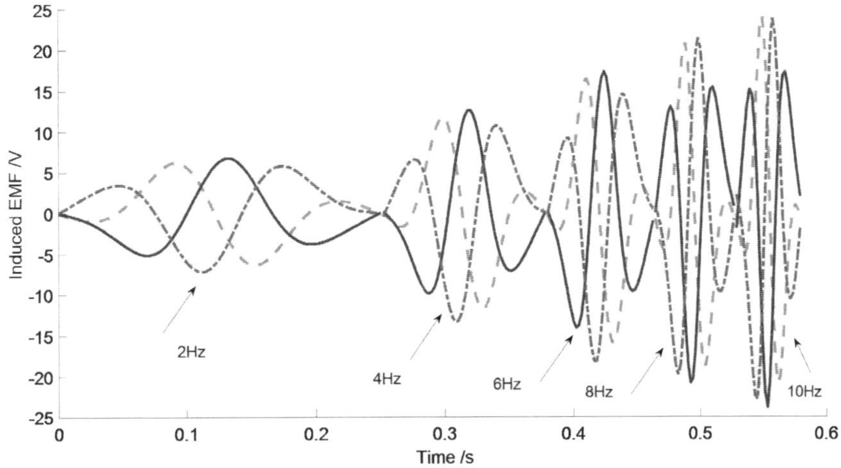

Figure 6. Three-phase Induced EMF under various frequency excitations

As illustrated in Figure 6, the amplitude of the induced EMF progressively increases with the rise in road excitation frequency, while the cycle period shortens at higher frequencies. This phenomenon can be primarily attributed to the enhanced operating speed of the actuator resulting from increased excitation frequency, which amplifies the amplitude of the three-phase induced EMF. Simultaneously, under the same amplitude condition (one pole pitch), the higher the actuator's speed, the shorter the cycle period becomes due to the compressed temporal duration required to complete motion within the fixed displacement range.The induced EMF frequency, amplitude and Induced EMF under various frequencies as illustrated in Table 2.

Table 2. The induced EMF frequency, amplitude and Induced EMF under various frequencies

Frequency(Hz)	Amplitude(m)	Induced EMF(V)
2	0.012	7.09
4	0.012	13.27
6	0.012	18.53
8	0.012	21.36
10	0.012	23.97

4.2. Variation Characteristics of the Actuator Force

Electromagnetic thrust force is a critical performance indicator for evaluating the actuator. In this paper, the steady-state electromagnetic force is analyzed using the finite element model described above. The actuator's synchronous motion speed is set to 0.36 m/s, and the excitation current is set to 3A. A three-phase symmetrical sinusoidal current is applied as the input, and calculation time is set to 400 ms. The calculated actuator electromagnetic thrust force is shown in Figure 7.

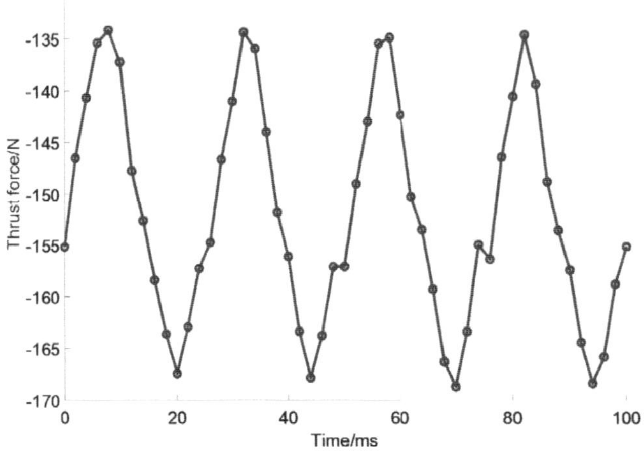

Figure 7. Electromagnetic thrust force of actuator

As shown in Figure 7, the actuator electromagnetic thrust force exhibits fluctuations under current excitation rather than remaining stable. These fluctuations arise from two main sources. On the one hand, Ripple force generated by the interaction between the stator magnetic field harmonics and the mover magneto-motive force harmonics, with the 3rd harmonic being the dominant component. On the other hand, Magnetic reluctance force generated by actuator.

In addition, this paper analyzes the variation of the actuator electromagnetic thrust force under input currents ranging from 0 A to 8 A. Due to the presence of force fluctuations, the average value is taken as the actual actuator electromagnetic thrust force, and the difference between the peak value and the average value is defined as the fluctuating force. The calculation results are shown in Figure 8 and Figure 9.

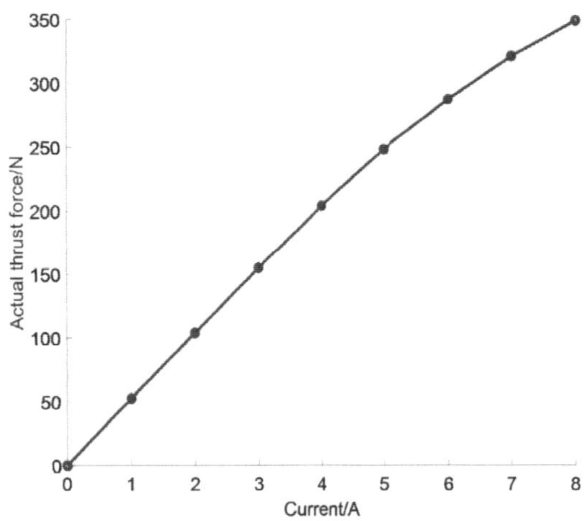

Figure 8. Actual electromagnetic thrust force

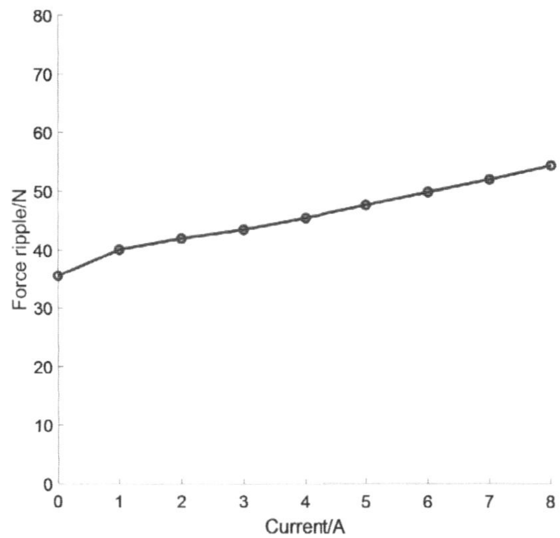

Figure 9. Fluctuating force

As shown in Figure 8, the actual electromagnetic thrust force increases gradually with the excitation current, exhibiting an approximately linear relationship. From Figure 9, it can be observed that the fluctuating force also increases with the current, and the rate of increase becomes more pronounced at higher input excitation currents.

5. Conclusions
For this paper, a structure of 12-slot 14-pole radial magnetized cylindrical permanent magnet actuator applied to vehicle active suspension was proposed and investigated. Through the actuator performance analysis, it has been discovered that the actuator electromagnetic thrust force exhibits fluctuations under current excitation rather than remaining stable. The electromagnetic thrust force and the fluctuating force increase gradually with the excitation current,which the fluctuating force reached 53N while the actuator's synchronous motion speed is 0.36 m/s, the excitation current is set to 3A.

Acknowledgements
This work was supported by the Key Research and Development Promotion Special Project (Science and Technology Research Project) of Henan Province(Grant No.232102240048)

References
[1] Damien Sammier, Olivier Sename, Luc Dugard. Skyhook and H_{∞} Control of Semi-active Suspensions:Some Practical Aspects[J]. Vehicle System Dynamics ,2003,39(4):279-308.
[2] Yagiz N, Sakman LE. Fuzzy Logic Control of a Full Vehicle Without Suspension Gap Degeneration[J]. International Journal of Vehicle Design, 2006,42(1-2):198-212.
[3] Yu Fan, Zhang Yongchao, Zhang Guoguang. Review of Vehicle Electromagnetic Suspension Technology [J]. Automotive Engineering, 2012, 34(7): 569-573.
[4] Ismenio Martins, Jorge Esteves,Gil D Marques,et al.Permanent-Magnets Linear Actuators Applicability in Automobile Active Suspensions[J]. IEEE Transactions on Vehicular Technology, 2006,55(1):86-94.
[5] Bart LJGysen, Jeroen LG Janssen. Design Aspects of an Active Electromagnetic Suspension System for Automotive Applications [J]. IEEE Transactions on Industry Applications, 2010,45(5):1589-1597.
[6] Babak Ebrahimi, Mir Behrad Khamesee, Farid Golnaraghi. A Novel Eddy Current Damper: Theory and Experiment[J]. Journal of Physics D: Applied Physics, 2009,42:1-6.
[7] Deng Zhaoxiang, Lai Fei. Research on Electromagnetic Linear Actuator for Active Suspension of Vehicles [J]. Journal of Mechanical Engineering, 2011, 47(14): 121-128.
[8] Lin Jiongkang, Cheng Ka Wai Eric, Zhang Zhu, et al. Active Suspension System Based on Linear Switched Reluctance Actuator and Control Schemes[J]. IEEE Transactions on Vehicular Technology, 2013,62(2):562-572.

ResPTFE: Self Supervised Learning-Based Train Bearing Fault Diagnosis in Labelled Data-Starved Scenarios

Yuan Xu, Haimeng Sun, Deqiang He*, Jinxin Wu

Guangxi Key Laboratory of Manufacturing System & Advanced Manufacturing Technology, School of Mechanical Engineering of Guangxi University, Nanning 530004, China

*Corresponding author's e-mail: hdqlqy@gxu.edu.cn

Abstract. Fault diagnosis of train traction motor bearings (TTMB) is a key component to ensure the safe operation of trains. However, due to the high cost of acquiring labelled data in different fault modes of TTMB, the lack of labelled data leads to the poor performance of most existing intelligent diagnosis methods. To this end, a self-supervised TTMB fault diagnosis model based on residual parallel time-frequency encoder (ResPTFE) is proposed in this paper. Firstly, a time-domain data enhancement method is proposed to generate diverse fault samples. Secondly, a time-frequency encoder combining residual network with squeezing and excitation attention mechanism is constructed, which can effectively extract key time-frequency features. Subsequently, a cross-correlation regularised loss function is designed to enhance the correlation between corresponding features and ensure the independence between different features by constraining the interaction between frequency domain and time domain features. Finally, the experimental results show that ResPTFE is more suitable for TTMB fault diagnosis in the case of scarce labelled data.

1. Introduction

With the increasing popularity of high-speed trains, the demand for efficient travelling has been met. However, compared to ordinary trains, the complex operating conditions of high-speed trains accelerate the wear and tear of key components, especially the train traction motor bearings (TTMB). As a key component in high-speed train operation, the condition of TTMBs directly affects the safety and reliability of trains. Therefore, it is crucial to carry out fault diagnosis research on TTMB to ensure the safe operation of high-speed trains [1-3].

With the rapid development of artificial intelligence and big data technologies, intelligent fault diagnosis methods based on deep learning have gradually gained attention and achieved good results [4-5]. Initially, most deep learning diagnostic methods rely on supervised learning frameworks, e.g., He et al. [6] proposed a method for intelligent detection of weld quality of underground car bodies using supervised learning. Such methods usually rely on sufficient labelled data to achieve high diagnostic performance. However, due to the complexity of TTMB fault data and the high cost of manual labelling, the lack of labelled samples limits their application in TTMB fault diagnosis.

To address the decline in diagnostic accuracy caused by insufficient labeled data, researchers have gradually explored intelligent diagnosis methods based on self-supervised learning (SSL). SSL can learn the feature structure of fault signals from unlabelled data through predefined excuse tasks, and achieve efficient fault diagnosis using only a small amount of labelled data. Zou et al [7] proposed a method combining discrete wavelet transform and improved deep belief network to achieve high-precision TTMB fault diagnosis without the need for a large amount of labelled data. Nie et al [8] proposed a model combining SSL and sparse filtering, which significantly improves the diagnostic performance with limited labelled data. However, most of the self-supervised models still need to construct negative samples for comparative learning, which is computationally intensive and that the labelled data usually accounts for more than 10% of the total dataset.

To address the aforementioned issues, this study proposes a self-supervised fault diagnosis model based on the residual parallel time-frequency encoder (ResPTFE), which utilizes unlabeled data for training to learn the time-domain and frequency-domain features of fault signals. This model does not require the construction of negative sample pairs; instead, it predicts the corresponding time-domain representations in the feature space based on the frequency-domain representations. To ensure a high correlation between time-domain features and their corresponding frequency-domain features while maintaining the independence of different features, a cross-correlation regularization loss function is designed to fully learn the potential fault representations in the unlabeled data. Ultimately, only 1.3% of the total dataset of labeled data is used to fine-tune the pre-trained model, achieving high-precision TTMB fault diagnosis. The main contributions of this paper are as follows:

(1) A time-domain data augmentation method is proposed to generate diverse signal samples, enhancing the model's ability to recognize complex fault patterns.

(2) An encoder based on the residual squeeze and excitation network (ResSeNet) is constructed to effectively extract key fault features in both time and frequency domains, significantly reducing computational complexity by incorporating depthwise separable convolutions.

(3) A cross-correlation regularization loss function is designed to maximize the correlation between the corresponding time-domain and frequency-domain features while ensuring the independence and sparsity of different features, effectively reducing feature redundancy.

2. Methodology

2.1. Data enhancement module

Data augmentation can be used to diversify the raw data so as to simulate different working conditions and enhance the model's understanding of data diversity and invariance, which is particularly suitable for SSL frameworks that rely on unlabelled data. Therefore, this paper proposes an enhancement module for time-domain signals containing seven signal enhancement methods: Gaussian SNR noise, Shift, TimeMask, Perturbed weighted average, Sliding window, Amplitude perturbation and Gain.

2.2. A SSL framework based on ResSeNet

The structure of the self-supervised framework is shown in Fig. 1. It primarily consists of a temporal encoder, a frequency encoder, temporal projection heads and frequency projection heads, and prediction heads. The unannotated temporal data is represented in the frequency domain using fast fourier transform (FFT), followed by data augmentation in both the temporal and frequency domains. The augmented temporal and frequency data are separately input into the temporal and frequency encoders constructed by ResNet, and then projected into a low-dimensional space through the projection heads. Finally, the prediction head completes the contrastive prediction of the temporal and frequency domains.

Journal of Physics: Conference Series **3024** (2025) 012018 doi:10.1088/1742-6596/3024/1/012018

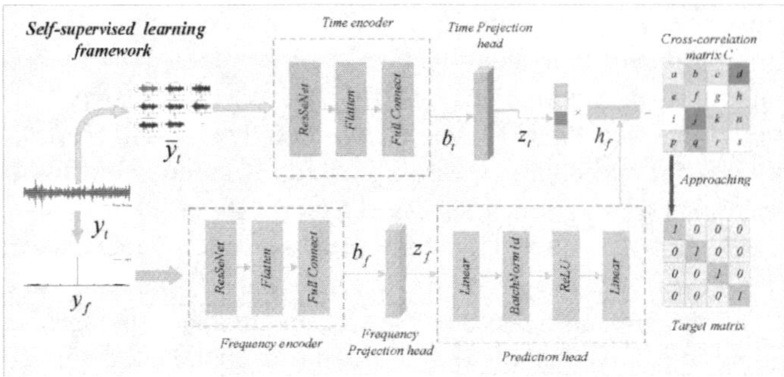

Figure 1. SSL frame diagram

2.2.1. Encoder structure. This paper propose a ResSeNet-based encoder for efficient feature extraction. The main structure of ResSeNet is shown in Fig.2, which introduces the squeeze and excite attention mechanism (SEAM) in the residual block and uses the depthwise separable convolution instead of the standard convolution, which significantly reduces the number of parameters and computation, thereby improving the efficiency of the model.

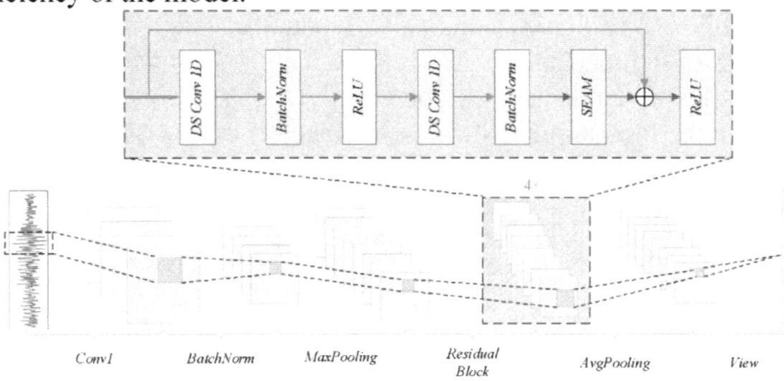

Figure 2. Network structure of ResSeNet

2.2.2. SEAM structure. The Squeeze and Excitation (SE) module serves as an independent and easily portable computational unit, effectively enhancing important features by computing and weighting the inter-channel correlations, demonstrating significant effectiveness in image recognition tasks. This module can automatically learn the weights of each channel, enhancing key features and suppressing interfering features, thereby improving the model's discriminative power.

Since the SE module is typically used to process two-dimensional images or matrix inputs, and the fault bearing vibration data used in this paper consists of one-dimensional sequential data. The SE module has been adapted to handle one-dimensional sequence inputs, as shown in Fig. 3. Firstly, Squeeze is performed on the incoming data, and a channel descriptor is constructed by doing a global average pooling of U. The information contained in each channel is embedded in the global distribution. Then it is to activate the data output from the previous step through two levels of full connectivity via Excitation after activation by Sigmoid function, outputting weight data in the range of (0-1), and then multiplying the weights with the C channels activated by the data to use as input to the next level. The important information is emphasised by the different rationing of the channel weights to improve the directionality of the feature extraction.

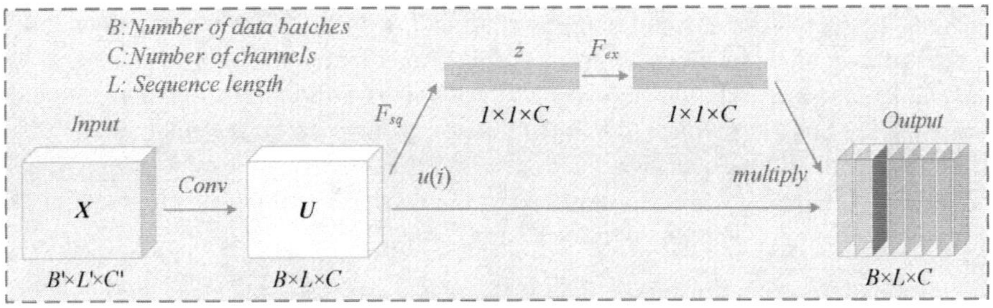

Figure 3. A schematic of SEAM's structure

2.2.3. Cross-correlation regularization loss function. In this paper, a cross-correlation regularisation loss function is designed to measure the relationship between time-domain features and frequency predictions so that the inter-correlation matrix between frequency predictions and temporal features approaches the unit matrix, i.e., the diagonal elements converge to 1 and the non-diagonal elements converge to 0. Meanwhile, the L_1 regularisation is introduced to promote the sparsity of the model parameters by minimising the sum of the differences in the absolute values between the prediction matrix and the unit matrix. This regularisation strategy enhances the adaptability and interpretability of the model under new data and supports efficient feature selection.

The loss function is defined as follows:

$$\zeta = \sum_{i=1}^{N}(1-C_{ii})^2 + \lambda\sum_{i=1}^{N}\sum_{j=1,j\neq i}^{N} C_{ij}^2 + \alpha\sum_{i=1}^{N}\sum_{j=1}^{N}\left|C_{ij} - I_{ij}\right| \tag{1}$$

The first term in Eq. (3) denotes the squared difference loss of the diagonal elements, the second term denotes the squared difference loss of the non-diagonal elements and is multiplied by a constant measuring the uncorrelated contribution, λ , and the third term denotes the loss of the L_1 regularisation and is multiplied by the regularisation parameter, α . C_{ij} denotes the elements in the cross-correlation matrix, where N is the dimensions of the frequency domain prediction and time domain features, i and j are the indexes of the dimensions. I_{ij} is the unitary matrix, and C is the inter-correlation matrix between frequency domain predictions and time domain features in the batch dimension:

$$I_{ij} = \begin{cases} 1 & i = j \\ 0 & i \neq j \end{cases} \tag{2}$$

$$C = \frac{1}{N}\frac{h_f^T}{\|h_f\|_2}\frac{z_t}{\|z_t\|_2} \tag{3}$$

where $\|\cdot\|$ denotes the l_2 -paradigm, which is used to ensure that h_f and z_t are always in comparable ranges and that C converges to an $N \times N$ unit matrix.

2.3. The fault diagnosis process of ResPTFE

This paper proposes a TTMB fault diagnosis method based on ResPTFE, which enhances the model's recognition accuracy under conditions of limited labeled data samples. Fig. 4 illustrates the specific diagnosis process of ResPTFE:

Phase 1. Vibration data of seven types of faulty bearings are collected and processed.

Phase 2. Firstly, an FFT is performed on a portion of the preprocessed time-series data to map the unlabelled time-domain data to the frequency domain in order to capture more spectral features. Next, a series of data enhancement operations are performed on the time domain signals. Subsequently, the enhanced time-domain data and frequency-domain data are input to the corresponding time encoder and frequency-domain encoder, respectively. As the model performs a predefined self-supervised task, the encoder progressively learns robust representations through feature extraction to enhance the model's generalisation ability in downstream tasks. Complementarity and consistency of features are further enhanced by cross-correlation regularisation of the loss function to minimise the discrepancy between the frequency domain prediction and the time domain representation.

Phase 3. The pre-trained time encoder is retained and the model is fine-tuned using a small amount of labelled data. By connecting the classifier head at the end of the time-domain encoder, the model is able to efficiently perform the diagnostic task of TTMB faults.

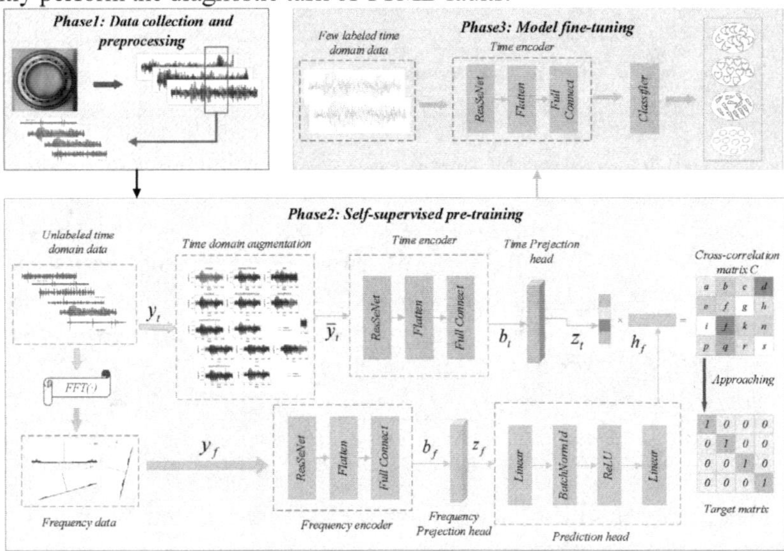

Figure 4. Brief flow chart of ResPTFE

3. Experimental

In this section, a series of experiments are conducted on ResPTFE using the collected data to verify its effectiveness and superiority. To ensure the fairness of the experiments and avoid random results, all experiments are run five times.

3.1. Data collection and processing

To better simulate the fault conditions of the TTMB during actual operation, a specific traction motor drive-end bearing was custom-made in a 1:1 ratio, model SKF NU 216 ECM/C4VA3091, with an inner diameter of 80mm and an outer diameter of 140mm. Seven different types of faults were artificially created on the inner ring, outer ring and rolling elements of the bearing using wire cutting technology, including: rolling element fault (RF), inner ring fault (IF), inner ring and rolling element fault (IR), inner ring and outer ring fault (IO), outer ring fault (OF), outer ring and rolling element fault (OR), and inner ring and outer ring and rolling element fault (IOR). The above fault categories are assigned labels 0-6, and subsequent data collection will be carried out at bearing speeds of 1500 rpm and 2100 rpm.

Different types of faulty bearings were mounted on an experimental bench as shown in Fig. 5 for data collection, and vibration sensors were mounted on the bearing caps to collect vibration data. A sampling frequency of 20 kHz was used to collect the vibration data during the operation of the faulty bearing. After data collection, the sliding window technique was applied to segment the time series data to generate fixed-length samples. The window size is set to 1024 and the sliding step size is set to 512. There are 2300 samples for each type of fault at different rotational speeds. To ensure a balanced sample size, the fault samples were randomly divided into a training set (1770 unlabelled samples and 30 labelled samples), a validation set (300) and a test set (200). It is worth noting that each type of fault dataset contains only 30 labelled samples for model tuning, which is 1.3% of the total number of samples.

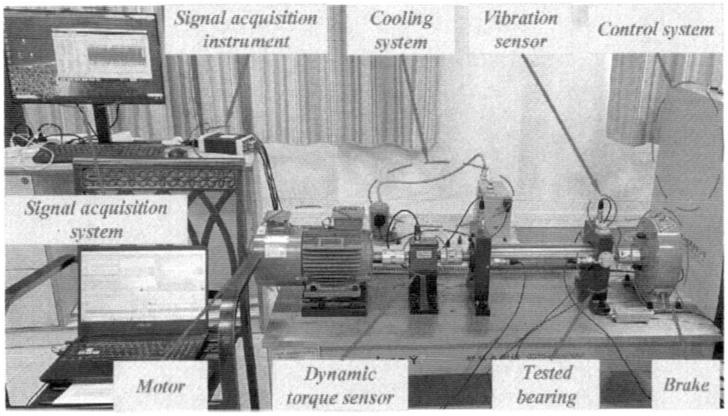

Figure 5. Train traction motor bearing failure simulation test bench

3.2. Comparison with several advanced models

To verify the diagnostic performance of ResPTFE, this study selected the following advanced models for comparison: (1) TFPred [9] (Time-frequency prediction); (2) TFAI [10] (Time-frequency alignment and Interaction); (3) SSMN [11] (Semi-supervised meta-learning network). Except for SSMN, which is a semi-supervised model, the other models are all SSL models. All four models were experimentally validated on the aforementioned dataset.

The aforementioned models were tested at different speeds, with each group of experiments repeated five times. Fig. 6 shows the accuracy and average accuracy of the five groups of experiments. The average accuracy of ResPTFE in the two scenarios is 2.57% and 0.78% higher than that of TFPred, indicating that the improvement strategy enhances the model's performance in TTMB fault diagnosis. Additionally, in different speed scenarios, ResPTFE's classification accuracy is higher than that of SSMN by 5.67% and 2.62%, with SSMN performing the worst among these models, indicating that semi-supervised learning methods are not advantageous in fault diagnosis tasks with limited labeled data.

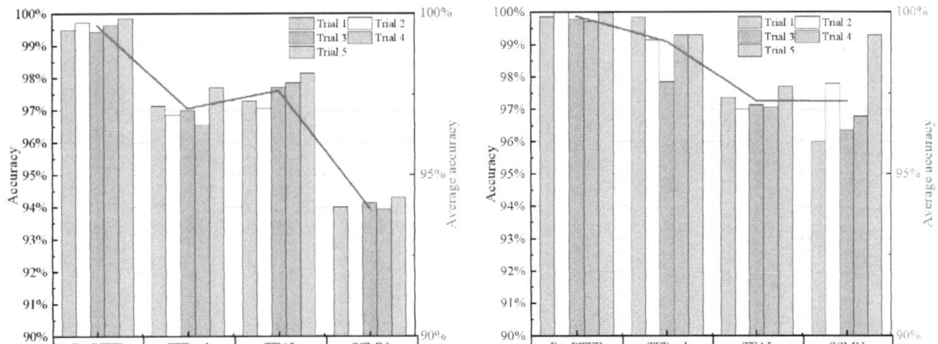

	ResPTFE	TFPred	TFAI	SSMN
Trial 1	99.5%	97.14%	97.29%	94%
Trial 2	99.71%	96.86%	97.07%	93.43%
Trial 3	99.43%	97%	97.71%	94.14%
Trial 4	99.64%	96.57%	97.86%	93.93%
Trial 5	99.85%	97.71%	98.14%	94.29%
—	99.63%	97.06%	97.61%	93.96%

(a) 1500rpm

	ResPTFE	TFPred	TFAI	SSMN
Trial 1	99.86%	99.85%	97.36%	96%
Trial 2	100%	99.14%	97%	97.79%
Trial 3	99.79%	97.86%	97.14%	96.36%
Trial 4	99.71%	99.29%	97.07%	96.79%
Trial 5	100%	99.29%	97.71%	99.29%
—	99.87%	99.09%	97.26%	97.25%

(b) 2100rpm

Figure 6. Accuracy statistics of five experimental results of each model under two rotational speeds

3.3. Visualisation of feature extraction

To gain a deeper understanding of the feature extraction effect of ResPTFE, the features extracted by ResPTFE at two rotational speeds are visualised using the nonlinear dimensionality reduction algorithm T-SNE, which exhibits that samples of the same fault modes are clustered together, while the different modes have clear classification boundaries. The results are shown in Fig. 7, and the T-SNE visualisation shows that the features extracted by ResPTFE have a large interclass spacing, a small intraclass spacing, and a clear distinction between different fault classes.

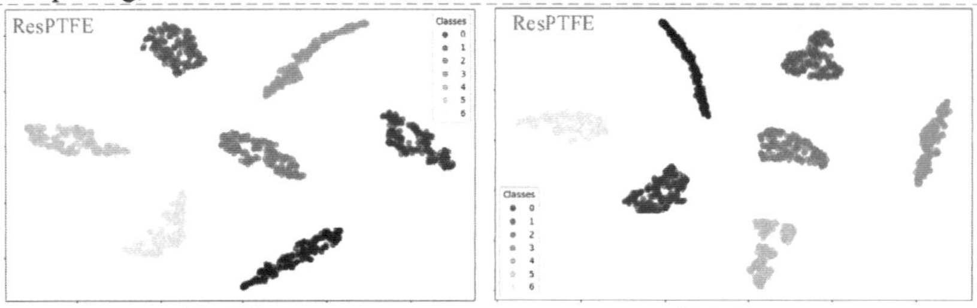

(a) 1500rpm (b) 2100rpm

Figure 7. Visualization of features extracted by ResPTFE

4. Conclusion

To address the difficulties in fault diagnosis caused by the scarcity of labeled data for TTMB, this paper proposes a SSL model named ResPTFE. Firstly, a set of data augmentation methods was developed to expand the data scale. Subsequently, an encoder based on ResSeNet was constructed and experimentally validated for its efficiency in extracting both time-domain and frequency-domain features. In addition, a cross-correlation regularization loss function is designed to measure the relationship between time domain features and frequency domain prediction, so that the model can better learn the potential features in the signal and improve the subsequent diagnosis effect. Combined with Ranger optimizer and ReduceLROnPlateau learning rate scheduler, the stability and training efficiency of the model were improved. On a simulated TTMB fault testing platform built in the laboratory, various fault state data were collected, and the performance of ResPTFE was systematically evaluated using multiple comparative models. Experimental results show that under different speed conditions, compared with other advanced models, ResPTFE has higher accuracy and stability in fault identification, as shown in Fig. 6, the classification accuracy can reach 100% in some scenarios.

Acknowledgements

The research was supported by the National Natural Science Foundation of China [Grant Number 52072081], the National Natural Science Foundation of China [Grant Number U22A2053], Guangxi Science and Technology Plan Project [Grant Number Guike AD25069080, 2025GXNSFBA069131], Foundational Research Capacity Enhancement Program for Young and Middle-Aged Teachers in Guangxi Higher Education Institutions (2025KY0060, 2021KY0987), Innovation Project of Guangxi Graduate Education [Grant No. YCBZ2024013] and Major Talent Project in Guangxi Zhuang Autonomous Region.

References

[1] Zou X, He D, Jin Z et al 2023 Intelligent diagnosis method of bearing fault based on ICEEMDAN and Ghost-IRCNN Proceedings of the Institution of Mechanical Engineers, Part C: Journal of Mechanical Engineering Science vol 237 no 13 pp 3115-3130
[2] He D, Zhang Z, Jin Z et al 2025 RTSMFFDE-HKRR: A fault diagnosis method for train bearing in noise environment Measurement vol 239 pp 115417
[3] Jin Z, Sun Y 2023 Bearing fault diagnosis based on VMD fuzzy entropy and improved deep belief networks Journal of Vibration Engineering & Technologies vol 11 no 2 pp 577-587
[4] He Y, He D, Lao Z et al 2024 Few-shot fault diagnosis of turnout switch machine based on flexible semi-supervised meta-learning network Knowledge-Based Systems vol 294 pp 111746
[5] Yuan Q, Jin Z, Jia S et al 2019 An improved opportunity maintenance model of complex system Advances in Mechanical Engineering vol 11 no 4 pp 1687814019841517

[6] He D, Ma R, Jin Z et al 2023 Welding quality detection of metro train body based on ABC mask R-CNN Measurement vol 216 pp 112969
[7] Zou Y, Zhang Y, Mao H 2021 Fault diagnosis on the bearing of traction motor in high-speed trains based on deep learning Alexandria Engineering Journal vol 60 no 1 pp 1209-1219
[8] Nie G, Zhang Z, Shao M et al 2023 A novel study on a generalized model based on self-supervised learning and sparse filtering for intelligent bearing fault diagnosis Sensors vol 23 no 4 pp 1858
[9] Chen X, Yang R, Xue Y et al 2024 TFPred: Learning discriminative representations from unlabeled data for few-label rotating machinery fault diagnosis Control Engineering Practice vol 146 pp 105900
[10] Fu D, Liu J, Zhong H et al 2024 A novel self-supervised representation learning framework based on time-frequency alignment and interaction for mechanical fault diagnosis Knowledge-Based Systems vol 295 pp 111846
[11] Feng Y, Chen J, Zhang T et al 2022 Semi-supervised meta-learning networks with squeeze-and-excitation attention for few-shot fault diagnosis ISA Transactions vol 120 pp 383-401

Elastic dynamics equation of robot mechanism with Groove Pin Motion Pair with Clearance

Chen zhenglong[1], He liang[1, 2, *], Wang Yaping[1], Su Jiaqiang[1], Yang Lihua[1], Ganwei Cai [1, 3[ORCID:0000-0002-1936-5048]]

[1] School of Construction Machinery, Hunan Sany Polytechnic College, Changsha,410129, P.R. China

[2] College of Mechanical and Electrical Engineering, Central South University, Changsha,410083, P.R. China

[3] College of Mechanical Engineering, Guangxi University, Nanning Guangxi 530004, P.R. China

*Corresponding author's e-mail: hel85@sany.com.cn

Abstract. The topological structure and degrees of freedom of the metamorphic mechanism can be changed. In the field of robotics, especially in medical robots, construction robots, military robots, and industrial robots, it has a very promising application prospect. The study of elastic dynamics of robot mechanisms with metamorphic kinematic pairs with clearance has not yet been studied. In this paper, on the basis of the continuous contact theory with metamorphic kinematic pairs with clearance, a finite element model of the metamorphic kinematic pairs with clearance is proposed, and the elastic dynamics equation of the metamorphic groove Pin Pair with clearance is established according to the KED theory. It lays a certain foundation for the research of vibration response, motion stability, bifurcation and chaos of the robotic mechanism with metamorphic kinematic pairs with clearance.

1. Introduction

The metamorphic kinematic pairs has the characteristic of changing the constraint number, so that the metamorphic mechanism can change the topological structure and freedom, which is suitable for many requirements of robots, especially in the fields of aerospace robots, medical robots, industrial robots, autonomous excavators (construction robots), military robots and so on[1, 2]. The clearance theory of fixed constrained motion pair is not applicable to the metamorphic motion pairs, so the relevant clearance theory of metamorphic kinematic pairs need to be studied[3]. The elastic dynamics of robot mechanism is an important branch of robot mechanism. With the increasing speed and decreasing quality of robot mechanism, the elastic vibration of robot mechanism becomes more and more serious, and the residual vibration of robot mechanism seriously affects its work efficiency. The mechanism researchers have been studying the elastic dynamics of robot mechanism for a long time and have achieved a lot of results [4].

Journal of Physics: Conference Series **3024** (2025) 012019 doi:10.1088/1742-6596/3024/1/012019

The clearance of kinematic pairs has many adverse effects on the kinematic and dynamic performance of the mechanism, which has been studied for a long time [3]. Continuous contact model, two-state model and three-state model have been widely recognized and applied[5,6,7]. The continuous contact model does not reflect the impact between the components of the kinematic pair, it is very convenient for mechanism analysis[3]. In this regard, Furuhashi[8], Erkaya[9, 10]and others have made good contributions. People have accumulated good experience for the research of variable constrained kinematic pairs with clearance by studying the traditional fixed constrained kinematic pairs with clearance for a long time.

Literature [3,11] describes the structure and working principle of the metamorphic linkage with grooved pin motion pair, wherein the elastic force of the spring conforms to the force gradient design criteria of the metamorphic linkage[12].

The two configurations and four stages of a motion cycle of linkage with the groove pin motion pair were introduced in detail in literature[3]. In this study, the finite element model of the metamorphic kinematic pair with clearance is proposed, and the elastic dynamic equation of the metamorphic mechanism is established, which provides a theoretical basis for the calculation of the elastic dynamic response of metamorphic mechanism with motion pair with clearance and the study of its nonlinear dynamic performance.

2.The Element model of Finite Element of the Grooved Pin Metamorphic Motion Pairs with Clearance

Based on the continuous contact theory of the clearance of metamorphic mechanism[5]and the KED theory, the finite element model of the clearance of the grooved pin metamorphic kinematic pairs can be established.

On the basis of the original mechanism of metamorphic kinematic pairs with clearance, a clearance beam(BC) element is added to transform the mechanism into ideal mechanism with no clearance, as shown in Figure 1. The massless elastic beam element BC represents the clearance of the motion pair, and the original pin is the slider at node C. The mass of this pin is included in the mass matrix as the concentrated mass. Figure 2 is the schematic diagram of the clearance element, u1、u2、u3、u4、u5、u6、u7、u8 are the coordinates of the local coordinate system of the clearance element, U2、U3、U4、U5、……are the coordinate of the global coordinate system, θ is the Angle between the local coordinate system and the global coordinate system. The node displacements u2 and u6 of B and C reflect the change of clearance. According to literature [13], the contact stiffness coefficient is

$$k = \frac{2ED_{al}^{D/2}}{\sqrt{\pi}(1-\mathrm{D})}\left(al^{\frac{1-D}{2}} - ac^{\frac{1-D}{2}}\right) \tag{1}$$

See the reference[13] for the parameters in the formula.

Journal of Physics: Conference Series **3024** (2025) 012019 doi:10.1088/1742-6596/3024/1/012019

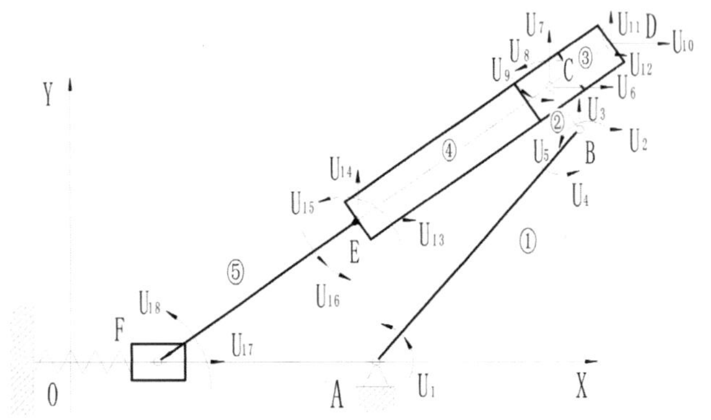

Figure 1. New mechanism with no clearance after adding imaginary link

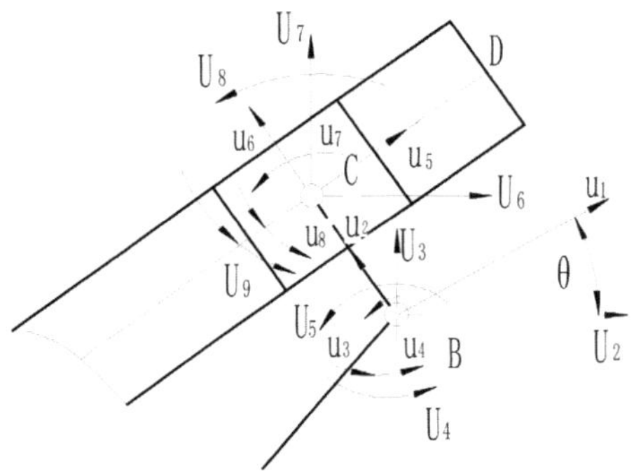

Figure 2. Local coordinates and global coordinates of gap element

The elastic deformation of the clearance element is u6-u2, so the deformation energy of the element is

$$
\begin{aligned}
V &= \frac{1}{2}k(u_6 - u_2)^2 = \frac{1}{2}k(u_2^2 + u_6^2 - 2u_2u_6) \\
&= \frac{1}{2}\begin{bmatrix} 0 \\ u_2 \\ 0 \\ 0 \\ 0 \\ u_6 \\ 0 \\ 0 \end{bmatrix}^T
\begin{bmatrix}
0 & 0 & 0 & 0 & 0 & 0 & 0 & 0 \\
0 & k & 0 & 0 & 0 & -k & 0 & 0 \\
0 & 0 & 0 & 0 & 0 & 0 & 0 & 0 \\
0 & 0 & 0 & 0 & 0 & 0 & 0 & 0 \\
0 & 0 & 0 & 0 & 0 & 0 & 0 & 0 \\
0 & -k & 0 & 0 & 0 & k & 0 & 0 \\
0 & 0 & 0 & 0 & 0 & 0 & 0 & 0 \\
0 & 0 & 0 & 0 & 0 & 0 & 0 & 0
\end{bmatrix}
\begin{bmatrix} 0 \\ u_2 \\ 0 \\ 0 \\ 0 \\ u_6 \\ 0 \\ 0 \end{bmatrix} \\
&= \frac{1}{2}\mathbf{u}^T \bar{\mathbf{k}} \mathbf{u}
\end{aligned}
\tag{2}
$$

Where $\bar{\mathbf{k}}$ is the stiffness matrix of the clearance element.

The clearance element is different from the beam element in that it only has the concentrated mass located at node C. The kinetic energy of the clearance element is

$$T=\frac{1}{2}\dot{\mathbf{u}}^{T} m_C \dot{\mathbf{u}}=\frac{1}{2}m_C(\dot{\mathbf{u}}_5^2+\dot{\mathbf{u}}_6^2)$$

$$=\frac{1}{2}\begin{bmatrix}0\\0\\0\\0\\\dot{u}_5\\\dot{u}_6\\0\\0\end{bmatrix}^T\begin{bmatrix}0&0&0&0&0&0&0&0\\0&0&0&0&0&0&0&0\\0&0&0&0&0&0&0&0\\0&0&0&0&0&0&0&0\\0&0&0&0&m_C&0&0&0\\0&0&0&0&0&m_C&0&0\\0&0&0&0&0&0&0&0\\0&0&0&0&0&0&0&0\end{bmatrix}\begin{bmatrix}0\\0\\0\\0\\\dot{u}_5\\\dot{u}_6\\0\\0\end{bmatrix} \quad (3)$$

$$=\frac{1}{2}\mathbf{u}^T\bar{\mathbf{m}}\mathbf{u}$$

Where $\bar{\mathbf{m}}$ is the mass matrix of the clearance element.

3. Elastodynamic Equations of the Linkage with Groove Pin Joint with Clearance

On account of the continuous contact theory of the motion pair with clearance [3], the linkage with groove pin joint clearance is transformed into a non-clearance mechanism (fig. 1), and the slot DE is divided into two beam elements DC and EC of varying lengths, and D, E and C are the element nodes. The length changes of these two elements can be easily obtained from the position change
s of the slide block in the kinematic study of the mechanism. The metamorphic mechanism is divided into 5 elements, AB, BC, CD, CE and ED, in which BC is the clearance element and CD and CE are the variable length element. In order to avoid the stiffness matrix being singular, the joint elastic displacement at node A of the AB beam is only curvature U_1 according to the cantilever beam hypothesis.

In the non-clearance mechanism (fig. 1), except for the clearance element BC as a special finite element element, the other elements are general beam elements, and the conversion matrix [R] between the element coordinates and the overall coordinates of each element, and the coordinate coordination matrix between the local number of element i and the system number of element i is Bi, so the equivalent mass matrix and stiffness matrix of each element in the overall coordinates are respectively

$$\mathbf{m}_i=\mathbf{R}_i^T\bar{\mathbf{m}}\mathbf{R}_i \quad (4)$$
$$\mathbf{k}_i=\mathbf{R}_i^T\bar{\mathbf{k}}\mathbf{R}_i$$

The coordinate coordination matrix between the element i local number and the system number is Bi, then the mass matrix and stiffness matrix of the unit system number are respectively

$$\mathbf{M}_i=\mathbf{B}_i^T\mathbf{m}_i\mathbf{B}_i \quad (5)$$
$$\mathbf{K}_i=\mathbf{B}_i^T\mathbf{k}_i\mathbf{B}_i$$

Therefore, the mass matrix and stiffness matrix of the metamorphic linkage are respectively

$$\mathbf{M} = \sum_{i=1}^{5} \mathbf{M}_i \qquad (6)$$
$$\mathbf{K} = \sum_{i=1}^{5} \mathbf{K}_i$$

Therefore, without considering the damping effect, the motion equation of the linkage with groove pin motion joint clearance is

$$\mathbf{M}\mathbf{U}\ddot{\mathbf{U}} + \mathbf{K}\mathbf{U} = \mathbf{P} - \mathbf{M}\ddot{\mathbf{U}}_r \qquad (7)$$

Where, P is the generalized force vector, which consists of the torque acting on the driving bar AB to point A and the spring force acting on node F. $\ddot{U}_r$ is the vector of the acceleration of the rigid body at each node

4. Conclusion

With the characteristic of configuration transformation, the metamorphic linkage has a wide application prospect in the field of robotics. The clearance of the joint has an important effect on the dynamic performance of this kind of robot linkage, but no one has done in-depth research on the metamorphic joint so far, which seriously affects the study on the performance of the robot linkage and its engineering application. With the increase of the speed of the robot linkage, the study of its kinematic elastic dynamics (KED) has been paid more and more attention, but the elastodynamic performance of the robot linkage with metamorphic kinematic joints with clearance has not been studied. In this paper, based on the continuous contact theory of the clearance of the metamorphic joints, the corresponding finite element model of the clearance of the grooved pin joints is proposed, and the elastic dynamics equation of the metamorphic linkage is established according to the KED theory, which lays a foundation for the study of the elastic dynamics of the metamorphic linkage with clearance. It has a certain significance for the theoretical research and engineering application of robot linkage.

5. Acknowledgments

This work was supported in part by the Natural Science Foundation of Changsha City (kq2208085) , in part by the National Natural Science Foundation of China (51765005) and in part by the Guangxi Key Laboratory of Special Engineering Equipment (2411KFYB03).

References

[1] Pan Y C, Cai G W, Wang H Z, et al. 2014 Topological analysis of configuration evolution and biological modeling of anovel type of electric loading mechanism with metamorphic functions(Journal of Mechanical Engineering, vol 50)1 pp 38-46

[2] Li D L, Zhang Z H, et al. 2010 Overview and prospects of metamorphie mechanism(Journal of Mechanical Engineering vol 46)13 pp 14-21

[3] Hu J X, He L, Wang Y P, Yang L H, Cai G W 2025 Continuous Contact Model of Metamorphic Groove Pin Pair with Clearance(Applied Mathematics vol 16) 1 pp 15-41

[4] Thompson B S and Sung C K 1984 A Variational Formulation for the Dynamic Viscoelastic Finite Element Analysis of the Robotics Manipulators Constructed from Composite Materials(ASME Journal of Mechanisms, Transmissions and Automation in Design vol 106) pp183-190

[5] Earles S W E and Wu C L S 1972 Motion Analysis of a Rigid-Link Mechanism with Clearance at a Bearing Using Lagrangian mechanics and Digital Computation (Conference on Mechanisms, London, England)

[6] Furuhahi T， Morita N, Matsuura M 1978 Research on Dynamics of Four-bar Linkage with Clearances at Turning Pairs (Bulletin of JSME 21)

[7] Bahgat B M， Osman M O M, Sankar T S 1981 An Approach for Dynamic Analysis of Mechanical Systems with Multiple Clearance Using Lagrangian Mechanics (ASME paper 81-DET-3)
[8] Furuhashi T, Morita M 1978 Research on Dynamics of Four-Bar Linkage with Clearances at Turning Pairs: 4th Report, Forces Acting at Joints of Crank-Lever Mechanism (Bulletin of JSME vol 21)158 pp1299-1305
[9] Erkaya S and Uzmay I 2009 Optimization of transmission angle for slider-crack mechanism with joint clearance (Structural and Multidisciplinary Optimization vol 37)5 pp 493-508
[10] Erkaya S and Uzmay I 2008 A neural-genetic (NN-GA) approach for optimizing mechanisms having joints with clearance (Multibody System Dynamics vol 20) pp 69-83
[11] Sun L L 2010 Research on the cell modes and cell processes of the variational cell mechanism (Shenyang: Northeastern Universit)
[12] Li S j, Wang H G, Dai J S 2014 The Equivalent Resistance Gradient Model of Metamorphic Mechanismsand the Design Method (Journal of Mechanical Engineering vol 50) 1 pp 18-23
[13] Zhang X L， Huang Y M， Fu W P， et al. Fractal model of normal contact stiffness between rough surfaces (Chinese Journal of Applied Mechanic vol 17) 2 pp 31-36.

Intelligent Test System Design for Aerospace Equipment Harnesses

Ping Sun[1], Yong Zhang[2,3*], Nan He[1], Gongping Zhou[1], Mei Li[1]

[1.] Department of Mathematics and Computer Technology, Guilin Normal University, Guilin, Guangxi Province, 541199, P.R. China

[2.] School of Mechanical Engineering, Guilin University of Aerospace Technology, Guilin, Guangxi Province, 541004, P.R. China

[3.] Guangxi Key Laboratory of Special Engineering Equipment and Control, Guilin, Guangxi Province, 541004, P.R. China

[*]Corresponding author's e-mail: zy@guat.edu.cn.

Abstract. Aiming at the current aerospace wiring harness, which is prone to breaks, short circuits, insulation failures, contract issues, and other quality problems due to the complex structure and numerous connections, an intelligent wiring harness test system was developed by combining the current development trend of intelligent manufacturing industry and Internet of Things technology. Experimental verification shows that the system greatly improves the detection efficiency of the wiring harness test, reduces the labor intensity of the wiring harness test, effectively reduces the test errors introduced by human factors in the control component products, and ensures the quality of the products and the production schedule.

1. Introduction

A wiring harness is a kind of component that bundles multiple wires or cables together according to a certain rule and layout, connecting various electronic components on the product, consisting of wires, insulating sheaths, insulating wrapping materials, terminals, etc., and plays a role in transmitting the signals and electrics in various electrical systems. With the rapid development of the intelligent manufacturing industry and the continuous upgrading of the Internet of Things technology, electronic control units have been applied in large quantities, and the demand for wiring harnesses has been increasing, which are used to connect the control components composed of a variety of sensors, actuators, or micro-controllers and other components, and to exchange data with the cloud servers or local networks. The quality of the wiring harness directly affects the stability and reliability of the system.

In recent years, China's rapid development of aerospace technology, with the progress of national defense science and technology and space technology breakthroughs, aerospace mission high-frequency launches, the corresponding control components of the equipment is more and more intelligent, precision, complexity, by the use of environmental factors in the occasion, the wiring harness failure to become an aerospace equipment contact failure of one of the important reasons, such as some of the occasions of the temperature difference and humidity is relatively large, some of the occasions have a relatively Large noise and mechanical vibration, these factors directly lead to the acceleration of the micro-vibration wear of the equipment[1], resulting in unqualified contact resistance between the wiring harnesses, leading to product failure, according to statistics, the failure caused by the cable parts accounted for about 20% of the total failure rate of a certain type of product[2]. In the aerospace model product testing process, to simulate the use of harsh climatic conditions, the equipment needs to be a long time in a wide range of temperatures (-55 ° C ~ 125 ° C) full-load operation, the risk of sudden failure increases, it is difficult to distinguish the direction of the many wiring harnesses for troubleshooting, for the protection of aerospace system products safe and reliable operation and on-time delivery, has a very important value. If we use a multimeter or a low resistance meter to measure the contact resistance of every single output of the equipment to analyze and judge the quality of the wiring harness by using the traditional mode of manual point-by-point lapping, it is not only inefficient but also has a high misdetection rate. Therefore, this paper develops a set of intelligent wiring harness test systems to realize the automatic test of the contact resistance of the wiring harness.

2. Overall System Design

Aerospace field control components from the design, assembly, and test to the delivery process, wiring harness conductivity test is a very important work, it is the principle of the product's various outputs through the electrical connector and other connectors lead out, using a multimeter or low resistance meter and other instruments to manually measure the resistance or voltage between the nodes, when the node conductive, under the action of the current, usually produces a small voltage drop When the node conducts, under the action of the current, there is usually a small voltage drop, and the contact resistance between the conductors is very small or close to 0. When the node does not conduct, the obstruction encountered when the current passes through is increased accordingly, so that the contact resistance between the conductors is very large or close to infinity. In this way, it can determine whether there is a path between the nodes, but the efficiency is very low. Compared with civil electronic equipment, the use of national defense electronic equipment has a more complex environment, and the temperature drift of components has a great impact on its test results[3 ~ 4]. The intelligent test system can automatically test the contact resistance of aerospace product bundles in a wide temperature range, quickly identify quality problems such as circuit breakage, short-circuiting, bad insulation, poor contact, etc., and ensure that the products are in the whole process of scientific research and production, electrical continuity The electrical continuity of the product during the whole process of research and production is ensured to meet the requirements of the technical conditions. The overall architecture of the system is shown in Figure 1. The external interface of the software is shown in Figure 2.

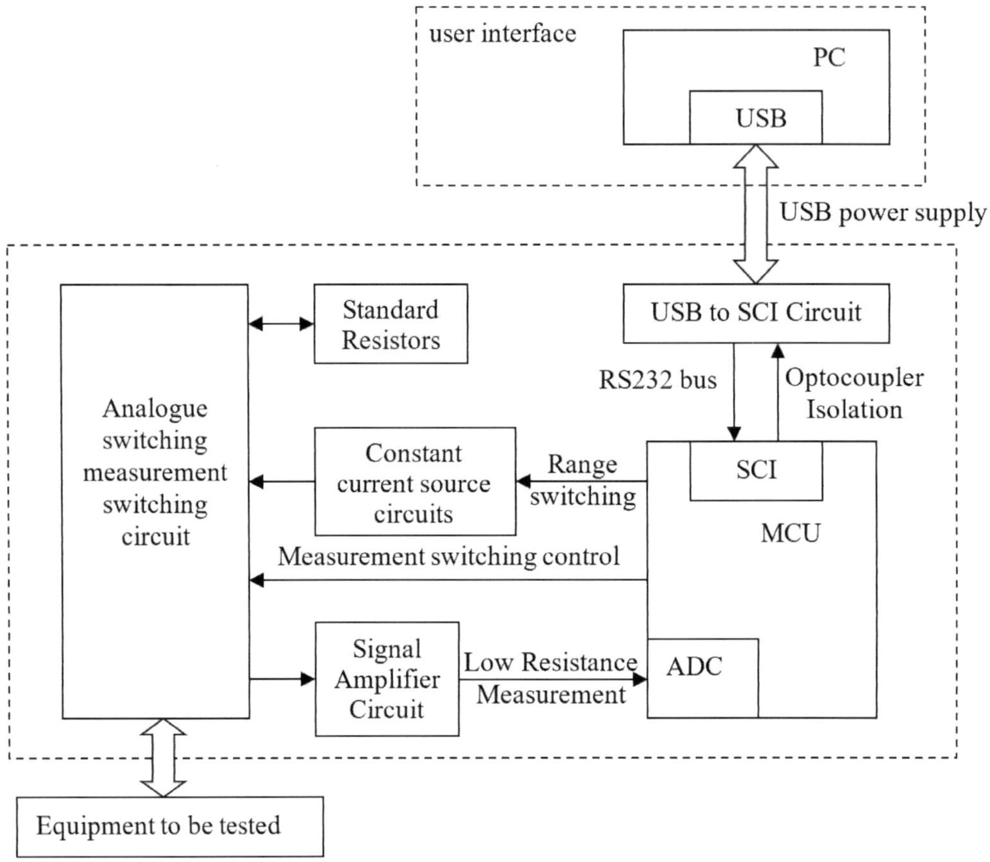

Figure 1. Overall system architecture

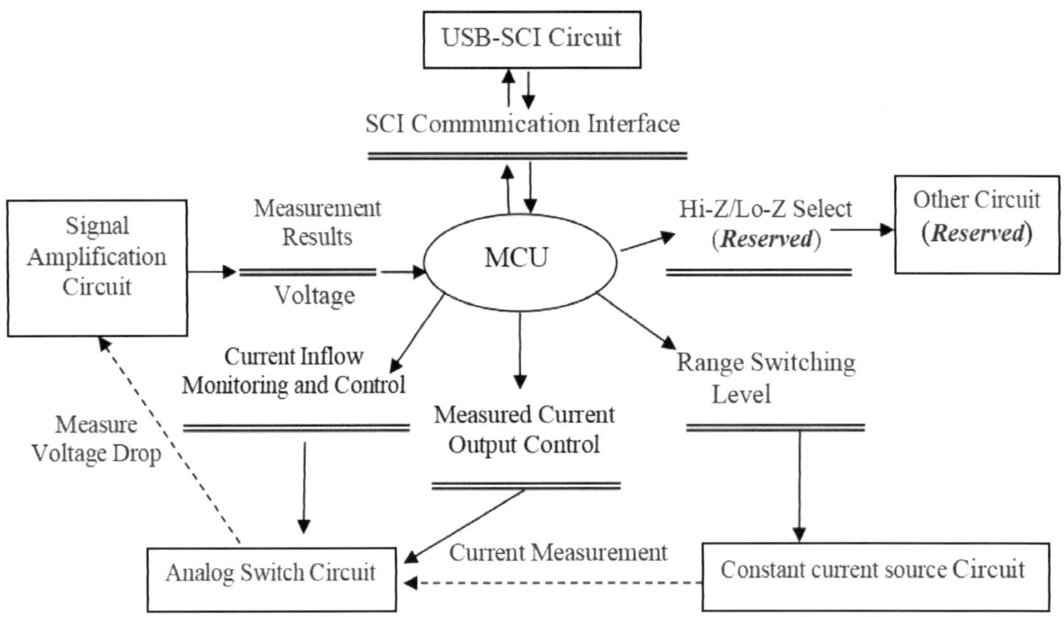

Figure 2. Schematic diagram of software external interface

3. Communication Protocol Design

The communication protocol is the soul and essence of the bus, which determines whether the intelligent test system and the equipment to be tested can communicate normally, the system selects RS232 bus communication mode, and the communication protocol refers to the character mode of Modbus communication protocol, the format is 1 bit start bit, 8 bit data bit, 1 bit stop bit, no parity bit. The system packages the sent data to form a data frame in the form of a string[5], i.e., in the original data, each byte uses two ASCII characters to represent its high four bits and low four bits respectively, data 0~ 9 is represented by the character '0'~'9', data 10~15 is represented by the character 'A'~'F' to represent it. The data 10 15 is represented by the characters 'A' and 'F' and starts with ':' (colon) and ends with '\r' (carriage return) and '\n' (line feed), where '\r' and '\n' are used as end characters. ' and '\n' have ASCII code values of 0x0D and 0x0A, respectively. The raw data (identified as Data0~DataN) is packed in the format shown in Table 1.

Table 1. Data Packaging Format

Raw data	Data0	Data1	Data2	...	DataN
	0x02	0x3A	0x64	...	0xF2
Packed strings	:023A64...F2\r\n				

In the communication protocol, it is stipulated that Data0 of the original data is the device node number, and to facilitate future expansion, its value is agreed to be 0xAA here. At the same time, it is stipulated that Data1 is the function code, which is used for describing the control commands received by the system, and the description of the function code is shown in Table 2:

Table 2. Function code Data1 and its description

Function code	Identifiers	Descriptive
0x0C	GET_ONOFF_MESSAGE	access control
0x0D	CMD_LINK_DEVICE	Online command to check if communication is normal
0x04	CMD_LOW1_ADJUST	Calibrated low resistance 1 stop
0x07	CMD_LOW2_ADJUST	Calibrated low resistance 2 stops
0x08	CMD_LOW1_MEASURE	Measurement from low resistance 1
0x0E	CMD_LOW2_MEASURE	Measurement from low resistance 2

The last byte DataN can be used as an additional checksum byte so that the checksum (only the lower eight bits are considered) is 0xFF after Data0 ~ DataN is summed up. The addition of the checksum byte effectively determines whether the data frame has been corrupted in the process of sending and receiving. The main functions realized in the software design part are: the intelligent test system receives the data frames sent by the upper computer through the RS232 bus, analyzes and configures them according to the format of the communication protocol, and controls the system circuit to realize the functions.

4. Hardware Design

USB cable interface for external equipment power supply, the output voltage is 5V, the output current can be up to 500mA, and intelligent test system operating current is about 65mA, so the system can not be connected to an external power supply, the use of USB cable direct power supply, you can drive the entire test system circuit. At present, many PC chassis shell is charged, the most common reason is poor contact or poor grounding, followed by dry weather caused by static electricity accumulation. When the case is charged, the current is often conducted to the USB port through the metal casing, which may damage the measurement circuitry on the one hand, and on the other hand, may pose a certain threat to the safety of the operator and the electronic equipment [6]. To avoid the risk of indirect electric shock, the circuits are electrically isolated in the design, which can effectively reduce noise interference and signal distortion and improve the stability and performance of the circuit after being isolated by the DC-DC power module.

The four-terminal measurement method is a classic measurement method for measuring low resistance, which can effectively reduce or eliminate the influence of wire resistance on the result when measuring small resistance, which is also the reason why the four-terminal measurement method is invariably adopted in the market measuring standard apparatus to measure the standard resistance of low and medium values [7~8]. Based on the traditional four-terminal measurement method, the intelligent test system utilizes the ADC part in Fig. 3, which is realized by the ADC that comes with the microcontroller, to measure the contact resistance of the wire harness by combining the "four-terminal measurement method" and the "on-site calibration" method.

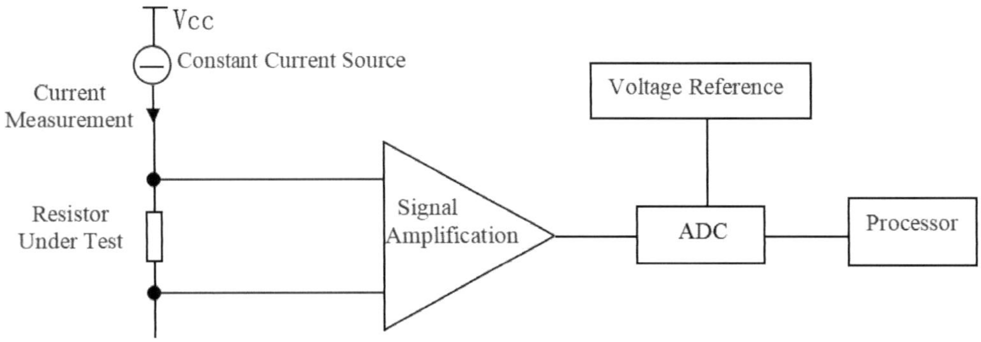

Figure 3. The measurement principle of the four-terminal method

The measurement current is provided by the constant current source circuit, and the resistance to be measured can be zero resistance, reference resistance, or target resistance from the device to be measured. To obtain a measurement ADC of the resistance to be measured, it is necessary to complete the line switching of the four terminals.

The parameters involved in calculating the target resistance are shown in Table 3.

Table 3. Parameters involved in the calculation of resistance

Markings	Descriptive
ZeroADC	AD conversion value obtained by the ADC module when switching to zero resistance.
RefRes	Reference resistor, 2Ω precision alloy gold foil resistor in this design
RefADC	AD conversion value obtained by the ADC module when switching to the reference resistor.
ObjADC	AD conversion value obtained by the ADC module when switching to the target resistor.

Before each series of measurements, a field calibration is performed to obtain the ZeroADC and Before each series of measurements, a field calibration is performed to obtain the ZeroADC and RefADC to eliminate the effects of circuit parameter differences, temperature, and other factors. The target resistance ObjRes can be calculated based on the following scaling equation.

$$\frac{ObjRes}{ObjADC - ZeroADC} = \frac{RefRes}{RefADC - ZeroADC} \tag{1}$$

Adjusted to

$$ObjRes = \frac{(ObjADC - ZeroADC) \bullet RefRes}{RefADC - ZeroADC} \tag{2}$$

Theoretically, when using this method to measure low resistance, the measured value can reach a high degree of accuracy; in addition, with this measurement method, only the 2Ω reference circuit for calibration has a special accuracy requirement (at least 0.5% accuracy), and the rest of the resistive components in the circuit only require the conventional 5% accuracy, which can greatly reduce the component cost.

5. Software Design

Intelligent test system lower computer software is prepared in C language after initialization is completed, waits for the microcontroller to finish receiving a frame of data, and judges whether the unpacking checksum is normal, if normal, then parses and executes the received commands, and after the execution of the commands, enable the reception of the new data, and empty the reception buffer, wait for the reception of the next frame of data, and so on. The main program flow is shown in Figure 4.

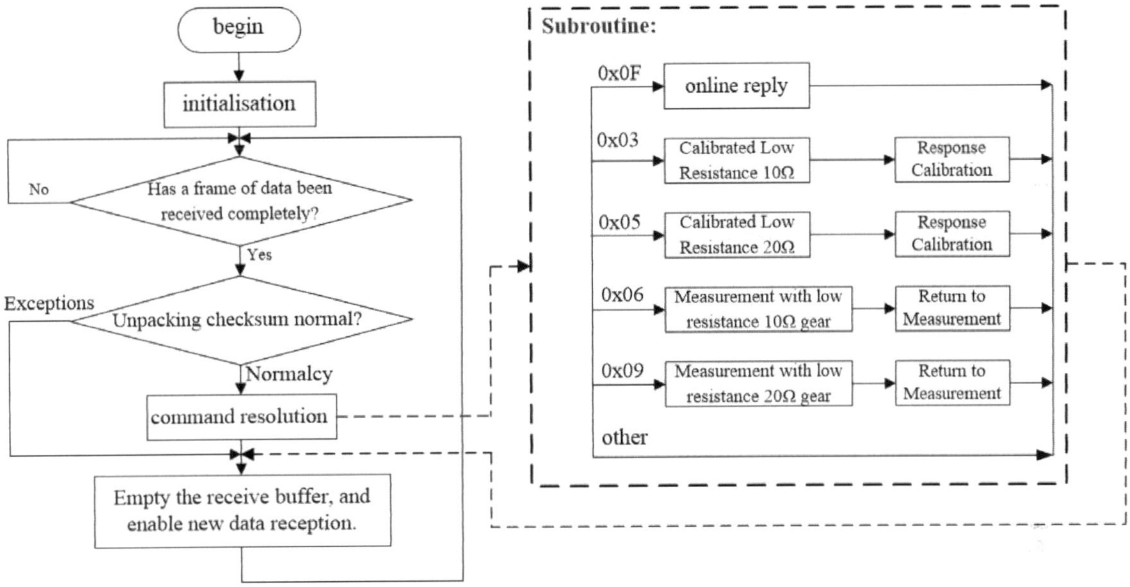

Figure 4. Microcontroller main program workflow

Intelligent test system host computer software interface shown in Figure 5, the user can based on the actual test requirements, configure the current inflow, current outflow, and gear, manual measurement configuration of the wiring harness to be measured, but also can be set up in advance a series of operations, or select the configuration file has been saved, by the system to automatically open the serial port, on-line testing, and the calibration of the 0 resistance and the reference resistance and so on, the entire process of measurement spends a The whole measurement process takes very short time. At the same time, the user can real-time test data, observe the progress of measurement, and terminate the current operation. This greatly improves work efficiency. In automatic mode, the user can pre-set a series of operation measurements, or select the saved configuration file from the template library. After the test starts, the system automatically controls the operation of the test process, including opening the serial port, online testing, calibrating the low-resistance 10 Ω file, calibrating the low-resistance 20 Ω file, making measurements according to the enabled measurements, and updating the results of the measurements, and closing the serial port until the test is completed, and the whole measurement process generally takes a few seconds to a few tens of seconds. The whole measurement process usually takes a few seconds to tens of seconds. During the measurement process, the user can observe the measurement progress and test data in real-time, or artificially terminate the current measurement operation. This mode is suitable for the standard process of testing tasks, the user does not need to customize the process of repetitive work, to greatly simplify the workload and improve the efficiency of the effect. Measurement results are stored as Excel files.

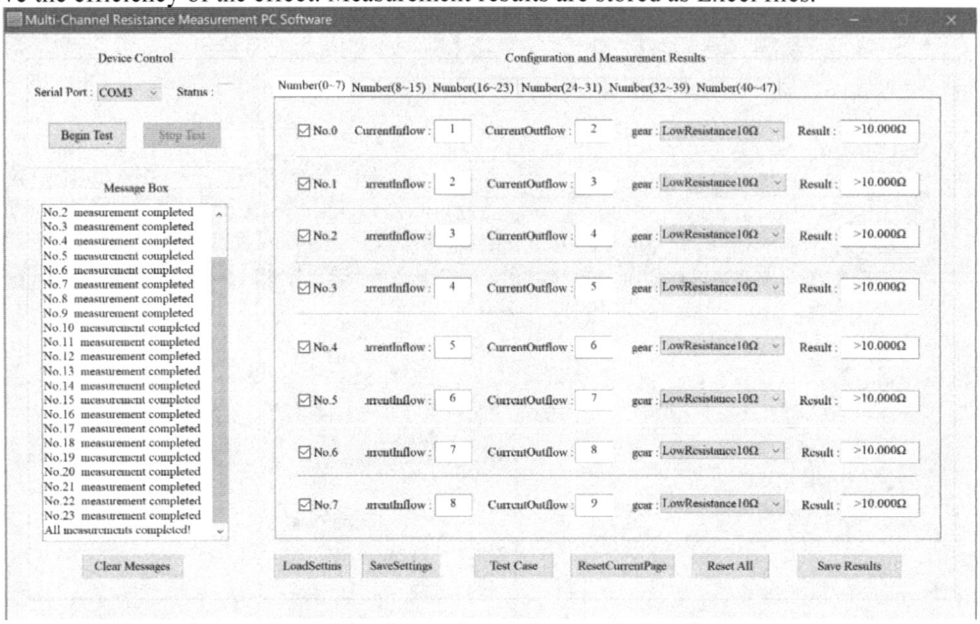

Figure 5. Interface of the debugging software of intelligent test system

6. Test Validation

To verify the function of the wiring harness intelligent test system, the ZX54 type laboratory DC resistance box (accuracy of 0.5 level) is used for measurement and verification when the measured resistance is 10Ω or less, the 10Ω file of the intelligent test system is used for verification at room temperature when the measured resistance is 10Ω~20Ω, the 20Ω file of the intelligent test system is used for verification, and the test results are shown in Table 4.

Table 4. Resistance measurement results

Resistance box test value (Ω)	Measuring gear	Test System Test Value (Ω)	Absolute error (mΩ)	Relative error (%)
0.100	10 Ω	0.105	5	5.0
0.200	10 Ω	0.204	4	2.0
0.300	10 Ω	0.303	3	1.0
0.500	10 Ω	0.503	3	0.6
0.700	10Ω	0.706	6	0.8
0.900	10Ω	0.901	1	0.1
1.000	10 Ω	1.003	3	0.30
5.000	10 Ω	5.002	2	0.04
9.000	10 Ω	9.008	8	0.08
10.000	20Ω	10.016	16	0.16
11.000	20Ω	11.017	17	0.15
12.000	20Ω	12.016	16	0.13
18.000	20Ω	18.017	17	0.09

Resistance box test value (Ω)	Measuring gear	Test System Test Value (Ω)	Absolute error (mΩ)	Relative error (%)
19.900	20Ω	19.918	18	0.09

From the measurement data, it can be seen that the maximum absolute error of measurement for 10Ω gear is 8 mΩ, and the maximum absolute error of measurement for 20Ω gear is 18 mΩ, which can meet the measurement requirements on many occasions.

To further verify the accuracy of the intelligent test system, the CMR-128 microohmmeter (micro resistance meter, the national military standard production line special low-resistance meter equipment, accuracy of 0.05%) produced by Prime Macro Electronic Enterprises Co., Ltd. was selected to measure and verify the test system, with an ambient temperature of -55℃~125℃, and since most of the tests are mΩ level resistors, the 10Ω file of the system was selected for testing. The results were obtained as shown in Table 5 below.

Table 5. Resistance measurement results (with 10Ω gear)

Low Resistance Meter Test Value(Ω)	Test temperature	Measured value of test system (Ω)	Absolute error (mΩ)	Relative error (%)
1.532	-55°C	1.530	2	0.13
5.024	-20℃	5.021	3	0.06
0.0367	-10℃	0.039	2	5.45
0.0911	40℃	0.090	1	1.09
0.160	100℃	0.163	3	1.9
0.2186	125℃	0.227	8	4.57

From the measurement data, it can be seen that the maximum absolute error of measurement of 10Ω gear in -55℃~125℃ ambient temperature is 4mΩ, which can meet the measurement requirements on many occasions.

7. Concluding Remarks

Currently, this intelligent testing system has been implemented in the internal control components of a military-industrial organization. It enables the testing of continuity status between multiple pins in control component products and measures the actual contact integrity of cable harnesses, achieving an accuracy of up to 1%. The system can rapidly and accurately identify quality issues caused by faults such as open circuits, short circuits, poor contact, or miswiring in harnesses. It overcomes the limitations of traditional detection methods that rely on low-resistance meters or multimeters for sequential single-channel measurements, offering the following advantages: high efficiency, reliability, simple structure, user-friendly operation, strong versatility, easily expandable testing capacity, and cost-effectiveness.In summary, the design of this intelligent testing system significantly enhances the efficiency of harness testing while reducing labor intensity. It effectively minimizes human-induced testing errors in aerospace equipment, ensuring product quality and production timelines. This innovation holds significant practical implications for improving R&D efficiency in aerospace product development and advancing industry progress.

Acknowledgments

The authors are grateful to the financial support from the Fund Project: Project for Enhancing Young and Middle-aged Teacher's Research Basis Ability in Colleges of Guangxi in 2023, "Research on Data Transmission and Receiving System of Internet of Things Based on CAN Bus" (2023KY0983), Leader: Sun Ping.

References

[1] WANG Xiaojun, ZHOU Xingpeng, WANG Yi, ZHAO Yongjie. Design of a high-precision insulation resistance meter with a broad range[J]. *Journal of Electronic Measurement and Instrument*, 2021, 26(2): 155-160.

[2] Mengyuan LI, Danmin LI, Jiwei ZHANG, Xianglong CHEN. Research on predictive maintenance technology for aerospace equipment under big data background[J]. China New Telecommunications, 2023(2): 25-28.

[3] Peng WU, Yizhe ZHANG, Jingqian WEN, Chunyang WANG, Pei GAO. Discussion on key issues in digital manufacturing of aviation wire harness[J]. Aeronautical Manufacturing Technology, 2017(14): 97-100.

[4] Shouxian XIA. IoT system applied to civil aerospace equipment production[J]. Automation and Information Engineering, 2021(2): 39-41.

[5] Yi LUO, Kun YANG, Lin DONG, Yanhui ZHU, Chunxue SHANG. Research on dynamic measurement method of contact resistance for switch contacts[J]. Chinese Journal of Scientific Instrument, 2015(1): 49-54.

[6] Dongya CHEN, Yujie QIU, Chunjuan WEI, Qiangqiang YANG. Research on Modbus communication protocol consistency of smart energy meters on the distribution side[J]. Ship Electronic Engineering, 2024(1): 37-38.

[7] Xinyu JING. Design of a precision micro-resistance measurement system using programmable constant current method[J]. Electronic Technology Design and Application, 2016(10): 40-42.

[8] Wentian ZHAN, Haijun YANG, Hongyue ZHANG. Research on automatic continuity testing method for EMU cables[J]. Science and Technology Forum, 2016(8): 110-111.

Muscle Synergy-driven TimesNet Method for Continuous Recognition of Lower Limb Motion Patterns

Hao Zhou[1,2#], Xiaohui Li[2#], Yinghu Peng[2], Xueyan Lyu[2], Dingxun Jin[2] and Lin Wang[2,3*]

[1] Southern University of Science and Technology, Shenzhen 518055, China

[2] Shenzhen Institutes of Advanced Technology, Chinese Academy of Sciences, Shenzhen 518055, China

[3] Shenzhen Lower Limb Intelligent Rehabilitation Engineering Research Center, Shenzhen 518055, China

[#]Hao Zhou and Xiaohui Li contributed equally to this manuscript.

[*]Corresponding author's e-mail: lin.wang1@siat.ac.cn

Abstract. Accurate recognition of human lower limb motion patterns is essential for the effective control of intelligent assistive robots. However, most existing methods in recent studies relied on the complete gait cycle as input, limiting their ability to perform continuous recognition throughout the entire motion process. To address this limitation, this study proposed a muscle synergy-driven TimesNet model with low computational complexity to enable efficient and accurate continuous recognition of lower limb motion patterns. Surface electromyography (sEMG) signals were collected from 11 muscles of the dominant legs of 15 subjects during five types of lower limb movements. The 11-channel preprocessed sEMG signals were decomposed into 8 muscle synergies using principal component analysis (PCA). To comprehensively extract features from the one-dimensional time series data, a deep learning model known as TimesNet was employed to capture both intra-period and inter-period information from the synergy data. The proposed method achieved an average recognition accuracy of 0.91, with computational cost measured at 0.98 billion FLOPs, demonstrating state-of-the-art performance with low computational complexity. These results indicated that the proposed method enables efficient and accurate continuous recognition of lower limb motion patterns. This study provided a more practical and real-time applicable solution for motion pattern monitoring in the control of assistive robotic devices, such as exoskeletons and intelligent prostheses.

1. Introduction

Lower limb motion pattern recognition is critically important for controlling assistive robots, such as powered exoskeletons and intelligent prostheses. Accurate and continuous motion recognition can provide real-time feedback to the robot controller, aiding in trajectory planning and switching between assistance modes

In the past few years, some methods have been proposed for lower limb motion patterns recognition based on machine learning algorithms. Shi [1] extracted features from 3 muscles, and then achieved the recognition of 5 lower limb motion patterns based on the scale unscented Kalman filter (SUKF) and neural network (NN). Vijayvargiya [2] integrated convolutional neural network (CNN) and long short-term memory (LSTM) to recognize 3 movements based on surface electromyography (sEMG) signals.

Tu [3] achieved in accurate recognition of 5 movements based on CNN and kernel extreme learning machine (KELM) model. In our previous study [4], an efficient and accurate recognition model integrating Gramian Angular Fields and multiscale fusion of ResNet was proposed, in order to extract the features of sEMG data more comprehensively. However, these studies focused on the lower limb motion patterns recognition for the entire gait cycle, with the recorded data of the entire gait cycle as the inputs of the proposed models, which would cause significant time delay in the control of assistive robots. Additionally, the computational complexity of the model proposed in these studies was high, reducing the recognition efficiency.

As for the choice of sensors, sEMG enables real-time analysis with high temporal resolution, and facilitates the detection of subtle muscle activation changes, allowing for the anticipation of movement intentions even before observable motion occurs [5, 6]. It would not lead to information delays, as seen in inertial measurement unit (IMU), or potential issues such as occlusion, limited camera capture range, and data privacy concerns commonly associated with vision-based sensors [6]. These characteristics underscore the unique role of sEMG in enhancing the accuracy and responsiveness of lower limb motion pattern recognition system, and sEMG has been widely utilized in recent studies of motion pattern recognition [1-3, 7, 8]. Additionally, by decomposing complex muscle activation patterns into a set of fundamental and low-dimensional synergies, muscle synergy analysis simplifies the representation of neuromuscular control strategies [9]. This not only reduces computational complexity but also enhances the interpretability of the underlying physiological mechanisms. In motion patterns recognition, muscle synergy analysis could facilitate the identification of shared movement primitives across different tasks and conditions, providing a physiologically meaningful framework for distinguishing between similar motion patterns [10].

Therefore, this study aimed to propose a muscle synergy-driven deep learning method, with lower computational complexity, to achieve the efficient and accurate continuous recognition of lower limb motion patterns.

2. Material and Methodology

2.1. *Participant information*

15 healthy subjects (22 ~ 35 years) participated in this study. The subjects' preferred legs were the right ones and all of them were free from lower limb mobility impairments. Written informed consent was provided by the subjects before the experiment. The experimental protocol received ethical approval from the Human Ethics Committee of the Shenzhen Institutes of Advanced Technology, Chinese Academy of Sciences (Number: SIAT-IRB-200315-H0478).

2.2. *Experiment procedure*

2.2.1. Data acquisition. Raw sEMG signals were recorded from 11 muscles in the subject's dominant leg by a wearable 16-channel sEMG acquisition system, with the sampling frequency as 2000 Hz and bandwidth as 0-500 Hz. The 11 selected muscles included gluteus medius (GM), rectus femoris (RF), vastus lateralis (VL), vastus medialis (VM), biceps femoris (BF), semitendinosus (ST), peroneus longus (PL), tibialis anterior (TA), medial gastrocnemius (MG), lateral gastrocnemius (LG), and soleus (SOL). The placement of the electrodes followed the Surface Electromyography for the Non-Invasive Assessment of Muscles (SENIAM) [11].

Shown in Figure 1. (a), the subjects were required to perform 5 different motion patterns, including standing, walking, running, squat, and jump, each repeated 5 times. There were a 30-second rest between repetitions and a 5-minute rest between different patterns to prevent muscle fatigue.

2.2.2. Data preprocessing. The preprocessing process was shown in Figure 1. (b). To eliminate power line interference, motion artifacts, ECG interference, and Gaussian white noise [6], the raw sEMG signals were denoised using a notch filter (50 Hz), a 4th-order Butterworth bandpass filter (20-450 Hz), and a wavelet threshold filter (the db4 wavelet, decomposition level-6, and soft thresholding). The denoised data were rectified and normalized to [0, 1]. Then, the activation state data were extracted from the sEMG signals.

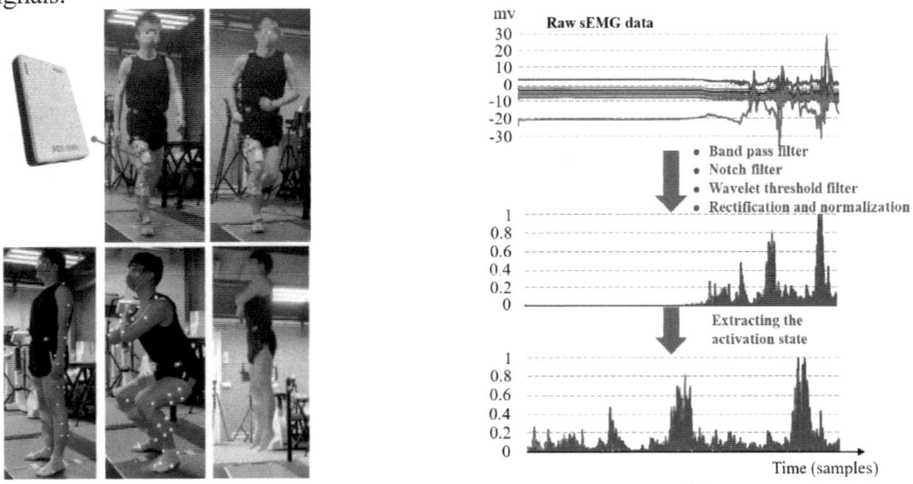

Figure 1. (a) the wearable sEMG signal acquisition device and the five daily motion patterns. (b) the data preprocessing process.

2.2.3 Class label assignment. The preprocessed data were segmented by a sliding window with the length as W and the stride as S, shown in Figure 2. There was a label of motion pattern for sEMG data at each time sample. The sliding window was labeled according to the motion pattern at the last time sample in the window. Each sliding window served as a data sample in this study.

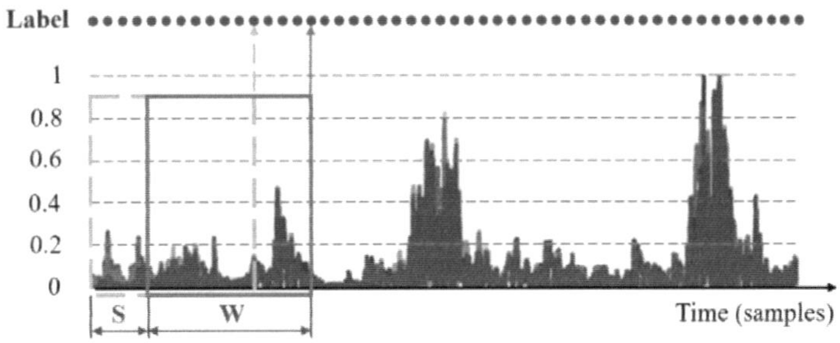

Figure 2. The class label assignment in this study.

2.3. Muscle synergy-driven recognition method

2.3.1. Muscle synergy. The principal component analysis (PCA) [12] serves as a decomposition method, usually applied to the muscle synergy analysis [13]. In this study, based on PCA, the preprocessed sEMG data $A^{c\times t}$ was decomposed to $Y^{z\times t}$ and $W^{c\times z}$, with $Y^{z\times t}$ as the data in the synergy space:

$$A^{c\times t} = W^{c\times z}Y^{z\times t} + E^{c\times t} \tag{1}$$

where $Y^{z\times t}$ is the activation coefficient matrix, containing the muscles activation information of the synergies, $W^{c\times z}$ refers to the basis vectors of the synergy space, and $E^{c\times t}$ is a residuals matrix. c is the number of recorded sEMG channels, t refers to the time samples, and z is the number of synergies.

The number of synergies was determined by the variance accounted for (VAF). It's suggested that the average VAF exceeds 0.9, with only a slight increase in the average VAF when an additional synergy is added [14].

$$VAF = 1 - SS_{res} / SS_{tot} = 1 - \left\| A - \hat{A} \right\|^2 / \left\| A - \bar{A} \right\|^2 \tag{2}$$

where SS_{res} is the sum of squared residuals, SS_{tot} is the total sum of squares, $\|\cdot\|$ denotes the ℓ_2 operator norm, A is original sEMG signals, $\bar{A}$ is the mean of A, and $\hat{A}$ is the reconstructed sEMG signals.

2.3.2. TimesNet. In this study, TimesNet [15] model was adapted to achieve the recognition of the lower limb motion patterns. It has been proved to achieve state-of-the-art in time series forecasting and classification tasks [15]. Considering the limitations of 1D time series in feature extraction, TimesNet transformed it into a set of 2D tensors based on the multi-periodicity in 1D time series, in order to comprehensively extract both intraperiod- and interperiod-information.

The architecture of TimesNet in this study is shown in Figure 4, which is composed of two TimesBlocks. For the *l*-th layer of TimesNet, the input is X_{1D}^{l-1}, and the output is X_{1D}^{l}, with the TimesBlock defined as:

$$X_{1D}^{l} = TimesBlock(X_{1D}^{l-1}) + X_{1D}^{l-1} \tag{3}$$

X_{1D}^{0} = Embedding (Y^T) ∈ $R^{t\times d_{\text{mdoel}}}$ served as the input of TimesNet. In each TimesBlock, at first, according to the top-*k* amplitude values in the amplitude-frequency plot based on Fast Fourier Transform (FFT), the most significant *k* frequencies $\{f_1,\cdots,f_k\}$ were selected for the input X_{1D}^{l-1}, coupled with *k* period lengths $\{p_1,\cdots,p_k\}$, $p_i = \lceil t/f_i \rceil$. X_{1D}^{l-1} was reshaped into a set of 2D tensors $X_{2D}^{l} = \{X_{2D}^{l,1}, X_{2D}^{l,2},\cdots,X_{2D}^{l,k}\}$, with $X_{2D}^{l,i} \in R^{p_i\times f_i\times d_{\text{mdoel}}}, i\in\{1,\cdots,k\}$, shown in Figure 3.

Then, a parameter-efficient Inception block [16] was used to extract features from X_{2D}^{l}, thus getting the learned 2D representations $\hat{X}_{2D}^{l} = \{\hat{X}_{2D}^{l,1},\cdots,\hat{X}_{2D}^{l,k}\}$. At last, $\hat{X}_{2D}^{l}$ was reshaped back 1D tensors $\hat{X}_{1D}^{l} = \{\hat{X}_{1D}^{l,1},\cdots,\hat{X}_{1D}^{l,k}\}, \hat{X}_{1D}^{l,i} \in R^{t\times d_{\text{mdoel}}}$, which was then fused from *k* 1D tensors to one 1D tensor $X_{1D}^{l} \in R^{t\times d_{\text{mdoel}}}$ based on adaptive aggregation strategy.

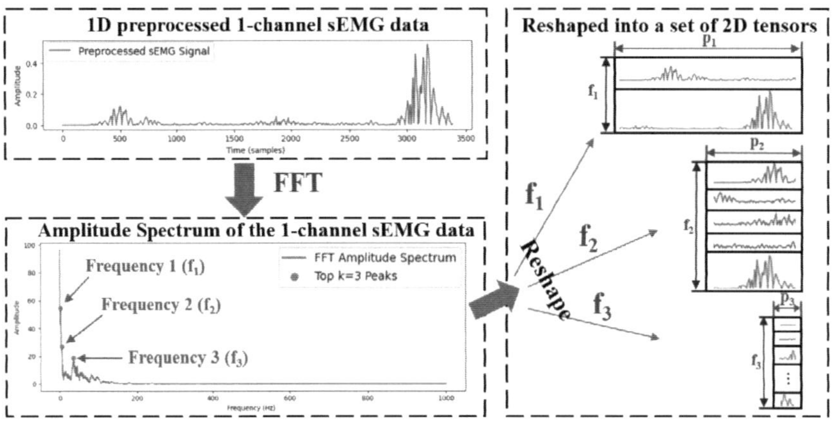

Figure 3. 1D preprocessed sEMG data were reshaped into a set of 2D tensors based on Fast Fourier Transform (FFT), taking 1-channel sEMG data and k=3 as an example.

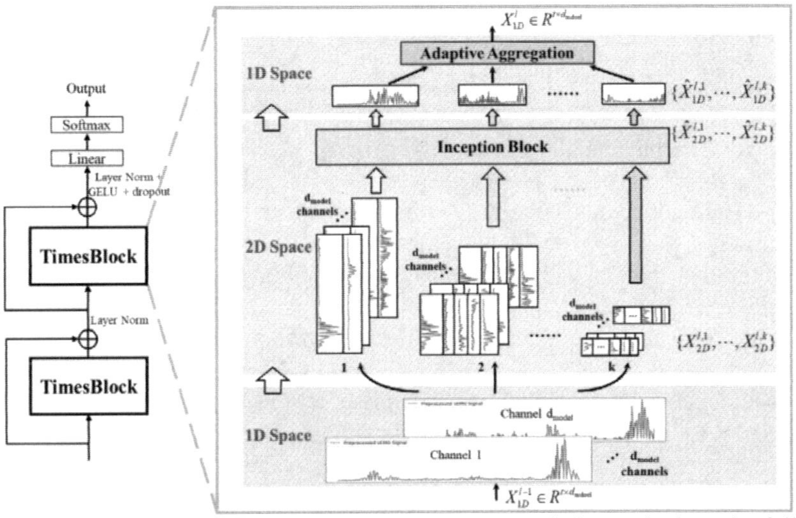

Figure 4. The architecture of TimesNet, composed of 2 layers of TimesBlock and 1 fully connected layer.

2.4. *Configurations and Performance metrics*

2.4.1. Configurations. The sEMG data after dimensionality reduction in the sliding window served as the input of TimesNet. The length of the window W was set to [500, 1000, 1500, 2000] time samples, namely [250, 500, 750, 1000] ms, with the stride S as 100 time samples.

The cumulative VAFs with different numbers of synergies across all the motion patterns were calculated. Across the 15 subjects, the averaged VAFs with 8 and 9 synergies were 0.954 ± 0.017 and 0.976 ± 0.011 (mean ± 1 std), respectively. Therefore, according to the standard mentioned in section 2.3.1, the number of synergies was finally set to 8.

There were totally 375 data files in the dataset. After randomly shuffling the data files, they were divided into training, validation, and test sets in a 7:1:2 ratio. During training, the 5-fold cross validation, early stopping strategy, cosine annealing decay strategy, Adam optimizer, and cross-entropy loss function (CEL) were implemented. The initial learning rate was set to 0.0002, batch size was 32, the max epoch was 150, and the patience of early stopping was 9 epochs. As for the TimesNet, dropout was 0.1, k was 5, and d_{model} was 16. The training and inference process were conducted in an AMD Ryzen Threadripper PRO 5975WX 32-Cores 3.60 GHz Processor and an NVIDIA GeForce RTX 3090.

2.4.2. Performance metrics. 5 metrics were selected to assess the recognition performance of the model, including the *accuracy*, *weighted precision*, *weighted F_1-Score*, *weighted recall*, and *weighted AUC*.

$$accuracy = n^+ / n \tag{4}$$

$$weighted\ precision = \sum_{i=1}^{m} w_i P_i \tag{5}$$

$$weighted\ F_1 - Score = \sum_{i=1}^{m} w_i F_{1i} \tag{6}$$

$$weighted\ recall = \sum_{i=1}^{m} w_i R_i \tag{7}$$

$$weighted\ AUC = \sum_{i=1}^{m} w_i AUC_i \tag{8}$$

where n^+ is the number of data samples correctly recognized by the model, n is the total number of the data samples (sliding windows), m is the number of recognized categories, and w_i is the weight of the *ith* category, $w_i = n_i/n$, with n_i as the number of the samples for the *ith* category. With the One-vs-Rest scheme applied to the *ith* category, P_i, F_{1i}, R_i, and AUC_i refer to the *precision*, *F_1-Score*, *recall*, and *AUC* for the *ith* category. Additionally, Floating Point Operations (FLOPs) per forward inference (batch size

= 1) and the total parameters of the model (Params) are used to assess the computational complexity.

3. Results and Discussion

3.1. Recognition performance based on the muscle synergy-driven method

The recognition performance of the proposed muscle synergy-driven TimesNet method is shown in Figure 5 and Table 1. The *accuracy*, *weighted precision*, *weighted recall*, *weighted* F_1 and *weighted AUC* of the proposed method were 0.91, 0.92, 0.91, 0.91, and 0.98, respectively. With the exception of the *weighted AUC*, the other metrics were approximately 3.4% higher than those of the TimesNet model without the muscle synergy strategy.

According to the confusion matrix in Figure 5, the recognition accuracies for jump and running were lower compared to other motion patterns, primarily due to the model misclassifying jumps as walking and running as walking to some extent. These misclassifications might be attributed to the similarities in muscle coordination between the misclassified motion patterns. Additionally, the computational complexity of the proposed method was low, with FLOPs and Params as 0.98 billion and 1.42 million respectively (W = 750 ms).

Overall, the results demonstrated that the muscle synergy-driven method can achieve accurate continuous recognition of target motion patterns while maintaining low computational complexity.

Table 1. The recognition performance of TimesNet with PCA and without PCA (W = 750 ms).

Model	Accuracy	Weighted-precision	Weighted-recall	Weighted-F_1	Weighted-AUC	FLOPs (M)	Params (M)
TimesNet+PCA	**0.91**	**0.92**	**0.91**	**0.91**	**0.98**	**982.8**	**1.42**
TimesNet	0.88	0.89	0.88	0.88	0.98	983.1	1.42

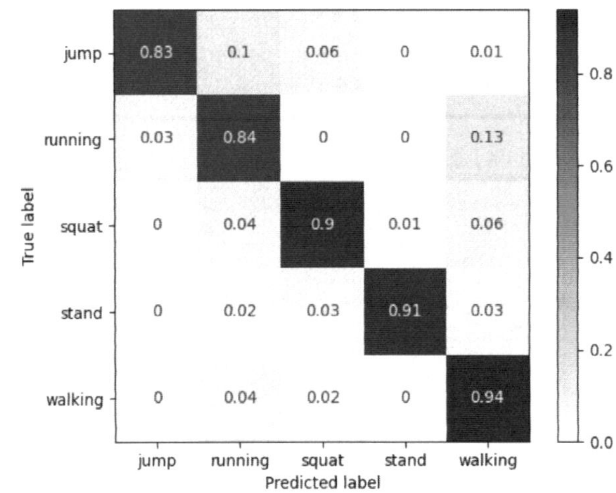

Figure 5. The confusion matrix of the proposed method for motion recognition (W = 750 ms).

3.2. Impact of the sliding window's length on the recognition performance

The sliding windows of varying lengths encapsulate different amounts of historical information, which may influence the model's performance. Consequently, the effect of the sliding window length on performance was analyzed in this study. The performance of the muscle synergy-driven TimesNet method with different sliding window lengths (W) is presented in Table 2. The results indicated that as the sliding window length increased, the model's recognition performance initially improved, then declined, with the best performance observed at a window length of 750 ms. This could be attributed to the insufficient historical information captured by shorter windows and the information redundancy introduced by longer windows.

Table 2. The recognition performance of the proposed method with different sliding window length *W*.

W (ms)	Accuracy	Weighted-precision	Weighted-recall	Weighted-F_1	Weighted-AUC	FLOPs (B)	Params (M)
250	0.84	0.88	0.84	0.85	0.97	**0.35**	**1.34**
500	0.88	0.91	0.88	0.89	0.97	0.64	1.38
750	**0.91**	**0.92**	**0.91**	**0.91**	**0.98**	0.98	1.42
1000	0.86	0.90	0.86	0.86	0.97	1.36	1.46

3.3. *Methods comparison*

The architecture design of TimesNet enhanced both inference efficiency and accuracy. Specifically, instead of directly extracting features from the 1D time series data (such as ResNet34 [17]), the 1D time scrics was initially reshaped into 2D data using FFT at the start of the TimesBlock. This design allowed for a more comprehensive extraction of both intra-period and inter-period information [15]. The comparison of proposed method with some other models in recognition performance was shown in Table 3. The muscle synergy-driven TimesNet outperformed the other models in all the assessment metrics.

Moreover, there were only 2 layers of TimesBlock in this study, leading to low computational complexity, thus achieving more efficient continuous recognition than the other models listed in Table 3. In contrast, the transformer-based models [18-20] were limited by the self-attention mechanism, where the dot product could lead to high computational demands. Shown in Table 3, the FLOPs and Params of the proposed method were 0.98 billion and 1.42 million respectively, lower than those of ResNet34 and Transformer-based models.

Additionally, most existing methods in recent studies relied on the complete gait cycle as input [1-4], lacking the capability for continuous recognition throughout the motion process. Using data from a complete gait cycle as input would introduce significant delay in the control of wearable assistive devices, hindering real-time feedback and reducing control accuracy. In contrast, this study achieved efficient and accurate continuous motion patterns recognition based on the proposed muscle synergy-driven TimesNet model. It provided a more practical method for real-time monitoring of lower limb motion patterns in clinical setting and assistive wearable robot applications.

Table 3. The comparison of the proposed method with the other state-of-the-art models in the recognition performance (*W* = 750 ms).

Models	Accuracy	Weighted-precision	Weighted-recall	Weighted-F_1	Weighted-AUC	FLOPs (B)	Params (M)
ResNet34 [17]	0.53	0.51	0.53	0.42	0.72	1.95	21.28
Transformer [19]	0.46	0.50	0.46	0.43	0.84	14.09	10.17
Informer [20]	0.71	0.69	0.71	0.68	0.94	9.52	18.58
PatchTST [18]	0.74	0.81	0.74	0.74	0.95	13.78	11.58
Proposed method	**0.91**	**0.92**	**0.91**	**0.91**	**0.98**	**0.98**	**1.42**

3.4. *Limitations and future work*

However, several limitations of this study should be acknowledged. First, the transitional phase between different motion patterns was not considered, as continuous recognition was only achieved during the stable phase of the gait. Second, the sample size in this study was relatively small, which might have limited the ability to train a more robust model. In the future, efficient and accurate continuous recognition of lower limb motion patterns will be achieved across both transitional and stable phases of the gait. Additionally, data from a larger cohort of healthy subjects and patients with lower limb dysfunction will be collected to train a more generalized and robust model.

4. Conclusion

Based on the multi-channel sEMG data, this study proposed a muscle synergy-driven TimesNet method, achieving in the accurate continuous recognition of the lower limb motion patterns with lower computational complexity. The proposed method could be utilized to provide real-time feedback to the controller of the assistive robot, such as powered exoskeleton, and intelligent prosthesis, so as to achieve accurate control with low time delay.

Acknowledgements
This work was supported in part by Shenzhen Sustainable Development Sci-Tech project (Grant KCXFZ20230731093501003), in part by Shenzhen Science and Technology Program (Grant KQTD20210811090217009), and in part by Shenzhen Strategic Emerging Industry Support Plans (Grant XMHT20230115002).

References

[1] Shi X *et al.*, "Feature Extraction and Classification of Lower Limb Motion Based on sEMG Signals," *IEEE Access,* vol. 8, pp. 132882-132892, 2020.
[2] Vijayvargiya A *et al.*, "Hybrid Deep Learning Approaches for sEMG Signal-Based Lower Limb Activity Recognition," *Mathematical Problems in Engineering,* vol. 2022, p. 3321810, 2022.
[3] Tu J *et al.*, "Lower limb motion recognition based on surface electromyography," *Biomedical Signal Processing and Control,* vol. 81, p. 104443, 2023.
[4] Zhou H *et al.*, "Integration of multiscale fusion of residual neural network with 2-D gramian angular fields for lower limb movement recognition based on multi-channel sEMG signals," *Biomedical Signal Processing and Control,* vol. 99, p. 106807, 2025.
[5] Cheng J *et al.*, "A framework for daily activity monitoring and fall detection based on surface electromyography and accelerometer signals," *IEEE journal of biomedical and health informatics,* vol. 17, no. 1, pp. 38-45, 2012.
[6] Vijayvargiya A *et al.*, "Human lower limb activity recognition techniques, databases, challenges and its applications using sEMG signal: an overview," *Biomedical Engineering Letters,* vol. 12, no. 4, pp. 343-358, 2022.
[7] Kyeong S *et al.*, "Surface electromyography characteristics for motion intention recognition and implementation issues in lower-limb exoskeletons," *International Journal of Control, Automation and Systems,* vol. 20, no. 3, pp. 1018-1028, 2022.
[8] Tao Y *et al.*, "Multi-channel sEMG based human lower limb motion intention recognition method," in *IEEE/ASME AIM*, 2019, pp. 1037-1042.
[9] Steele K M *et al.*, "Muscle synergies and complexity of neuromuscular control during gait in cerebral palsy," *Developmental Medicine & Child Neurology,* vol. 57, no. 12, pp. 1176-1182, 2015.
[10] Lanzani V *et al.*, "Kinematic–Muscular Synergies Describe Human Locomotion with a Set of Functional Synergies," *Biomimetics,* vol. 9, no. 10, p. 619, 2024.
[11] Stegeman D *et al.*, "Standards for surface electromyography: The European project Surface EMG for non-invasive assessment of muscles (SENIAM)," *Enschede: Roessingh Research and Development,* vol. 10, pp. 8-12, 2007.
[12] Abdi H *et al.*, "Principal component analysis," *Wiley interdisciplinary reviews: computational statistics,* vol. 2, no. 4, pp. 433-459, 2010.
[13] Naik G R *et al.*, "Principal component analysis applied to surface electromyography: a comprehensive review," *IEEE Access,* vol. 4, pp. 4025-4037, 2016.
[14] Roh J *et al.*, "Robustness of muscle synergies underlying three-dimensional force generation at the hand in healthy humans," *Journal of neurophysiology,* vol. 107, no. 8, pp. 2123-2142, 2012.
[15] Wu H *et al.*, "Timesnet: Temporal 2d-variation modeling for general time series analysis," in ICLR, 2023.
[16] Szegedy C *et al.*, "Going deeper with convolutions," in *IEEE CVPR*, 2015, pp. 1-9.
[17] He K *et al.*, "Deep residual learning for image recognition," in *IEEE CVPR*, 2016, pp. 770-778.
[18] Nie Y *et al.*, "A time series is worth 64 words: Long-term forecasting with transformers," *arXiv preprint arXiv:.14730,* 2022.
[19] Vaswani A, "Attention is all you need," *arXiv:1706.03762,* 2017.
[20] Zhou H *et al.*, "Informer: Beyond efficient transformer for long sequence time-series forecasting," in *AAAI*, 2021, vol. 35, pp. 11106-11115.

Design of Motor Speed Controller Based on FPGA

Yiding Wang[1], Li Gao[2], Mingxin Gu[3], Guoguo Wu[3], Peng Lu[3*]

[1] Chengdu SIWI Power Electronic Technology Co., Ltd., Chengdu, Sichuan Province, 610000, China

[2] State Key Laboratory of Elcctronic & Electrical Products (Chongqing), Chongqing Customs Technology Center, Chongqing, 400020, China

[3] School of Intelligent Manufacturing Engineering, Chongqing University of Arts and Science, Yongchuan 402160, China

*Corresponding author's e-mail: lupeng@cqwu.edu.cn

Abstract. In the field of modern industrial automation, the accuracy and stability of motor speed control technology are essential to ensure production efficiency and product quality. The purpose of this research is to design and implement an FPGA motor speed controller based on a PID control algorithm to meet the needs of high-precision motor control. This design analyzes the principle of the PID control algorithm, including the function of proportion, integral, and differential three links and their adjustment methods. By optimizing the parameters of the PID controller, the high precision control of motor speed is realized. In the design process, this study elaborated on the FPGA implementation of PID controller-specific flow and key technology. The results show that the controller can quickly respond to the change in motor speed and output the corresponding PWM signal to adjust the motor speed.

1. Introduction

With the continuous development of industry, intelligent equipment has been more widely used in various fields such as machinery, robots, petroleum, coal mines, etc. Traditional motor control technology has developed rapidly to meet the needs of actual production. Traditional motor speed control is difficult to meet the needs of precise speed control of different intelligent equipment, and various control methods have been developed for different control needs. Thus better control of the motor[1,2].

The field of motor speed control has become an important field of competition and cooperation among countries. By increasing the research and application of high-precision motors, countries not only enhance their international competitiveness but also promote the continuous innovation and development of motor control technology. This design can be widely used in different control scenarios of the model. According to different expected speeds, the parameters of the PID [3-5]link can be adjusted in time and the target speed can be changed in time, to achieve accurate control for different production processes and production needs and cope with different control objects. Reduce energy consumption and reduce production costs.

2. Preliminary

Motor speed control is an important part of automatic production, which determines the precision and efficiency of motion control and has great significance for improving social production efficiency and reducing production costs. In this paper, the motor speed control design based on PID is applied to the FPGA development board. The current motor speed is calculated by measuring the pulse width of PWM. The PID algorithm compares the current speed with the input desired speed and quickly adjusts the pulse width of PWM to make the current speed consistent with the target speed, achieving the purpose of motor speed control [6-8].

The real-time motor speed is calculated by processing the pulse data from the stepper motor through FPGA, and the PID algorithm reads the current speed information of the motor. Comparing the desired speed with the current speed, the error value is calculated, which reflects the difference between the current speed and the target speed. The control parameters are calculated according to the PID algorithm, including the error value multiplied by the scale coefficient to get the scale term, which is used to rapidly reduce the error; The error value is accumulated and multiplied with the integral coefficient to get the integral term, which is used to eliminate the static difference and improve the stability of the system. The differential term is obtained by multiplying the rate of error change with the differential coefficient, which is used to predict the error change and correct it in advance to reduce overshoot. These three terms are weighted together to obtain the PWM duty cycle correction. Finally, this correction is added to the current PWM duty cycle to obtain a new PWM duty cycle, which is used to control the power supply of the motor to adjust its speed. The whole process is repeated until the motor reaches the desired speed or a preset control time. In summary, the overall scheme design of the appropriate system as shown in Figure 1 is determined.

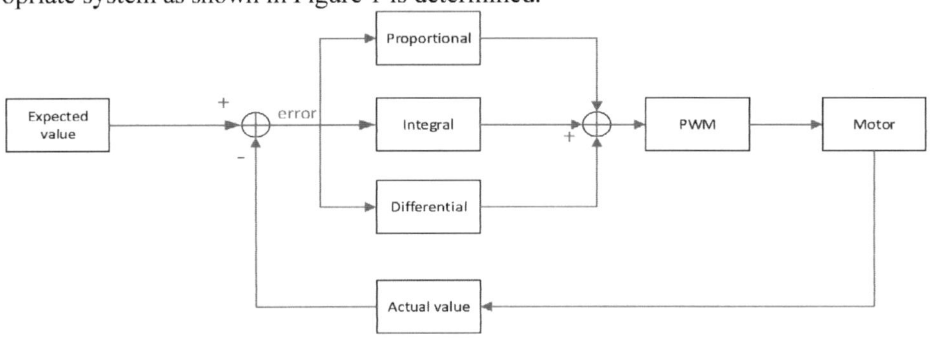

Figure 1. PID Schematic diagram of PID control

3. Methods

In contrast to traditional embedded control, FPGA can be reprogrammed in real-time at the hardware level to fulfill diverse tasks, endowing it with flexibility and the ability to adapt to evolving requirements and standards. Concurrently, FPGA is equipped with parallel processing capabilities, enabling it to execute multiple tasks and thus providing enhanced computational speed and responsiveness. This characteristic allows FPGA to exhibit remarkable performance in applications demanding high-performance computing, such as image processing, radar signal processing, and medical signal processing. Although embedded systems can also achieve parallel processing, they are generally constrained by hardware, and their performance may not rival that of FPGA. Given the multiple advantages of FPGA, it is selected as the control core in this paper for the calculation and control of the motor.

3.1. Hardware circuit design

This design takes FPGA (EP4CE15F23C8) as the main control core, with its powerful logic processing ability and high programmable flexibility, providing a solid hardware foundation for the motor control system. Combined with the TB6612FNG motor drive module, the module ensures the effective drive of the DC deceleration stepper motor with its excellent drive performance and high stability. It is particularly important that the design uses a DC gear motor with an encoder, through the encoder real-time accurate capture and feedback of the motor operating state information, such as speed and position, to achieve precise control of the motor parameters. The whole system integrates precision control logic and a closed-loop feedback mechanism to ensure the stability and accuracy of motor operation, which provides solid theoretical and technical support for further exploration in the field of automation application.

Motor selection. The performance of the motor stands as a pivotal factor determining the overall performance of the equipment. In this regard, the motor speed controller plays an indispensable role in optimizing motor performance. Through precise control of the motor speed, the motor speed controller enables the motor to better conform to the working requirements of the equipment, thereby enhancing the stability and reliability of the equipment. The MG370 encoder motor is selected for this design.

3.1.1. Controller selection. Compared with traditional embedded control, FPGA has outstanding characteristics at the hardware level. It has a strong parallel data processing capability, which means that it can handle multiple tasks at the same time, greatly improving the efficiency of data processing. FPGA plays a key role, it is responsible for the PID control algorithm and PWM pulse width calculation. Through precise programming, the FPGA can achieve precise control of the motor speed, so that the motor can run stably at the preset speed. This precise control not only improves the response speed of the system but also enhances the stability of the system. PID algorithm can quickly respond to the change in motor speed, and the accurate calculation of PWM pulse width ensures that the motor can get the appropriate drive signal. This series of advantages allows the motor control algorithm can be quickly executed to achieve real-time control of the motor speed and bring more efficient and reliable solutions to the motor control field.

3.1.2. Determination of motor control strategy Among many control algorithms, PID control stands out for its unique advantages. Compared with auto-disturbance rejection control and fuzzy control, the PID control algorithm is simpler and less difficult to implement. It can adjust parameters flexibly in the face of continuous signal or discrete signal. After long-term practice tests, PID control has excellent stability, and after decades of development, has been quite mature.TB6612FNG is a highly efficient and stable motor driver chip from Toshiba Semiconductor. It uses a dual full bridge drive design and can drive two motors at the same time, each drive channel has two logic input pins, an output pin, and a PWM pin. This allows the user to control the operating state of the motor by adjusting the level state of the logic input pin, and achieve precise speed regulation of the motor through the PWM pin.

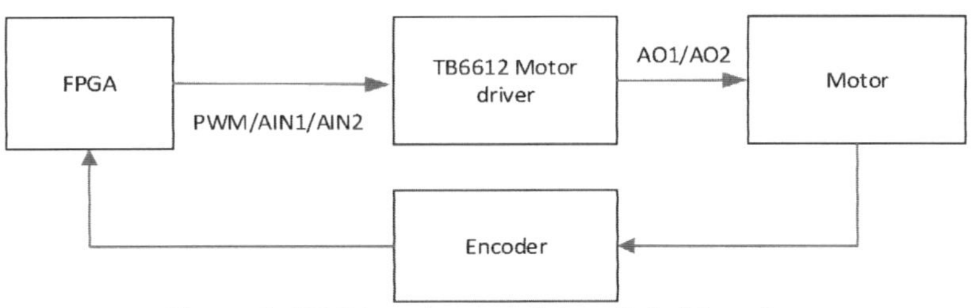

Figure 2. FPGA and motor drive digital interface

3.2. Algorithm and software implementation

3.2.1. Discrete PID algorithm The PID algorithm is the most commonly used control algorithm in industrial production because of its simple and practical characteristics. Its core idea is to control by calculating the function relationship of a proportional-differential-integral error value. The formula of the PID algorithm is a core part of the control theory, which is used to adjust the output of the system according to the error between the input and the expected output. However, the specific implementation method can be different due to different application scenarios and needs. The formula is shown in Formula 1.

$$u(t) = Kp * e(t) + Ki * \int e(t)dt + Kd * de(t)/dt \quad (1)$$

But it involves many fields such as industrial automation, robot control, aerospace, automotive control, and temperature control. By detecting and adjusting the error of the controlled object in real time, the discrete PID controller can realize the control effect of high precision, high stability, and reliability. Its formula is shown in the following formula 2, in which the effects of each variable are shown in Table.

$$u(k) = Kp * e(k) + Ki * \sum e(i) + Kd * (e(k) - e(k - 1)) \quad (2)$$

Where, u(k) is the control quantity, k_p, k_i, and k_d respectively are the proportion coefficient, integral coefficient, and differential coefficient.

3.2.2. Influence of PID parameters In this experiment, positional PID is used to change the parameters of proportional P, integral I, and differential D successively using control variables to determine the influence of the three links on the dynamic performance of the system.

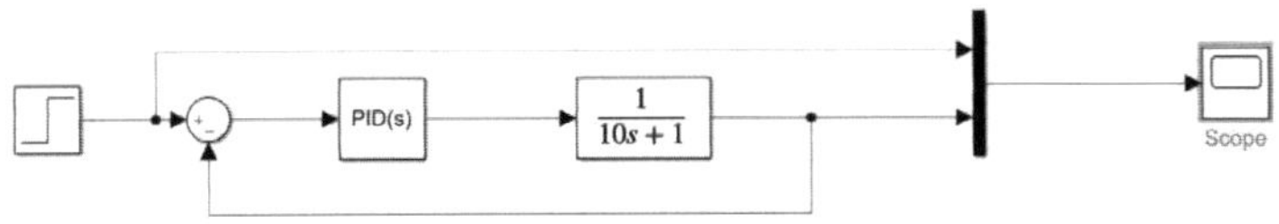

Figure 3. System simulation diagram

The results show that the increase of K_p reduces the rise time of the system, but the overshoot of the system also increases. K_i and K_d in a special ratio can eliminate the overshoot problem caused by K_p. When this relation holds, the system hardly oscillates and can reach the target speed quickly.

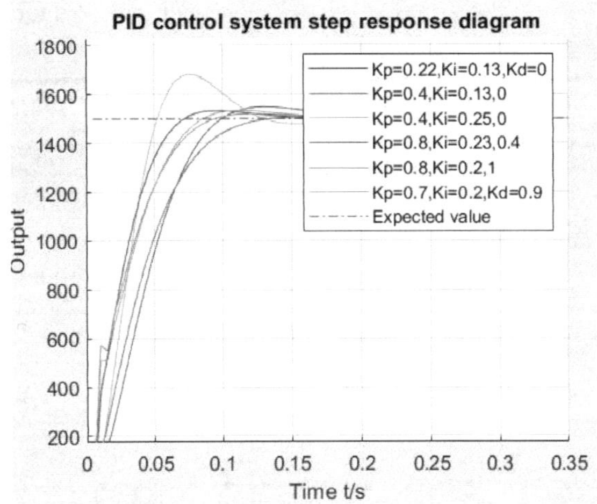

Figure 4. Influence curve of PID parameters on the system

3.2.3. Experimental verification The PID controller first obtains the error between the current state of the controlled object and the set value, and then calculates the output signal through the proportion, integral, and differential parts respectively. The output signals of these three parts are synthesized to form the final controller output signal, which is transmitted to the actuator to affect the state of the controlled object. By constantly adjusting the output signal, the PID controller can make the output value of the controlled object gradually close to the set value, and realize the stable control of the system. As shown in Figure 5, the target value is 800 RPM and y is the actual value. The horizontal coordinate is time minutes, and the vertical coordinate is revolutions. It can be seen that the system response time is fast and no overshoot.

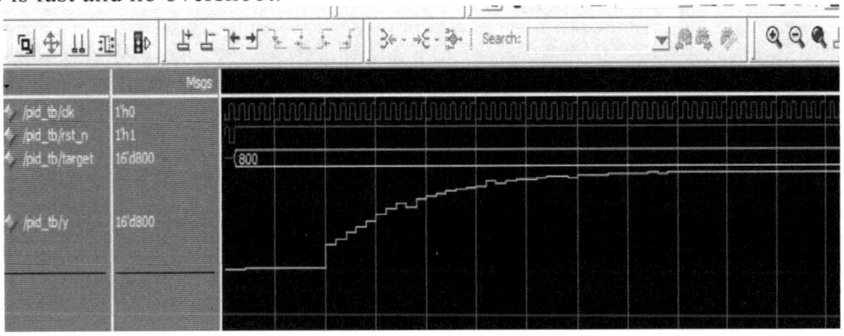

Figure 5. Motor speed test curve

4. Conclusion

In this design, the hardware design for the FPGA board is created, and the PID control algorithm and PWM signal generation module are realized. Then, the curve change of adjusting duty cycle was observed through the Modelsim simulation and finally verified on the development board. Using the oscilloscope to test the output PWM duty cycle is consistent with the target duty cycle previously set, which verifies the correctness of the design.

Acknowledgments

This research was funded by the Chongqing Education Commission Project (KJQN202201323), Yongchuan District Project (2023yc-jckx20093), Natural Science Foundation Project of Yongchuan District, Chongqing (Grant No. 2023yc-jckx20081).

References

[1] Altınkaya H, Ekmekci D. Tuning of PID Controller in PLC-Based Automatic Voltage Regulator System Using Adaptive Artificial Bee Colony – Fuzzy Logic Algorithm[J]. Electronics, 2024,13(24):5039-5039.

[2] Taş G, Özdamar M. Estimation of PID parameters of BLDC motor system by using machine learning methods[J]. Signal, Image and Video Processing, 2024,19(1):140-140.

[3] Lu W. Optimization and design of electromechanical control automation based on dual motor control algorithm[J]. Frontiers in Mechanical Engineering, 2024,101485041-1485041.

[4] Liu F, Rao L, He Z, et al. Robust PID Control for Upper Limb Exoskeleton Robot based on the Beetle Antennae Search Algorithm[J]. IAENG International Journal of Applied Mathematics, 2024,54(12):

[5] Shekher V, Sisodiya A, Sinha K A, et al. Optimal tuning of PID controller for V/f control of linear induction motor using artificial biological intelligence[J]. Franklin Open,2024,9100183-100183.

[6] He X, Yu Q, Pan X, et al. Improved beluga whale optimization-based variable universe fuzzy controller for brushless direct current motors of electric tractors[J]. Computers and Electrical Engineering, 2024,120(PC):109866-109866.

[7] Proaño P, Díaz R, Chillán C, et al. Sliding Mode Control Proposed Using a Clegg Integrator for Speed Control of a Three-Phase Induction Motor †[J]. Engineering Proceedings, 2024,77(1):8-8.

[8] Kong W, Zhang H, Yang X, et al. PID control algorithm based on multistrategy enhanced dung beetle optimizer and back propagation neural network for DC motor control.[J]. Scientific reports, 2024,14(1):28276.

Vibration Isolation and Control Based on Backstepping Control for Magnetic Adjustable Stiffness Fast Steering Mirror

Jin Luo[1,2,3,4], Jiachen Li[1,2,3,4], Luyao Zhang[1,2,3], Li Miao[1,2,3,4*], Jiuqiang Deng[1,2,3,4], Jinying Li[1,2,3,4], Yao Mao[1,2,3,4]

[1] National Key Laboratory of Optical Field Manipulation Science and Technology, Chinese Academy of Sciences, Chengdu 610209, China

[2] Key Laboratory of Optical Engineering, Chinese Academy of Sciences, Chengdu 610209, China

[3] Institute of Optics and Electronics, Chinese Academy of Sciences, Chengdu 610209, China

[4] University of Chinese Academy of Sciences, Beijing 100049, China

[*]Corresponding author's e-mail: miaoli@ioe.ac.cn

Abstract. Fast steering mirrors (FSMs) are vital for achieving precise beam alignment, yet mitigating vibrations poses significant challenges in demanding settings like vehicle-mounted systems. To resolve this problem, we propose a combined approach of passive vibration isolation and active control to enhance vibration isolation and control performance. A magnetic adjustable stiffness mechanism (MASM) is incorporated into the FSM, allowing it to be tuned to quasi-zero stiffness (QZS) to achieve QZS vibration isolation. Additionally, the adjustable stiffness FSM is controlled in a closed-loop using backstepping control (BSC). Simulation results demonstrate that this combined passive and active approach improves the vibration isolation and control performance of the FSM.

1. Introduction

Fast steering mirrors (FSM) have important applications in precision beam control, including optoelectronic measurement, adaptive optics, laser communication, and astronomical observation [1-3]. The FSM controls the direction of the light beam by rotating the mirror surface through a voice coil motor or piezoelectric ceramics [4]. When the FSM operates in complex environments, such as vehicle-mounted scenarios, there are higher requirements for vibration suppression.

Within the domain of passive vibration attenuation, the lowest frequency that a system can mitigate is determined by its natural frequency scaled by a factor of $\sqrt{2}$ [5], which depends on the stiffness of the system. Therefore, the lower the system's stiffness, the lower the minimum frequency that can be suppressed, leading to better vibration isolation performance. To enhance the vibration suppression ability of the FSM, the stiffness should be reduced. However, reducing stiffness also leads to a decrease in the load-bearing capacity of the FSM. As a result, conventional FSMs typically cannot achieve good passive vibration isolation performance.

To tackle this challenge, this study introduces the integration of a magnetic adjustable stiffness mechanism (MASM) into the FSM to modify its rigidity, adjusting it to quasi-zero stiffness (QZS). This approach ensures that the FSM maintains its load-bearing capacity while achieving near-zero stiffness, thereby improving passive vibration isolation performance.

Additionally, the paper designs a backstepping control (BSC) for the FSM equipped with the MASM. This not only solves the nonlinearity of the stiffness introduced by the MASM but also enhances the active vibration suppression capability, ensuring that the FSM meets tracking control requirements[6].

The vibration isolation ability of the FSM is a combination of passive isolation and active suppression. Methods that rely solely on active control algorithms to suppress vibration have an upper limit to their effectiveness. However, the best vibration isolation effect can be achieved only through the integration of passive isolation. This is the core idea of the FSM vibration isolation method presented in this paper, which combines both passive and active approaches.

This study first explores the MVSM by developing a model for the vehicle-mounted adjustable stiffness FSM in Section 2. It then presents the design of the BSC for the adjustable stiffness FSM in Section 3. The performance of the proposed approach is evaluated in Section 4, focusing on vibration suppression and precision control, before providing a summary and conclusion in Section 5.

2. Model establishment

2.1. Modeling of the Magnetic MASM

For an FSM, the rotational rigidity is mainly controlled by its flexible hinge, and this hinge's stiffness remains unchanged after the manufacturing process. To further improve vibration isolation performance, MASM is proposed [7].

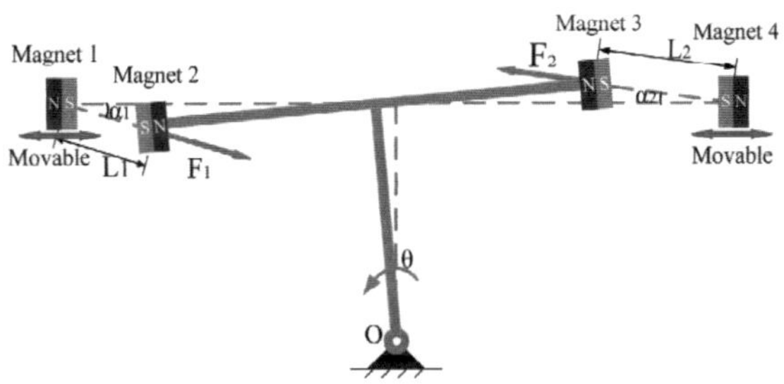

Figure 1. Diagram depicting the MASM configuration

As depicted in the structural illustration of the MVSM in Figure 1, two perpendicular and fixed rods can rotate around the hinge's point *O*. At both ends of the rods, two cylindrical magnets, Magnet 2 and 3, are fixed, while two opposing cylindrical magnets, Magnet 1 and 4, are located at the ends of Magnet 2 and 3. These two magnets can move along the horizontal direction. When the rods rotate by an angle θ, the force F_1 between Magnet 1 and 2, and the force F_2 between Magnet 4 and 3, produce torques at point *O* By adjusting the positions of Magnet 1 and 4, the stiffness of the MASM can be modified.

The interaction force between the pair of cylindrical magnets has been explored in [8] and [9], with the equation provided below:

$$F_{mag} = \frac{J_1 J_2}{2\mu_0} \sum\nolimits_{i=1}^{2} \sum\nolimits_{j=3}^{4} a_1 a_2 a_3 f_z' [-1]^{i+j}, \tag{1}$$

where the intermediate expression is

$$f_z' = K(a_4) - \frac{1}{a_2} E(a_4) + \left[\frac{a_1^2}{a_3^2} - 1 \right] \Pi\left(\frac{a_4}{1-a_2} \middle| a_4 \right), \tag{2}$$

where

Journal of Physics: Conference Series **3024** (2025) 012023 doi:10.1088/1742-6596/3024/1/012023

$$
\begin{aligned}
&a_1 = z_i - z_j,\\
&a_2 = \frac{(r_1 - r_2)^2}{a_1^2} + 1,\\
&a_3^2 = (r_1 - r_2)^2 + a_1^2,\\
&a_4 = \frac{4r_1 r_2}{a_3^2},\ 0 < a_4 \le 1.
\end{aligned} \tag{3}
$$

The functions $K(m)$, $E(m)$, and $\Pi(m)$ denote the complete elliptic integrals of the first, second, and third types, respectively.

The torques applied by forces F_1 and F_2 at the rotation center point O are given as follows [7]:

$$
\begin{aligned}
M = {} & F_{mag}(L_1)\sqrt{a^2 + 2ab_1 + b_1^2 + h^2}\sin\left(\arctan\left[h/(a+b)\right]\right.\\
& \left.-\arcsin\left[\left(a + h\tan(\theta/2)\right)\sin\theta / L_1\right]\right)\\
& -F_{mag}(L_2)\sqrt{a^2 + 2b_2 + b_2^2 + h^2}\sin\left(\arctan\left[h/(a+b_2)\right]\right.\\
& \left.+\arcsin\left[\left(a - h\tan(\theta/2)\right)\sin\theta / L_2\right]\right).
\end{aligned} \tag{4}
$$

The torque M can be expanded using a Taylor series, simplifying equation (4):

$$
M = k_n\theta + \gamma\theta^3 \tag{5}
$$

Where k_n represents the adjustable stiffness of the MASM. By adjusting the positions of Magnet 1 and Magnet 4, k_n can be changed. When Magnet 1 and 2, and Magnet 3 and 4 repel each other, k_n becomes a adjustable negative value; conversely, it becomes a adjustable positive value. γ is the nonlinear stiffness coefficient.

When the MASM is applied to the FSM, the physical configuration is shown in Figure 2. With the flexible hinge's rotational stiffness denoted as k_n, the FSM's total stiffness is given by $kn+kh$. By adjusting the stiffness of the MASM, when $k_n + k_h \approx 0$ and $k_n + k_h > 0$, the FSM exhibits quasi-zero stiffness.

Figure 2. Physical model of MASM applied to FSM

2.2. *Modeling of Adjustable Stiffness FSM*

The vehicle-mounted adjustable stiffness FSM is shown in Figure 3. The displacement disturbance θ_d from the road surface is transmitted to the base of the FSM through the vehicle suspension, resulting in a vibration θ_w at the base of the FSM, which further affects the mirror displacement θ_m. J_w, c_w and k_w denote the rotational inertia, damping coefficient, and stiffness of the vehicle suspension, respectively. J_m and c_m denote the rotational inertia and damping of the FSM, and $F(t)$ corresponds to the motor's driving input. The mathematical description of the vehicle-mounted adjustable stiffness FSM is as follows:

$$J_m\ddot{\theta}_m = -c_m(\dot{\theta}_m - \dot{\theta}_w) - (k_h + k_n)(\theta_m - \theta_w) - \gamma(\theta_m - \theta_w)^3 + F(t), \tag{6}$$

$$J_w\ddot{\theta}_w = -c_w(\dot{\theta}_w - \dot{\theta}_d) - k_w(\theta_w - \theta_d) + c_m(\dot{\theta}_m - \dot{\theta}_w) + (k_h + k_n)(\theta_m - \theta_w) + \gamma(\theta_m - \theta_w)^3 - F(t), \tag{7}$$

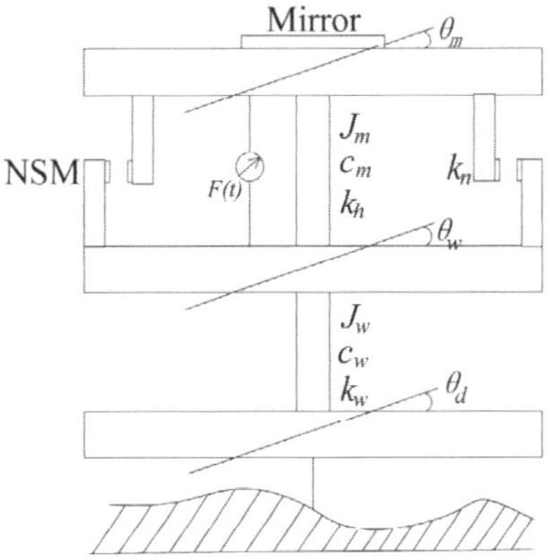

Figure 3. Vehicle-mounted adjustable stiffness FSM

3. Design of backstepping control

A BSC is designed for the vehicle-mounted adjustable stiffness FSM. By rewriting equation (6), we obtain:

$$\ddot{\theta}_m = -a\theta_m - b\dot{\theta}_m + u + d, \tag{8}$$

Where $a = (k_h + k_n)/J_m$, $b = c_m / J_m$, and u represent the control rates, $u = F(t)/J_m$, and d represents the total disturbance. Define the state variables as $x_1 = \theta_m$, $x_2 = \dot{\theta}_m$, and the equation is given by:

$$\begin{cases} \dot{x}_1 = x_2 \\ \dot{x}_2 = -ax_1 - bx_2 + u + d \end{cases}. \tag{9}$$

Let the tracking error be,

$$z_1 = x_1 - x_{1d}, \tag{10}$$

and define the Lyapunov function V_1:

$$V_1 = \frac{1}{2}z_1^2, \tag{11}$$

$$\dot{V}_1 = z_1\dot{z}_1 = z_1(\dot{x}_1 - \dot{x}_{1d}). \tag{12}$$

To ensure system stability, $\dot{V}_1 \le 0$, i.e., it must be negative definite. Observing equation (12), the requirement is satisfied if $\dot{x}_1 - \dot{x}_{1d}$ approaches $-k_1 * z_1$ $(k_1 > 0)$

$$\dot{V}_1 = -k_1 z_1^2. \tag{13}$$

At this stage, x_{2d} is designated as the desired tracking value for x_2, and since $\dot{x}_1 = x_2$, we have:

$$x_{2d} = \dot{x}_{1d} - k_1 z_1. \tag{14}$$

The tracking error z_2 is:

$$z_2 = x_2 - x_{2d}, \tag{15}$$

$$\dot{z}_2 = \dot{x}_2 - \dot{x}_{2d} = -ax_1 - bx_2 + u + d - \ddot{x}_{1d} + k_1(\dot{x}_1 - \dot{x}_{1d}). \tag{16}$$

Next, define the Lyapunov function V_2:

$$V_2 = \frac{1}{2}z_1^2 + \frac{1}{2}z_2^2. \tag{17}$$

$$\dot{V}_2 = z_1\dot{z}_1 + z_2\dot{z}_2 = z_1(x_2 - \dot{x}_{1d}) + z_2\dot{z}_2 = -k_1 z_1^2 + z_2(z_1 + \dot{z}_2)\cdot \tag{18}$$

To ensure system stability, $\dot{V}_2 < 0$, i.e., it must be negative definite. Let

$$z_1 + \dot{z}_2 = -\mathrm{k}_2 z_2\ , \tag{19}$$

where $\mathrm{k}_2 > 0$, yielding:

$$\dot{V}_2 = -k_1 z_1^2 - k_2 * z_2^2\ . \tag{20}$$

By calculating equation (19),

$$z_1 - ax_1 - bx_2 + u + d - \ddot{x}_{1d} + k_1(\dot{x}_1 - \dot{x}_{1d}) = -k_2 x_2 + k_2\dot{x}_{1d} - k_1 k_2 z_1, \tag{21}$$

the control rate is obtained as:

$$u = -k_2 x_2 + k_2\dot{x}_{1d} - k_1 k_2 z_1 - k_1(\dot{x}_1 - \dot{x}_{1d}) + \ddot{x}_{1d} - z_1 + ax_1 + bx_2 - d. \tag{22}$$

To address the issue of uncertain disturbance $\tilde{d}$, adaptive control is introduced.

Since $\tilde{d}$ is unknown, its estimated value is used instead in equation (22) as the actual value. Replace $\tilde{d}$ with $\hat{\tilde{d}}$ in The control rate u. The error between them is $z_3 = \tilde{d} - \hat{\tilde{d}}$. The prerequisite for using $\hat{\tilde{d}}$ instead of $\tilde{d}$ is that the estimated value $\hat{\tilde{d}}$ converges to the true value. This transforms the problem into solving for the estimated value $\hat{\tilde{d}}$. An adaptive law is constructed using the system's state or output variables. A proper Lyapunov function is selected to ensure that the error approaches zero, with its expression given below:

$$V_3 = \frac{1}{2}z_1^2 + \frac{1}{2}z_2^2 + \frac{1}{2\lambda}z_3^2, \tag{23}$$

$$\dot{V}_3 = z_1\dot{z}_1 + z_2\dot{z}_2 + \frac{1}{\lambda}z_3\dot{z}_3 = -k_1 z_1^2 + z_1 z_2 + z_2(-ax_1 - bx_2 + u + d - \dot{x}_{2d}) + \frac{1}{\lambda}z_3\dot{z}_3. \tag{24}$$

Since $\tilde{d}$ is unknown but constant, we have $\dot{z}_3 = \dot{\hat{\tilde{d}}}$, $\dot{\tilde{d}} = 0$. Substituting this into equation (24):

$$\begin{aligned}\dot{V}_3 &= -k_1 z_1^2 + z_1 z_2 + z_2(-ax_1 - bx_2 + u + z_3 + \hat{\tilde{d}} - \dot{x}_{2d}) + \frac{1}{\lambda}z_3\dot{\hat{\tilde{d}}} \\ &= -k_1 z_1^2 + z_1 z_2 + z_2(-ax_1 - bx_2 + u + \hat{\tilde{d}} - \dot{x}_{2d}) + z_3(z_2 + \frac{1}{\lambda}\dot{\hat{\tilde{d}}}).\end{aligned} \tag{25}$$

Equation (25) may be expressed as $\dot{V}_3 = \dot{V}_2 + z_3 * (z_2 + \frac{1}{\lambda} * \dot{\hat{\tilde{d}}})$.

The control rate u is given by equation (22), $\dot{V}_2 < 0$, Therefore, it is necessary to ensure $\dot{V}_3 < 0$, which leads to $z_2 + \frac{1}{\lambda} * \dot{\hat{\tilde{d}}} = 0$, yielding $-\lambda * z_2 = \dot{\hat{\tilde{d}}}$. Finally, when u is given by equation (22), $\dot{\hat{\tilde{d}}} = -\lambda * z_2$, and the closed-loop system variables z_1, z_2, z_3 all converge to 0, ensuring the system is asymptotically stable.

4. Simulation verification

To verify the performance of the BSC and the FSM equipped with the MASM, Simulations were carried out in MATLAB. Table 1 summarizes the simulation parameters.

Table 1. Simulation Parameters

Parameters	Value
J_m	2.75×10^{-4} kg·m^2
k_h	13.4 N·m/rad
k_n	-13.3 N·m/rad
γ	1.47×10^{-3} N/m^3
c_m	7.1×10^{-3}N·m·s/rad
J_w	5×10^{-4} kg·m^2
k_w	13.4 N·m/rad
c_m	7.4×10^{-3}N·m·s/rad
k_1	55
k_2	90
λ	2

Figure 4 highlights the effect of a road surface step disturbance θ_d on the FSM. The control input $\theta_r = 0$, and at t = 1s. a step disturbance of $\theta_d(t \geq 1) = 1$. is applied. The stiffness of the FSM is adjusted to quasi-zero stiffness (QZS), with the QZS FSM stiffness being k_h+k_n = 0.1 Nm/rad. The proposed BSC+QZS method results in smaller mirror displacement θ_m, with reduced oscillation amplitude and fewer oscillations, and no steady-state error. The BSC+QZS method demonstrates better disturbance rejection performance against the step disturbance from the road surface.

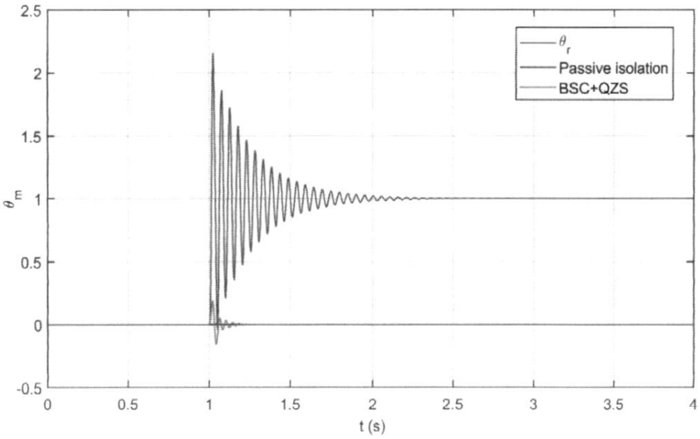

Figure 4. Effect of road surface step disturbance θ_d on FSM

Figure 5 shows the step response of the QZS FSM based on BSC, where $\theta_r(t \geq 1) = 1$, The overshoot of the step response is 5.3%, and the time required to reach 90% of the steady-state value is 31 ms, indicating excellent step response performance.

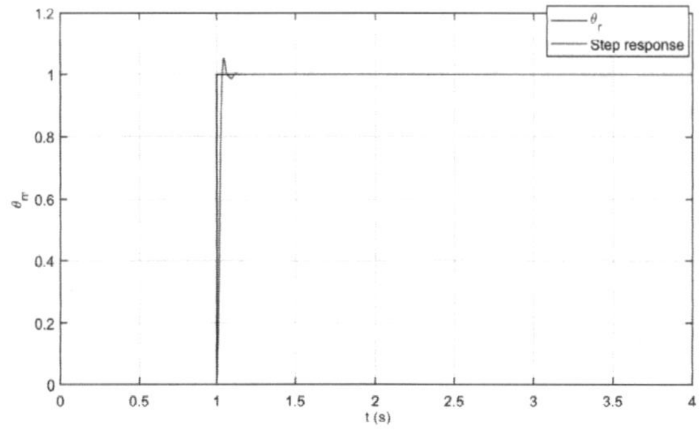

Figure 5. Step response of QZS FSM based on BSC

Figure 6 shows the effect of FSM stiffness on vibration isolation and control performance. By adjusting the stiffness of the MASM, the overall stiffness of the FSM is modified. $\theta_r = \sin(4\pi t)$, At $t = 1$s a road surface step disturbance of $\theta_d(t \geq 1) = 1$ is applied. The figure demonstrates that as the total stiffness k_h+k_n of the FSM decreases, the suppression effect on the road disturbance improves, and the tracking performance at the initial application of the control signal also improves.

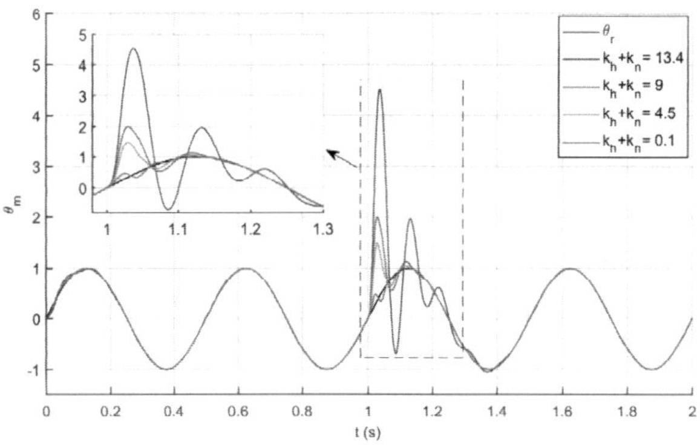

Figure 6. Effect of FSM stiffness on vibration isolation and control performance

Figure 7 demonstrates the impact of sinusoidal disturbances on the tracking accuracy of the QZS FSM system. The control input is $\theta_r = \sin(4\pi t)$, and the input disturbance force is $\theta_{disturb} = 0.2\sin(100\pi t)$. As shown in Figure 7, the system exhibits good tracking performance while effectively suppressing disturbances.

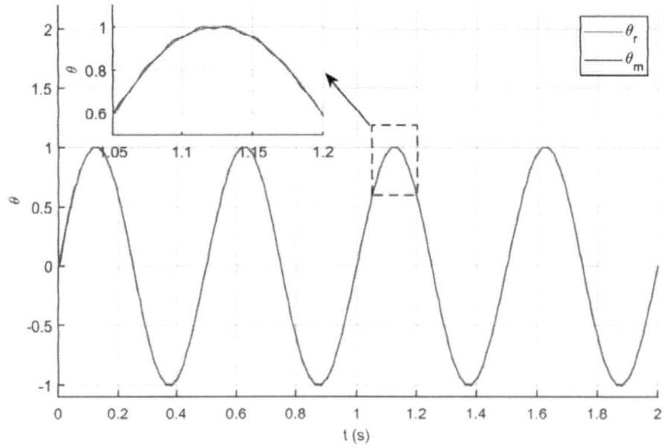

Figure 7. Effect of sinusoidal disturbances on QZS FSM tracking performance

5. Conclusion

This study develops a BSC strategy for vehicle-mounted adjustable stiffness FSMs integrated with MVSM. significantly improving both vibration isolation capability and dynamic control precision. The MVSM is applied to the FSM to adjust its stiffness, allowing the FSM to achieve QZS. A BSC is developed specifically for the adjustable stiffness FSM. Simulation results demonstrate that the proposed QZS+BSC method more effectively suppresses road surface step disturbances compared to passive vibration isolation, exhibiting excellent step response performance. The results also confirm that the lower the FSM stiffness, the better the suppression of road surface step disturbances, and the method shows good suppression effects against sinusoidal disturbances.

Acknowledgments

This work was supported by the Special Research Assistant Program, Chinese Academy of Sciences, China (Grant No. E329691C21) and Sichuan Science and Technology Program, China (Grant No. 2024NSFSC1481).

References

[1] Zhang B, Nie K, Chen X, and Mao Y, et al 2022 Development of sliding mode controller based on internal model controller for higher precision electro-optical tracking system *Actuators* vol 11 no 1 p 16 (MDPI)

[2] Zhang W, Yuan J, Yan C, Gao Z, and Dong Y 2021 Multi-objective optimization design of natural frequency of two-degree-of-freedom fast steering mirror system *IEEE Access* vol 9 pp 33689–33703

[3] Schlarp J, Csencsics E, and Schitter G 2019 Optical scanning of a laser triangulation sensor for 3-D imaging *IEEE Trans. Instrum. Meas.* vol 69 no 6 pp 3606–3613

[4] Luo J, Mao Y, Dai W, Yang F, Zhang L, and Miao L 2023 Design and analysis of variable-stiffness fast-steering mirror *Actuators* vol 13 no 1 p 5 (MDPI)

[5] Davoodi E, Safarpour P, Pourgholi M, and Khazaee M 2020 Design and evaluation of vibration reducing seat suspension based on negative stiffness structure *Proc. Inst. Mech. Eng., Part C: J. Mech. Eng. Sci.* vol 234 no 21 pp 4171–4189

[6] Li J, Zhuang S, Wang H, Deng J, and Mao Y 2024 Design of backstep control based on a Softsign linear–nonlinear tracking differentiator for an electro-optical tracking system *Photonics* vol 11 no 2 p 156 (MDPI)

[7] Luo J, Zhang L, Miao L, and Mao Y 2024 A new magnetic negative stiffness mechanism: A case study of fast steering mirror for quasi-zero stiffness vibration isolation *Adv. Mech. Eng.* vol 16 no 10 p 16878132241288091

[8] Zhu T, Cazzolato B, Robertson WS, and Zander A 2015 Vibration isolation using six degree-of-freedom quasi-zero stiffness magnetic levitation *J. Sound Vib.* vol 358 pp 48–73

[9] Robertson W, Cazzolato B, and Zander A 2011 A simplified force equation for coaxial cylindrical magnets and thin coils *IEEE Trans. Magn.* vol 47 no 8 pp 2045–2049

Optimization Study of Precise Control of Commercial Vehicle Cooling System Based on MAP Feedforward

Rongjiang Tang, Le Zhao, Quanhui Liu*, Wenwen Wang, Xin Jing

School of Mechanical and Electrical Engineering, Guilin University of Electronic Technology, Gui Lin, Guangxi, 541004, PR China

*Corresponding author's e-mail: 691319937@qq.com

Abstract. Aiming at the problems of overcooling, overheating, the insufficient control accuracy of water pump and fan, and wastage of power consumption of traditional thermal management systems of commercial vehicles under transient conditions, this paper constructs a thermal management system control strategy model based on GT-Suite and Simulink platform and carries out experimental validation of the model under the C-WTVC cyclic conditions, and the results show that the model has a high degree of accuracy. In this paper, the MAP feed-forward control strategy is adopted to query the MAP of heat dissipation demand through the engine speed and torque, combined with the ambient temperature correction algorithm to get the corrected heat dissipation demand and calculate the target speeds of the water pump and the fan based on the theory of minimum power allocation, and at the same time, introduce the fuzzy PID feedback control to realize the precise regulation. Simulation results show that the optimized thermal management system improves the coolant temperature control accuracy by 88.9%, reduces the total power consumption of the water pump and fan by 53.4% under the C-WTVC cycle condition, which effectively reduces the fuel consumption, and provides the theoretical basis and methodological support for the energy-saving optimization of thermal management system for commercial vehicles.

1. Introduction

With the increasing number of motor vehicles, the continuous consumption of large amounts of petroleum resources has triggered a severe energy shortage and aggravated the degree of environmental pollution. Heavy commercial vehicles with large-displacement diesel engines, which account for 51% of the total fuel consumption of the vehicle [1], require more heat to be carried away by the thermal management system. The internal coolant flow of the engine, the airflow of the radiator, and the heat dissipation efficiency together determine the efficiency of the thermal management system.

The thermal management system control strategy coordinates the control of each controllable component with the aim of making the thermal management system efficiency optimal as a whole rather than optimizing the control of a certain component. At present, the traditional thermal management system mainly suffers from such defects as long warm-up time, overcooling or overheating under transient operating conditions, and waste of component power consumption [2-3]. Aiming at the above problems, related scholars have studied the water pump [4], fan [5] and thermostat [6] of the thermal management system, which effectively improves the efficiency of the thermal management system and reduces the warm-up time. With the progress of science and technology, some scholars began to use simple rule-based control to control the thermal management components, and gradually, a variety of more advanced control algorithms have been applied to thermal management systems, such as PID control [7], fuzzy control [8], sliding-mode control, and adaptive control. Meanwhile, multivariate cooperative control has also been applied to thermal management systems, such as composite control with system thermal balance state information as feedforward, MAP+PID control or model predictive control (MPC) control [9]. However, the related control algorithms for pumps and fans do not achieve absolute accuracy, and there are still some shortcomings in terms of operational stability.

Aiming at the deficiencies existing in the important components of the thermal management system mentioned above, this paper proposes an overall control strategy scheme for the thermal management system, which adopts the joint MAP feedforward and fuzzy PID feedback control for the water pump and fan in order to improve the control accuracy of the water pump and fan, and to reduce the unnecessary power consumption, and finally conducts a joint simulation of the thermal management control system, and the results show that the thermal management system of this paper has a very good optimization.

2. Thermal management system simulation platform construction

2.1. Model building

The thermal management system model in this paper is built into the Vehicle module in the GT-SUITE platform to control the vehicle's operating status in real-time and simulate the vehicle's operation under the C-WTVC cycle conditions. Table 1 shows the relevant parameters of the whole vehicle and engine.

Table 1. Vehicle and engine main configuration parameters

Vehicle		Engines	
Sports event	Parameter value	Sports event	Parameter value
Driving method	6x4 Front Rear Drive	Engine displacement (L)	11
Overall weight (kg)	8800	Cylinder number	6
Full load weight (kg)	49000	Compression ratio	17
Wheelbase (mm)	3300+1350	Intake method	Supercharged intercooler
Rear axle main reduction ratio	3.7	Rated speed (r/min)	1900
Windward area (m²)	10.17	Maximum output power (kw)	339
wind resistance	0.53	Maximum torque （N·m）	2200

Pump module in the model library is used to simulate the pump work, then the pump shaft power calculation formula is shown in equation (1).

$$p_w = \frac{\dot{V}_w \cdot \Delta p_w}{\eta_w} \tag{1}$$

Where p_w for the pump shaft power, kW; $\dot{V}_w$ for the coolant flow, m^3/s ; Δp_w for the pump coolant inlet and outlet pressure difference, kPa; η_w for the pump efficiency.

The fan module is used in the model to simulate the fan operation, and the main input parameters include the fan's structural parameters. The flow and pressure ratios at one reference data point and the pressure ratio at zero flow at the same fan speed are input to the model, and the flow rates at other fan speeds are extrapolated according to the law of similarity, and the extrapolation formula is shown in equation (2).

$$\dot{V}_N = \dot{V}_{ref} \frac{N}{N_{ref}}(P_R - 1) = (P_{Rref} - 1)(\frac{N}{N_{ref}})^2 \tag{2}$$

where $\dot{V}_N$ and $\dot{V}_{ref}$ are the volume flow rate of the fan at the actual and reference rotational speeds, m^3/s, respectively; P_R and P_{Rref} are the pressure-to-lift ratios of the actual and reference rotational speeds, respectively; and N and N_{ref} are the actual and reference rotational speeds of the fan, $r/\min$, respectively.

2.2. Experimentation and model validation

Figure 1 shows the vehicle test equipment site for China Automobile Research Corporation (CARC) to test the rotating hub environmental chamber, which can simulate the vehicle's operating status under road cycle driving conditions and can test the vehicle's economy, dynamics and other value-added tests. The temperature, humidity, and air pressure in the environmental chamber meet the technical requirements stipulated by the regulations, which ensures the effective implementation of the test. The C-WTVC (Adapted World Transient Vehicle Cycle) cycle was chosen for the test, which is based on the international heavy commercial vehicle transient cycle (WTVC) standard, and some speed segments were adapted to form a complete driving cycle with three scenarios, including high speed (432 seconds), highway (468 seconds) and urban (900 seconds). The total duration of the complete driving cycle is 1,800 seconds.

Figure 1. Prototype and hub environment chamber test environment

The results of simulation and test comparison are shown in Fig. 2, and it can be seen that this model is able to operate smoothly according to the set C-WTVC cycle conditions.

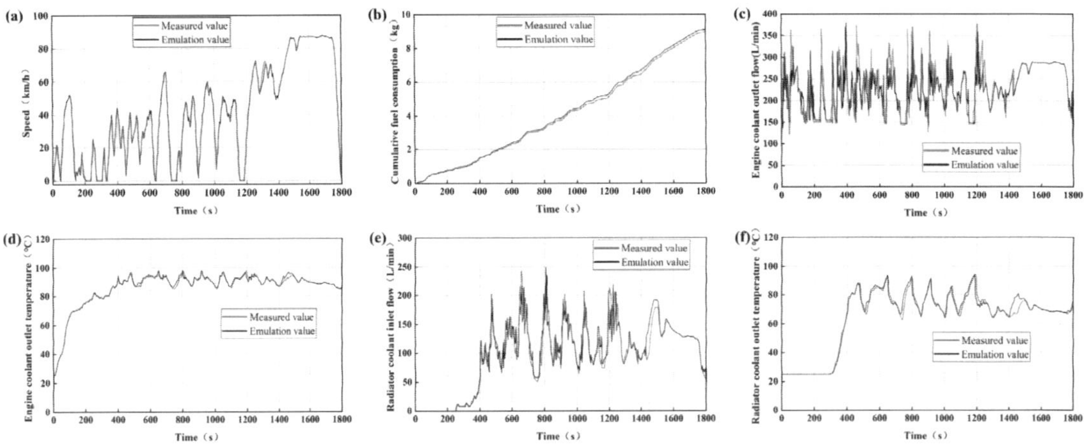

Figure 2. Comparison of simulation and test results

After the simulation yields the relevant data of the engine, based on the rotational speed and torque of the engine at any moment under the C-WTVC condition, the value of the heat dissipation of the engine at a steady state at that moment can be interpolated in the engine heat dissipation demand MAP shown in Fig. 3.

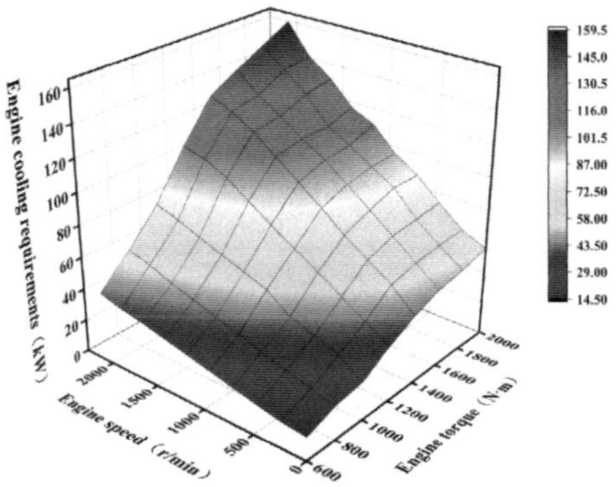

Figure 3. Engine Cooling Requirements MAP

3. Thermal management system control strategy

The overall control strategy of this paper is divided into the warm-up phase and the cooling phase. When the engine starts, the water pump runs at the lowest speed, and then the system determines whether the thermostat is open or not. After the thermostat is open, it is fuzzy-controlled to determine whether it is fully open. After the thermostat is fully open to go into the cooling stage, then the water pump and fan joint MAP feed-forward and fuzzy PID feedback control to determine whether the coolant temperature is lower than the target temperature; when lower than the target temperature after returning to the cycle of fuzzy control.

3.1. Fuzzy PID controller design

Fuzzy PID control is highly robust, especially suitable for dealing with time-varying, hysteresis nonlinear and other complex conditions, and has a very fast response speed. The calculation formula of the control quantity u(t) is shown in equation (3).

$$u(t) = K_p e(t) + K_i \int_0^t e(t)dt + K_d \frac{de(t)}{dt} \tag{3}$$

In the formula, r is the set control target value, y is the actual output value, and the difference between the two values, e (t), is the input of the PID controller after the proportional, integral, and differential three links will be controlled u output to the controlled object until the system output value reaches the set control target value.

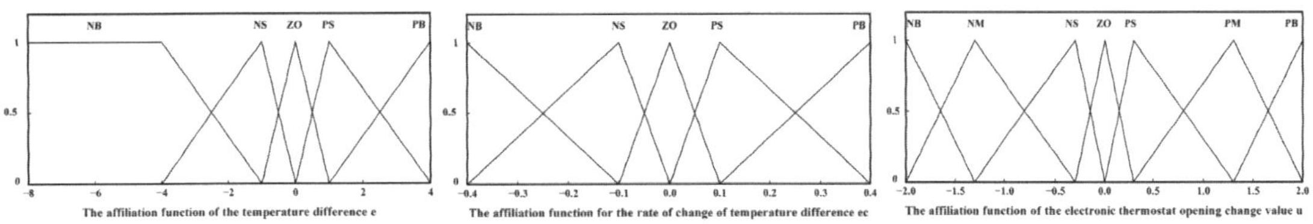

Figure 4. Input and thermostat opening change value affiliation function

A fuzzy PID controller consists of a PID controller and a fuzzy controller. The input of the fuzzy controller takes the difference between the engine outlet temperature and the target temperature e and its rate of change ec, and after the fuzzy controller blurring, fuzzy reasoning, and de-fuzzing, the outputs are proportional incremental coefficients ΔK_p , integral incremental coefficients ΔK_i , and differential incremental coefficients ΔK_d. The use of the three incremental coefficients in real-time adjusts the PID controller's three parameters to enhance the adaptivity of the PID controller, and finally, the output of the PID controller is the corrected heat dissipation. The affiliation functions for the temperature difference e, the rate of change of temperature difference ec and the value of change of thermostat opening u are shown in Fig. 4, respectively.

3.2. *Combined MAP feed-forward and fuzzy PID feedback control of water pump and fan*

3.2.1. *MAP feed-forward control*

MAP feed-forward control is based on a type of open-loop control, a type of control that compensates for disturbances or changes in a given value, allowing the thermal management system components pump and fan to quickly and accurately calculate a target speed that meets the current cooling requirements based on the real-time engine operating conditions. The heat dissipation performance of the radiator at different coolant flow rates and airflow rates can be obtained by offline calibration of the radiator coolant inlet temperature of 95°C and intake air temperature of 30°C by the Design of Experiment (DOE) method.

In the case of the same cooling performance of the radiator, there are different pump speeds and fan speed combinations that can meet the cooling conditions; there must be a pump and fan power consumption of the minimum speed. The theoretical formula for the minimum power consumption of pumps and fans for the same cooling requirement is:

$$\begin{cases} \min(P_c) = \min(P_p + P_f) \\ s.t.Q_{eng}(n, T_r) = Q_{rad}(\dot{m}_c, \dot{m}_a, \Delta T_{c-a}) \\ s.t.\dot{m}_{c-\min} \le \dot{m}_c \le \dot{m}_{c-\max} \\ s.t.\dot{m}_a \le \dot{m}_a \le \dot{m}_{a-\max} \end{cases} \tag{4}$$

Where: P_c, P_p, P_f are the total power consumption of thermal management system components, pump power consumption, fan power consumption, kW; Q_{eng} is the heat dissipation demand of the engine in different operating conditions, kW; n is the engine speed, r/min; T_r is the engine torque, $N \cdot m$; Q_{rad} is the heat dissipation of the radiator, kW; ΔT_{c-a} is the temperature difference between the coolant side of the radiator and the air side; $\dot{m}_c$, $\dot{m}_a$ are the mass flow rate of the coolant and the air through the radiator, kg/min; and is the minimum coolant mass flow rate; and is the maximum allowable flow rate of coolant and air respectively, kg/min; $\dot{m}_{c-\min}$ is the minimum coolant mass flow rate; $\dot{m}_{c-\max}$ and $\dot{m}_{a-\max}$ are the maximum allowable flow rates of coolant and air, respectively.

Correcting for the heat dissipation of the radiator, the formula is calculated as:

$$\frac{Q_{rad}}{\Delta T_{in}} = kA \approx C \tag{5}$$

$$\frac{Q_a}{\Delta T_{in-a}} = \frac{Q_{eng}}{\Delta T_{in-r}} \tag{6}$$

Where: Q_a is the corrected heat dissipation requirement of the heat sink based on the ambient temperature; k is the convective heat transfer coefficient; A is the heat transfer area of the radiator, m^2; C is a constant quantity, kW/K; ΔT_{in} is the logarithmic mean temperature difference between the hot and cold fluids, K; ΔT_{in-r} is the temperature difference between the coolant of the radiator and the inlet of the air at the current ambient temperature, K; and ΔT_{in-a} is the calibrated temperature difference between the radiator coolant and the air inlet at a specific ambient temperature, K.

Correcting the effect of ambient temperature on the heat dissipation demand by the above formula, the speed of the pump and fan calculated by the theory of minimum power distribution is output according to the heat dissipation demand.

3.2.2. Fuzzy PID feedback control

In the fuzzy PID control system, the coolant control target temperature of the thermal management system is set to 93°C. The domain of the fuzzy controller input quantity temperature difference e is set to be [-4,4], the domain of the temperature difference rate of change ec is set to be [-0.4,0.4], and the domains of the output quantities ΔK_p, ΔK_i, and ΔK_d are set to be [-3,3], [-0.4,0.4], and [-0.1,0.1] respectively, which will be the input, The domain of the output quantities is divided into seven fuzzy subsets of NB, NM, NS, ZO, PS, PM, and PB.

After completing the fuzzy inference on ΔK_p, ΔK_i, ΔK_d according to the fuzzy rules, the centre of gravity method is used to find out the determined value of the incremental coefficients of the PID controller parameters and finally complete the adaptive adjustment of the PID controller parameters to realize the correction of the radiator's heat dissipation.

4. Analysis of joint simulation results

The joint simulation model of this paper is shown in Fig. 5. Under the cold start conditions of the C-WTVC standard test cycle, the initial coolant temperature and ambient temperature are both set to run the system at 25°C, while the target control temperature of the engine coolant outlet is set to 93°C in the control strategy model. The comparison graphs before and after the optimization of the engine coolant outlet temperature are shown in Figure 6, respectively. After entering the cooling stage, the optimized coolant temperature curve oscillation is significantly weaker than that before optimization, which better matches the heat dissipation demand of the engine, reduces the deviation of the coolant temperature from the target temperature, greatly improves the control accuracy of the coolant temperature, and then enhances the ability of the coolant temperature to follow the target temperature, so that the engine is kept at the optimal temperature for operation.

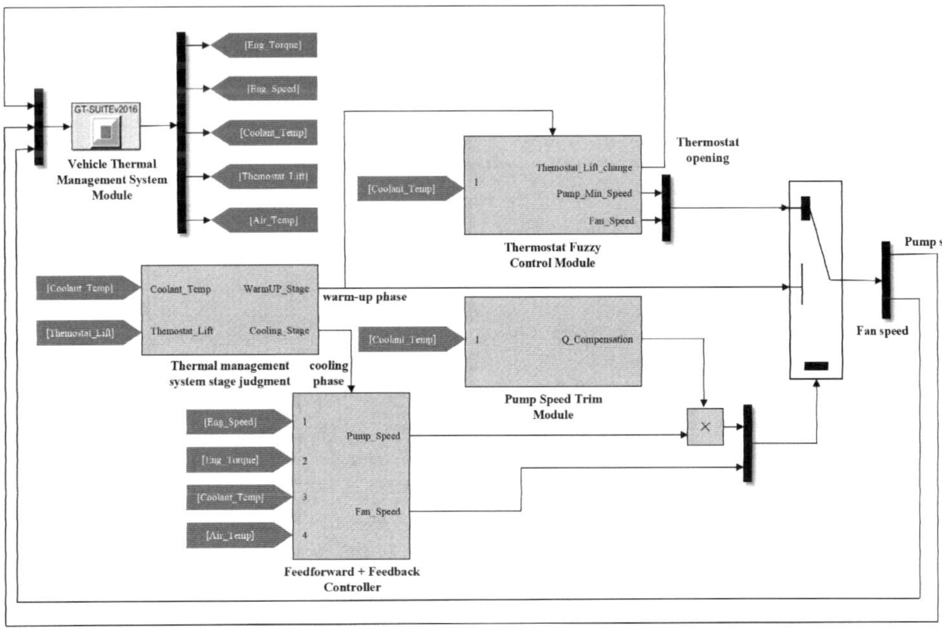

Figure 5. Joint Simulation Thermal Management System Control Strategy Modeling

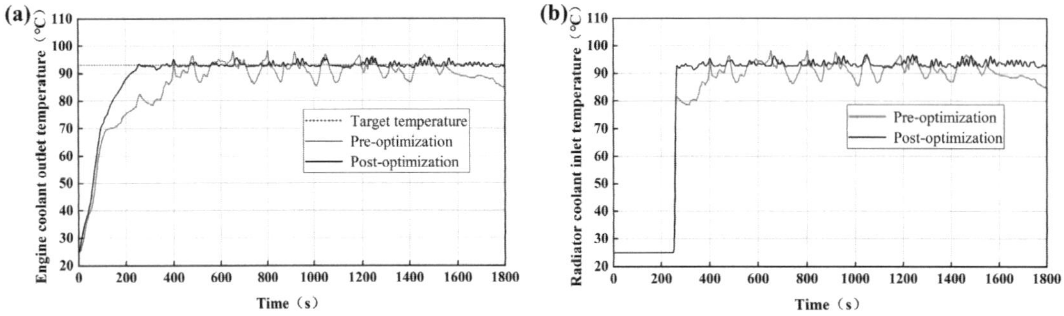

Figure 6. Comparison chart before and after coolant temperature optimization

Subsequently, the temperature control performance of the thermal management system's optimized control strategy under the C-WTVC cycle condition after the thermostat starts working from 252 to 1800s before optimization and after the thermostat enters the cooling stage from 348 to 1800s after optimization are counted respectively, as shown in Table 1.

Table 2. Temperature control performance related statistics

Sports event	$\tau_{\pm 1℃}$	$T_{\max} - T_{set}$	$T_{set} - T_{\min}$	$\Delta\overline{T}$
Original vehicle not optimized	29%	5.1℃	7.4℃	-1.89℃
Post-optimization	82%	2.6℃	1.2℃	0.21℃

Where $\tau_{\pm 1℃}$ is the percentage of time in which the absolute value of the difference between the engine coolant outlet temperature and the target temperature is less than 1 ℃ during the statistical time; Tmax and Tmin are the highest and lowest values of the coolant temperature; Tset is the target value of the temperature; $\Delta\overline{T}$ is the average temperature difference between the engine coolant outlet temperature and the target temperature. The statistical results show that the optimized $\tau_{\pm 1℃}$ increases by 53%, and the temperature control effect is more obvious. The temperature oscillation range is low from [-7.4,5.1] to [-1.2,2.6], and the oscillation range is reduced by 69.6%. The average temperature deviation was reduced from -1.89°C to 0.21°C, and the control accuracy was improved by 88.9%.

Fig. 7 compares the water pump and fan speed before and after optimization. The fan speed after optimization is substantially lower than that before optimization, and the pump and fan's power consumption decreases from 965.2kJ and 2003.7kJ to 394.8kJ and 989.5kJ, respectively. The total power consumption is reduced by 1584.6kJ, which is a reduction of 53.4%.

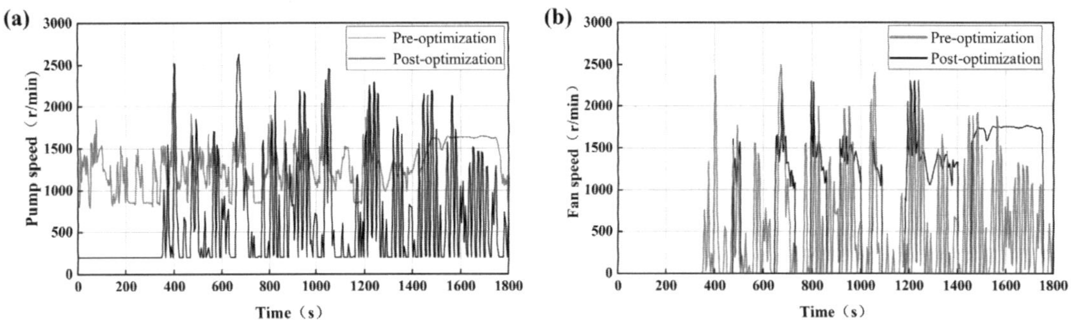

Figure 7. Comparison chart before and after coolant temperature optimization

5. Conclude

In this paper, MAP feed-forward and fuzzy PID feedback control are used for the water pump and fan to find the heat dissipation demand MAP based on the engine conditions to solve the heat dissipation demand, correct the heat dissipation demand at different ambient temperatures through the heat balance and heat transfer equations, and design a fuzzy PID controller based on the target temperature difference of the coolant and the rate of change of the temperature difference to carry out the feedback control and compensate for the heat dissipation demand, and after determining the heat dissipation demand Match with the speed MAP that minimizes the total power of the water pump and fan to output the speed of the water pump and fan for more rapid and accurate control of the coolant temperature. The final results show that under the C-WTVC cycle conditions of the whole vehicle, the optimized thermal management system has improved the coolant temperature control accuracy by 88.9%, while the total power consumption of the water pump and fan has been reduced by 53.4%, saving a certain amount of fuel consumption.

Acknowledgments

Science and Technology Major Project of Guangxi (Grant No. GuiKeAA24206038), Guangxi Key Research and Development Plan (Grant No. GuiKeAB24010262).

References

[1] Liu YT, Cao and Liu CP 2022 Progress of Energy and Power System Technologies for Commercial Vehicles Under China's Carbon Peaking and Carbon Neutrality Goals. Chinese Journal of Automotive Engineering, 4:p478-494.

[2] Kang J, Cho J, Park S 2023 Investigation of friction loss characteristics of engine pistons for different engine operating conditions. International Journal of Automotive Technology, 24(2): 503-511.

[3] Xu B, Li X 2021 A Q-learning based transient power optimization method for organic Rankine cycle waste heat recovery system in heavy duty diesel engine applications. Applied Energy, 286: 116532.

[4] Xie Y, Ou J, Li W, et al 2022 An intelligent eco-heating control strategy for heat-pump air conditioning system of electric vehicles. Applied Thermal Engineering, 216: 119126.

[5] Cui N 2017 Research Summary on Fan Driving Method of Construction Machinery Cooling System. Mechanical Engineering & Automation, 2 : p 211-213.

[6] Ghasemi Zavaragh H, Kaleli A, Solmuş İ, et al 2021 Experimental analysis and evaluation of thermostat effects on engine cooling system. Journal of Thermal Science, 30: 540-550.

[7] Zhao R, Qin D, Chen B, et al 2022 Thermal management of fuel cells based on diploid genetic algorithm and fuzzy PID. Applied Sciences, 13(1): 520.

[8] Ling SM 2017 Fuzzy PID adaptive control and Simulation of cooling water temperature of marine diesel engine cylinder liner.Ship Science and Technology,39(12):p93-95.

[9] Smith R D, Alleyne A G 2023 Iterative Learning Model Predictive Control for Thermal Management of Urban Air Mobility Vehicles. IEEE Conference on Control Technology and Applications (CCTA). IEEE, 2023: 948-954.

Sugarcane Field Terrain Inversion and Height Recognition Based on 3D-LiDAR and Improved RANSAC Algorithm

Meiqi Shi[1], Yanzhou Li[1*], Yingchun Pan[2], Linfei Lu[1], Jin Wei[1]

[1] College of Mechanical Engineering, Guangxi University, Nanning 530000, China

[2] Guangxi LiuGong Agricultural Machinery Co., Ltd, Liuzhou 545000, China

*Corresponding author's e-mail: lyz197916@126.com

Abstract. This study proposes a method for recognizing sugarcane field ground height based on 3D-LiDAR and an improved RANSAC algorithm, aiming to accurately extract and identify the height of sugarcane growth points from the sugarcane field environment. First, ROI region extraction and Gaussian filtering preprocessing were performed on the experimental bench data collected by LiDAR. Based on the improved RANSAC algorithm and combined with the point cloud morphology characteristics, a multi-model parallel fitting strategy was adopted to simultaneously fit both ground and sugarcane models. The terrain obscured by sugarcane leaves was reconstructed through cubic polynomial fitting. Traversing the Z-direction minimum of the sugarcane point cloud and labeling the sugarcane growth points, morphological filtering was used to remove the residual noise, and finally calculating the ground height of the sugarcane growth points. The results of indoor simulations and sugarcane field experiments show that the ground height recognition algorithm can accurately identify the boundary point between sugarcane and the ground. The average errors between terrain inversion height and actual measured height are 0.09cm and 0.1cm, respectively. This provides reliable technical support for the adaptive adjustment of sugarcane harvester header height.

1. Introduction

Sugarcane is an important perennial crop widely used in the production of sugar, alcohol, biofuels, and other products. In China, the Guangxi and Yunnan regions are the main sugarcane-growing areas, contributing approximately 80% of the national sugarcane production [2,3]. However, these regions are mostly hilly, with rugged terrain and steep slopes, creating significant obstacles for the mechanization of sugarcane harvesting. Currently, the mechanization rate of sugarcane harvesting in China remains relatively low. One significant reason is the uneven terrain, which complicates the accurate control of the cutting point at the base of the sugarcane stalks, thus affecting harvesting efficiency and quality [4]. Therefore, accurately obtaining terrain information of the regions to be harvested in sugarcane fields is crucial for achieving autonomous adjustment of the sugarcane harvester's cutting height

There are two types of terrain detection methods currently applied: contact and non-contact. Contact methods rely on direct interaction between mechanical equipment and the terrain, sensing and adapting to terrain changes via ground-touch sensors or suspension systems [5]. Suggs et al. developed a floating bottom cutter that adjusts its height to adapt to variations in the sugarcane field by incorporating a raised circular hub on the bottom plane of the cutter [6]. Pasi et al. used a towed soil wheel and the average value of multiple ultrasonic sensor measurements to obtain surface undulation information [7]. Xie et al. designed an optimal state feedback LQR controller to control the lifting of the cutting header by adjusting the torque [8]. Ding et al. designed an arc-shaped swing arm terrain detection device [9]. Non-contact terrain perception utilizes advanced sensing technologies such as computer vision, LiDAR, and ultrasound to obtain terrain information without direct contact with the ground. Wu et al. proposed the use of computer vision methods to obtain cutting height and features [10]. Long et al. proposed an indirect measurement method based on vehicle inclination angle and header inclination angle [11]. Zhao et al. employed ground-penetrating radar to measure the vertical distance between the ground and the antenna [12].

In summary, recent years have witnessed a rise in research on acquiring ground height information using contact profiling devices both domestically and internationally, but this method suffers from poor time response and difficult device installation. Non-contact methods largely remain in the theoretical research stage. In the complex sugarcane field environment, vision technology is easily affected by lighting and environmental conditions, while ultrasonic sensing technology, commonly used for close-range measurements, is greatly influenced by sensor resolution, signal-to-noise ratio, and other factors. In contrast, LiDAR can provide high-precision terrain information and is less affected by lighting, achieving measurement accuracy at the millimeter level.

Therefore, this study proposes a method for ground segmentation and sugarcane growth point height recognition using 3D-LiDAR as an environmental perception sensor. The method first performs ROI extraction and Gaussian filtering preprocessing on the 3D point cloud of the sugarcane field. It then uses an improved RANSAC algorithm and a cubic polynomial terrain inversion algorithm to segment the ground and mark the positions of sugarcane growth points. Morphological filtering is applied to further remove residual sugarcane leaf points, and the ground height of sugarcane growth points is finally calculated. The process of the ground height identification method for sugarcane growth points is illustrated in Figure 1.

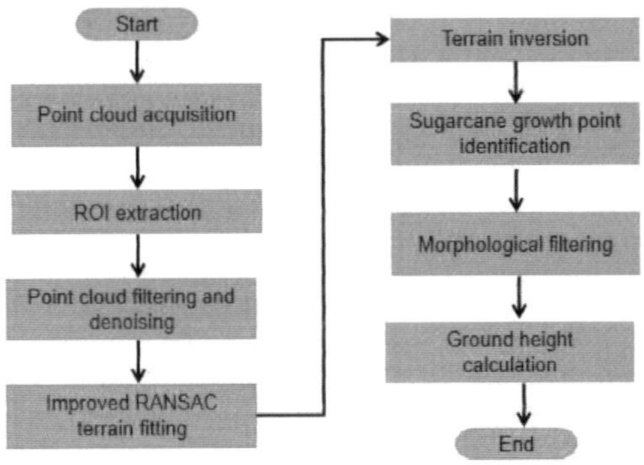

Figure 1. Flowchart of the ground height identification method for sugarcane growth points.

2. Materials and Methods

2.1. Data Collection

The data collection for this study was divided into two stages: indoor simulation and field experiments. First, a simulated sugarcane field test bench was constructed in the laboratory to collect data and verify the effectiveness of data processing algorithms in a controlled environment. Subsequently, further data collection was conducted in actual sugarcane field environments to test the robustness of the algorithm in complex scenarios.

2.1.1. Data collection equipment This study used the RS-Helios-5515 LiDAR model for data collection. As shown in Figure 2, the LiDAR features 32 laser beams with a wavelength of 905nm and a maximum range of 150 m. It offers a 360° horizontal scanning range and a vertical scanning span from -55° to +15°, with a maximum scanning frequency of 20 Hz. Through Ethernet communication, it provides real-time feedback on the three-dimensional coordinates, reflectivity, and other information of the measured object.

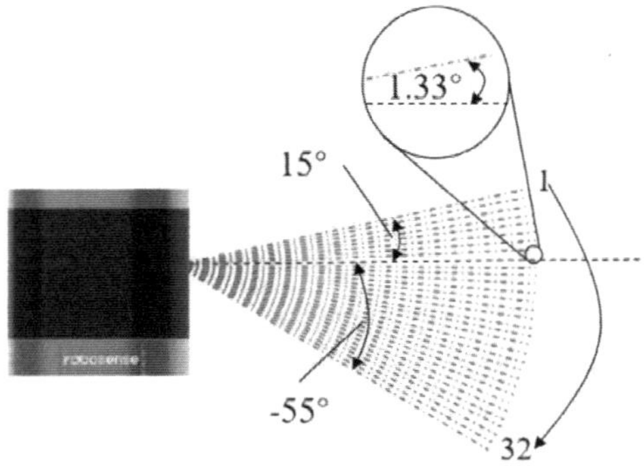

Figure 2. Schematic diagram of LiDAR sensor.

2.1.2. Indoor simulation test data collection The actual sugarcane field environment was simulated in the laboratory to evaluate the impact of interference conditions, like sugarcane leaves and weeds, on the ground segmentation of sugarcane fields scanned by LiDAR.

Table 1. Test bench environment simulation.

Item	Description
Soil moisture	20%
Terrain height variation	68~78cm
Ground obstructions	Sugarcane leaves, weeds
Test platform dimensions	Length 2 m, width 1 m, height 0.6 m
Sugarcane planting configuration	Adjustable row and plant spacing, 4 rows with 12 sugarcane plants

In order to simulate the sugarcane harvester's real-world operation in a single-line arrangement, the LiDAR was mounted on the table situated to the test platform's left side. Single-row mode refers to the sugarcane harvester harvesting only one row of sugarcane during operation, which helps improve harvesting accuracy and efficiency. To guarantee that the conditions of the indoor simulation experiment matched those of the sugarcane field trials, the LiDARwas positioned at an elevation of 81.2 cm, with a horizontal offset of 85 cm relative to the test stand. The LiDAR acquired data at a rate of 10 Hz.

Figures 3 and 4 illustrate the test bench's layout and experimental environment. Figure 3 presents the overall arrangement of the test bench, while Figure 4 provides a detailed view of the distribution of environmental obstructions.

Figure 3. Test Bench Layout.

(a) **(b)** **(c)**

Figure 4. Experimental environment setup. (a) Ground unobstructed. (b) Weed cover. (c) Sugarcane leaves and weed cover.

2.2. ROI segmentation and extraction

The indoor test bench point cloud data scanned by LiDAR includes other environmental information within the room. To specifically extract the sugarcane field test bench section, it is necessary to segment and extract the ROI (Region of Interest) [13]. Common methods for ROI segmentation include threshold segmentation, clustering, and region growth. Compared to other methods, the region growing algorithm offers higher accuracy and better robustness against noise and local irregularities in point cloud data.

First, select the initial seed points of the test bench area in CloudCompare. These seed points are located in the central area of the sugarcane test bench. Then, the growth conditions were defined, and through multiple repeated experiments, the optimal values for the distance threshold $d_{\max}$ and the normal vector angle difference threshold $\theta_{\max}$ were determined to be 5 and 10, respectively. Initialize the seed point set S and the processed point set P. Add the initial seed points to S. Select a point p from S, remove it from S, and add it to the processed point set P. Find the points adjacent to point p. If q meets the conditions of Formula 1 and Formula 2, add q to S. Repeat this process until S is empty [14].

$$\|p-q\| \le d_{\max} \tag{1}$$

$$\cos^{-1}\left(\frac{n_p \cdot n_q}{\|n_p\|\|n_q\|}\right) \le \theta_{\max} \tag{2}$$

Utilizing CloudCompare's segmentation function to separate the grown ROI from the entire room point cloud, thereby generating an independent point cloud file. By following the above steps, the target area can be effectively extracted from the complex point cloud data, ensuring the pertinence and accuracy of data processing. The extracted results are presented in Figure 5.

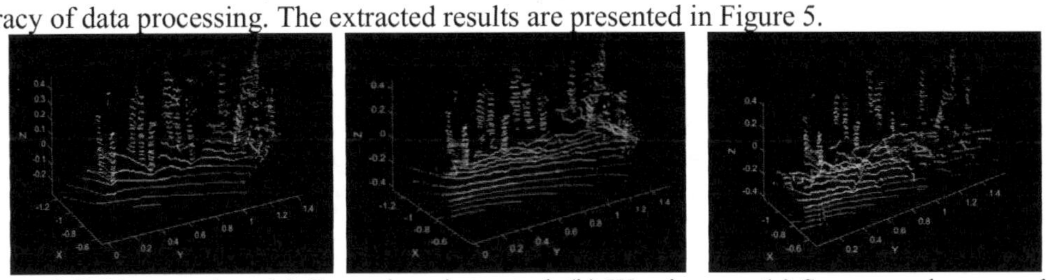

Figure 5. ROI extraction. (a) Ground unobstructed. (b) Weed cover. (c) Sugarcane leaves and weed cover.

2.3. Point cloud filtering and denoising

Due to the influence of instrument vibration, environmental factors, and the surface properties of the measured objects, noise is inevitably present in the point cloud data [15]. Especially during field data collection, the movement of the sugarcane harvester, visual obstructions, and obstacles can introduce outliers into the point cloud model. These noise points and outliers significantly affect the accuracy of subsequent point cloud processing operations, such as ground point cloud segmentation and height recognition.

Gaussian filtering was used to denoise the data. Gaussian filtering performs a statistical analysis on the neighborhood of each point and identifies points that exceed the normal range as outlier noise points to be removed [16]. The process of Gaussian filtering is shown in Figure 6.

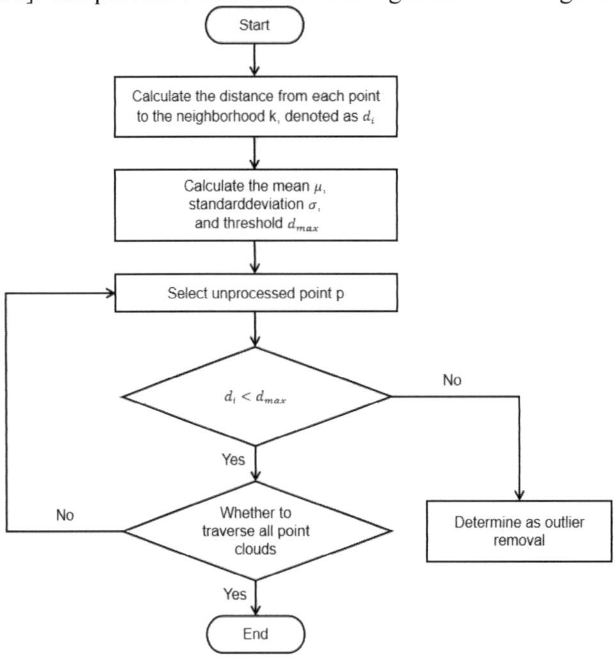

Figure 6. Gaussian filtering flowchart.

(1) The total number of point clouds in the statistical dataset is n . Calculate the average distance between each point P and its nearest k points, and aggregate them into d_i.

$$d_i = \frac{1}{k}\sum_{j=1}^{k}\left\|p_i - p_j\right\| \tag{3}$$

(2) Calculate the mean μ and standard deviation σ of point set d_i according to Formulas 4 and 5.

$$\mu = \frac{1}{n}\sum_{i=1}^{n} d_i \tag{4}$$

$$\sigma = \sqrt{\frac{1}{n-1}\sum_{i=1}^{n}(d_i - \mu)^2} \tag{5}$$

(3) After obtaining the mean μ and standard deviation σ , calculate the maximum threshold $d_{\max}$ of the Gaussian distribution according to Formula 4, where a represents the proportional coefficient.

$$d_{\max} = \mu + \alpha \times \sigma \tag{6}$$

Compare the average distance d_i of each point P with the maximum threshold $d_{\max}$. If a point's d_i is greater than the maximum threshold $d_{\max}$, classify this point as a noise point and remove it [17].

2.4. Sugarcane growth point identification

2.4.1. Improved RANSAC algorithm In situations where the ground is undulating and covered with sugarcane leaves and weeds, ground extraction mainly adopts plane fitting, rule-based segmentation, and clustering methods. These methods yield good results when dealing with simple scenes, but under the conditions of ground occlusion, model fitting accuracy decreases, computational complexity is high, and the segmentation effect is poor.

The Random Sample Consensus (RANSAC) algorithm is a robust model fitting method that includes the basic principles of random sampling, model fitting, evaluation model, and iterative optimization [18]. Figure 7 illustrates the use of RANSAC for fitting ground and sugarcane point clouds. In the figure, the ground point cloud obscured by sugarcane leaves cannot be identified, and some sugarcane point clouds are filtered out. This highlights the problem of the RANSAC algorithm's low computational efficiency and insufficient model fitting accuracy in handling occlusion and dense point cloud data. The traditional RANSAC method usually requires 500-1000 iterations to converge on complex terrain surfaces, which is costly and affects the real-time performance.

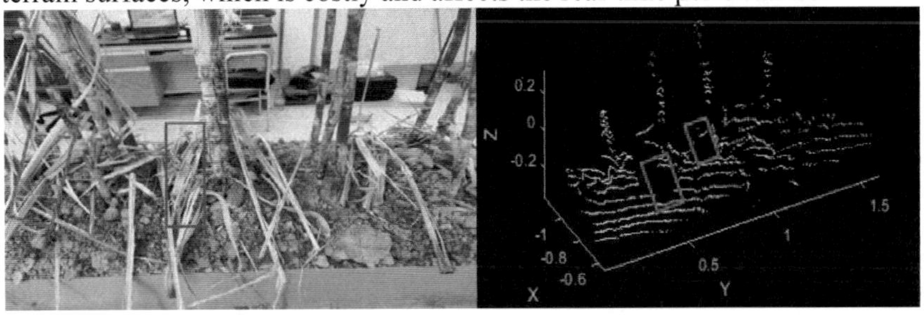

Figure 7. RANSAC point cloud iteration.

Therefore, this paper proposes an improved RANSAC algorithm based on an intelligent sampling strategy of point cloud shape features to enhance algorithm performance. As depicted in Figure 8, the ground point cloud forms a smooth and continuous linear structure, forming a horizontally distributed collection of points that are relatively flat and continuous. In contrast, sugarcane point clouds exhibit a vertical cylindrical shape, which is significantly different from ground point clouds that are vertically oriented and have obvious height characteristics. The weed point clouds appear as scattered point sets with low density and irregular distribution. The sugarcane leaf point clouds are clusters of point clouds connected to sugarcane plants, with high density and irregular linear shape.

Figure 8. Point cloud shapes.

Based on the traditional RANSAC algorithm, an intelligent sampling strategy is introduced to sample ground points and sugarcane points based on point cloud shape features.This intelligent selection of sample point sets reduces the number of iterations and improves computational efficiency. First, select a small subset of sample points from LiDAR point cloud data as the initial sample set. Adopt a multi-model parallel fitting strategy to calculate a ground model and a sugarcane model based on these points. Then apply these two models to all data points to verify the matching between each data point and the model. If the data points match the model, they are included in the model; otherwise, they are excluded [19]. This process is repeated multiple times, and the model with the highest number of matching sample points is chosen as the final fitted model, representing the segmented ground and sugarcane.

Compared with the traditional RANSAC method, this method reduces the number of iterations to 150-300 by adaptive thresholding and intelligent sampling strategy, and reduces the computational cost by 50%-70%. Experiments on a dataset containing 10,000 points show that the processing time of the traditional RANSAC method is 1.2 seconds, while the proposed method takes only 0.45 seconds, which is a computational efficiency improvement of about 62.5%, and significantly improves the real-time and applicability of terrain reconstruction.

2.4.2. Terrain inversion When obstacles like sugarcane leaves cover the ground, certain areas may become undetectable. Therefore, it is necessary to use terrain inversion methods to recover the terrain information of the obstructed area, thereby achieving accurate reconstruction of the obstructed ground [20]. Interpolation methods are suitable for handling flat ground. Statistical methods can handle noise and outliers but have limited model fitting capability for complex terrain changes. In contrast, cubic polynomial fitting is well-suited for modeling uneven surfaces. Its mathematical characteristics and adaptability allow it to effectively model terrains with moderate undulations while mitigating the risks of underfitting or overfitting.

For the ground points extracted by the improved RANSAC algorithm and unaffected from the influence of sugarcane foliage, a cubic function is applied to create a smooth surface model that accurately reflects height changes. First, set the polynomial degree to *n*=3 and use the method of least error to approximate the ground points, yielding the cubic equation below:

$$P(x,y) = a_3x^3 + a_2x^2 + a_1x + b_3y^3 + b_2y^2 + b_1y + c \quad (7)$$

Where $a_3, a_2, a_1, b_3, b_2, b_1, c$ represent the fitting coefficients, x and y refer to the horizontal coordinates of the points, and $P(x,y)$ indicates the vertical coordinates of the ground point. By solving the fitting parameters of the above polynomials using the least squares method, a ground model for the occluded region can be constructed.

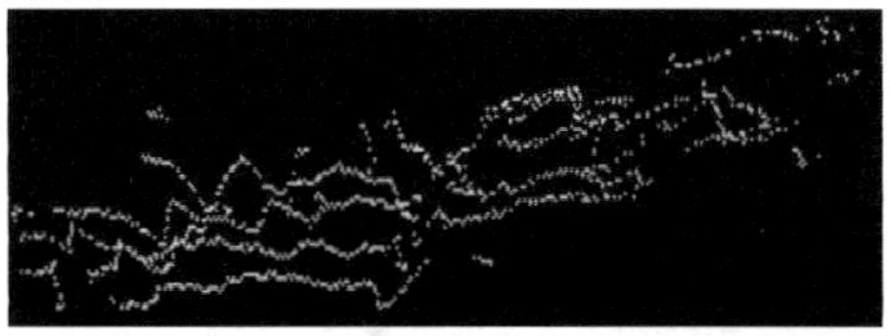

Figure 9. Sugarcane leaf occlusion.

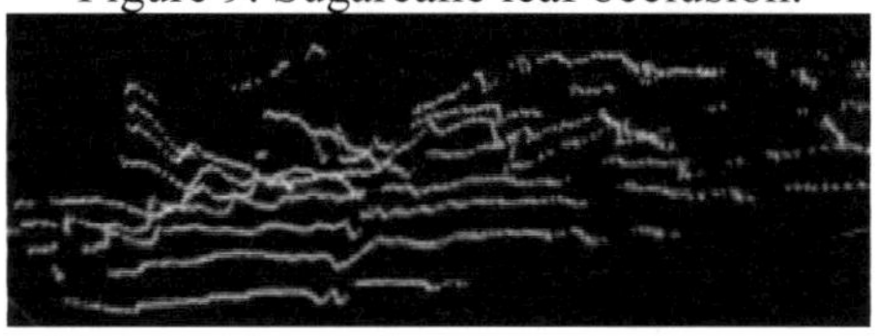

Figure 10. Terrain inversion.

Because of the occlusion from the leaves, the point cloud shown in Figure 9 was not accurately captured and displayed, resulting in a missing terrain contour. To address this issue, Figure 10 illustrates the results of recovering the occluded terrain through a cubic polynomial fit, which produces a continuous, seamless ground contour. The terrain reconstruction process constructs a seamless curve using available clear terrain data, which in turn fills in the previously missing terrain areas. The reconstructed terrain model is able to more realistically reflect the height variations and morphological features of the ground.

2.4.3. Identifying sugarcane growth points In order to determine where the sugarcane meets the ground, a starting minimum Z value is defined. Then, every point in the sugarcane structure is traversed to see whether its Z value is lower than the current minimum recorded value. If the condition is satisfied, the minimum Z value is updated and the coordinates of the point are recorded as the current minimum point. After traversing all the points, the finalized minimum point is the lowest position at the base of the sugarcane model. Next, this lowest point is mapped to the ground, with the location labeled in the ground point cloud to indicate the junction position between the sugarcane and the ground, and the relevant part of the sugarcane point cloud is removed.

2.5. Height recognition

In the ground segmentation stage, the ground has been extracted and the positions of the sugarcane have been marked. To accurately obtain the ground level at the growing points of sugarcane, morphological processing methods and height conversion are applied based on the three-dimensional coordinates (x,y,z) data obtained by LiDAR. This allows for precise identification of height information [21].

Use the dilation operation first to expand the area of ground points and fill in small depressions, thereby making the ground point cloud data smoother and more continuous [22]. The expansion operation expands the structural elements and increases the pixel points in the ground point cloud data, thereby eliminating local ground fluctuations caused by weed and sugarcane leaf noise.

$$Dilation(A) = A \oplus B \tag{8}$$

Where A is the original ground point cloud data, and B is the structural element of the expansion operation.

Since the image has been dilated, the resulting ground point cloud is larger than the original. Therefore, the erosion operation is used to remove prominent noise points from the ground point cloud data, reduce unnecessary pixels, eliminate interference caused by abnormal data, make the ground point cloud data clearer and more reliable, and restore it to the original size [23].

$$Erosion(A) = A \ominus B \tag{9}$$

The installation height of the LiDAR serves as a reference and is used to convert the height information of the extracted ground points into the elevation of the sugarcane growth points. As illustrated in equation (10), the ground height is determined by a difference between the LiDAR's mounting height and the extracted ground point Z value.

$$H_S = H_L - z_0 \tag{10}$$

Where H_S is the ground height of the sugarcane growth points, H_L is the installation height of the LiDAR, and z_0 is the Z-direction coordinate value of the sugarcane growth points acquired by the LIDAR.

3. Results

3.1. Indoor simulation experiment results

3.1.1. Ground Segmentation in Simulation Experiment The results of ground segmentation for indoor experimental bench data are shown in Figure 11. For the ROI extracted point clouds, the red box represents noise, and the arrows indicate the point cloud data after Gaussian filtering. By comparing these two sets of images, it is evident that noise has been effectively removed, and the point cloud data has become clearer and smoother. The red dot signifies the boundary between the ground and sugarcane. It can be seen from the diagram that the ground cloud has a continuous linear structure, which represents the contour of the surface.

(a) **Ground unobstructed:** In the absence of interference, the ground point cloud displays a clear continuous linear structure with smooth and distinct ground contours.

(b) **Ground with weeds:** After adding weed interference, although some parts of the ground are obstructed, the ground point cloud still maintains continuity. In areas with dense weeds, some noise points may appear, but they do not significantly affect the overall structure of the ground point cloud.

(c) **Ground with weeds and sugarcane leaves:** When the ground is disturbed by both weeds and sugarcane leaves, the continuity of the ground point cloud is greatly affected. However, through terrain inversion, the occluded ground point cloud can still be effectively reconstructed, resulting in a complete ground contour.

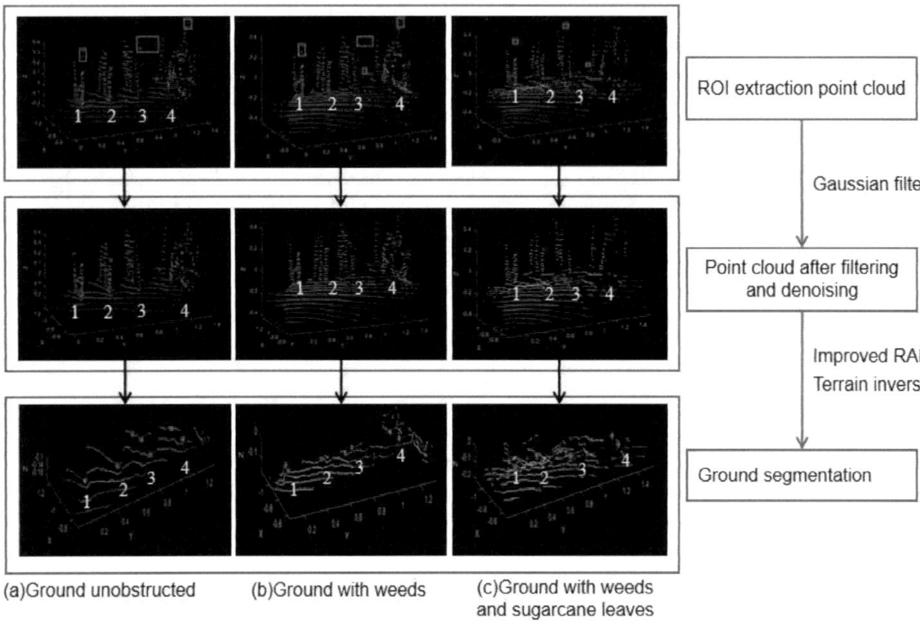

Figure 11. Ground segmentation process in simulation experiment.

3.1.2. Analysis of terrain inversion and height identification results In order to verify the accuracy of the terrain inversion and the identification of the ground, the ground height obtained by the inversion is compared with the real ground height, including the case where the ground is obstructed.

The LiDAR sensor was mounted and positioned at 81.2 cm high, which is used as a reference level. As shown in Figure 12 (c), the dividing point between the ground and the sugarcane determined through LiDAR is 8.83 cm lower than the reference level, which corresponds to a true height of 72.37 cm. In order to validate this result, the true terrain elevation was measured at the identical sugarcane growing point using a tape gauge, as depicted in Figure 12 (d). The height covered by the sugarcane leaves was 74.61 cm, while the true ground elevation was about 72.33 cm.

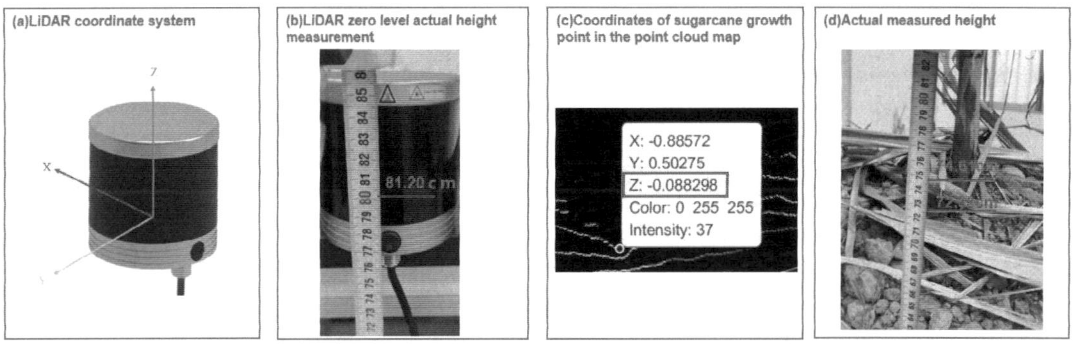

Figure 12. Measuring the height of the sugarcane growth point in simulation experiment.

To more precisely assess the impact of terrain inversion, a comparison was made between the true ground elevation, the height obscured by sugarcane leaves and weeds, and the height variations before and after the inversion process. The detailed results are presented in Table 2.

Table 2. Comparison of terrain inversion and actual measurement heights.

Obstruction Type	Measurement Point	Sugarcane leaves and weeds cover height (cm)	Actual measured ground height (cm)	Height before terrain inversion (cm)	Height of terrain inversion (cm)
None	1	/	70.95	70.87	70.90
	2	/	77.46	77.35	77.37
	3	/	70.92	70.83	70.85
	4	/	73.50	73.56	73.56
Weed obstruction	1	74.50	70.51	72.36	70.43
	2	76.85	73.76	74.50	73.88
	3	80.26	75.30	78.17	75.47
	4	79.48	77.39	78.25	77.28
Weeds and sugarcane leaves obstruction	1	80.25	77.16	79.42	77.07
	2	74.61	72.33	73.58	72.37
	3	72.15	69.85	71.37	69.94
	4	75.20	74.63	74.95	74.56

From the comparison data in Table 2, it is evident that the terrain inversion method effectively corrects ground height when obstructed by sugarcane leaves and weeds, resulting in inverted ground heights of sugarcane growth points closely matching the actual measured heights. In scenarios without obstruction, the difference between the height before terrain inversion and the actual measured height is minimal. Terrain inversion fine-tunes these values, resulting in a closer approximation to the actual height. When obstructed by weeds and sugarcane leaves, the ground height before terrain inversion tends to be higher, primarily due to initial height overestimation caused by these obstacles. Terrain inversion further optimizes and corrects for these interference effects, enhancing the accuracy of the final height identification results.

Figure 13 presents the statistical results of terrain height recognition errors for sugarcane growth points. Prior to terrain inversion, the recognition error was relatively large, and the data points were more dispersed, averaging 0.98 cm in error. After the terrain inversion processing, there is a significant improvement in the consistency of the altitude determined by the point cloud map and the actual altitude measured. The error was reduced to ± 0.17cm, and the mean error was reduced to 0.09 cm. These findings suggest that the enhanced RANSAC and terrain inversion algorithms have effectively boosted the accuracy and stability of terrain height recognition in obstructed environments.

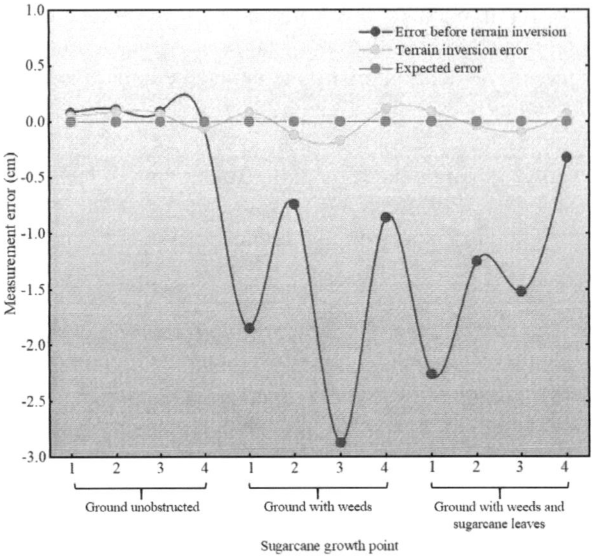

Figure 13. Statistics of height error for the growing points of sugarcane.

3.2. Field experiment

To validate the effectiveness of identifying sugarcane growth point terrain heights using improved RANSAC and terrain inversion algorithms under real field conditions, field data collection was conducted in April 2024 at the sugarcane plantation in Fusui and Chongzuo, Guangxi. Since sugarcane harvesters typically operate in single rows, the LiDAR was positioned 82.65 cm above the ground on the left side of the sugarcane harvester header to capture comprehensive data from the harvesting side of the sugarcane. During data collection in the sugarcane field, the harvester moved at a speed of 1.5 km/h, with the LiDAR scanning at a frequency of 10 Hz.

3.2.1. Ground segmentation of sugarcane field The ground segmentation results of the field data are depicted in Figure 12, where red dots indicate the border between the ground and sugarcane. The figure illustrates that the field environment is more complex than the indoor test bench environment, yet the segmentation algorithm can still effectively extract the ground point cloud.

(a) Sugarcane field ground segmentation: In the actual sugarcane field environment, the terrain is variable and the vegetation is dense, including obstructions such as growing weeds and fallen sugarcane leaves. Despite these challenges, the algorithm accurately identifies and extracts the ground point cloud, with red dots clearly representing the boundary points between the ground and sugarcane.

(b) Ground segmentation with fallen sugarcane: Fallen sugarcane increases the difficulty of ground segmentation due to its proximity to the ground. As illustrated in Figure 14, the distinction between the fallen sugarcane data points and the ground data points remains well-defined. The red dots precisely outline the separation between the sugarcane and the ground.

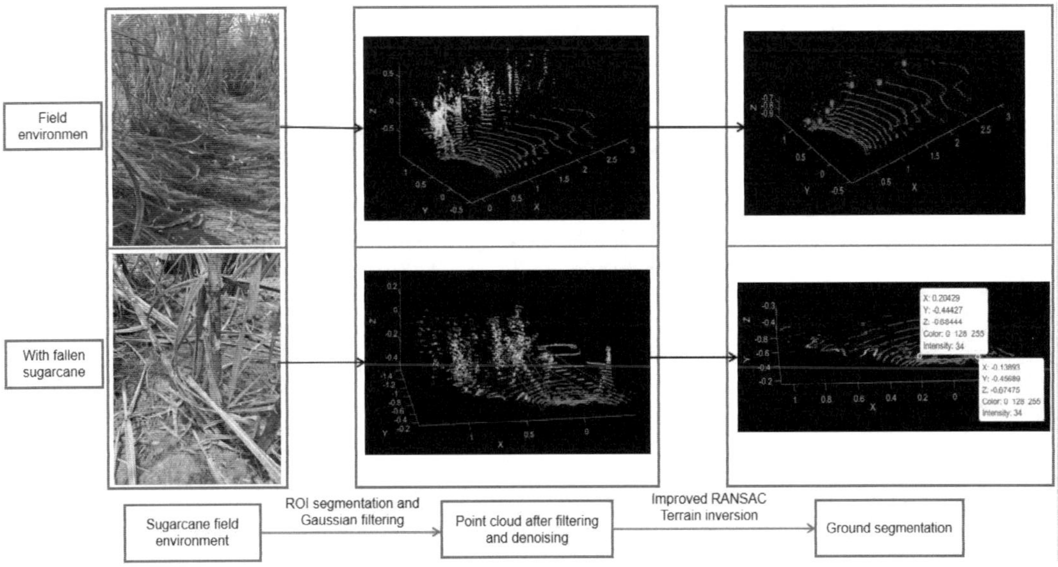

Figure 14. Field ground segmentation results.

3.2.2. Analysis of terrain inversion and height identification results In the field experiment, the significant ground occlusion resulting from complex terrain and dense vegetation necessitated detailed measurements and a comparative analysis was conducted to validate the precision of terrain reconstruction and ground height estimation in the sugarcane field.

As illustrated in Figure 15(a), the true height of the zero horizontal plane measures 82.65 cm. As observed in Figure 15(b), the dividing point separating the sugarcane from the ground is positioned 66.68 cm beneath the zero horizontal plane, resulting in a computed actual height of 15.97 cm. To validate this outcome, a measuring instrument was utilized to determine the actual height of the contact point where the sugarcane meets the ground in the field, as illustrated in Figure 15(c). The measured ground height approximates 15.84 cm, while the height obscured by sugarcane leaves measures 17.13 cm.

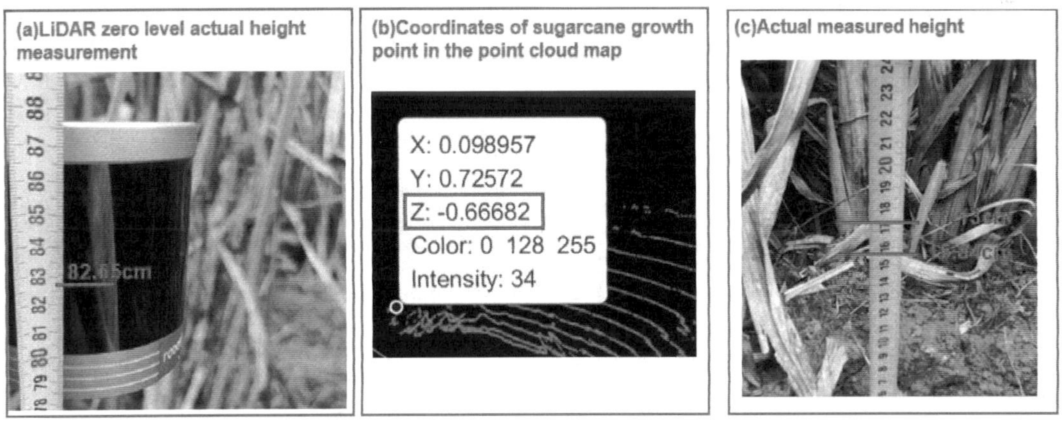

Figure 15. Measurement of terrain height at the sugarcane growth point.

Table 3 presents the results of terrain inversion and height identification in sugarcane fields. By comparison, it is clear that in intricate field conditions, the terrain inversion approach successfully adjusts ground height, even when obstructed by elements like sugarcane leaves and weeds.

Table 3. Comparison of sugarcane field terrain inversion and actual measured heights.

Measurement Point	Sugarcane leaves and weeds cover height (cm)	Actual measured ground height (cm)	Height before terrain inversion (cm)	Height of terrain inversion (cm)
1	17.13	15.84	16.73	15.97
2	15.35	11.55	13.84	11.58
3	13.79	12.5	12.97	12.61
4	15.20	14.13	15.06	14.25
5	14.68	12.00	13.85	11.93
6	19.54	14.35	17.62	14.23
7	14.90	14.80	14.88	14.71
8	16.55	13.47	15.06	13.36

Table 4 also includes a detailed analysis of the terrain inversion results concerning fallen sugarcane. Although fallen sugarcane complicates ground segmentation, the terrain inversion method adeptly identifies and restores ground height.

Table 4. Comparison of terrain inversion and actual measured heights for fallen sugarcane.

Measurement Point	Sugarcane leaves and weeds cover height (cm)	Actual measured ground height (cm)	Height before terrain inversion (cm)	Height of terrain inversion (cm)
1	16.90	14.28	15.25	14.21
2	16.35	15.22	15.93	15.17
3	14.55	11.35	13.71	11.43
4	15.26	12.41	14.68	12.28
5	15.97	15.65	15..88	15.76
6	20.00	13.47	18.26	13.32
7	18.64	14.15	16.32	14.25
8	17.38	12.76	15.94	12.64

Figure 16 illustrates the statistical results of terrain height recognition errors for sugarcane growth points in a sugarcane field. Prior to terrain inversion, data points exhibited a relatively scattered distribution, characterized by substantial fluctuation in error amplitude, averaging 1.61 cm. Subsequent to terrain inversion processing, the error between estimated and actual measured heights significantly decreased, with error ranging within ±0.15 cm and average error reduced to 0.10 cm. This underscores terrain inversion's high accuracy and stability in sugarcane field conditions, facilitating effective real-time identification of ground heights for sugarcane growth points during harvesting.

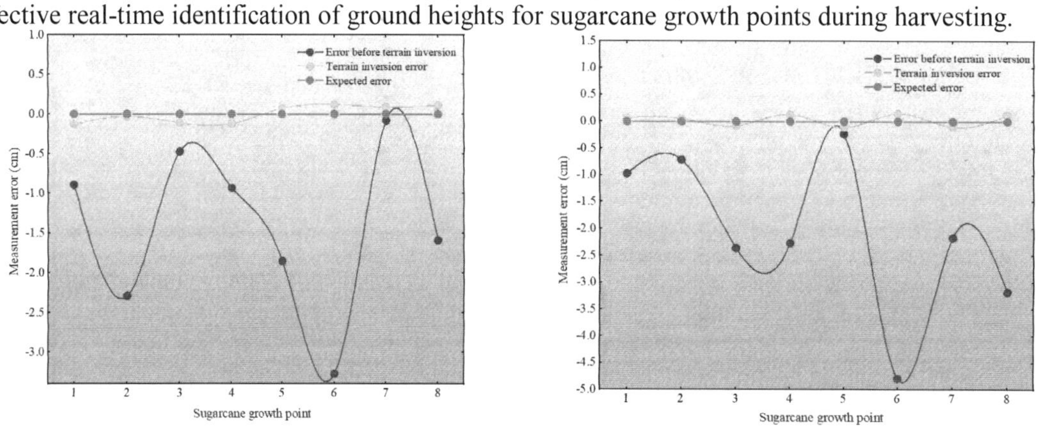

(a) (b)

Figure 16. Statistical chart terrain height error of sugarcane growth points in sugarcane field. (a)Field environment. (b)Fallen sugarcane.

4. Discussion

This study utilizes 3D-LiDAR technology to enhance ground segmentation and height recognition methods using RANSAC and terrain inversion. The terrain height is accurately extracted under challenging environmental conditions, including interference from sugarcane leaves and weeds, and its effectiveness and robustness are validated through indoor simulations and field experiments. The approach exhibits high accuracy in identifying the terrain height of sugarcane growth points, thereby offering crucial technical support for dynamically adjusting the header height of sugarcane harvesters to accommodate ground undulations.

Utilizing LiDAR technology enables obtaining high-precision terrain information, overcoming the slow response and complexity of traditional contact terrain detection methods, and compensating for the limitations of non-contact methods in practical applications. In contrast to Zhao's ground penetrating radar-based stone detection method and Wu's computer vision approach, LiDAR offers millimeter-level accuracy and is less affected by environmental light. While ground profiling wheels and ultrasonic ranging methods have demonstrated efficacy under specific conditions, they are unsuitable for the complex and variable environments of sugarcane fields.

However, this study also acknowledges certain limitations. LiDAR performance may be impacted to some extent during foggy or rainy weather conditions. Future research could explore multi-sensor data fusion techniques, including integrating LiDAR data with computer vision, ultrasound, and other sensor inputs, to enhance the stability and accuracy of terrain information across diverse environments.

5. Conclusions

This study proposes a ground segmentation and height recognition method for sugarcane fields based on 3D-LiDAR, offering a reliable basis for achieving precise control of sugarcane harvester cutting height. The main conclusions are as follows:

(1) ROI extraction and Gaussian filtering preprocess 3D-LiDAR sugarcane field point cloud data to reduce the impact of interference factors on point cloud processing.

(2)An improved RANSAC algorithm and a cubic polynomial terrain inversion algorithm were proposed to segment ground point clouds. The ground point cloud exhibits a continuous linear shape, while the sugarcane point cloud displays a vertical cylindrical shape. This significant shape difference provides a basis for extracting ground and sugarcane point clouds. When the terrain is obstructed by weeds and sugarcane leaves, cubic polynomial fitting is applied to reconstruct the blocked terrain, enhancing the accuracy of terrain segmentation.

(3)Traverse the minimum value in the Z-direction of the sugarcane point cloud to label the sugarcane growth points. Apply morphological filtering technology to remove residual sugarcane leaf points, retain the main structure of the ground, and determine the ground elevation at the sugarcane growth points.

(4)The results of the indoor simulation experiment indicate that the difference between the terrain inversion height and the actual measured height is within a range of ±0.17 cm, with the mean error being 0.09 cm. Field experiments in sugarcane fields demonstrate that the method for identifying the height of sugarcane growth points performs equally well, with a mean error of 0.10 cm between the elevation of the terrain inversion and the actual height measured. The sugarcane field ground segmentation and height recognition method for sugarcane growth points proposed in this article basically meets the accuracy requirements, providing technical support for the sugarcane harvester to adaptively adjust the height of the cutting header with the ground undulation.

Funding

This research was funded by the Guangxi Major Science and Technology Special Fund of China (Guike AA22117007, Guike AA22117005), Beihai Science and Technology Development Plan Project of China (2023174001), and National Key Research and Development Program of China (2022YFD2301100).

References

[1] López-Ortega M G, Guadalajara Y, Junqueira T L, et al. Sustainability analysis of bioethanol production in Mexico by a retrofitted sugarcane industry based on the Brazilian expertise[J]. Energy, 2021, 232: 121056.
[2] Omprakash S. Assessment of sugarcane industry: Suitability for production, consumption,and utilization[J]. Annals of Agrarian Science, 2018, 16(4): 389-395.
[3] Qu Y. The current situation and countermeasures of full mechanization of sugarcane production in China [J]. Modern Agricultural Equipment, 2019, 40 (2): 3-8.
[4] Bai J, Ma S, Yang G, et al. Testing and Analysis of the Extractor of a Sugarcane Chopper Harvester[J]. Transactions of the ASABE, 2020, 63(2): 251-257.
[5] Dong Z. Research on the profile and distance measurement of sugarcane harvester blades [D]. Guangxi: Guangxi University, 2022.
[6] Suggs C, Abrams C. Automatic Hydraulic Height Control Mechanism for Rotary Cutting Harvesting Equipment[J]. 1972, 15(3): 428-432.
[7] Suomi P, Oksanen T. Automatic working depth control for seed drill using ISO 11783 remote control messages[J]. Computers and Electronics in Agriculture, 2015, 116(C): 30–35.
[8] Xie Y, Ashley Greer. Header height control of a combine harvester system[J]. Dynamic Systems and Control Conference, 2010, 11(2): 135-139.
[9] Ding Z, Ma S, Wang F, et al. Development and Testing of an Electro-Hydraulic System for Height Control of Sugarcane Harvester Basecutter[J]. Journal of the ASABE, 2022, 65(2): 305-312.
[10] Wu Y, Shang X, Zhang C, et al. Obtaining Operating Parameters of Intelligent Leaf Vegetable Harvester under Natural Light [J]. Zhejiang Agricultural Journal, 2017, 29 (11): 1930-1937.
[11] Long Z, Xiang Y, Li Y, et al. An adaptive height adjustment system for cutting heads based on inclination sensors [J]. Journal of China Agricultural University, 2021, 26 (11): 200-208.
[12] [Zhao L, Zhang J, Jiao S, et al. Ground surface detection method using ground penetrating radar signal for sugarcane harvester base-cutter control[J]. Biosystems Engineering, 2022, 219: 103-123.
[13] Ding Y, Gong W, Cheng Z, et al. Model test and engineering application of landslide surface vertical deformation measurement based on multi-phase UAV images[J]. Geoscience and Technology Bulletin, 2023. 42(02): 267-278.
[14] Moyano J, León J, Nieto-Julián J, et al. Semantic interpretation of architectural and archaeological geometries: Point cloud segmentation for HBIM parameterisation[J]. Automation in Construction, 2021. 130: 103856.
[15] Hu X, Wei X, Sun J. A Noising-Denoising Framework for Point Cloud Upsampling via Normalizing Flows[J]. Pattern Recognition, 2023. 140:109569.
[16] Wen G, Zhang H, Guan Z, et al. Bilateral filter denoising of Lidar point cloud data in automatic driving scene[J]. Infrared Physics & Technology, 2023. 131:104724.
[17] Li Q, Zhu H. Performance evaluation of 2D LiDAR SLAM algorithms in simulated orchard environments[J]. Computers and Electronics in Agriculture, 2024. 221: 108994.
[18] Deng S, Xu Q, Yue Y, et al. Individual tree detection and segmentation from unmanned aerial vehicle-LiDAR data based on a trunk point distribution indicator[J]. Computers and Electronics in Agriculture, 2024. 218:108717.
[19] Goetz K, Soitinaho R, Oksanen T. Ploughing furrow recognition for onland ploughing using a 3D-LiDAR sensor[J]. Computers and Electronics in Agriculture, 2023. 210: 107941.
[20] Luo H, Xu Q, Cheng Y, et al. 3-D deformation inversion: A MATLAB toolbox for automatically calculating SAR-derived 3-D deformation maps of glacier, landslide, and land subsidence[J]. Environmental Modelling & Software, 2024,178:106074.

[21] Meng X, Wang L, Silván-Cárdenas J, et al. A multi-directional ground filtering algorithm for airborne LIDAR[J]. ISPRS Journal of Photogrammetry and Remote Sensing, 2009. 64(1): 117-124.
[22] Li X, Wang X, Wei T. Adaptive morphology and multiscale combination for remote sensing image segmentation of vegetation regions[J]. Advances in Lasers and Optoelectronics, 2022. 59(24): 240-246.
[23] Wu J, Wei X, Wang H, et al. Weld extraction method for tank trucks based on edge detection[J]. Electric Welding Machine, 2021. 51(07): 93-98+118.

Human-Robot Interaction Control Based on Motion Intent Estimation

Luming Yang[*][0009-0001-6144-5586], **Yina Wang**[0000-0003-3574-8302]

[1] School of Electrical Engineering, Shenyang University of Technology, Shenyang, Liaoning Province, 110870, P.R. China

[*]Corresponding author's e-mail: ylm456258@163.com

Abstract. This paper presents an adaptive impedance control framework for human-robot cooperation under dual uncertainties of unknown human motion intent and uncertain robot dynamics. Human motion intention is denoted by the desired trajectory within the kinematic model of the human limb. Obtaining this trajectory poses notable challenges due to the highly nonlinear and time-varying dynamics inherent in human limb modeling. To address this issue, a novel online estimation architecture is developed by integrating Radial Basis Function Neural Networks (RBFNN). This adaptive impedance controller integrates estimated motion intent to enable real-time tracking of a predefined impedance reference model. This developed framework's active collaboration with its human partner can be demonstrated by experimental results. This work provides novel theoretical insights and practical contributions to the field of human-robot interaction.

1. Introduction

Accelerated advancement of robot systems, coupled with advancements in intelligent control and machine learning, has propelled robotic systems into mission-critical applications ranging from medical rehabilitation to industrial manufacturing [1]. Human-robot collaboration (HRC) capitalizes on the synergistic relationship between human cognitive flexibility and robotic precision, emerging as a transformative paradigm for complex task execution [2]. Central to HRC effectiveness is the implementation of physical human-robot interaction (PHRI) systems that exhibit both compliance and adaptability, particularly in dynamic environments where human intent and task requirements evolve unpredictably [3]. Despite technological progress, existing control architectures struggle to reconcile the conflicting demands of robust trajectory tracking and responsive human adaptation, prompting urgent methodological innovations.

Current PHRI control strategies face fundamental limitations when deployed in real-world collaborative scenarios. Force control methodologies, while theoretically capable of direct interaction force regulation [4], prove impractical due to their acute sensitivity to environmental disturbances and reliance on precise dynamic models. Hybrid position/force control schemes, despite their spatial decoupling advantages [5], inherently treat interaction forces as disturbances to be suppressed rather than collaborative cues to be interpreted. Conventional impedance control techniques [7-13] partially address these issues through adjustable force-position relationships but remain constrained by their offline parameter calibration requirements and performance degradation during abrupt intent transitions [15]. Even advanced variants like position-based impedance control [14], which employs virtual trajectory generation, fail to adapt impedance parameters dynamically in response to real-time human behavior. These limitations collectively underscore a critical gap: the absence of intent-aware control mechanisms capable of bridging human cognitive signals with robotic actuation.

Parallel challenges persist in human motion intent estimation, the cornerstone of proactive collaboration. Kinematics-driven approaches [16-17] oversimplify intent derivation by neglecting individual biomechanical variations and sensorimotor adaptation. Momentum conservation-based methods [18] suffer from unrealistic physical assumptions that collapse under multi-contact conditions, while filter-based techniques [19-20] sacrifice temporal resolution through excessive state discretization. Crucially, existing estimators predominantly focus on single-degree-of-freedom motions or discrete intent states, failing to capture the multidimensional, time-varying nature of natural human movements – a deficiency that fundamentally limits their clinical and industrial applicability.

The following order is used to organize the remaining parts of this paper. Section 2 describes in detail the specific human-robot collaborative system and the interactive information media involved, and designs a robot trajectory tracking controller based on the Backstepping Control. In Section 3, we provide an in-depth description of the human limb motion model and the neural network motion intention estimation method. In Section 4, we integrate the estimation results of motion intent into the impedance control of the robot and establish a control block diagram of the entire system. The proposed method's effectiveness is verified in Section 5 through extensive experimental studies.

2. Description of systems

Consider the human-robot interaction task shown in Figure 1, where the user completes walking training with the assistance of a robot. In this task, it should be ensured that the robot moves freely and flexibly under the guidance of the user, minimizing the user's energy consumption and providing the user with a comfortable operating experience. For secure collaboration, this can be achieved by limiting the position and speed of the robot

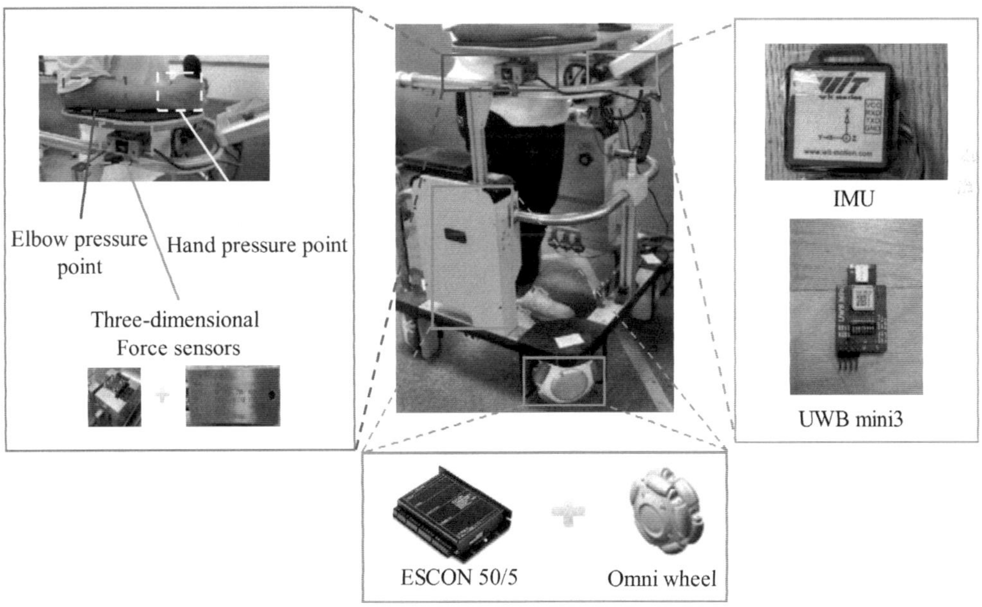

Figure 1. The intelligent walking-aid robot

2.1. Omni-directional mobile lower limb rehabilitation robot

The robot is composed of a power supply system, an interactive screen, an emergency braking, a support plate and an omnidirectional mobile chassis. In passive mode, the user selects a preset trajectory via the touch screen to assist walking. The omnidirectional moving chassis consists of four independently driven omnidirectional wheels, which allows the robot to achieve a variety of movements in the plane.

Journal of Physics: Conference Series **3024** (2025) 012026 doi:10.1088/1742-6596/3024/1/012026

In order to realize human-robot interaction, inertial measurement units, three-dimensional force sensors, ultrasonic sensors and UWB indoor positioning sensors are installed on the robot. All sensors communicate with the host computer through the serial port, and their specific working principle is shown in Figure 2.

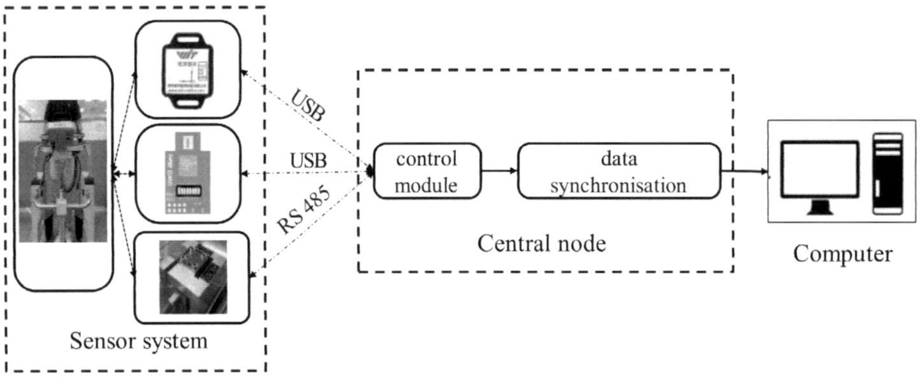

Figure 2. Structure of Motion Information Acquisition System

2.2. Robot Kinematics Modelling

A kinematic model of the robot can be created based on the following assumptions:
1) The wheels do not slip;
2) The robot's geometric center is where its center of gravity is located.

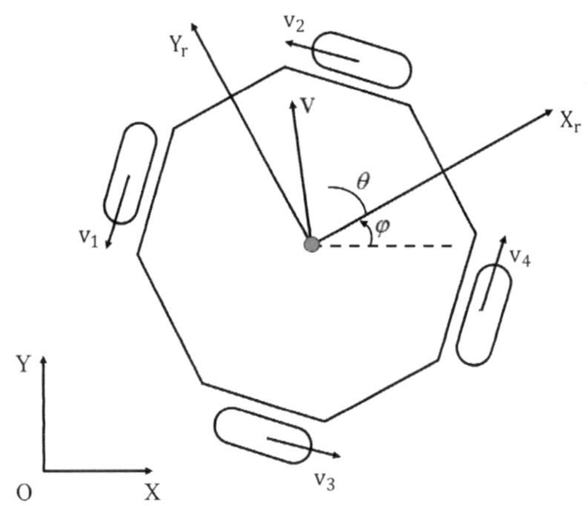

Figure 3. Structure of the four-wheeled omnidirectional mobile robot

As shown in Figure 3, $X = [x(t)\ y(t)\ \varphi(t)]^T$ represents the position of the mobile robot, where x、y represents the position of the robot in the global coordinate system, and φ represents the attitude angle of the robot.

According to the principle of vector synthesis, the kinematic model of the mobile robot is shown in equation (1):

$$\omega = \frac{1}{r} J_r v_r, J_r = \begin{bmatrix} -\sin\theta & \cos\theta & 1 \\ \cos\theta & \sin\theta & -1 \\ -\sin\theta & \cos\theta & -1 \\ \cos\theta & \sin\theta & 1 \end{bmatrix} \tag{1}$$

$\omega = [\omega_1\ \omega_2\ \omega_3\ \omega_4]^T$ and $v_r = [\dot{X}_r\ \dot{Y}_r\ \dot{\varphi}]^T$ represents the rotational speed of the omnidirectional wheel and the speed in the robot coordinate system, respectively.

2.3. Robot Motion Controller

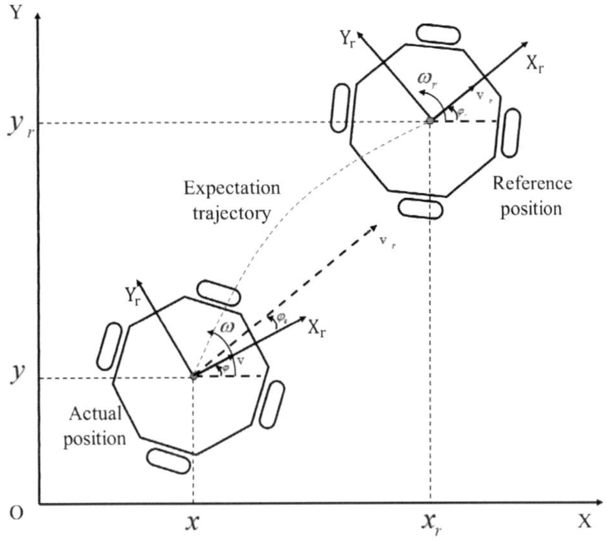

Figure 4. Trajectory Tracking Schematic

Referring to Figure 4, for the desired position $X_{dr} = [x_r\ y_r\ \varphi_r]^T$ and the desired velocity $V_r = [v_r\ \omega_r]^T$ in the world coordinate system, the corresponding errors can be expressed in the robot coordinate system as shown in Equation (2), where $R(\varphi)$ represents the coordinate transformation matrix.

$$X_e = \begin{bmatrix} x_e \\ y_e \\ \varphi_e \end{bmatrix} = R(\varphi)\begin{bmatrix} x_r - x \\ y_r - y \\ \varphi_r - \varphi \end{bmatrix} \tag{2}$$

The corresponding differential equation for the position error can be obtained by deriving the above equation:

$$\begin{bmatrix} \dot{x}_e \\ \dot{y}_e \\ \dot{\varphi}_e \end{bmatrix} = \begin{bmatrix} y_e\omega - v_x + v_r\cos\varphi_e \\ -x_e\omega - v_y + v_r\sin\varphi_e \\ \omega_r - \omega \end{bmatrix} \tag{3}$$

We assume here that the Lyapunov function is

$$V_1 = \frac{1}{2}(x_e^2 + y_e^2 + \varphi_e^2) \tag{4}$$

Find its derivative as shown in the following equation (5), which is required to be less than zero when the system is asymptotically stabilized.

$$\dot{V}_1 = x_e\dot{x}_e + y_e\dot{y}_e + \varphi_e\dot{\varphi}_e \tag{5}$$

$$\dot{x}_e = -k_x x_e\ , \dot{y}_e = -k_y y_e\ , \dot{\varphi}_e = -k_\varphi \varphi_e \tag{6}$$

When k_x、k_y、$k_\varphi > 0$, the derivative of the Lyapunov function is less than zero and the system is asymptotically stable.

Substituting into the position error differential equation yields the control rate as follows:

$$
\begin{aligned}
v_x &= y_e\omega + v_r\cos\varphi_e + k_x x_e \\
v_y &= -x_e\omega + v_r\sin\varphi_e + k_y y_e \\
\omega &= \omega_r + k_\varphi \varphi_e
\end{aligned}
\tag{7}
$$

3. Estimation of movement intentions

In human-robot collaboration, comprehending each other's motion intentions is of vital significance. To enable effective communication, we can make use of various types of sensors, such as force sensors and position sensors. Sensors act as interfaces in human-robot interaction. This study develops methods to estimate human motion intent from robotic sensory data.

3.1. *Human Limb Movement Modelling*

We use a human limb motion model to formulate the motion intention of a human user in this part. As presented in reference [21], the general model adopted herein for depicting the dynamics of limb motion ought to incorporate mass - damped stiffness characteristics:

$$-M_{HL}\ddot{x} - C_{HL}\dot{x} + G_{HL}(x_{H\_d} - x) = f \tag{8}$$

The inertial mass, damping, and spring/stiffness matrices for the human limb are represented by M_{HL}, C_{HL}, and G_{HL}, respectively, as diagonal arrays. The trajectory that the human partner's motor intention plans, referred to as x_{H_d}, is x_{H_d}.

Previous research has validated that damping and spring/stiffness properties typically govern the kinematic modeling of human limbs. The following simplified model is the result of our work:

$$-C_{HL}\dot{x} + G_{HL}(x_{H\_d} - x) = f \tag{9}$$

Assumption 1: The unknown functions C_{HL} and G_{HL} refer to x and $\dot{x}$, namely $C_{HL}(x,\dot{x})$ and $G_{HL}(x,\dot{x})$.

Assumption 2: Human motion intent x_{H_d} is inferred from interaction f, position x, and velocity $\dot{x}$.

$$x_{H\_d} = F_{H\_d}(f,\dot{x},x) \tag{10}$$

Here, $F_{H\_d}(\cdot)$ is a function that is not known.

In (9), the time-varying and uncertain nature of G_{HL} and C_{HL} makes function $F_{H\_d}(\cdot)$ non-linear. During cooperation, the human partner may alter limb impedance at any time, complicating the estimation of motor intention x_{H_d}. We address this with a neural network approach. It can uncover system info, map unknowns, and approximate functions.

3.2. *Neural network based motion intention estimation*

This work employs RBFNN to model the estimation function for human intent. The structure of this neural network is generally expressible as follows:

$$
\begin{gathered}
y(\sigma, R_{in}) = \sigma^T H(R_{in}),\ \sigma, H(R_{in}) \in R^M \\
H(R_{in}) = [h_1(R_{in}), h_2(R_{in}), \cdots, h_M(R_{in})]^T \\
h_k(R_{in}) = \exp\left[\frac{-(R_{in}-\mu_k)^T(R_{in}-\mu_k)}{\tau_k^2}\right] \\
k = 1,2,\cdots,M
\end{gathered}
\tag{11}
$$

The RBF neural network receives input from $y(\sigma, R_{in})$ which is a continuous function related to R_{in}. M stands for the number of hidden-layer nodes. The vector $\mu_k = [\mu_{k,1}\ \mu_{k,2}\ \cdots\ \mu_{k,n}]^T$ represents the center of the receptive field for the hidden-layer nodes. The parameter τ_k stands for the width of the Gaussian basis function. Moreover, the weight vector of the hidden-layer nodes is denoted as σ.

With the RBF neural network, the human partner's movement intention and our estimate of it are given by the following equation:

$$x_{H\_d,i} = \hat{\sigma}_i^T H_i\left(R_{in\_i}\right) + \varepsilon_i,\ \hat{x}_{H\_d,i} = \hat{\sigma}_i^T H_i(R_{in\_i}) \tag{12}$$

Here, $(\cdot)_i$ is the i-th element of $(\cdot)$. $R_{in\_i} = [f_i\ x_i\ \dot{x}_i]^T$ is the network input. ε_i represents the error that occurs during the estimation process, and the value $\hat{\sigma}_i$ is what we obtain as an approximation of the ideal weight σ_i. Empirically, the error will be minimized when M is sufficiently large.

Notably, the function $F_{Hd}(\cdot)$ fitted by the neural network in this paper is continuous, a generally reliable assumption. This is because, in typical human-robot cooperation scenario, human users adjust limb impedance in a smooth and stable manner. The discontinuity of function will be explored in future research.

From the gathered data, we calculate $H_i(R_{in})$. The direction-propagation algorithm is employed to obtain $\hat{\sigma}_i$ in (12). Given prior discussion, our control aim is to make the robot move "actively" to the human-predefined position, minimizing the interaction force f_i. Thus, $\hat{\sigma}_i$ can be adjusted online following the fastest-decreasing path according to the loss function below.

$$E_i = \frac{1}{2} f_i^2 \tag{13}$$

Again, we can get:

$$\hat{\sigma}_i(t) = -\beta_i' \frac{\partial E_i}{\partial \hat{\sigma}_i} = -\beta_i' \frac{\partial E_i}{\partial f_i} \frac{\partial f_i}{\partial x_{H\_d,i}} \frac{\partial x_{H\_d,i}}{\partial \hat{\sigma}_i} = -\beta_i' f_i \frac{\partial f_i}{\partial x_{H\_d,i}} \frac{\partial x_{H\_d,i}}{\partial \hat{\sigma}_i} \tag{14}$$

Note that the learning rate β_i' is a positive integer.

Based on (9) and (12) in the preceding section, we can compute:

$$\frac{\partial f_i}{\partial x_{H\_d,i}} = G_{HL,i},\ \frac{\partial f_i}{\partial \hat{\sigma}_i} = H(R_{in\_i}) \tag{15}$$

Bringing equations (15) into (14), respectively, we obtain:

$$\dot{\hat{\sigma}}_i(t) = -\beta_i f_i H_i(R_{in\_i}) \tag{16}$$

Here, $\beta_i = \beta_i' G_{HL}$. Since G_{HL} is an undetermined parameter within the framework of the human limb movement model, we substitute it with a suitable parameter β_i.

From the above, we can obtain the update rate of $\hat{\sigma}_i$, where dt stands for the sampling interval used in the robot system.

$$\sigma_i(t) = \sigma_i(t-1) - \beta_i f_i(t-1) H(R_{in\_i}(t-1))dt \tag{17}$$

Notably, the update of $\hat{\sigma}_i$ occurs online, enabling human users to alter their motion intentions at any time. In practical use, we set specific switching conditions to determine the adaptability of weight parameters, simplifying calculation and enhancing system robustness. The figure 5 below shows simulation results of motion intention prediction for different numbers of hidden-layer nodes.

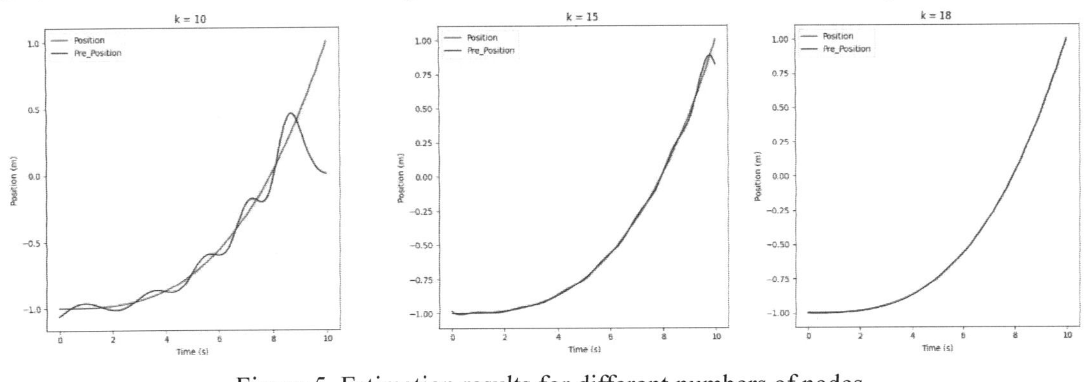

Figure 5. Estimation results for different numbers of nodes

4. Impedance Controller

In the prior section, we estimated the human motion intention. As neural-network estimation has errors and may reach local minima during loss-function iteration, x_{H_d} and $\hat{x}_{H\_d}$ aren't identical. Position controllers are unsuitable for tracking robot position to estimate intent x_{H_d}. This compensates, via impedance control, part of the error between x_{Hd} and $\hat{x}_{H\_d}$, minimizing user energy consumption when guiding the robot. The control block diagram of the system is presented in Figure 6.

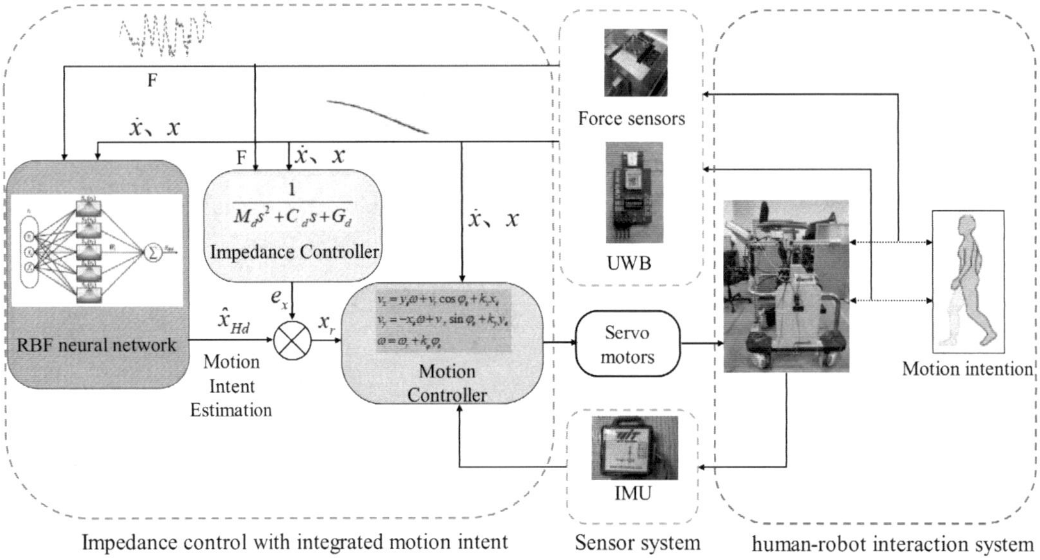

Figure 6. The control block diagram of the system.

In pre-defined rehabilitation training tasks, we presume that the robot's motion trajectory is pre-specified and achievable through control. The human partner in the human-robot cooperation job this research examines determines the necessary motion, which is unknown for controller design. In earlier studies, researchers used impedance control to regulate the robot's movement so that it could adjust to the stresses exerted by its human partner. Similarly, the target impedance model governs robotic dynamics as detailed below:

$$M_x(\ddot{x} - \ddot{x}_r) + C_x(\dot{x} - \dot{x}_r) + K_x(x - x_r) = f \tag{18}$$

Here , x_r represents the initial position of the robot's motion. Regarding M_x, C_x and K_x denote to the required inertia matrix, damping matrix, and stiffness matrix.

Impedance control modulates robot position based on interaction force f. The human operator perceives this as interacting with a virtual mass-spring-damper system characterized by inertia M_x, damping C_x, and stiffness K_x, moving from initial position x_r to target x. Large deviations between x and x_r require greater effort to manipulate the robot.

Under various conditions, x_r estimation relies on robotic prediction of human motion intent. For instance, in the case of a human shaking hands with a robotic arm, even though the robot cannot precisely forecast the actual human movement, x_r can be estimated based on a fundamental grasp of human hand - shaking behavior. Nevertheless, this empirical approach is somewhat inflexible and fails to ensure optimal performance. Thus, in the third section of this paper, we put forward human limb movement models to predict users' movement intentions and incorporate them into the existing impedance control. Therefore, the impedance controller can be expressed in the frequency domain as:

$$E(s) = \frac{F(s)}{M_x s^2 + C_x s + K_x} \tag{19}$$

Journal of Physics: Conference Series **3024** (2025) 012026 doi:10.1088/1742-6596/3024/1/012026

5. Experiments

In this part, we verify the proposed method via experiments. The experiment utilizes a four - wheel omnidirectional mobile rehabilitation robot in the laboratory, as depicted in Fig. 1. In the experiment, the human user's upper limb interacts physically with the robot through the pallet and handle. The interaction force between human and robot is measured directly by a 3D force sensor under the pallet. For the experimental environment setup, we use UWB indoor localization to acquire the robot's real - time position and calculate the robot's 2D - plane moving velocity by differentiating position data. The experimental environment is shown in Figure 7. As the user's true movement intention cannot be obtained during the experiment, we substitute it with the marked trajectory on the ground.

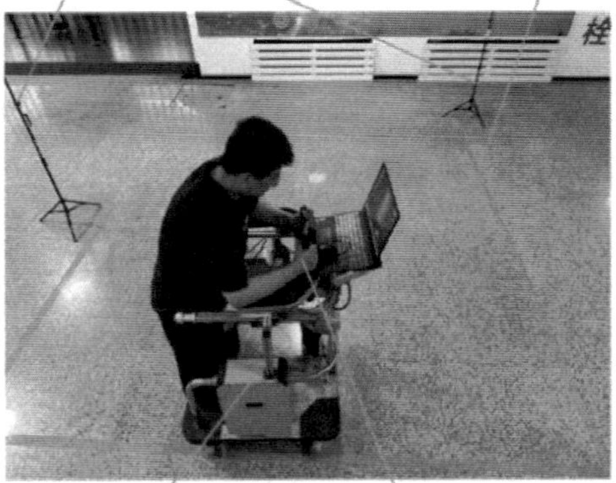

Figure 7. Experiment scene

We consider two scenarios with distinct motion intentions here. In the first scenario, the human subject moves anticlockwise along a rectangular path, resulting in a desired trajectory composed of discrete waypoints. For the second experimental case, the human subject intends to move anticlockwise along a circular path with a radius of 1.5 meters, resulting in a dynamically changing motion trajectory. For comparison in both scenarios, we apply impedance control alone and impedance control integrated with intent recognition to the robot respectively. Additionally, zero-stiffness impedance control is employed as a baseline comparison. The impedance parameters defined in Equation (20) are set as $M_x = 0.02$, $C_x = 1.0$ and $K_x = 0$. The neural network architecture incorporates 17 hidden-layer nodes. Similarly, adjusting these parameter values may be implemented to optimize control performance.

Fig. 8 presents the first scenario outcomes. Figure 8a shows robot trajectories under impedance control and the proposed method, with the 'target trajectory' denoting the human operator's intended path. Empirical results demonstrate the presented approach outperforms impedance-based control in overall position tracking accuracy. Neural network estimation performance is shown via motion intention estimates. While Fig.8 reveals both methods roughly track expected trajectories, Fig.8b highlights significantly reduced interaction force with the proposed method. Results demonstrate that while both impedance control and the proposed approach enable effective human-robot collaboration, the latter reduces human energy consumption significantly.

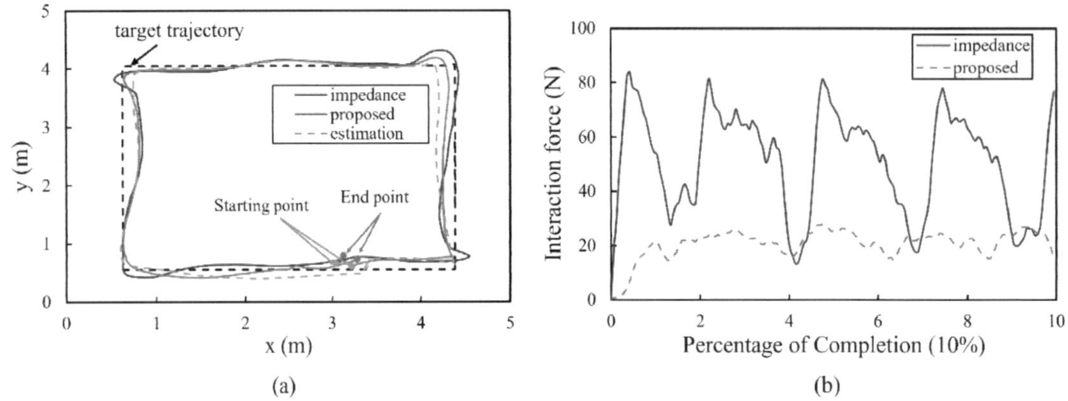

Figure 8. Trajectory tracking and Interactive force, during discrete target motion

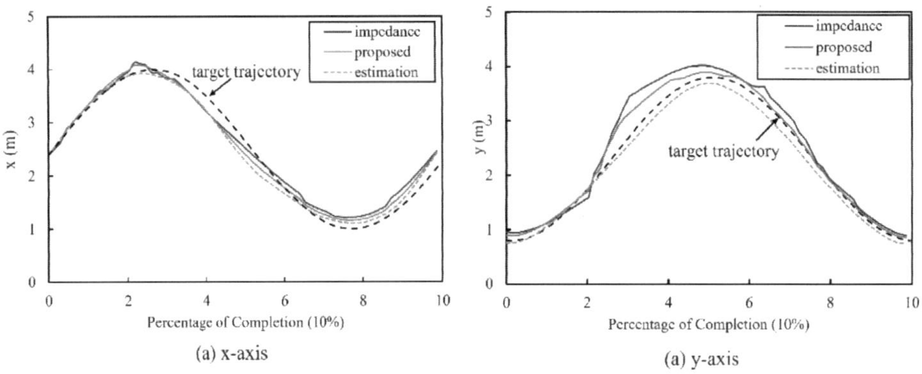

Figure 9. Trajectory tracking, for time-dependent trajectories

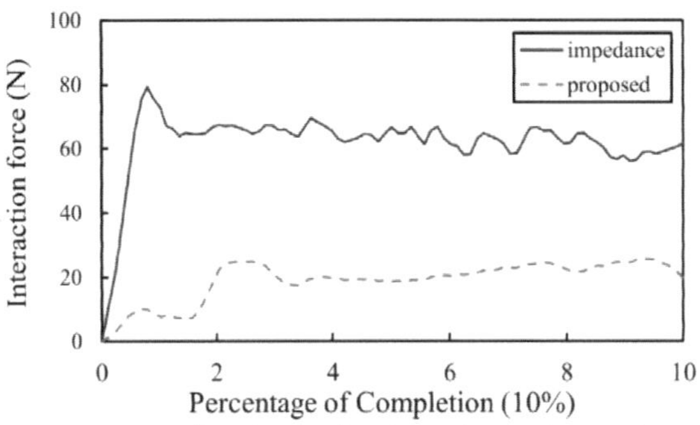

Figure 10. Interactive force, for time-dependent trajectories

Contrary to discrete motion in the initial case, real-world scenarios involve dynamically changing trajectories. The second task results are shown in Figs. 9 and 10. Figure 9(a-b) presents 2D position changes under different controls. Figure 10 demonstrates faster responses with the proposed method, requiring 65N external force to initiate motion, imposing significant physical load. Two strategies enhance impedance control: reducing M_x and C_x to improve compliance, though empirical evidence shows target inertia cannot be arbitrarily reduced while substantial damping ensures stability. Alternatively, human operators can increase limb stiffness to dominate system dynamics, but this becomes impractical for high-mass robots due to inherent inertia. Constant impedance controllers fail to actively track time-varying human motion. Compared to impedance control, the proposed method reduces external moment significantly (e.g., 20N in Fig. 10). Earlier findings confirm target position achievement with reduced force during motion intent modifications. These results validate the proposed method by estimating human motion intention through interaction force, position, and velocity parameters.

Table. 1 Tracking error under different trajectories

	Circle			**Rectangle**		
	Proposd	Impedance	estimation	Proposed	Impedance	estimation
MSE(m)	0.008	0.009	0.009	0.009	0.014	0.014
Mean Error(m)	0.06	0.08	0.09	0.08	0.1	0.09
Std Error(m)	0.03	0.05	0.04	0.04	0.06	0.06

To analyze the results more clearly and quantitatively, the position errors in different scenarios, obtained from the first two sets of experiments, are presented in Table 1 respectively. In both scenarios, the method we proposed outperforms impedance control alone. Evidently from the table, impedance control combined with intent recognition also minimizes the average error in trajectory tracking.

6. Conclusion

This paper explores human - robot collaboration, observing human users' motion intentions via the desired trajectories estimated by the established human limb motion model. To tackle the issue of the unknown human limb model, a radial basis neural network is proposed to estimate the parameters of the human model. Estimated motion intentions are incorporated into the robot's impedance control system, enabling it to actively follow the human partner. Experimental results confirm the method's effectiveness.

References

[1] He, W., Li, Z., & Chen, C. L. P. 2017, "A survey of human-centered intelligent robots: issues and challenges", IEEE/CAA Journal of Automatica Sinica, 6602 - 609.

[2] Medina, J. R., Lawitzky, M., Mörtl, A., Lee, D., & Hirche, S. 2011, An experience - driven robotic assistant acquiring human knowledge to improve haptic cooperation, Proceedings of the 2011 IEEE/RSJ International Conference on Intelligent Robots and Systems (IROS), IEEE, pp. 2416–2422.

[3] Yang, C., Huang, K., Cheng, H., Li, Y., & Su, C. - Y. 2017, Haptic identification by ELM–controlled uncertain manipulator, IEEE Trans. Syst. Man Cybern.: Syst., 47(8), 2398–2409.

[4] Newman, W. S. 1992, “Stability and performance limits of interaction controllers”, Trans. ASME, J. Dyn. Syst., Meas., Control, vol. 114, pp. 563–570.

[5] Dupree, K., Liang, C. - H., Hu, G., & Dixon, W. E. 2008, Adaptive Lyapunov - based control of a robot and mass–spring system undergoing an impact collision, IEEE Trans. Syst. Man Cybern. Part B (Cybern.), 38(4), 1050–1061.

[6] Raibert, M. H., & Craig, J. J. 1981, Hybrid position/force control of manipulators, J. Dyn. Syst. Meas. Control, 102(127), 126–133.

[7] Hogan, N. 1985, “Impedance control: An approach to manipulation—Part I: Theory; Part II: Implementation; Part III: Applications”, Trans. ASME, J. Dyn. Syst., Meas., Control, vol. 107, no. 1, pp. 1–24.

[8] Jung, S., & Hsia, T. C. 2000, “Robust neural force control scheme under uncertainties in robot dynamics and unknown environment”, IEEE Trans. Ind. Electron., vol. 47, no. 2, pp. 403–412.

[9] Lynch, K. M., Liu, C., Sorensen, A., Kim, S., Peshkin, M., Colgate, J. E., Tickel, T., Hannon, D., & Shiels, K. 2002, “Motion guides for assisted manipulation”, Int. J. Robot. Res., vol. 21, no. 1, pp. 27–43.

[10] Lee, E., Park, J., Loparo, K. A., Schrader, C. B., & Chang, P. H. 2003, “Bang - bang impact control using hybrid impedance/time - delay control”, IEEE/ASME Trans. Mechatronics, vol. 8, no. 2, pp. 272–277.

[11] Buerger, S. P., & Hogan, N. 2007, "Complementary stability and loop shaping for improved human - robot interaction", IEEE Trans. Robot., vol. 23, no. 2, pp. 232–244.
[12] Hirata, Y., Wang, Z., Fukaya, K., & Kosuge, K. 2009, "Transporting an object by a passive mobile robot with servo brakes in cooperation with a human", Adv. Robot., vol. 23, pp. 387–404.
[13] Patel, R. V., Talebi, H. A., Jayender, J., & Shadpey, F. 2009, "A robust position and force control strategy for 7 - DOF redundant manipulators", IEEE/ASME Trans. Mechatronics, vol. 14, no. 5, pp. 575–589.
[14] Roy, J., & Whitcomb, L. L. 2002, Adaptive force control of position/velocity controlled robots: theory and experiment, IEEE Trans. Robot. Autom., 18(2), 121–137.
[15] Iqbal, K., & Zheng, Y. F. 1999, "Arm - manipulator coordination for load sharing using predictive control," in Proc. IEEE Int. Conf. Robot. Autom., pp. 2539–2544.
[16] Burdet, E., & Milner, T. E. 1998, "Quantization of human motions and learning of accurate movements," Biol. Cybern., vol. 78, pp. 307–318.
[17] Corteville, B., Aertbelien, E., Bruyninckx, H., Schutter, J. D., & Brussel, H. V. 2007, "Human - inspired robot assistant for fast point - to - point movements", in Proc. IEEE Int. Conf. Robot. Autom, pp. 3639–3644.
[18] Erden, M. S., & Tomiyama, T. 2010, "Human - intent detection and physically interactive control of a robot without force sensors", IEEE Trans. Robot., vol. 26, no. 2, pp. 370–382.
[19] Wang, Z., Peer, A., & Buss, M. 2009, "An HMM approach to realistic haptic human - robot interaction" in Proc. 3rd Joint Eurohapt. Conf. Symp. Hapt. Interf. Virt. Environ. Teleoperat. Syst., pp. 374–379.
[20] Wakita, K., Huang, J., Di, P., Sekiyama, K., & Fukuda, T. 2013, "Human walking intention based motion control of an omnidirectional - type cane robot", IEEE/ASME Trans. Mechatronics, vol. 18, no. 1, pp. 285–296.
[21] Rahman, M., Ikeura, R., & Mizutani, K. 2002, "Investigation of the impedance characteristic of human arm for development of robots to cooperate with humans", JSME Int. J. Series C, vol. 45, no. 2, pp. 510–518.

Design and Implementation of an Autonomous Sensing Intelligent Steering Wheel with Hands-Off Detection

Pengyang Fu[1], Xi Li[1*], Yuanhua Chen[1], Jin Feng[1], Jie Lai[2]

[1] School of Automotive Engineering, Guilin University Of Aerospace Technology, Guilin Guangxi, 541000, P.R. China

[2]Guangxi Yuchai Power Co.,ltd. Yulin Guangxi, 541000, P.R. China

*Corresponding author's e-mail: 2021050320204@stu.guat.edu.cn

Abstract. The research background of this paper is based on the urgent need for vehicle intelligence and safety. With the rapid development of intelligent driving technology, as the core component of vehicle control, the intelligent steering wheel is the core component of vehicle control, its performance directly affects driving safety and user experience. However, the existing smart steering wheel technology has significant shortcomings in terms of sensor sensitivity and control algorithm accuracy, such as traditional systems due to single sensor configuration or limited algorithm processing capacity, as a result, the detection accuracy is insufficient and the response speed is slow, which is difficult to meet the real-time requirements in complex driving environments. In addition, the lack or insufficiency of fatigue driving monitoring and warning functions further increases driving safety risks. According to statistics, the proportion of traffic accidents caused by fatigued driving worldwide is as high as 20%-30%[1]. Therefore, by optimizing the sensor configuration and improving the control algorithm, an intelligent steering wheel system with high precision and high response speed is designed and implemented. The system adopts multi-sensor fusion technology, combined with a capacitance sensor, torque sensor, and biometric sensor, which can detect the driver's hand state, grip force change, and physiological characteristics in real-time, and through the artificial intelligence algorithm to achieve fatigue driving monitoring, early warning and active intervention functions. The experimental results show that the system has excellent performance in different driving environments, which lays a solid foundation for improving driving safety and user experience. Future research will further optimize the sensor module, explore more intelligent functions, and promote its deep integration and collaboration with other intelligent driving systems.

1. Introduction

1.1. Background and significance of the study

The research background of intelligent steering wheels mainly stems from the increasing demand for vehicle safety and intelligence in modern society. With the progress of science and technology, especially the rapid development of sensor technology control algorithms and artificial intelligence technology[2], the design and implementation of intelligent steering wheels have become possible. The smart steering wheel can not only monitor the driving state of the driver in real-time but also automatically intervene to assist the driver in controlling the vehicle in an emergency, thus greatly improving driving safety. In addition, the smart steering wheel can also work with other intelligent systems of the vehicle to achieve a more intelligent driving experience such as self-driving and self-parking.

1.2. Objectives and tasks

The goal of this study is to design and implement an autonomous sensing smart steering wheel for off-hand detection. By optimizing the sensor configuration and improving the control algorithm, the detection accuracy and response speed of the system are improved. Specifically, the main tasks of the study include: designing high-sensitivity sensor modules to ensure accurate recognition of driver's hand movements in diverse driving environments; developing efficient control algorithms, and improving the performance of the system, the real-time processing and analysis of sensor data are realized, to quickly respond to the driver's operation changes. A system prototype is built and verified by experiments to ensure the stability and reliability of the system.

2. Second, the overall design of the system

2.1. Design requirements

The design of the intelligent steering wheel needs to meet the following main requirements: first, the system needs to have high sensitivity and high accuracy of the hand-off detection function to accurately determine whether the driver holds the steering wheel with both hands in a variety of driving environments; Secondly, the system should have the ability of quick response, which can warn the driver quickly and take corresponding measures when the driver leaves the vehicle Thirdly, the system should have a good user interface so that drivers can intuitively understand the system status and warning information Finally, the system needs to have good stability and reliability to ensure that the long-term use of the process will not fail.

2.2. System architecture

The system architecture is mainly composed of three parts: a Sensor Module, a control unit, and an actuator. The sensor module is responsible for collecting the driver's hand movement and steering wheel state data; the control unit processes and analyzes the sensor data to judge the driver's Operation State, when necessary, the actuator is responsible for carrying out the commands of the control unit, such as issuing warnings, adjusting the vehicle state, etc.. In addition, the system also includes a user interface module to enable drivers to interact with the system.

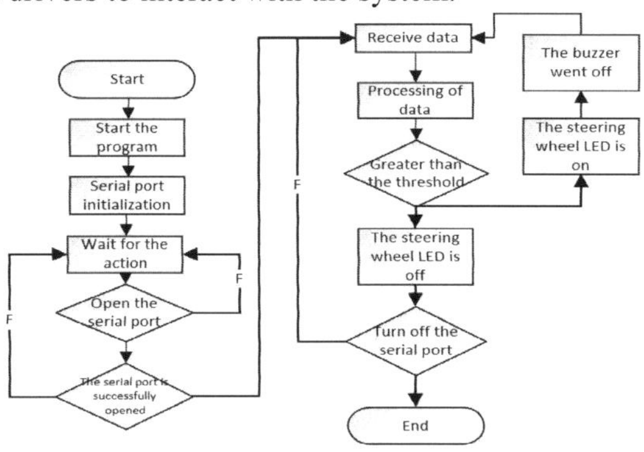

Figure 2-1. steering wheel program flow chart

In the design of the sensor module, this study will use a variety of types of sensors, such as capacitance sensor, Angle detection sensor data fusion, which can improve the accuracy and robustness of the detection. The control unit will use STM32F103C8T6 embedded chip to ensure the real-time and accuracy of data processing and analysis. The user interface module includes an LED indicator and buzzer, enabling the driver to intuitively understand the state of the system and to set and adjust when the driver is tired and his eyes are closed.

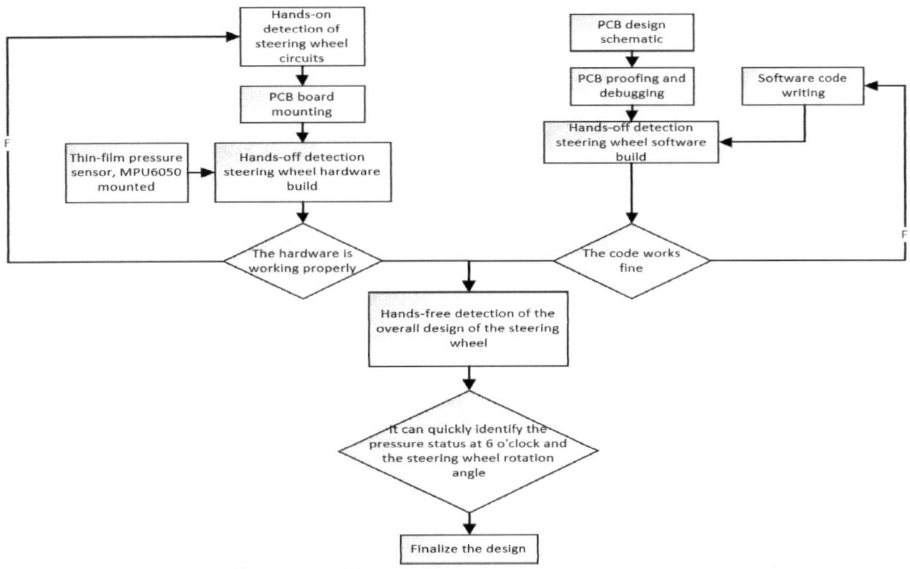

Figure 2-2. flow chart of out-of-hand detection steering wheel

3. Hardware design

3.1. Sensor module design

In terms of sensor design, this study focuses on optimizing key performance parameters. The sensitivity and anti-interference capability of capacitive sensors is enhanced through the use of high-precision sensing chips and optimized layout structures, while environmental compensation algorithms are developed to eliminate interference from temperature and humidity. The detection accuracy and response speed of torque sensors are improved by optimizing installation positions, calibration methods, and data processing algorithms. The accuracy and comfort of biometric sensors are enhanced by refining sensing chip precision and designing lightweight wearable devices.

The performance of capacitive pressure sensors primarily depends on the dielectric layer thickness parameter, which directly affects the plate spacing d. As shown in Figure 3-1, according to the capacitance formula $C=\varepsilon A/d$, the dielectric layer thickness is inversely proportional to the capacitance value[3]. An increase in thickness leads to a larger electrode spacing, reducing the capacitance value. This characteristic provides a theoretical foundation for sensor design.

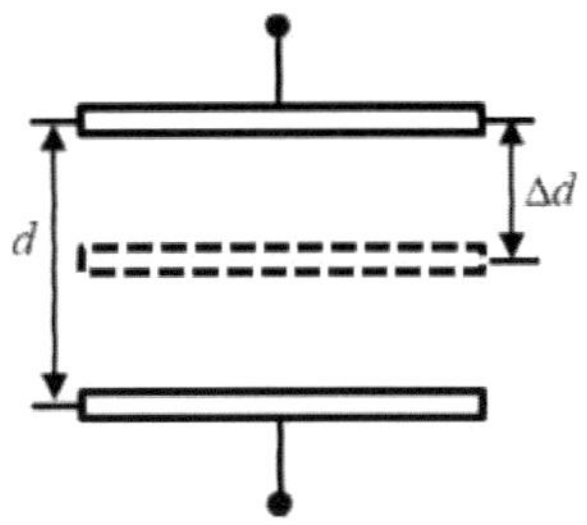

Figure 3-1. schematic diagram of capacitive sensor with varying dielectric layer thickness

In this study, four force sensors with different dielectric layer thicknesses of 250 MM, 500 MM, 750 MM, and 1000 μm were designed. The theoretical capacitance of each sensor is calculated by the capacitance formula, and the actual capacitance of the sample is measured by the LCR tester. The results are as follows.

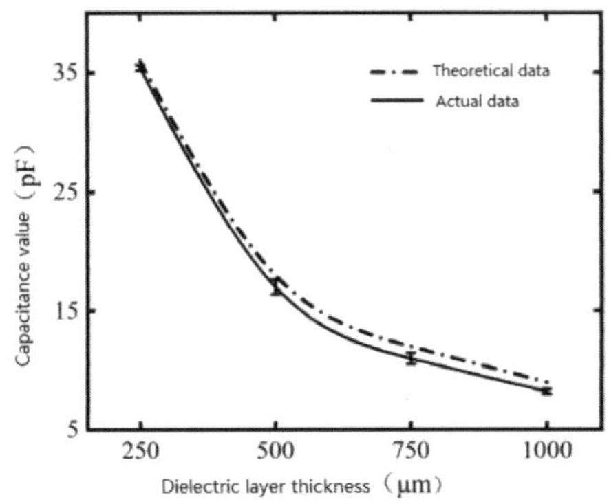

Figure 3-2. capacitance diagram of force sensors with different thickness

Experimental results demonstrate an inverse relationship between dielectric layer thickness and sensor sensitivity: reduced thickness enhances capacitance variation and improves sensitivity, enabling more precise detection of driver hand force. However, excessively thin layers risk structural failure. Consequently, a 500 μm dielectric layer is selected for the hands-off monitoring force sensor design.

Pressure values serve as a key detection metric, directly indicating hand-wheel contact. As illustrated in Figure 3-2, the hands-off detection system records grip pressure during simulated driving, accurately identifying grip states when pressure values exceed a defined threshold. Dynamic grip force fluctuations during operation are effectively captured. Analysis reveals a positive correlation between vehicle speed and average grip force, reflected in rising pressure values. While speed acceleration does not significantly affect driver tension, lane-changing maneuvers increase tension due to collision risks, elevating grip force. These behavioral patterns are consistently quantified through pressure value variations.

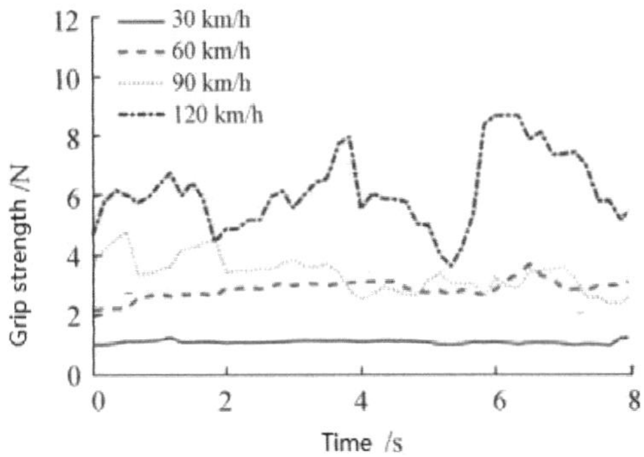

Figure 3-3. grip diagram under different speeds in free lane change

To monitor and display driver control states in real-time, multiple flexible force sensors are strategically distributed across different regions of the steering wheel to accommodate various grip patterns. Proper hand positioning enhances control and reaction capabilities, significantly reducing accident risks. These sensors detect hand contact states and transmit data to the hands-off detection system, ensuring accurate monitoring. A zonal layout strategy is adopted for sensor placement. Sensors at the 9 and 3 o'clock positions monitor hand grip presence, while those at the 6 and 12 o'clock positions detect contact areas. This multi-region configuration enables precise tracking of hand positions and applied forces, providing comprehensive grip state assessment.

Given the variability of hand placement, single-point sensor arrangements are insufficient. A multi-sensor network is implemented, with six pressure sensors uniformly distributed on the steering wheel to ensure complete monitoring coverage and reliable data acquisition.

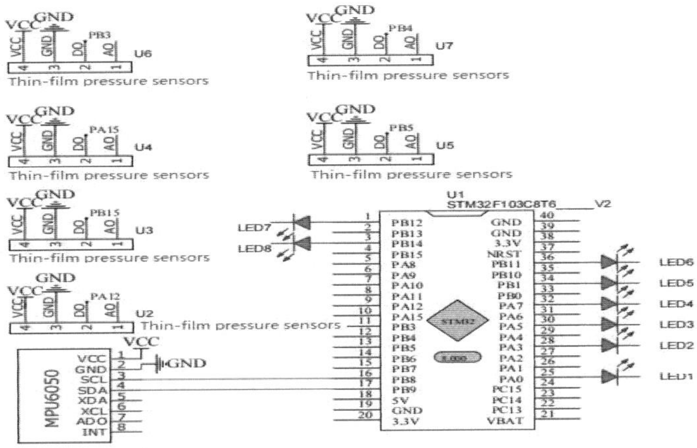

Figure 3-4. design schematic diagram

3.2. *Design of control unit*

The control unit serves as the central processing module of the intelligent steering wheel system, integrating sensor data acquisition, processing, and driver state analysis. Built around the STM32F103C8T6 microcontroller (ARM Cortex-M3 architecture, 72MHz clock frequency), the system delivers high-performance real-time control with low power consumption and compact integration.

System reliability is enhanced through the FreeRTOS real-time operating system, enabling efficient task scheduling and resource management. Critical data integrity is ensured via on-chip Flash memory and external EEPROM/FRAM, safeguarding system parameters and operational records. A dual-bus communication architecture, incorporating CAN and LIN protocols, facilitates robust data exchange with vehicle subsystems. Leveraging hardware-software co-design, dynamic power optimization, and a modular framework, the system supports scalable functionality and OTA updates, meeting the demands of intelligent steering systems in complex automotive environments.

4. Software design

4.1. *MPU6050 angle detection*

The MPU6050 sensor initialization and data acquisition process is implemented on an STM32 microcontroller via I2C communication. The MPU_Init function configures the sensor by resetting (writing 0x80 to MPU_PWR_MGMT1_REG), waking (writing 0x00), setting gyroscope range to ± 2000dps (MPU_Set_Gyro_Fsr(3)), accelerometer range to ±2g (MPU_Set_Accel_Fsr(0)), and sample rate to 100Hz (MPU_Set_Rate(100)). It disables interrupts (MPU_INT_EN_REG = 0x00), FIFO (MPU_FIFO_EN_REG = 0x00), and I2C master mode (MPU_USER_CTRL_REG = 0x00), verifies communication by reading the device ID register (MPU_DEVICE_ID_REG), and sets the clock source to PLL X-axis reference (MPU_PWR_MGMT1_REG = 0x01).

Dynamic configuration of gyroscope and accelerometer ranges is supported, such as MPU_Set_Gyro_Fsr(3) for ±2000dps and MPU_Set_Accel_Fsr(0) for ±2g. The digital low-pass filter (LPF) is configured via MPU_Set_LPF, e.g., setting LPF to 50Hz for a 100Hz input (MPU_Set_LPF(rate/2)). Data acquisition includes temperature, gyroscope, and accelerometer readings. Temperature is read from MPU_TEMP_OUTH_REG and MPU_TEMP_OUTL_REG, converted using the formula: temperature = 36.53 + (raw value / 340). Gyroscope and accelerometer data are read from MPU_GYRO_XOUTH_REG and MPU_ACCEL_XOUTH_REG (6-byte data converted to 16-bit signed integers).

4.2. Thin-film pressure sensors

In order to read the input state and control the output of GPIO pins, the system uses HAL library function to operate. The status of the input pin is collected by the HAL () function, which reads the level status of the pin on the specified GPIO port and may return either GPIO (high level) or GPIO (low level) . The system configuration input pins include: GPIOB, GPIOA, GPIOA, GPIOB, GPIOB, and GPIOB.

The output pin is controlled by the HAL () function, whose logic is as follows: when a GPIO (low-level) input pin is detected, the corresponding output pin is set to a low level; When a GPIO (high-level) input pin is detected, the corresponding output pin is set to a high level[4]. System-configured output pins include: GPIOB, GPIOB, GPIOA, GPIOA, GPIOA, and GPIOA. This mapping relationship between input and output pins ensures that the system can adjust the output signal in real-time according to the input state.

Table 4-1. Pin mapping

Input pins	Output pins	Description of the function
GPIOB_PIN_15	GPIOB_PIN_11	Detect GPIOB_PIN_15, control GPIOB_PIN_11
GPIOA_PIN_12	GPIOB_PIN_1	Detect GPIOA_PIN_12, control GPIOB_PIN_1
GPIOA_PIN_15	GPIOA_PIN_7	Detect GPIOA_PIN_15, control GPIOA_PIN_7
GPIOB_PIN_3	GPIOA_PIN_5	Detect GPIOB_PIN_3, control GPIOA_PIN_5
GPIOB_PIN_4	GPIOA_PIN_3	Detect GPIOB_PIN_4, control GPIOA_PIN_3
GPIOB_PIN_5	GPIOA_PIN_0	Detect GPIOB_PIN_5, control GPIOA_PIN_0

4.3. Ui Design

To improve the interactivity and user experience of the user interface, this research will develop the user interface based on the touch screen. Through the touch screen driver can be intuitive to set and adjust to improve the convenience and accuracy of operation[5].In the real driving environment, the application of a touch screen can reduce the use of physical buttons, simplify the operation process, provide clearer and more intuitive information display, and reduce the risk of driver distraction. The user interface will also develop an intelligent prompt function to provide personalized prompts and suggestions based on the driver's operating habits and driving environment. For example, when the system detects that the driver has maintained the same driving posture for a long time, the display screen will prompt the driver to adjust the posture to prevent fatigued driving.

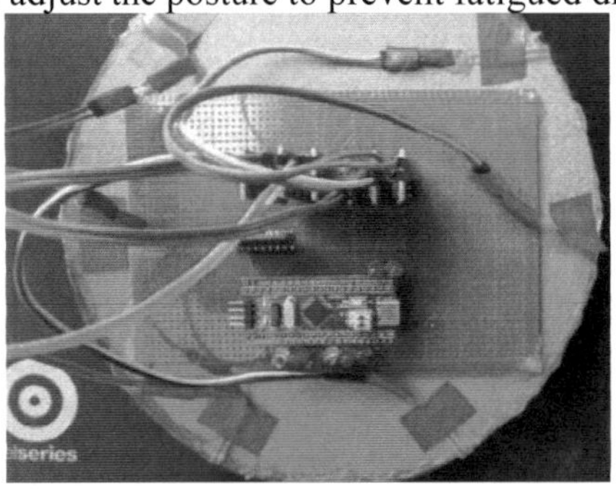

Figure 4-1. Photographs of the real thing

5. Experimental Verification

5.1. Functional testing

The functional tests mainly include hand-off detection, fatigue driving monitoring, user interaction, and other functional tests. By simulating different driving scenarios and operation behaviors, it is verified whether the intelligent steering wheel system can accurately detect the driver's hand movements and operation states. For example, in different temperature and humidity environments, whether the system can quickly detect the driver holding the steering wheel state, and send a warning and signal to ADAS to take protective measures such as automatic deceleration measures. Here is the capacitance sensor on the temperature and humidity of the experimental test.

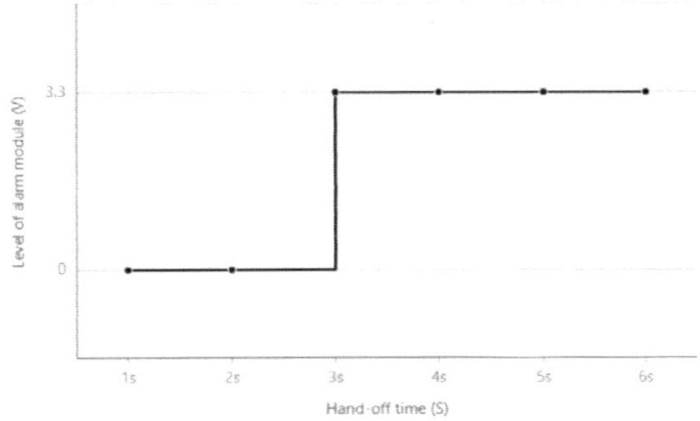

Figure 5-1. alarm module level

5.2. performance analysis

The performance analysis focuses on systematically evaluating the detection accuracy, response speed, and operational stability of the intelligent steering wheel system. Experimental data is used to validate system performance across diverse driving scenarios and environmental conditions, including variations in humidity and temperature[6]. Key metrics include detection precision under varying conditions, response latency in emergencies, and long-term operational reliability.

A rigorous evaluation framework is established, employing quantitative indicators such as error rate and accuracy for detection precision, millisecond-level response delays for system reactivity, and mean time between failures (MTBF) for stability assessment. This methodology provides a comprehensive and objective performance evaluation, offering reliable data for system optimization.

Environmental temperature poses significant challenges to sensor performance. Under extreme conditions, ranging from -30 ° C in winter to 70-90 ° C in vehicle interiors during summer, sensor sensitivity exhibits an inverse relationship with temperature. Experimental results (Figure 5-2) demonstrate that while capacitance decreases with rising temperatures, the 0.5% variation caused by temperature effects is negligible compared to the 10% change induced by external forces, confirming minimal temperature impact on sensor performance.

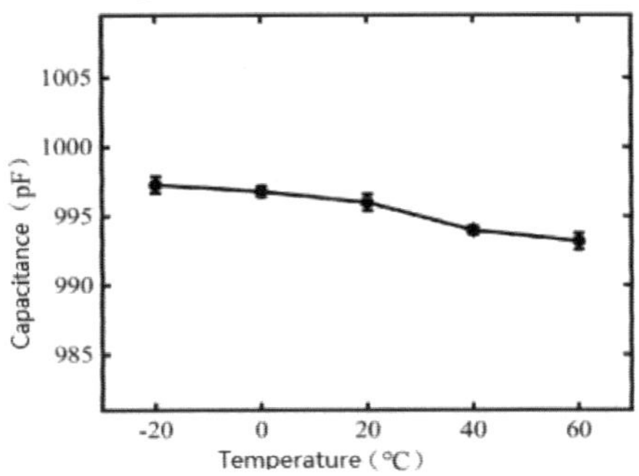

Figure 5-2. the change of sensing capacitance of thin film with temperature

In addition to the temperature factor, environmental humidity is also an important parameter affecting the performance of the sensor. Gaseous water molecules will affect the conductivity of the sensor, thus changing the key performance indicators such as electron transmission rate, sensitivity, and response time. In this study, a simulated experimental environment (shown in Figure 5-3)

Journal of Physics: Conference Series **3024** (2025) 012027 doi:10.1088/1742-6596/3024/1/012027

consisting of a closed acrylic test chamber, a humidifier, and a humidity tester was constructed. The experimental data show that the humidity of the test chamber is 15% at the ambient temperature of 30 °C, and the humidity can reach 70% under the simulated rainy day condition of 10 °C. However, the capacitance of the sensor always fluctuates around 997 PF, and there is no significant change[7]. This result indicates that humidity changes such as rain and snow have little effect on the sensor performance.

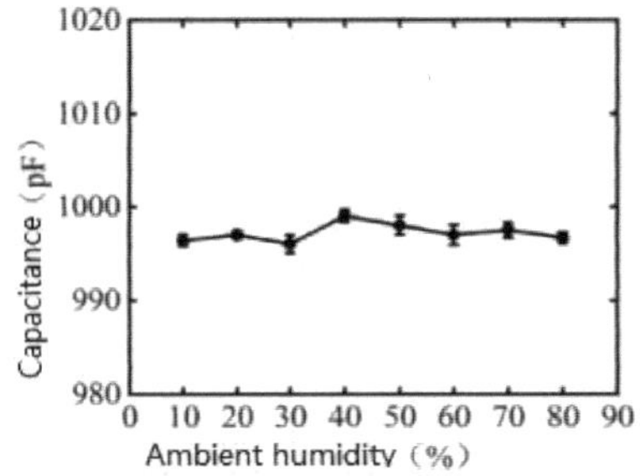

Figure 5-3. image of film sensing capacitance changing with humidity

6. Conclusion and prospect

6.1. Summary of research

This study successfully developed and implemented a set of out-of-hand detection autonomous induction intelligent steering wheel systems. Through the optimization of the sensor layout and the improvement of the control algorithm, the system shows high sensitivity and high accuracy in the off-hand detection function. The experimental results show that the system can accurately identify the driver's hand state in a variety of driving scenarios, and issue warnings or take corresponding measures in time, which significantly improves driving safety. At the same time, the system is also equipped with a friendly user interface, and stable operation, to provide drivers with more intelligent driving support.

References

[1] Yeong D J, Velasco-Hernandez G, Barry J, et al. Sensor and sensor fusion technology in autonomous vehicles: A review[J]. Sensors, 2021, 21(6): 2140.
[2] Yu L ,Yan Z ,Linyi X , et al.Roll angle of autocollimator measurement method based on hollow cube corner reflector[J].Optoelectronics Letters,2022,18(02):91-96.
[3] Sakamaki, R., and M. Horibe. "Precision Adjustment of Probe-Tilt Angle With RF Signal Detection Technique." IEEE Transactions on Instrumentation and Measurement (2020).
[4] Bi C, Huang J, Xing G, et al. Safewatch: A wearable hand motion tracking system for improving driving safety[J]. ACM Transactions on Cyber-Physical Systems, 2019, 4(1): 1-21.
[5] Wang Z, Kiourti A, Lee R. A proximity sensor for the steering wheel based on leaky coaxial cable[J]. IEEE Sensors Journal, 2020, 20(24): 14799-14808.
[6] Li Y. Anti-fatigue and collision avoidance systems for intelligent vehicles with ultrasonic and Li-Fi sensors[C]. IEEE 3rd International Conference on Information Communication and Signal Processing. IEEE, 2020: 203-209.
[7] Kim S K, Lee G H, Jeon C, et al. Bimetallic Nanocatalysts Immobilized in Nanoporous Hydrogels for Long-Term Robust Continuous Glucose Monitoring of Smart Contact Lens[J]. Advanced Materials, 2022, 34(18): 2110536.
[8] Zhang D, Zhong Y, Wu Y, et al. Liquid metal elastomer with flytrap-inspired pillar structure for stress sensing[J]. Composites Science and Technology, 2021, 216: 109066.

Research of Immune-Ant Colony Algorithm for the Detection of Weld Pool Image

Dukun Ding*

Dongguan Polytechnic, No.3 Daxue Road, Songshan Lake District, Dongguan, Guangdong, 523808, P.R. China

*Corresponding author's e-mail: dukunding@163.com

Abstract. This paper has researched a new immune-ant colony algorithm for weld pool detection.Firstly, the tungsten inert gas welding experiment system has been setup and welding images can be captured in real time. The tungsten inert gas welding experiment system is firstly built up, which can be used for images real time capturing. Based on the experiment system, the immune-ant colony algorithm has been designed.In the end, several testing experiments have been performed, which dictated that the performer of the researched algorithm is better. In testing experiments, after about 35 times iterative computations, a global optimum value for image segmentation is determined,which is 40% faster convergence compared to standalone ant colony algorithms. And wonderful weld pool segmentation images can be received. The edge of pool and weld seam can be easily extracted, which is helpful for weld automation.

1. Introduction

The weld technique is very important for metal connection. It is widely used in mechanic manufacturing, automobile industry, construction industry, and so on[1]. Nowadays, in some complicated curve welding, it is still operated by mankind, which is full of electric arc, spatter and poisonous gas. It is damage for workers' health. So it is very significant to realize the automatic of welding, especially for the complicated curve welding. Many sensors have been applied to the goal realization,which are the ultrasonic sensor,mechanic sensor, vision sensor, welding arc sensor, and so on[2-4]. And the most attractive method in the sensors above is the vision sensor,which has the merit of no touch and large information accessing. In the research field of weld pool detection, many intelligent algorithms have been researched, which are the DNA algorithm, the genetic algorithm,and so on. However, the algorithms are time costing and easy to fail into local optimality[5-6].

In the paper, the ant-colony algorithm has been studied for weld pool detection.In the first section of the algorithm, the immune computation is applied, which has the merit of rapid calculation. And in the latter part, the ant colony algorithm is used, which can receive a global optimum value. The new method has integrated the merit of the two algorithm,which can avoiding local optima and accelerate convergence. So the weld pool edge is extracted preferably. It has supplied a novel way for weld pool detection.

2. Setup of experimental system

In the paper, Tungsten Inert Gas weld is chosen to be researched welding technology. The hardware of experimental system is shown in fig.1, which includes the mechanic body, the CCD sensor, the

Journal of Physics: Conference Series **3024** (2025) 012028 doi:10.1088/1742-6596/3024/1/012028

industry computer, the images capture system, the motion control card, and so on. The mechanic part of experimental system includes the the portal frame and welding table. The welding table is driven by four step motors. There are six fixture on the workbench. The travel switches are installed along the gantry frame and weld workbench. They are used to prevent over travelling. Optical filter is installed in front of CCD sensor. The optical filter can filter weld arc disturbance. And in the welding experiments, weld pool images are captured in real time.

Based on the setup experimental hardware system, the software is designed. The software interface is shown in fig.2. It can control CCD sensor to capture image in real-time, decode the image, perform A/D conversion and save the data to industry computer hard disk. The software system also has the function of image procession, such as the image gray enhancement, median filtering operation, image edge detection, and so on.

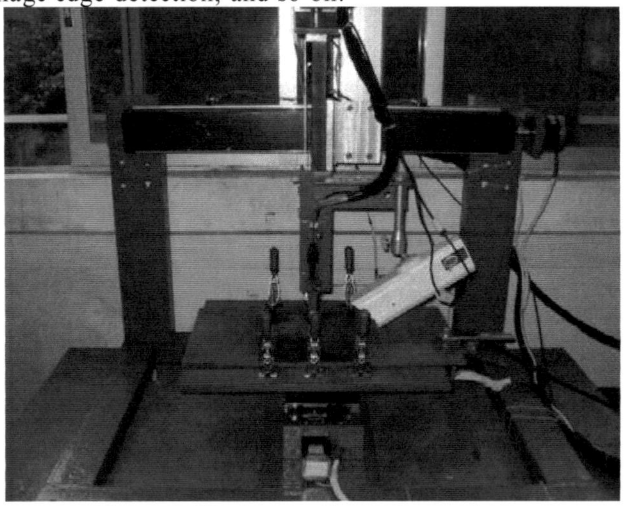

Figure 1. Hardware of experimental system

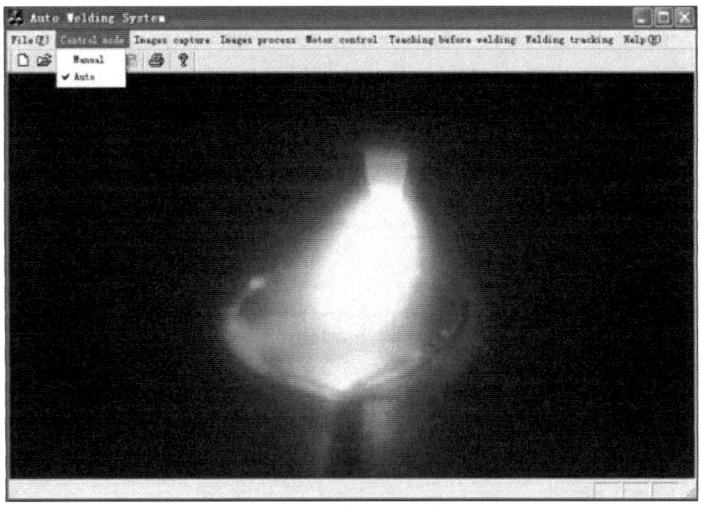

Figure 2. Software of system

Table 1. The parameters of CCD

Camera brand	CCD resolution	Frame rate (fps)	PSNR of Image (db)	SSIM of Image
MARS-170-662GTM	1600×1104	662	25	0.8

3. Designed of immune-ant colony algorithm

The ant colony algorithm originates from the simulation of real ant colony action. Individual action of some insects is very simple. However, group action of the insects are very complex. That is because each insect individual has one kind of substance for information transfer, which is called ectohormone.In the moving procession, one ant can leave the substance. When the other ant moves along the same path, it can sense the substance, which can be helped to point the moving direction for the ant.

The information of positive feedback phenomenon can be shown by the collective action, which is that the more ants move along one path, selected probability of the path is bigger. And the goal of searching food can be reached by the information communication between ant individual. The ant colony algorithm can be used for searching, which is the simulation of ant group action, which is no needs of any priori knowledge. In the beginning phase, the searching path is randomly chosen. As the more solution spaces is learned more, the searching becomes more regularity. And the final optimal solution can be reached[7-8]. Fig.3 shows working flow of the algorithm.

3.1. Set up of fitness function

Suppose k is the segmentation threshold of pool image. A data set of image pixel is C_0, and the range of data set is $[0,1,\ldots,k\text{-}1]$. C_1 is another image pixel data set, and the gray level of the set is $[k, k+1,\ldots,L\text{-}1]$. The weld pool image gray fitness function can be defined as following:

$$f(d_i) = \omega_0(\mu_0 - \mu_T)^2 + \omega_1(\mu_1 - \mu_T)^2 \tag{1}$$

$$\begin{cases} \omega_0 = \sum_{q=0}^{k-1} p_q(r_q) \\ \omega_1 = \sum_{q=k}^{L-1} p_q(r_q) \\ \mu_0 = \sum_{q=0}^{k-1} q p_q(r_q)/\omega_0 \\ \mu_1 = \sum_{q=k}^{L-1} q p_q(r_q)/\omega_1 \\ \mu_T = \sum_{q=0}^{L-1} q p_q(r_q) \\ p_r(r_q) = \dfrac{n_q}{n}, q = 0,1,2,\cdots,L-1 \end{cases}$$

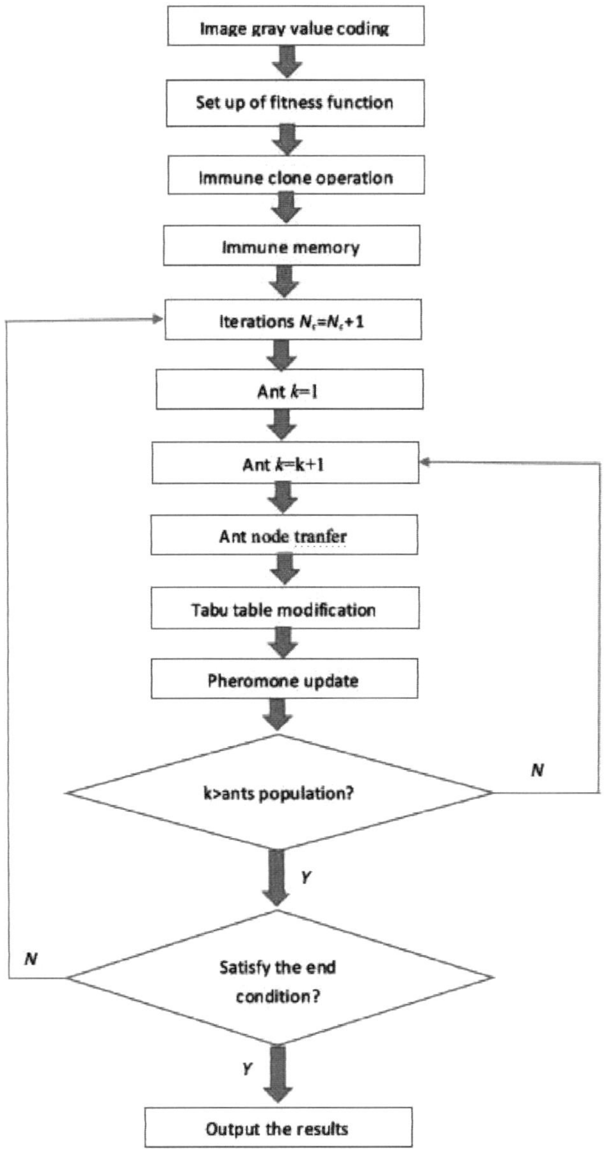

Figure 3. Work flow of the algorithm

3.2. *Immune clone operation*

The operation is to choose the high affinity antibodies to clone accord to the relative affinity of each antibody individual. So the antibody group can develop the direction of big affinity. The calculation formula of antibody relative affinity is:

$$p_i = \frac{f(d_i)}{\sum_{j=1}^{N} f(d_j)} \tag{2}$$

In the formula, p_i is the antibody relative affinity. d_i is the antibody individual. $f(d_i)$ is the affinity between antibodies and antigens, which is shown in formula(1).

3.3. *Immune memory*

When each iteration is completed, the judge operation of algorithm will be performed,which is used to estimate whether the fitness of current antibody generation is higher than their parent antibodies. If the result is "yes", then the antibody will join into the next antibody generation.

3.4. *Ant node transfer*

In the moving procession, the k ant moves to the direction according to the pheromone on the path. On t moment, the transfer probability of k ant from i node to j node is:

$$P_{ij}^{k}(t)=\begin{cases}\dfrac{\tau_{ij}(t)\eta_{ij}(t)}{\sum\limits_{s\in A}\tau_{is}(t)\eta_{is}(t)}, if \quad j\in A \\ 0, if \quad j\notin A\end{cases} \tag{3}$$

In the formula, A is the node where the k ant will move at the next moment. The A is $\{1,2, \ldots, l\text{-}1\} - \xi_k$. ξ_k is a node set where the k ant has moved and it will change with the evolution procession of algorithm. $f(d_i)$ is the fitness function. η_{ij} is the visibility factor, which is $\eta_{ij}=\dfrac{1}{f(d_i)}$.

3.5. *Update of pheromone on the path*

After n moments, the ant will complete one time circulation. The moving path pheromone can be adjusted by the following function.

$$\begin{cases}\tau_{ij}(t+n)=(1-\rho)\tau_{ij}(t)+\Delta\tau_{ij} \\ \Delta\tau_{ij}=\sum\limits_{k=1}^{m}\Delta\tau_{ij}^{k}\end{cases} \tag{4}$$

$\Delta\tau_{ij}^{k}$ is the k ant left its pheromone on the path (i,j) of formula(4). Its value is determined by the quality of path. The path is shorter, and the number of pheromone released by the ant will be more. $\Delta\tau_{ij}$ is the increase number of pheromone on path (i,j). ρ is the pheromone attenuation coefficient of ,and its range is [0,1). In most application, ρ is always less than 1, which can be used to reduce the infinite accumulation of pheromone along the path.

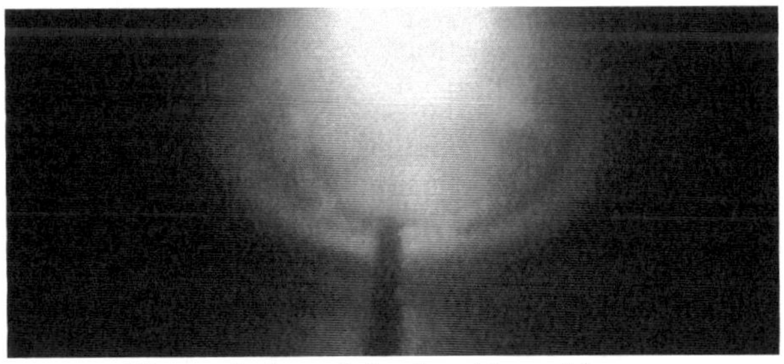

Figure 4. the original weld pool image

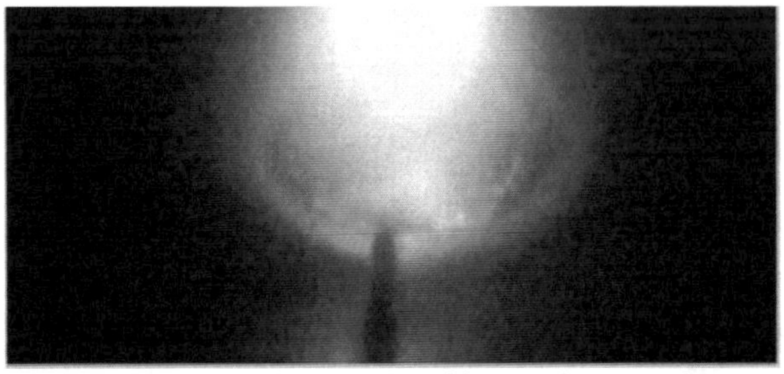

Figure 5. Pool image after median filter and gray enhancement

Figure 6. Processed by immune-ant colony algorithm

4. Conclusion

This paper has researched a new way for the pool detection in weld image, and it is based on the immune-ant colony algorithm. A tungsten inert gas welding experiment system has been built up and used to capture welding images. The weld images are firstly pretreated. Then the immune-ant colony algorithm has been designed. The new algorithm can overcome the weakness of immune and ant colony algorithm, which is local optimum and slow convergence. The final experiment result shows that the new algorithm has a better performance. 30% fewer iterations are required and a clear weld pool segmentation image can be received. So the pool and weld seam edge can be easily extracted, which is very helpful for weld automation.

Acknowledgments

The authors acknowledge 2024 Guangdong higher vocational colleges production and education integration innovation platform project (No.2024CJPT027), Guangdong Provincial Department of Education 2023 high-end equipment (intelligent robots, new materials) key special areas project (No. 2023ZDZX3092), Dongguan social development of science and technology key project (No. 20231800939952), intelligent terminal and intelligent manufacturing special project of Dongguan Polytechnic: Platform and team project (No. ZXA003), intelligent terminal and intelligent manufacturing special project of Dongguan Polytechnic (No. ZXD202307), Guangdong HUST Industrial Technology Research Institute, Guangdong Provincial Key Laboratory of Manufacturing Equipment Digitization project (No.2023B1212060012) and Guangxi manufacturing system and advanced manufacturing technique key lab project (No.22-035-4S018).

Reference

[1] Kim Chan Kyu, Kim Jin Hyeok and Hong Hyun Uk 2021 Behavior of weld pool convection and columnar-to-equiaxed grain transition in gas tungsten arc welds of ferritic stainless steels with different aluminum contents Journal of Materials Processing Technology vol 289 p116946

[2] Yang Hong, Li Yulong, Dong Yangping and Cui Qingbo 2019 Ultrasonic welding packaging of FBG and its bending sensing characteristics vol 40 Transactions of The China Welding Institution p69-75

[3] Gao Yanfeng and Wu Dong 2018 Effects of lateral wind on the rotating arc sensing and the weld seam tracking in a horizontal welding Process Transactions of The China Welding Institution vol 39 p36-40
[4] Huang Wensheng and Xu Hongli 2018 Development of six-DOF welding robot with machine vision Modern Physics Letters B,vol 32 p1840079
[5] Zhu Yanjun, Wu Zhisheng and Li Ke 2019 Welding deviation detection method based on weld pool image contour features China Welding vol 28 p35-44.
[6] Huang Junfen, Xue Long and Huang Jiqiang 2019 GMAW Penetration State Prediction Based on Visual Sensing Journal of Mechanical Engineering vol 55 p41-46.
[7] Yang Shuyin 2015 Pattern recognition and intelligent computing (Beijing: Electronic Industry) p1-50
[8] Kang Lishan Computational intelligence(Beijing: Science press) p25-90

Decoupling control of electromagnetic bearing rigid rotor system based on IGWO and linear active disturbance rejection inverse system

Youbing Li[1,2], Zhenning Zhu[1,2], Zhixian Zhong[1,2*]

[1]Key Laboratory of Advanced Manufacturing and Automation Technology (Guilin University of Technology), Guilin 541006, Guangxi, China

[2]College of Mechanical and Control Engineering, Guilin University of Technology, Guilin 541006, Guangxi, China

[*]Corresponding author's e-mail: 2005zhzhx@163.com

Abstract. This study proposes a decoupling control strategy integrating an improved grey wolf optimization algorithm (IGWO) and a linear active disturbance rejection inverse system for rigid rotor system control in electromagnetic bearings. Initially, a mathematical model of the radial four-degree-of-freedom electromagnetic bearing rotor system is developed, and the system is decoupled into four pseudo-linear subsystems via the inverse system method. To address the model dependency of the inverse system, a linear active disturbance rejection controller (LADRC) is designed to regulate the decoupled subsystems. Furthermore, the IGWO algorithm is implemented to rapidly optimize LADRC parameters, enhancing the controller's dynamic response. Comparative simulations and experiments against decentralized PID control demonstrate the proposed method's superior performance in decoupling accuracy, settling time, and disturbance rejection.

1. Introduction

With the advantages of no friction loss and strong controllability, electromagnetic bearings have important application value in flywheel energy storage, aerospace and other fields[1][2] Under low-speed conditions, the rotor's operational speed remains significantly below its critical threshold, permitting simplification to a rigid rotor model. Nevertheless, its dynamic behavior exhibits pronounced coupling and nonlinearity, with gyroscopic effects becoming increasingly prominent as rotational velocity rises [3]. Therefore, it is necessary to design control strategies to achieve decoupling control to improve the system performance of electromagnetic bearings.

In recent years, the linear decoupling control of micro-geometric decoupling and inverse system decoupling has become a research hotspot of linear decoupling. Micro-geometry decoupling needs to transform the control problem into a geometric domain, which involves complex nonlinear coordinate transformation, and its theoretical difficulty limits its application in electromagnetic bearings[4]. Conversely, the inverse system decoupling approach achieves full decoupling of the magnetic bearing rotor while significantly reducing computational complexity. Cheng et al.[5] suggested integrating the α-order inverse system (using state feedback) with the original system to form a pseudo-linear structure, while the internal model controller markedly enhances disturbance rejection capabilities.

However, the inverse system decoupling method relies heavily on precise system modeling, and its robustness requires further enhancement. In contrast, the active disturbance rejection controller (ADRC) demonstrates superior disturbance rejection capabilities with minimal dependence on accurate model parameters, which provides a new idea for decoupling control of electromagnetic bearing[6][7]. To streamline the parameter tuning process, Gao[8] proposed a linear active disturbance rejection controller (LADRC) that is more in line with the control requirements, which promotes the application of this technology. Based on finite element analysis and LADRC theory, Liang et al. [9] realized online self-tuning of key parameters of bearingless flux-switching permanent magnet motor and obtained good control performance, but the complexity of controller parameter tuning was high.

Nature-inspired optimization algorithms are extensively employed in parameter tuning owing to their adaptability and straightforward implementation. The Grey Wolf Optimizer (GWO) has demonstrated superior global search capabilities [10]. However, it exhibits diminished search efficiency and a propensity to converge prematurely to local optima. To enhance its performance, researchers have attempted to integrate other optimization algorithms to improve its capabilities. Cheng X et al. [11] The GWO algorithm was enhanced through integration with the Particle Swarm Optimization (PSO) algorithm, demonstrating a notable improvement in its overall performance.

The four-degree-of-freedom electromagnetic bearing model is first constructed, followed by reversibility analysis. The coupled system is decoupled into four pseudo-linear subsystems via the inverse system method. Given the model dependency of this approach, linear active disturbance rejection controllers are applied to each subsystem to enhance robustness and disturbance rejection. Subsequently, the Particle Swarm Optimization algorithm is integrated with the Grey Wolf Optimizer to develop the improved grey wolf optimization algorithm (IGWO), enabling rapid global optimization of LADRC parameters. Simulations and experiments validate the method's decoupling accuracy, dynamic response, and disturbance rejection performance.

2. Mathematical model of four-degree-of-freedom electromagnetic bearing rotor system

Figure 1 depicts the radial four-degree-of-freedom electromagnetic bearing structure. Point O marks the rotor's geometric center, with θ_x, θ_y, and ω indicating angular displacements around the X-, Y-, and Z-axes. Electromagnetic resultant forces F_{ay}, F_{ax}, F_{by} and F_{bx} act on bearings at ends A and B in the X and Y directions. Distances from the mass center to end bearings and sensors are defined by l_a, l_b and l_{sa}, l_{sb}.

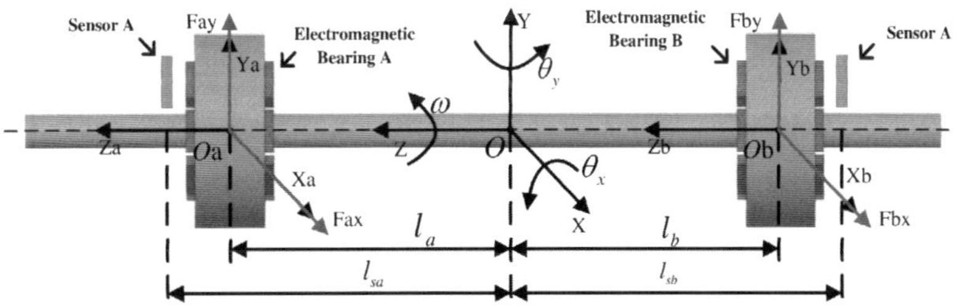

Figure 1. Rigid rotor structure diagram of four-degree-of-freedom electromagnetic bearing

After the rotor is disturbed, the external damping of the rotor is ignored. According to the rigid rotor dynamics theory, the motion equation at the center of mass is obtained as follows:

$$\begin{cases} m\ddot{x} = F_{ax} + F_{bx}, m\ddot{y} = F_{ay} + F_{by} - mg \\ J_x\ddot{\theta}_x + J_z\omega\dot{\theta}_y = -l_aF_{ay} + l_bF_{by}, J_y\ddot{\theta}_y - J_z\omega\dot{\theta}_x = l_aF_{ax} - l_bF_{bx} \end{cases} \tag{1}$$

Where, m denotes the rotor mass, and J_x, J_y, and J_z represent the rotor's moments of inertia about the X-, Y-, and Z-axes.

The displacement and rotation angle at the center of mass can be expressed by the displacement of the bearings at both ends, and the relationship is:

$$x = \frac{l_b}{l} x_a + \frac{l_a}{l} x_b, y = \frac{l_b}{l}, y_a + \frac{l_a}{l} y_b, \theta_x = \frac{y_b - y_a}{l}, \theta_y = \frac{x_a - x_b}{l} \tag{2}$$

Where, $l = l_a + l_b$, is the distance between the two electromagnetic bearings at both ends.

The characteristics of each pair of coils in the system are the same, and the electromagnetic force of the bearings at both ends can be expressed as:

$$F_{ax} = K_{sa} x_a + K_{ia} i_{xa}, F_{bx} = K_{sb} x_b + K_{ib} i_{xb}, F_{ay} = K_{sa} y_a + K_{ia} i_{ya}, F_{by} = K_{xb} y_b + K_{ib} i_{yb} \tag{3}$$

Where, K_{sa} and K_{sb} are the displacement stiffness coefficients of the electromagnetic bearings; K_{ia} and K_{ib} are the current stiffness coefficients; i_{xa}, i_{xb}, i_{ya} and i_{yb} represent the X- and Y-direction control currents at Electromagnetic Bearings A and B, respectively.

3. Design of control method for electromagnetic bearing rotor system

3.1. The reversibility judgment of electromagnetic bearing system

For the radial four-degree-of-freedom electromagnetic bearing rotor model, state variables are defined as *X*, input variables as *U*, and output variables as *Y*, expressed as follows:

$$\begin{cases} X = [x_1, x_2, x_3, x_4, x_5, x_6, x_7, x_8]^T = [x_a, x_b, y_a, y_b, \dot{x}_a, \dot{x}_b, \dot{y}_a, \dot{y}_b]^T \\ U = [u_1, u_2, u_3, u_4]^T = [i_{xa}, i_{xb}, i_{ya}, i_{yb}]^T, Y = [y_1, y_2, y_3, y_4]^T = [x_a, x_b, y_a, y_b]^T \end{cases} \tag{4}$$

The system's state equation, derived from Equations (1)-(4), is:

$$\begin{cases} \dot{x}_1 = x_5 \quad \dot{x}_2 = x_6 \quad \dot{x}_3 = x_7 \quad \dot{x}_4 = x_8 \\ \dot{x}_5 = R^a{}_x G_1 x_1 + R^a{}_x G_3 x_2 - M_a (x_8 - x_9) + R^a{}_i G_1 u_1 + R^b{}_i G_3 u_2 \\ \dot{x}_6 = R^a{}_x G_3 x_1 + R^a{}_x G_2 x_2 + M_b (x_8 - x_9) + R^a{}_i G_3 u_1 + R^b{}_i G_2 u_2 \\ \dot{x}_7 = R^a{}_x G_1 x_3 + R^a{}_x G_3 x_4 + M_a (x_6 - x_7) + R^a{}_i G_1 u_3 + R^b{}_i G_3 u_4 \\ \dot{x}_8 = R^a{}_x G_3 x_3 + R^a{}_x G_2 x_4 - M_b (x_6 - x_7) + R^a{}_i G_3 u_3 + R^b{}_i G_2 u_4 \end{cases} \tag{5}$$

Where, $R^a{}_x = \frac{K_{sa}}{m}$, $R^b{}_x = \frac{K_{sb}}{m}$, $R^a{}_i = \frac{K_{ia}}{m}$, $R^b{}_i = \frac{K_{ib}}{m}$, $G_1 = 1 + \frac{ml_a^2}{J_{xy}}$, $G_2 = 1 + \frac{ml_b^2}{J_{xy}}$, $G_3 = 1 - \frac{ml_a l_b}{J_{xy}}$, $M_a = \omega \frac{J_z l_a}{J_{xy} l}$, $M_b = \omega \frac{J_z l_b}{J_{xy} l}$, $J_{xy} = J_x = J_y$.

Calculating the derivative of the output concerning time, we can obtain:

$$J_u = [\ddot{y}_1, \ddot{y}_2, \ddot{y}_3, \ddot{y}_4]^T = \begin{bmatrix} R^a{}_x G_1 x_1 + R^b{}_x G_3 x_2 - M_a (x_7 - x_8) + R^a{}_i G_1 u_1 + R^b{}_i G_3 u_2 \\ R^a{}_x G_3 x_1 + R^b{}_x G_2 x_2 + M_b (x_7 - x_8) + R^a{}_i G_3 u_1 + R^b{}_i G_2 u_2 \\ R^a{}_x G_1 x_3 + R^b{}_x G_3 x_4 + M_a (x_5 - x_6) + R^a{}_i G_1 u_3 + R^b{}_i G_3 u_4 \\ R^a{}_x G_3 x_3 + R^b{}_x G_2 x_4 - M_b (x_5 - x_6) + R^a{}_i G_3 u_3 + R^b{}_i G_2 u_4 \end{bmatrix} \tag{6}$$

The corresponding Jacobian matrix is:

$$\frac{\partial J}{\partial U} = \begin{bmatrix} \frac{\partial \ddot{y}_1}{\partial u_1} & \frac{\partial \ddot{y}_1}{\partial u_2} & \frac{\partial \ddot{y}_1}{\partial u_3} & \frac{\partial \ddot{y}_1}{\partial u_4} \\ \frac{\partial \ddot{y}_2}{\partial u_1} & \frac{\partial \ddot{y}_2}{\partial u_2} & \frac{\partial \ddot{y}_2}{\partial u_3} & \frac{\partial \ddot{y}_2}{\partial u_4} \\ \frac{\partial \ddot{y}_3}{\partial u_1} & \frac{\partial \ddot{y}_3}{\partial u_2} & \frac{\partial \ddot{y}_3}{\partial u_3} & \frac{\partial \ddot{y}_3}{\partial u_4} \\ \frac{\partial \ddot{y}_4}{\partial u_1} & \frac{\partial \ddot{y}_4}{\partial u_2} & \frac{\partial \ddot{y}_4}{\partial u_3} & \frac{\partial \ddot{y}_4}{\partial u_4} \end{bmatrix} = \begin{bmatrix} R^a{}_i G_1 & R^b{}_i G_3 & 0 & 0 \\ R^a{}_i G_3 & R^b{}_i G_2 & 0 & 0 \\ 0 & 0 & R^a{}_i G_1 & R^b{}_i G_3 \\ 0 & 0 & R^a{}_i G_3 & R^b{}_i G_2 \end{bmatrix} \tag{7}$$

According to Equation (7), $\det\left(\frac{\partial J}{\partial U}\right) \neq 0$, so $\frac{\partial J}{\partial U}$ is a non-singular matrix. Therefore, the relative degree of the electromagnetic bearing system is $\alpha = (\alpha_1, \alpha_2, \alpha_3, \alpha_4) = (2, 2, 2, 2)$, and $\sum_{i=1}^{4} \alpha_i = 8 \leq n$, where

n denotes the system state variable count. Per inverse system theory, the system is invertible. The inverse system's input variables are $\varphi=[\varphi_1,\varphi_2,\varphi_3,\varphi_4]^T$, yielding:

$$\varphi=[\varphi_1,\varphi_2,\varphi_3,\varphi_4]^T=[\ddot{y}_1,\ddot{y}_2,\dot{y}_3,\ddot{y}_4]^T=\left[\ddot{x}_a,\ddot{x}_b,\ddot{y}_a,\ddot{y}_b\right]^T \tag{8}$$

Substituting Equation (8) into Equation (6), we can obtain the decoupled control quantity of the final inverse system for the electromagnetic bearing as follows:

$$\begin{cases} u_1=\dfrac{1}{R^a{}_i(G_1G_2-G_3^2)}(G_2\varphi_1-G_3\varphi_2)-\dfrac{R^a{}_x}{R^a{}_i}x_1+\dfrac{G_2M_a+G_3M_b}{R^a{}_i(G_1G_2-G_3^2)}(x_7-x_8) \\ u_2=\dfrac{1}{R^b{}_i(G_3^2-G_1G_2)}(G_3\varphi_1-G_1\varphi_2)-\dfrac{R^b{}_x}{R^b{}_i}x_2+\dfrac{G_3M_a+G_1M_b}{R^b{}_i(G_3^2-G_1G_2)}(x_7-x_8) \\ u_3=\dfrac{1}{R^a{}_i(G_1G_2-G_3^2)}(G_2\varphi_3-G_3\varphi_4)-\dfrac{R^a{}_x}{R^a{}_i}x_3-\dfrac{G_2M_a+G_3M_b}{R^a{}_i(G_1G_2-G_3^2)}(x_5-x_6) \\ u_4=\dfrac{1}{R^a{}_i(G_3^2-G_1G_2)}(G_3\varphi_3-G_1\varphi_4)-\dfrac{R^b{}_x}{R^b{}_i}x_4-\dfrac{G_3M_a+G_1M_b}{R^b{}_i(G_3^2-G_1G_2)}(x_5-x_6) \end{cases} \tag{9}$$

Equation (9) demonstrates that inverse system decoupling divides the radial four-degree-of-freedom electromagnetic bearing's rigid rotor into four independent single-input single-output pseudo-linear subsystems.

3.2. Linear active disturbance rejection controller (LADRC)

LADRC consists of a tracking differentiator (TD), linear extended state observer (LESO), and Linear state error feedback control law (LSEF). The TD is expressed as:

$$\begin{cases} x_1(k+1)=x_1(k)+hx_2(k), x_2(k+1)=x_2(k)-h(r^2x_1(k)+2rx_2(k)-r^2y_r) \\ e_1(t)=y_{\mathrm{r}}(t)-z_1(t), e_2(t)=\dot{y}_{\mathrm{r}}(t)-z_2(t) \end{cases} \tag{10}$$

Where, r is a speed factor; h is sampling step; $x_1(k+1)$ is the discrete tracking signal of y_{r}; $x_2(k+1)$ is the differential of y_{r}, that is, the speed input; $\dot{y}_{\mathrm{r}}$ is the differential of the input signal.

The linear extended state observer (LESO) can be expressed as:

$$\begin{cases} e(t)=z_1(t)-y(t), z_1(t+h)=z_1(t)+h(z_2(t)-\beta_1e(t)) \\ z_2(t+h)=z_2(t)+h(z_3(t)-\beta_2e(t)+b_0u), z_3(t+h)=z_3(t)+h(-\beta_3e(t)) \end{cases} \tag{11}$$

Where, β_1, β_2 and β_3 are the adjustable gains of the LESO, their values are related to the bandwidth w_0 of the observer, and there exists the following relationship: $\beta_1=3w_0,\beta_2=3w_0^2,\beta_3=w_0^3$.

The linear state error feedback control law (LSEF) is expressed as:

$$u_0(t)=k_pe_1(t)+k_de_2(t) \tag{12}$$

Where, k_p and k_d , which is related to the bandwidth w_{c} of the controller: $k_{\mathrm{p}}=w_{\mathrm{c}}^2$, $k_{\mathrm{d}}=2w_{\mathrm{c}}$.

It can be seen from the analysis that the LADRC control parameters have three parameters w_0, w_c and b_0 that need to be set. It is difficult to set the parameters by empirical trial and error method. This study employs the improved grey wolf algorithm (IGWO) for parameter tuning.

3.3. Design of improved grey wolf optimization algorithm (IGWO)

Grey wolf packs exhibit a hierarchical social structure comprising four tiers: $\alpha,\beta,\delta,\omega$. The α wolves act as leaders, embodying the optimal solution, while β and δ wolves support decision-making by representing the second- and third-best solutions. The ω wolves, occupying the lowest rank, follow the directives of higher-tier members. When prey is detected, the collective hunting behavior of the pack can be described as follows:

$$\vec{D}=|\vec{C}\cdot\vec{X}_p(t)-\vec{X}(t)|, \vec{X}(t+1)=\vec{X}_p(t)-\vec{A}\cdot\vec{D}, \vec{A}=2\vec{a}\cdot\vec{r}_1-\vec{a}, \vec{C}=2\cdot\vec{r}_2 \tag{13}$$

Where, $\vec{D}$ denotes the distance between the grey wolf and prey, t represents the current iteration, while

$\vec{X}_p(t)$ and $\vec{X}$ (t) correspond to the prey's and grey wolf's position vectors, respectively. Coefficient vectors $\vec{A}$ and $\vec{C}$, along with $\vec{a}$ (linearly decreasing from 2 to 0 during encircling), define the dynamics. Random vectors $\vec{r}_1$ and $\vec{r}_2$ are constrained within [0, 1].

The group's hunting behavior is described as:

$$\vec{D}_k = |\vec{C}_i \cdot \vec{X}_k - \vec{X}|, \vec{X}_i = \vec{X}_k - \vec{A}_i \cdot \vec{D}_k, \vec{X}_p(t+1) = (\vec{X}_1 + \vec{X}_2 + \vec{X}_3)/3 \tag{14}$$

Where, $k = \alpha$, β, δ; $i = 1, 2, 3$; $\vec{X}_p(t+1)$ represents the position vector of the prey when the grey wolves evolve to the $(t+1)$-th generation.

Upon encircling the prey, the Grey Wolf Optimizer (GWO) obtains the optimal solution via predatory strategies. When $|\vec{A}| \leq 1$, the pack concentrates on the prey, performing a local search; when $|\vec{A}| > 1$ the wolves disperse to escape local optima and perform a global search. However, GWO is prone to premature convergence and slow convergence speed. To enhance performance, the concept of Particle Swarm Optimization (PSO) is introduced. By substituting the position update mechanism of the particle algorithm for that of the grey wolves and dynamically adjusting the inertia factor w, a balance between local exploitation and global exploration capabilities is achieved. The range of variation for the inertia factor is set between [0.5, 1][11], with the velocity and position updates being represented as per Equation （15）:

$$v_i^{k+1} = w\left[v_i^k + c_1 r_1\left(x_1 - x_i^k\right) + c_2 r_2\left(x_2 - x_i^k\right) + c_3 r_3\left(x_3 - x_i^k\right)\right], x_i^{k+1} = x_i^k + v_i^{k+1} \tag{15}$$

Where, the position of the particle is denoted by x; v is the velocity of the particle; w stands for the inertia factor; c_1, c_2 are constants for acceleration; r_1, r_2 are random values within the range [0,1].

Therefore, the first equation in Equation (13) can be expressed as:

$$\vec{D}_k = |\vec{C}_i \cdot \vec{X}_k - w \cdot \vec{X}| \tag{16}$$

3.4. The realization of the overall control algorithm

Figure 2 illustrates the self-disturbance rejection control (ADRC) framework for the electromagnetic bearing rigid rotor system, combined with the control system employing the Improved Grey Wolf Optimizer (IGWO) to optimize Linear Active Disturbance Rejection Control (LADRC) parameters. Here, $y_{r1}, y_{r2}, y_{r3}, y_{r4}$ denote system reference inputs, y_1, y_2, y_3, y_4 represent four-degree-of-freedom outputs, φ_1, φ_2, φ_3, φ_4 define inverse system input variables, and u_1, u_2, u_3, u_4 correspond to electromagnetic bearing current inputs. Using the inverse system method, the system is decoupled into pseudo-linear subsystems (transfer function: $G_d(s) = 1/s^2$). On this basis, four LADRC controllers are cascaded to tune the decoupled system. The IGWO algorithm is then used for rapid tuning of the LADRC parameters w_0, w_c and b_0.

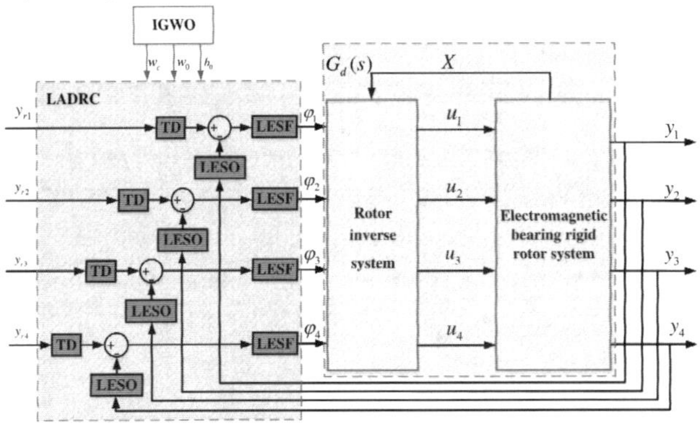

Figure 2. Principle block diagram of electromagnetic bearing control system

4. Simulation analysis

To validate the efficacy of the proposed approach, the IGWO algorithm is evaluated against GWO's parameter optimization performance, and the same population number (20), iteration number (20) and optimization parameter interval are set. Because the electromagnetic bearing needs to take into account both fast convergence and stable suspension, this paper uses the time multiplication absolute value integral error (ITAE) as the solution system fitness value for parameter optimization. The specific parameters of the rigid rotor system of the radial four-degree-of-freedom electromagnetic bearing are: Air gap $C_0 = 0.5\text{mm}$, $m = 10.01\text{kg}$, $l_a = l_b = 0.225\text{m}$, $J_x = J_y = 0.6370\text{kg}\cdot\text{m}^2$ $J_Z = 0.0076\text{kg}\cdot\text{m}^2$, $K_{sa} = K_{sb} = 1.20\times10^5\text{N/m}$, $K_{ia} = K_{ib} = 1.30\times10^2\text{N/A}$.

Under a 0.2mm step signal in the Xa direction of the electromagnetic bearing, Figure 3 compares the parameter optimization performance of IGWO and GWO. Results indicate that IGWO achieves lower fitness values and accelerates optimization speed by 43.75% compared to GWO. The LADRC parameters optimized by IGWO are: $w_0 = 1371.48$, $w_c = 231.05$ and $b_0 = 0.95$. The control parameters of the decentralized PID are also optimized by the IGWO method to obtain the following parameters: *Kp*=2551.7, *Ki*=100 and *Kd*=10470.

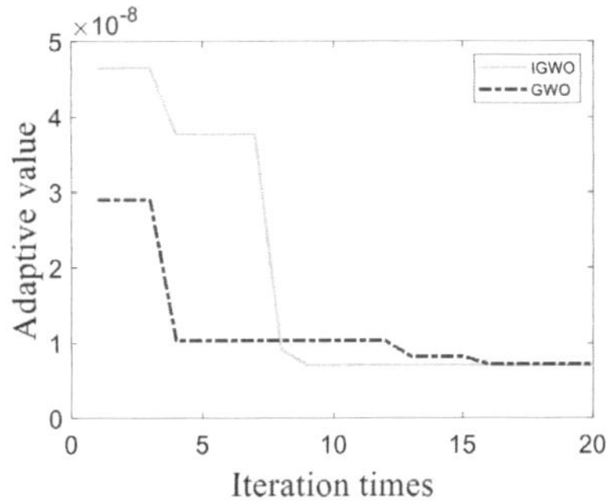

Figure 3. GWO and IGWO optimization parameter adaptation values

To validate the control method's efficacy, static levitation simulations are conducted and compared with conventional decentralized PID control. The Y-axis is designated as the gravitational direction, with the rotor's initial mass center position set to −0.5mm. Figure 4(a)-(b) displays the rotor levitation simulation results. During decentralized PID and LADRC parameter tuning, IGWO outperforms GWO, demonstrating superior control performance in the levitation process. Compared to decentralized PID, the proposed method achieves shorter settling times, zero overshoot, and enhanced levitation accuracy.

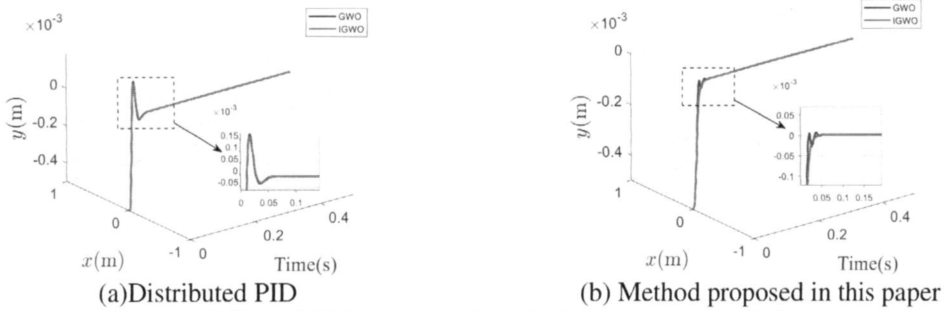

(a)Distributed PID (b) Method proposed in this paper

Figure 4. Effect of different control methods on rotor centroid floating

To assess the decoupling performance and anti-interference capability of the proposed method, 0.1mm and 0.2mm step responses are applied to the Xa and Ya directions at the reference position, while 50N and -40N disturbance forces are introduced at 2s. As depicted in Figure 5, the decentralized PID control exhibits significant overshoot and prolonged settling time. The proposed method reduces the overshoot in the Xa and Ya directions by 77.5 % and the adjustment time by 0.006 s. In terms of anti-interference ability, the method proposed in this paper can quickly suppress the disturbance in the

Xa and Ya directions, so that the rotor can be stably suspended at the reference position. In terms of decoupling ability, it can be seen that when Xa and Ya are given two degrees of freedom step signals, the decentralized PID control results in 0.76 nm and 1.52 nm displacement fluctuations in Xb and Yb directions due to structural coupling, respectively. The proposed method exhibits negligible impact, with position fluctuations confined to the 10^{-12} nm range. Simulations confirm its superior position tracking, anti-interference resilience, and effective static decoupling control.

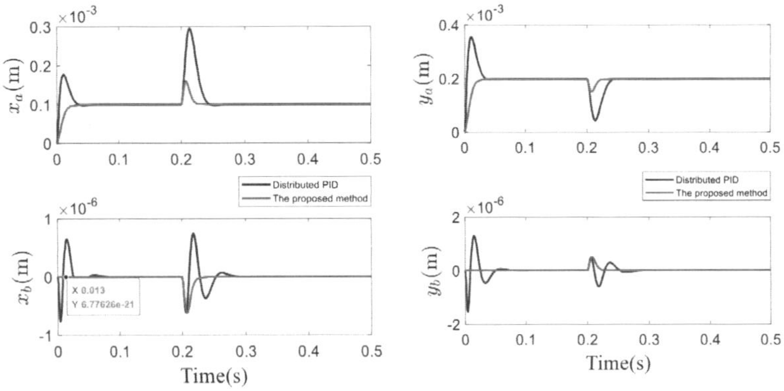

Figure 5. Displacement response of four degrees of freedom under step signal and disturbance

5. Experimental analysis

To validate the decoupling algorithm's control efficacy and engineering feasibility, Figure 6 illustrates the experimental platform and control flow of a four-degree-of-freedom electromagnetic bearing rigid rotor. The platform includes E50-05 A power amplifier, CWY-D0-810503-00-05-30-02 eddy current sensor (the signal of the front end is CWY-D0-810500-30-01-01-02), signal amplifier (self-made), Speedgoat real-time operating system, four-degree-of-freedom electromagnetic bearing test bench and computer. The test bench's physical parameters align with the simulation parameters. During testing, the displacement sensor tracks the rotor's four degrees of freedom and relays displacement data to the Speedgoat via the A/D interface. Subsequently, the control algorithm generates the control signal and outputs the voltage signal to the signal amplifier and power amplifier through the D/A interface. Finally, the power amplifier outputs the current signal to the electromagnetic bearing coil to control the movement of the rotor.

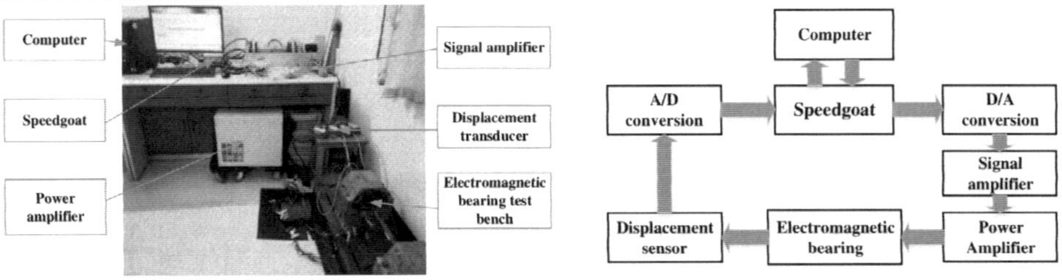

Figure 6. Electromagnetic bearing rigid rotor test bench and control flow chart

Figure 7 shows the experimental results of the displacement response of the rotor with different control methods. The floating experiment is carried out at t=0s, and a step signal of 0.1mm is applied to the Ya direction at t=10s. Affected by the manufacturing error of the test bench and the installation deviation of the bearing, the starting position in the y direction deviates from the nominal value of −0.5mm. Experiments show that after the proposed control method reaches the equilibrium position, the steady-state suspension fluctuation range of the four degrees of freedom is reduced by 20.78 % compared with the decentralized PID control. After 10 s, the step signal is applied in the Ya direction. While both control methods impact the Yb, Xa, and Xb degrees of freedom, the proposed strategy reduces positional fluctuations by 10.38%, 37.48% and 45.12%, respectively. Experimental validation confirms its enhanced ability to suppress coupling effects and resist external disturbances.

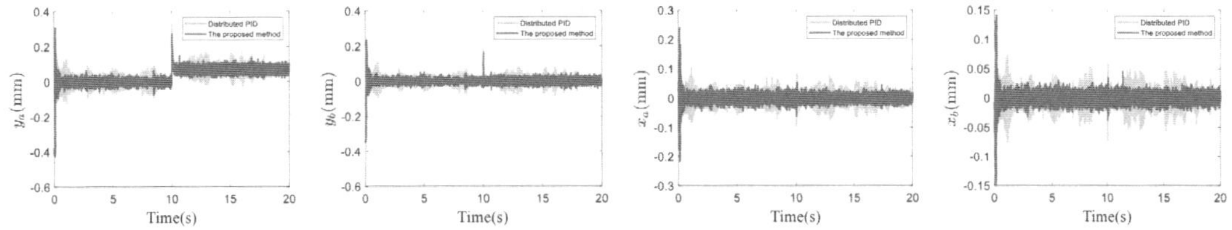

Figure7. Rotor floating and step response experiment

6. Conclusion

To address the multivariable and strong coupling nature of radial four-degree-of-freedom electromagnetic bearing rotor system, a hybrid decoupling control strategy combining IGWO and linear active disturbance rejection inverse systems is developed. Key findings include:

1)The inverse system method decouples the original system into four independent pseudo-linear subsystems, with model error and coupling disturbance dynamically compensated by LADRC, reducing dependence on the system mathematical model's accuracy.

2)The proposed IGWO converges 43.75% faster than traditional GWO, providing an efficient optimization method for complex control systems. Simulation and experimental analyses demonstrate the proposed method's superior signal tracking, enhanced anti-interference capability, and improved decoupling performance compared to decentralized PID control.

References

[1] Gao, J. Z. (2022). *Dynamic analysis and vibration control of magnetic levitation flywheel energy storage system* (Doctoral dissertation). North China Electric Power University, Beijing.

[2] Hawkins, L., Wang, Z., & Nambiar, K. (2018). Floating shock platform testing of a magnetic bearing supported chiller compressor: Measurements and simulation results. In *Turbo Expo: Power for Land, Sea, and Air* (Vol. 51142, p. V07BT34A042). American Society of Mechanical Engineers.

[3] Zhao, H. Y. (2018). *Analysis and decoupling control of high-speed motor rotor system supported by electromagnetic bearings* (Doctoral dissertation). Zhejiang University.

[4] Xie, J. J., Liu, G., & Wen, T. (2015). Composite compensation for load torque of active magnetic bearing in DGMSCMG. *Optics and Precision Engineering*, 23(8), 2211-2219.

[5] Cheng, X., Cheng, B., Deng, S., et al. (2020). State-feedback decoupling control of 5-DOF magnetic bearings based on α-order inverse system. *Mechatronics*, 68, 102358.

[6] Han, J. Q. (2008). *Active disturbance rejection control technology: Estimation and compensation for uncertain factors*. National Defense Industry Press.

[7] Li, B. L., Zeng, L., Zhang, P. M., et al. (2021). Sliding mode active disturbance rejection decoupling control for active magnetic bearings. *Electric Machines and Control*, 25(7), 129-138.

[8] Gao, Z. Q. (2013). Exploration of active disturbance rejection control theory. *Control Theory & Applications*, 30(12), 1498-1510.

[9] Liang, T., Zhou, Y., Chen, D., et al. (2024). Linear active disturbance rejection decoupling control of bearingless flux-switching permanent magnet motors based on system model assistance. *IEEE Transactions on Industrial Electronics*.

[10] Mirjalili, S., Mirjalili, S. M., & Lewis, A. (2014). Grey wolf optimizer. *Advances in Engineering Software*, 69, 46-61.

[11] Cao, Y., Gao, B. P., & Zhang, Z. H. (2021). A power grid fault diagnosis method based on PSO-GWO. *Electrical Measurement & Instrumentation*, 58(9), 35-40.

Static Modeling of Composite Continuum Robot Considering Coupling Effect

Jing Su[1,2], Tao Gao[1], Zechen Lin[3*], Yulu Luo[1], Rui Song[4], Fuxin Du[1,2*]

[1] School of Mechanical Engineering, Shandong University, Jinan 250061, China
[2] Key Laboratory of High-Efficiency and Clean Mechanical Manufacture of MOE, Shandong University, Jinan 250061, China
[3] Shandong Institute of Medical Device and Pharmaceutical Packaging Inspection, NMPA Key Laboratory for Quality Evaluation of Medical Materials and Biological Protective Devices, Shandong Province, Jinan, 250101, China
[4] School of Control Science and Engineering, Shandong University, Jinan 250061, China

*Corresponding author's e-mail: linzechen126@126.com, dufuxin@sdu.edu.cn

Abstract. Continuum robots, characterized by their flexible and hyper-redundant structure, exhibit exceptional adaptability for navigating complex anatomical regions, demonstrating significant potential in minimally invasive surgical applications. This paper proposes a static model for the composite continuum robot incorporating the coupling effect. The tension transmission characteristics are derived by analyzing the driving cables' frictional losses at nodal points. An accurate static model is established by considering both the end-load conditions of flexible beams and the coupling interactions between the concentric tube and notched continuum. Based on the coupling effect analysis and Euler-Bernoulli beam theory, deformation-driving relationships under various motion conditions are formulated for the composite continuum robot. Simulation and experimental verification demonstrate that the proposed static model achieves an average positioning error of 0.486 mm, exhibiting 1.846 times improvement in spatial accuracy compared with conventional constant curvature models. This static model provides theoretical foundations for precise control and clinical applications of composite continuum robots.

1. Introduction

Drawing inspiration from flexible biological structures like elephant trunks and octopus tentacles [1], continuum robots are distinguished by their capacity for continuous bending movements instead of relying on discrete rigid joints. Their highly redundant and compliant design enables them to be highly adaptable in confined spaces, surpassing traditional rigid robots' ability to navigate complex anatomical paths or irregular terrains [2]. This unique flexibility has made them attractive for minimally invasive surgery, where they can reach deep tissues through tiny incisions, and they have also shown potential in search and rescue missions and exploration work [3].

Continuum robots are classified according to their actuation strategies, each of which offers unique operational advantages. Cable-driven (CD) [4], [5] systems utilize coordinated cable tension to achieve enhanced maneuverability in confined spaces. Concentric tube (CT) [6] continuum robots employ nested pre-curved tubes to achieve submillimeter miniaturization, enabling access to extremely narrow anatomical regions [7]. Magnetic drive (MD) designs incorporate magnetic components to achieve

Journal of Physics: Conference Series **3024** (2025) 012030 doi:10.1088/1742-6596/3024/1/012030

precise wireless control via external magnetic fields, ensuring compact and reliable navigation [8]. Pneumatically driven (PD) structures achieve adaptive shape deformation by regulating air pressure [9], [10]. Hybrid actuation schemes effectively integrate multiple actuation mechanisms through synergistic actuation strategies [11]. Such as the composite transmission system combining concentric tube and cable actuation [12] and the rigid-flexible coupling actuation architecture of pneumatic artificial muscles and shape memory alloys (SMAs) [13]. This type of design significantly improves the robot's motion accuracy, load stability, and multimodal environmental adaptability in narrow cavities and demonstrates unique advantages in natural cavity surgery and narrow space detection.

Accurate modeling remains a significant challenge in continuum robotics, particularly for devices navigating constrained anatomical spaces. Traditional approaches, such as constant curvature (CC) [14] or piecewise constant curvature (PCC) approximation, strike a balance between computational efficiency and real-time control but ignore friction and material nonlinearity, which can lead to positioning errors. The Euler–Bernoulli beam (EB) model [15] and piecewise Euler–Bernoulli beam (PEB) model [16], while more accurate, are computationally difficult. Hybrid or simplified models, such as the pseudo-rigid body approximation model [17], address the trade-off between speed and accuracy. In addition, for hybrid-actuated continuum robots, the kinematic influence between different continua has not been explored.

This paper establishes an accurate static model by analyzing the tension transmission relationship and the coupling effect of the concentric tube continuum robot (CTR) on the notched continuum robot (NCR). Based on the coupling analysis and Euler-Bernoulli beam theory, the deformation drive relationship of the composite continuum robot (CCR) under different motion conditions is obtained. The key contributions of this paper are outlined as follows:

- This paper analyzes the force characteristics of the notched continuum while accounting for the frictional loss in the driving cable tension, which is equivalent to the deformation-driven behavior of a flexible notched beam.
- The impact of CCR motion on the NCR is analyzed, and a static model incorporating the coupling effect is developed using the Euler-Bernoulli beam theory.
- The experimental results indicate that the static model exhibits an average position error of 0.486 mm, which is 1.846 times greater than the position accuracy achieved by the CC model.

2. Structure of the robot

The composite continuum section is composed of concentric tube continuum and notched continuum nested together, as shown in Fig. 1. The external notched continuum can be bent and rotated, and its length is L_n. The inner concentric tube continuum can be rotated and fed. The length of the CTR's pre-bent part is L_t. The entire composite continuum robot has an overall feeding degree of freedom.

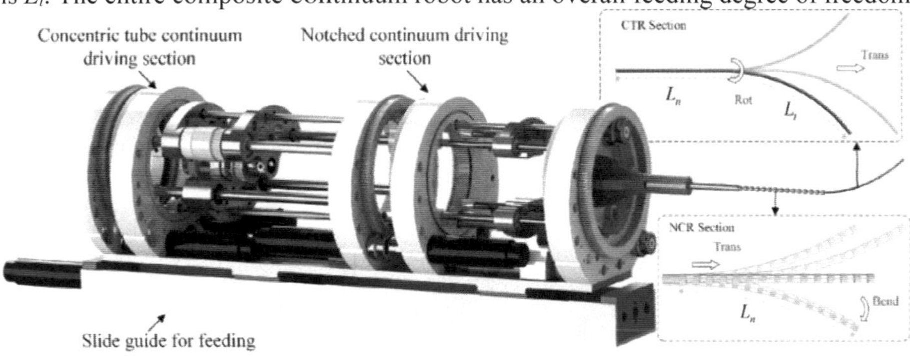

Figure 1. Overall structure of the CCR

3. Static modeling of composite continuum robot

A static model for the CCR is established by incorporating the influence of pre-bent sections in the concentric tube continuum. First, the frictional losses of driving cables at nodal points are analyzed to derive the force equilibrium equations and tension transmission coefficients. This enables the equivalent end-load conditions of flexible beams to be determined. Then, the overall deformation-driving relationships are obtained by evaluating the motion coupling effects of the CTR on the NCR. Finally, the static model of the CCR is resolved using the Euler-Bernoulli beam theory [18].

3.1. Tension transmission analysis

The frictional losses of driving cables at nodal points lead to non-uniform tension distributions, which directly affect the force-deformation behavior of flexible beams. Given the initial cable tensions, the driving forces acting on each flexible beam segment can be derived through tension transmission analysis. Where, only the driving effect of cables on motion deformation is considered, with cable tension losses attributed to frictional forces under a constant friction coefficient.

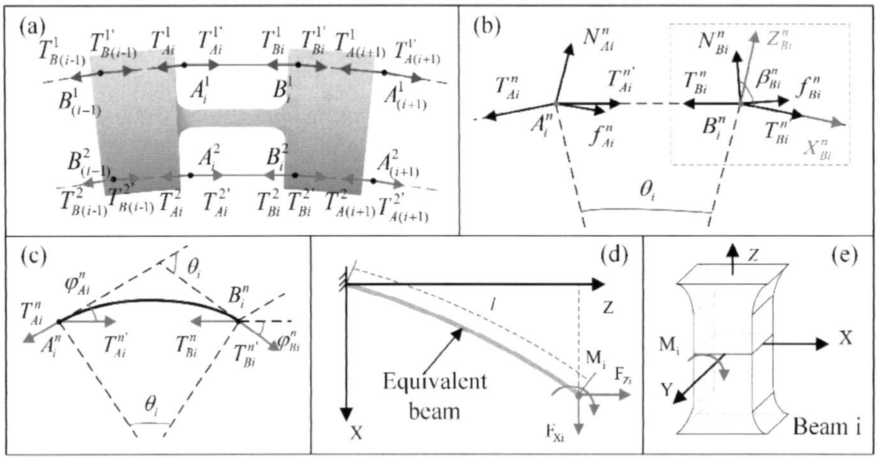

Figure 2. Force analysis of the notched continuum. (a) Tension transmission, (b) Force analysis, (c) Deflection angle analysis, (d) Force at the flexible beam, (e) Flexible beam structure

As shown in Fig. 2, since the force analysis at each node follows a similar methodology, taking the node B_i^n as an example, the equilibrium of four forces—cable tensions T_{Bi}^n and $T_{Bi}^{n'}$, frictional force f_{Bi}^n, and supporting force N_{Bi}^n —is established, as expressed in Eq.(1).

$$\mathbf{T}_{Bi}^n + \mathbf{T}_{Bi}^{n'} + \boldsymbol{f}_{Bi}^n + \mathbf{N}_{Bi}^n = 0 \tag{1}$$

Where n is the driving cables number (n=1,2), and i is the flexible beam number (i=1, 2, ...,14). Given that the tension loss in the driving cables results exclusively from frictional forces, the following relationship can be established:

$$\begin{cases} T_{Bi}^n = T_{Bi}^{n'} + f_{Bi}^n \\ f_{Bi}^n = \mu T_{Bi}^n \end{cases} \tag{2}$$

Where μ is the constant friction coefficient. For Node B_i^n, the local coordinate system is defined as shown in Fig.2(b). By resolving the vector Eq.(1) into scalar components, the following relationships are obtained:

$$\begin{cases} T_{Bi}^{n'} + f_{Bi}^n \cos\left(\beta_{Bi}^n - \pi/2\right) - N_{Bi}^n \sin\left(\beta_{Bi}^n - \pi/2\right) - T_{Bi}^n \cos\varphi_{Bi} = 0 \\ N_{Bi}^n \cos\left(\beta_{Bi}^n - \pi/2\right) + f_{Bi}^n \sin\left(\beta_{Bi}^n - \pi/2\right) - T_{Bi}^n \sin\varphi_{Bi} = 0 \end{cases} \tag{3}$$

Where β_{Bi}^n is the angle between the supporting force N_{Bi}^n and the coordinate system $\mathbf{X}_{Bi}^n$, and φ_{Bi} is the deflection angle of the driving cables.

As shown in Fig. 2(c) and Fig. 2(d), the two beams at each joint can be equivalently modeled as a single beam. A corresponding local coordinate system is defined, where the deflection angle of the i-th equivalent beam segment is represented by φ_i.. Assuming piecewise constant curvature, the curvature of the equivalent beam is considered uniform along its length, resulting in an arc-shaped bending profile. Accordingly, the relationship described in Eq.(4) can be derived.

$$\varphi_{Ai}^{n} = \varphi_{Bi}^{n} = \varphi_i = \arctan\left(r_i / z_i\right) = \theta_i / 2 \tag{4}$$

Where r_i and z_i are the coordinates of the equivalent beam, and θ_i is the bending angle of the ith flexible beam.

For the node B_i^n, the tension transmission coefficient is defined as t_{Bi}^{n} (corresponding to the direction of driving cables length reduction) or $t_{Bi}^{n\,'}$ (corresponding to the direction of driving cables length increase). The cases for node A_i^n and node B_i^n are analogous, and the expressions for t_{Bi}^{n} and $t_{Bi}^{n\,'}$ are given in Eq.(5).

$$\begin{cases} t_{Bi}^{n} = T_{Bi}^{n\,'} / T_{Bi}^{n}\left(t_{Bi}^{n} < 1\right) \\ t_{Bi}^{n\,'} = T_{Bi}^{n\,'} / T_{Bi}^{n}\left(t_{Bi}^{n\,'} > 1\right) \end{cases} \tag{5}$$

By substituting Eq.(5) into Eqs.(2) and (3), the expressions for the tension transmission coefficients t_{Bi}^{n} or $t_{Bi}^{n\,'}$ can be derived.

$$\begin{cases} t_{Bi}^{n} = \mu^2 - \mu^2 \cos\varphi_{Bi}^{n} - \sigma_1 + 1 \\ t_{Bi}^{n\,'} = \mu^2 - \mu^2 \cos\varphi_{Bi}^{n} + \sigma_1 + 1 \end{cases} \tag{6}$$

Where σ_1 is a symbol introduced to simplify the expressions for the tension transmission coefficients t_{Bi}^{n} and $t_{Bi}^{n\,'}$. The expression for σ_1 is provided in Eq.(7).

$$\sigma_1 = \frac{\sqrt{2}}{2}\mu \sin\varphi_{Bi}^{n} \sqrt{\frac{\mu^2 - \mu^2\cos\varphi_{Bi}^{n} + 2}{\cos\left(\varphi_{Bi}^{n} / 2\right)^2}} \tag{7}$$

As indicated by the above equations, the tension transmission coefficients t_{Bi}^{n} and $t_{Bi}^{n\,'}$ are determined solely by the deflection angle of the driving cables at the nodal points and the direction of cable movement. When the drive cables move in the forward direction (motion in the direction of cable length reduction), the relationship between the driving cables tensions satisfies Eq.(8).

$$\begin{cases} T_{Ai}^{n\,'} = k_{Ai}^{n} T_{Ai}^{n} \\ T_{Bi}^{n\,'} = k_{Bi}^{n} T_{Bi}^{n} \\ T_{Ai}^{n\,'} = T_{Bi}^{n} \\ T_{Bi}^{n\,'} = T_{A(i+1)}^{n} \end{cases} \tag{8}$$

When the driving cables undergoes reverse motion (motion in the direction of cables length increase), the coefficients k_{Ai}^{n} and k_{Bi}^{n} in the equation are replaced by $k_{Ai}^{n\,'}$ and $k_{Bi}^{n\,'}$, respectively. From Eq.(8), the vector relationship expressed in Eq.(9) can be derived.

$$\begin{cases} \mathbf{T}_{Ai}^{n\,'} + \mathbf{T}_{Bi}^{n} = 0 \\ \mathbf{T}_{Bi}^{n\,'} + \mathbf{T}_{A(i+1)}^{n} = 0 \\ \mathbf{T}_{B14}^{n\,'} + \mathbf{T}_{Aend} = 0 \end{cases} \tag{9}$$

The force acting on the i-th segment of the flexible beam is denoted as $\mathbf{F}_i$. The expression for $\mathbf{F}_i$ is provided in Eq.(10).

$$\mathbf{F}_i = \mathbf{T}_{Bi}^{n} + \mathbf{T}_{Bi}^{n\,'} + \mathbf{T}_{A(i+1)}^{n} + \mathbf{T}_{A(i+1)}^{n\,'} + \mathbf{T}_{B(i+1)}^{n} + \cdots + \mathbf{T}_{A14}^{n\,'} + \mathbf{T}_{B14}^{n} + \mathbf{T}_{B14}^{n\,'} + \mathbf{T}_{Aend} \tag{10}$$

By substituting Eq.(9) into Eq.(10), the force $\mathbf{F}_i$ acting at the i-th flexible beam can be derived as the tension of the driving cables at the end node of this segment, as shown in Eq.(11).

$$\mathbf{F}_i = \mathbf{T}_{Bi}^{n} \tag{11}$$

Considering the general case where both driving cables exert forces on the continuum structure, the expression for Force $\mathbf{F}_i$ acting at the i-th flexible beam segment is given by Eq.(12).

$$\mathbf{F}_i = \mathbf{T}_{Bi}^{1} + \mathbf{T}_{Bi}^{2} \tag{12}$$

When translating the tensions applied at the nodes to the beam's end, bending moments in varying directions are generated. Since the torsion and shear stiffness of the designed continuum structure are significantly greater than its bending stiffness, the torsional and shear deformations of the beam are neglected. As shown in Fig.2(d) and Fig.2(e), the tensions of the driving cables at the nodes are equivalently represented as deformation-driven actions at the beam's end. The force exerted at the *i*-th flexible beam is represented by Eq.(13).

$$\begin{cases} F_{Zi} = \sum_{n=1}^{2} \left| \mathbf{F}_i^n \right| \cos(\arctan(r_i, z_i)) \\ F_{Ri} = \sum_{n=1}^{2} \left| \mathbf{F}_i^n \right| \sin(\arctan(r_i, z_i)) \\ M_i^c = \frac{d}{2}(\left|\mathbf{F}_i^1\right| - \left|\mathbf{F}_i^2\right|) \end{cases} \tag{13}$$

Where $\mathbf{F}_i^1$ is the force exerted by driving cable 1 on the *i*-th beam segment, and $\mathbf{F}_i^2$ is the force from driving cable 2 on the same segment. By synthesizing the preceding equations, when the tension transmission coefficients are known, the resultant force at the *i*-th beam segment can be directly related to the initial tensions applied to the driving cables. Using an appropriate beam model to establish the relationship between beam deformation and its driving mechanisms, θ_i can be solved.

3.2. Coupling effect analysis

The pre-curved section of the CTR interacts with the NCR, creating a coupling effect that influences the movement of the corresponding notched segment. As the inner CTR rotates and the outer NCR bends, this mutual coupling modifies the curvature of the concentric tube continuum, which in turn generates a bending moment. This bending moment further impacts the curvature of the notched continuum, altering its deformation state. The bending moment M_i^t exerted by the concentric tube continuum on the *i*-th flexible beam segment of the notched continuum is expressed in Eq.(14).

$$M_i^t = E_t I_t (K - \overline{K_i}) \tag{14}$$

Where, E_t is the Young's modulus of the concentric tube continuum material, I_t is the moment of inertia of the concentric tube's cross-sectional area, K is the pre-curvature of the CTR, and $\overline{K_i}$ is the bending curvature after coupling with the *i*-th flexible beam segment, whose expression is given by Eq.(15).

$$\overline{K_i} = \theta_i / l \tag{15}$$

Where l is the length of the flexible beam segment. Based on the above equations, it can be obtained that

$$M_i^t = E_t I_t (K - \theta_i / l) \tag{16}$$

3.3. Euler-Bernoulli beam theory

The key to analyzing the motion of flexible beams in a continuum under deformation-driven actuation lies in the modeling of their structural mechanics. The Euler-Bernoulli beam theory, a widely recognized framework, applies under the following assumptions and conditions: (1) The beam material is homogeneous. (2) Beam deflection arises solely from bending, neglecting shear deformation and axial stretching. (3) The material behaves as a linear elastic medium, obeying Hooke's law.

In this study, the flexible beams within the continuum satisfy the Euler-Bernoulli assumptions: The displacement at any point during bending is small relative to the beam's dimensions. No pre-curvature is considered; the beam is assumed to start in a straight initial configuration. The constitutive equation for the Euler-Bernoulli beam is given in Eq.(17).

$$E_n I_n \frac{d\theta_i(s)}{ds} = M(s), s \in [0, L] \tag{17}$$

Where, E_n is the Young's modulus of the notched continuum material, I_n is the moment of inertia of the notched continuum beam's cross-sectional area (which remains constant during bending), s is an arbitrary point along the beam, $\theta_{i(s)}$ is the rotation angle of point on the *i*-th beam segment relative to the Zi-axis, and $M_{(s)}$ is the bending moment acting at points.

As analyzed in Section 3.2, the coupling effect of the CTR on each flexible beam segment of the NCR manifests as a bending moment exerted at the end of each segment. Therefore, the total bending moment M_i acting on the end of each flexible beam segment is the superposition of the bending moment M_i^t generated by the concentric tube continuum and the bending moment M_i^n induced by the driving cables.

$$M_i = M_i^t + M_i^n \tag{18}$$

Substituting Eq.(13) and Eq.(16) into Eq.(18) yields the expression for the total bending moment M_i.

$$M_i = E_t I_t (K - \theta_i / l) + \frac{d}{2}(|\mathbf{F}_i^1| - |\mathbf{F}_i^2|) \tag{19}$$

Given that the bending deformation of each flexible beam segment is assumed to follow a circular arc, the constitutive equation can be restated as shown in Eq.(20).

$$\theta_i = M_i l / E_n I_n \tag{20}$$

Based on Eq.(20), the relationship between the initial tension and the rotation angle θ_i can be established. When the initial tension is known, the bending angle of the flexible beam segment can be determined. The endpoint coordinates of each beam segment are also related to θ_i.

$$\begin{cases} r_i = 2\dfrac{l}{\theta_i}\sin\dfrac{\theta_i}{2}\sin\dfrac{\theta_i}{2} \\ z_i = 2\dfrac{l}{\theta_i}\sin\dfrac{\theta_i}{2}\cos\dfrac{\theta_i}{2} \end{cases} \tag{21}$$

3.4. Static model

When considering the coupling effects of the pre-curved sections of the CTR on the NCR, the bending angle of the flexible beam segment in the NCR can be determined given the tension applied to the driving cables and the feed length L_t of the CTR. Furthermore, with known actuation parameters ϕ_n and ϕ_t, the global coordinates of the CCR relative to {B0} can be derived in Fig.3.

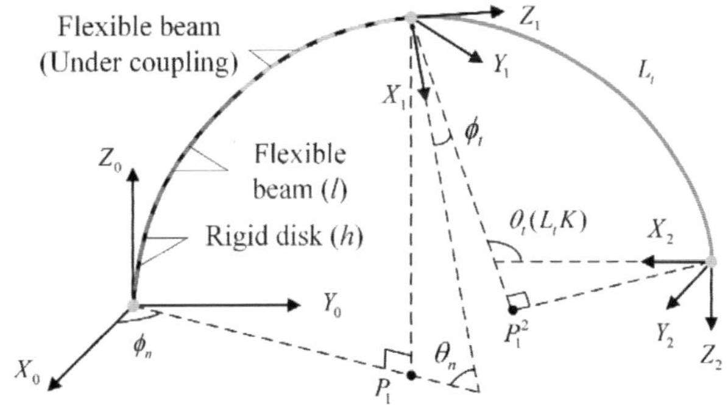

Figure 3. Coordinate system definition for the CC robot considering coupling effects

The coordinate systems for static analysis are established, as shown in Fig.3. The transformation matrix is expressed as Eq.(22).

$$T_0^2 = T_0' T_1' T_1 T_2' T_2 T_3' \cdots T_{12} T_{12}' T_{13} T_{13}' T_{14} T_{14}' T_1^2 \tag{22}$$

Where $T_0', T_1' \cdots T_{13}', T_{14}'$ are the transformation matrices. Similarly, $T_1, T_2 \cdots T_{13}, T_{14}$ denotes the transformation matrices linking the coordinate systems at the endpoints of each flexible beam segment. The explicit expressions for these transformation matrices are given in Eq.(23).

$$\begin{cases} T_0' = Trans(0,0,h)Rotz(\phi_n) \\ T_1' = T_i' = T_{14}' = Trans(0,0,h) \\ T_1 = T_i = T_{14} = Trans(r_i,0,z_i)Roty(\theta_i) \\ T_1^2 = Trans((1-\cos L_t K)\cos\phi_t / K, (1-\cos L_t K)\sin\phi_t / K, \sin L_t K / K)Rotz(\phi_t)Roty(\theta_t) \end{cases} \tag{23}$$

4. Simulation and experimental validation

4.1. Simulation

Simulations are performed in MATLAB by varying a single actuation parameter at a time, with structural parameters defined in Section 2. Under the conditions where the rotation angle ϕ_t of the concentric tube continuum is set to π, its feeding length L_t to 30 mm, and the rotation angle ϕ_n of the notched continuum to 0, the initial cable tension of the notched continuum is varied. The resulting posture configurations and end-effector trajectories are shown in Fig.4(a). When the feeding length L_t of the concentric tube continuum is set to 20 mm, the rotation angle ϕ_n of the notched continuum to 0, and the initial cable tension of the notched continuum held constant, the spatial posture configurations and end-effector trajectories of the composite continuum robot under varying rotation angles ϕ_t of the concentric tube continuum are shown in Fig.4(b).

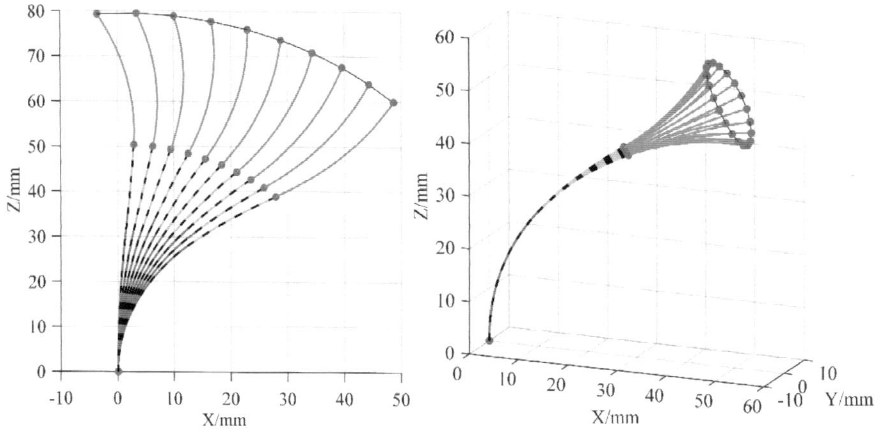

Figure 4. Shape and end-effector trajectory of the CCR. (a) ϕ_t, L_t, ϕ_n held constant; initial driving cables tension varies. (b) L_t, ϕ_n and initial driving cables tension held constant, ϕ_t varies.

4.2. Experiment

The prototype of the composite continuum robot platform is shown in Fig.5. To measure the robot's position, a magnetic tracker sensor was fixed at its tip. Unidirectional repeated bending motions were actuated by controlling the retraction and extension of the actuation cables.

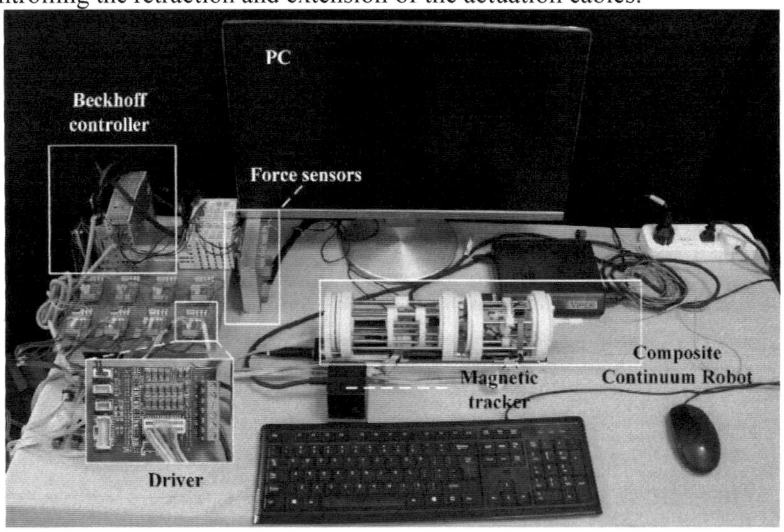

Figure 5. The experimental platform

Figure 6(a) shows the end-effector trajectories of the CCR under three conditions: the CC model, the proposed static model, and the actual experimental motion. As shown in Figs. 6(b) and 6(c), the experimental results indicate that the proposed static model achieves an average positional error of 0.486 mm, whereas the CC model shows a larger average error of 0.897 mm. This corresponds to the static model exhibiting 1.846 times greater positional accuracy compared to the CC model.

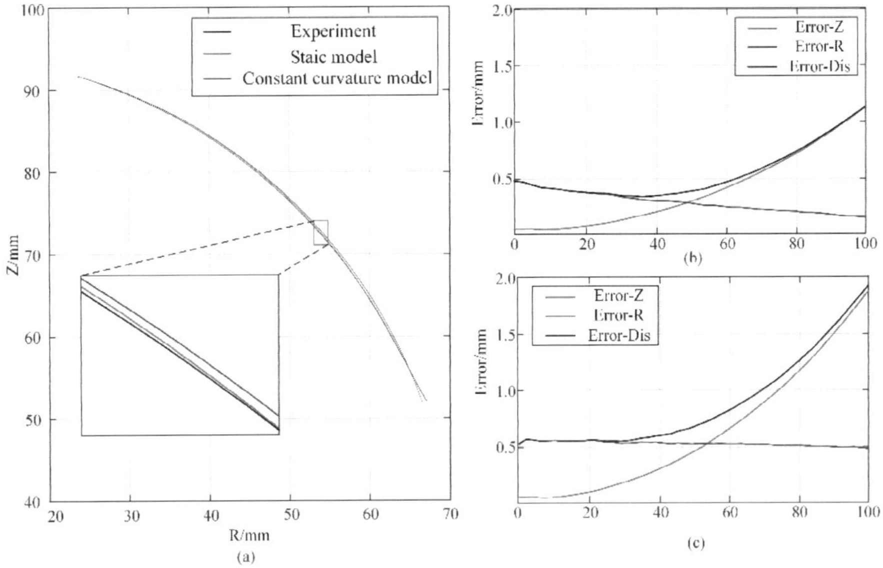

Figure 6. Trajectory and Position Error of the CCR. (a) End-effector trajectory. (b) The static model. (c) The CC model

5. Conclusion

This paper establishes a static model for the CCR that explicitly accounts for the coupling effect. The modeling process began by deriving the cable tension transmission relationships through an analysis of friction-induced losses in the actuation cables and their influence on the deformation of flexible beams. Then, the mutual coupling between the CTR and the NCR is modeled, revealing its significant impact on the deformation of the flexible beams. Leveraging Euler-Bernoulli beam theory, the flexible beams are modeled, and the static response of the CCR under various actuation conditions is derived by integrating the coupling effects. Simulation and experimental validation demonstrate that the proposed static model achieves significantly higher positional accuracy compared to the traditional constant-curvature model. Simulations involving variations in key actuation parameters, such as the feeding length, rotation angle of the concentric tube continuum, and cable tension, confirm their critical influence on the robot's spatial posture and end-effector trajectories. Experimental results quantified the model's performance: the static model exhibited an average positional error of 0.486 mm, representing a 1.846 times improvement in accuracy over the CC model. This work provides a theoretical foundation for the precise control and application of composite continuum robots.

Acknowledgments

This research was funded by the National Natural Science Foundation of China (Grant 52375020), the National Key Research and Development Program of China (Grants 2023YFB4705800 and 2022YFB4703000), and the Key Research and Development Program of Shandong Province, China (Grant 2022CXGC010503). Additional support was provided by the Shandong Provincial Postdoctoral Innovative Talents Program (Grant 238226), the Fundamental Research Funds for the Central Universities, and the Young Scholars Program of Shandong University, China.

References

[1] McMahan W Jones B Walker L Chitrakaran V Seshadri A and Dawson D 2004 ROBOTIC MANIPULATORS INSPIRED BY CEPHALOPOD LIMBS Proceedings of the Canadian Engineering Education Association (CEEA)

[2] Da Veiga T Chandler J Lloyd P Pittiglio G Wilkinson N Hoshiar A Harris R and Valdastri P 2020 Challenges of continuum robots in clinical context: a review Progress in Biomedical Engineering vol 2 no 3 p 032003

[3] Tonapi M Godage I and Walker I 2014 Next generation rope-like robot for in-space inspection 2014 IEEE Aerospace Conference pp 1-13

[4] Du F Zhang X Wu K Zhang G Zheng G Li Y and Song R 2024 A Sensor-Free Force Estimation Method for Notched Continuum Surgical Robot IEEE/ASME Transactions on Mechatronics pp 1-11

[5] Zhang X Du F Zhang G Wu K Zheng G Li Y and Song R 2024 Design and Modelling of Continuum Robot for Endoscopic Submucosal Dissection Surgery With Lifting Force Estimation The International Journal of Medical Robotics and Computer Assisted Surgery vol 20 no 5 p e2670

[6] Wang J Yang X Li P Song S Liu L and Meng M 2020 Design of a multi-arm concentric-tube robot system for transnasal surgery Medical & Biological Engineering & Computing vol 58 no 3 pp 497–508

[7] Webster R Romano J and Cowan N 2009 Mechanics of Precurved-Tube Continuum Robots IEEE Transactions on Robotics vol 25 no 1 pp 67–78

[8] Liu D Liu X Chen Z Zuo Z Tang X Huang Q and Arai T 2022 Magnetically Driven Soft Continuum Microrobot for Intravascular Operations in Microscale Cyborg and Bionic Systems vol 2022

[9] Wang Z Weng D Li Z Chen L Ma Y and Wang J 2023 A Magnetic-Controlled Flexible Continuum Robot with Different Deformation Modes for Vascular Interventional Navigation Surgery Actuators vol 12 no 6 p 247

[10] Tutcu C Baydere B Talas S and Samur E 2019 Quasi-static modeling of a novel growing soft-continuum robot The International Journal of Robotics Research vol 40 no 1

[11] Greer J Morimoto T Okamura A and Hawkes E 2017 Series pneumatic artificial muscles (sPAMs) and application to a soft continuum robot 2017 IEEE International Conference on Robotics and Automation (ICRA) pp 5503-5510

[12] Su J Zhang G Wei H Song R and Du F 2023 Compound continuum manipulator for surgery: Efficient static-based kinematics The International Journal of Medical Robotics and Computer Assisted Surgery vol 19 no 6 p e2561

[13] Zhang G Su J Du F Zhang X Li Y and Song R 2024 Composite continuum robots: Accurate modeling and model reduction International Journal of Mechanical Sciences vol 276 p 109342.

[14] Zhang G Du F Zhang X Wu K Zheng G Li Y and Song R 2024 Continuum Robots: A Real-Time Model-Based Data-Driven Nonlinear Controller IEEE Transactions on Industrial Electronics vol 71 no 12 pp 16176–16186

[15] Mishra A Mondini A Del Dottore E Sadeghi A Tramacere F and Mazzolai B 2018 Modular Continuum Manipulator: Analysis and Characterization of Its Basic Module Biomimetics vol 3 no 1 p 3

[16] Wang Z Jia Z Qian Sen Wang D Yu X and Liu X 2023 An Improved Static Model for Bidirectional Notched Continuum Robot Considering the Cable Tension Loss Journal of Mechanisms and Robotics vol 16 no 071006

[17] Sayadi A Cecere R and Hooshiar A 2024 Finite Arc Method: Fast-Solution Extended Piecewise Constant Curvature Model of Soft Robots with Large Variable Curvature Deformations Robotics Reports vol 2 no 1 pp 49–64

[18] Wu K 2022 A comprehensive static modeling methodology via beam theory for compliant mechanisms Mechanism and Machine Theory vol 169 p 104598

Research on Obstacle Avoidance Path Planning for Robotic Arms Using Improved RRT Algorithm

Fenglian Pan[1,2], Mingyu Zhao[2,3], Zhicheng Tan[2*], Guo Li[2], Yongjun Wang[2], Xing Yao[2]

[1] Guangxi Normal University, Guilin, Guangxi Zhuang Autonomous Region, 541004, P.R. China

[2] Guilin University of Aerospace Technology, Guilin, Guangxi Zhuang Autonomous Region, 541004, P.R. China

[3] Guilin University of Electronic Technology, Guilin, Guangxi Zhuang Autonomous Region, 541004, P.R. China

*Corresponding author's e-mail: tanzc@guat.edu.cn

Abstract. In robot path planning, the Rapidly-exploring Random Tree (RRT) algorithm is extensively employed for its effective search performance. However, the conventional RRT algorithm exhibits several limitations, including non-smooth paths, low search efficiency, and insufficient adaptability to dynamic environments. To tackle these issues, this paper introduces an improved RRT algorithm that integrates a target-biased sampling strategy, bidirectional expansion, adaptive step size adjustment, multi-objective optimization, and path smoothing. The target-biased sampling strategy dynamically adjusts the sampling probability, enhancing global exploration capability and accelerating convergence speed. The bidirectional expansion mechanism enables the search tree to expand concurrently from both the start and goal points, thereby substantially decreasing the search time. Adaptive step size adjustment modifies the step size based on the distance to the goal, improving the accuracy of path planning. Additionally, multi-objective optimization, combined with genetic algorithms, optimizes path length, smoothness, and motion time. Path smoothing is achieved through the application of cubic B-splines, which optimize path continuity and minimize unnecessary sharp turns. The experimental outcomes indicate that the enhanced algorithm surpasses conventional RRT and RRT* algorithms in search efficiency, path quality, and computational performance. This makes it a more effective approach for robot path planning in complex and dynamic environments.

1. Introduction

Path planning is a crucial issue in robotics and autonomous systems, particularly in complex environments. The Rapidly-exploring Random Tree (RRT) algorithm is widely recognized for its efficiency[1], but it suffers from issues like high randomness, non-smooth paths, and slow convergence, which hinder its performance, particularly in cases where both efficiency and precision are required. While the RRT* algorithm guarantees path quality through asymptotic optimality[2], it still faces challenges in real-time applications, especially in dynamic environments, due to high computational complexity and slow response to dynamic obstacles[3].

To overcome these limitations, this paper introduces an enhanced RRT algorithm that features several important innovations. The proposed approach integrates a target-biased sampling strategy, a bidirectional expansion mechanism, adaptive step size adjustment, multi-objective optimization, and path smoothing. The target-biased sampling strategy enhances exploration efficiency in the target region, while the bidirectional expansion mechanism speeds up the search process by growing trees from both the start and goal points simultaneously, adaptive step size adjustment optimizes exploration by dynamically modifying the step size according to the distance to the goal. Multi-objective optimization, implemented using genetic algorithms, balances path length, smoothness, and motion time, resulting in more practical and efficient paths. Additionally, path smoothing, achieved through cubic B-splines, eliminates redundant nodes and improves path continuity.

2. Related Works

Since its inception, the Rapidly-exploring Random Tree (RRT) algorithm and its variants have found extensive application in robot path planning, especially for autonomous obstacle avoidance in both static and dynamic environments. In their pioneering 2000 paper, S. M. LaValle and J. J. Kuffner laid the theoretical groundwork for the RRT algorithm.[1]. Subsequently, S. Karaman and E. Frazzoli proposed the RRT* algorithm, which demonstrates exceptional performance in high-dimensional spaces[2]. J. D. Gammell and colleagues further enhanced the field by presenting the Informed RRT* algorithm, enhancing search efficiency through the use of elliptical heuristic functions[4].

Yin et al[5]. proposed an enhanced Informed-RRT* method for robotic arm grasping planning, while Li et al.[6] enhanced the RRT algorithm to boost the obstacle avoidance capabilities of robotic arms, and Wan et al.[7] further optimized this process, achieving higher efficiency and reliability. Wu et al.[8] explored efficient path planning methods to increase the operational efficiency of robotic arms. Salzman O. et al.[9] presented the Asymptotically Optimal RRT algorithm, which notably enhanced both the speed and quality of motion planning. Additionally, M. Otte et al. proposed the RRTX algorithm, which supports rapid replanning in dynamic environments[10]. The Bi-AM-RRTStar algorithm, combining bidirectional search with attention mechanisms, demonstrated superior efficiency in dynamic environments[11]. Furthermore, Chen et al.[12] They introduced an enhanced path planning approach based on RRT-Connect, which overcomes the limitations of the traditional RRT-Connect algorithm and improves the robots' path planning abilities in dynamic environments.

Despite significant progress, most existing methods primarily focus on single-objective optimization, often neglecting the comprehensive optimization of path smoothness, motion time, and real-time replanning capabilities. The enhanced RRT algorithm introduced in this paper addresses these limitations through adaptive node selection, dynamic obstacle avoidance, multi-objective optimization, and path smoothing, thereby improving the efficiency and quality of path planning.

3. The Proposed Method

The overall procedure of the enhanced RRT algorithm is as follows: First, sampling points are generated using a goal-biased sampling strategy. Subsequently, a bidirectional expansion mechanism is employed, growing the path tree simultaneously from both the start and goal points. Next, an adaptive step-size adjustment strategy dynamically alters the step size. If no collision is detected, the step size is further adjusted to facilitate continuous growth; otherwise, the initial step size is retained for expansion. Once the two paths are connected to form an initial path, a multi-objective optimization strategy is applied for further refinement. Finally, redundant nodes are removed, and the path is smoothed using B-spline curves to obtain the final optimized path. The overall algorithmic flow is illustrated in Figure 1.

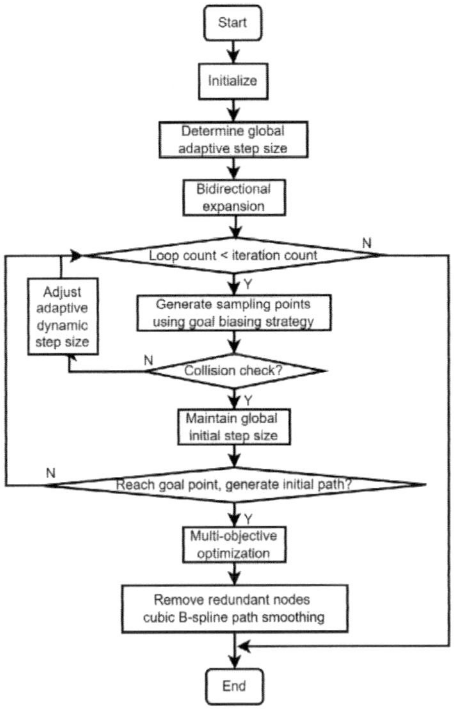

Figure 1. Improved RRT algorithm flowchart

3.1. Target Sampling Strategy Optimization

In the conventional RRT algorithm, sampling points are produced randomly, often leading to low search efficiency, especially in complex environments. The generated paths typically contain numerous redundant nodes, increasing path length and reducing smoothness. To boost the algorithm's performance for path planning in complex environments, this paper proposes an improved target-biased sampling strategy.

The core of this strategy is to increase the tendency toward the goal direction by introducing a target-biased probability parameter P_{bias} to control the selection of sampling points. Specifically, in each sampling iteration, the system produces a random number. If this number is less than or equal to the predefined P_{bias}, the goal point is chosen as the current sampling point, and the step size is enlarged. Otherwise, a point is randomly selected from the free space, and its validity is checked. If the point is valid, the predefined step size is used.

Furthermore, to further optimize this strategy, the value of P_{bias} can be dynamically modified according to the sampling success rate. The definition of P_{bias} is as follows:

$$P_{bias} = P_{min} + (P_{max} - P_{min}) \cdot exp\left(-\frac{k}{(T+1)^2}\right) \tag{1}$$

Here, P_{min} and P_{max} represent the minimum and maximum sampling probabilities, respectively, k is an environment complexity factor that controls the rate of probability adjustment, and T denotes the current sampling iteration. When the sampling success rate is high, P_{bias} is reduced to strengthen global exploration. Conversely, when the success rate is low, P_{bias} is increased to accelerate convergence toward the goal.

By incorporating the target-biased sampling strategy and the adaptive probability adjustment method, the algorithm's global exploration capability is significantly enhanced, improving search efficiency in the target region. Particularly in complex environments, this approach reduces the number of invalid samples, increases convergence speed, and improves the smoothness of the generated paths.

3.2. *Bidirectional Expansion Mechanism*

The traditional RRT algorithm expands unidirectionally from the start point toward the goal, which often results in low search efficiency, particularly in complex environments where the distance between the two points is large. To address this limitation, this paper introduces a bidirectional expansion mechanism, where the path tree expands concurrently from both the start and goal points. This method greatly decreases the number of iterations needed to find a feasible path, thereby enhancing overall performance.

In this method, two trees expand in parallel following standard RRT rules. During each iteration, new nodes are checked for collisions. If the two trees are sufficiently close and no obstacles are present between them, a feasible path is formed. This simultaneous exploration increases the likelihood of quickly finding a valid path while balancing global exploration with local refinement. By minimizing unnecessary expansions, the bidirectional mechanism improves path planning efficiency and generates smoother, more practical paths, especially in complex environments.

3.3. *Adaptive Step Size Adjustment*

In traditional RRT-based path planning, a fixed step size is typically used, which significantly impacts performance. A large step size may cause the expansion to overshoot the target region or collide with obstacles, while a small step size reduces search efficiency and increases computation time. To tackle this problem, this paper introduces an adaptive step size adjustment strategy that dynamically alters the step size according to the distance between the current node and the goal, thereby improving both search efficiency and accuracy.

The fundamental concept of this method is to modify the step size based on the distance to the goal: a larger step size is used when the distance is large to rapidly approach the goal, while a smaller step size is employed when the distance is small to ensure precision and safety. The step size adjustment strategy is defined as follows:

$$s = \begin{cases} s_{min}, & d < d_{min} \\ s_{min} + \dfrac{d - d_{min}}{d_{max} - d_{min}} \cdot (s_{max} - s_{min}), & d_{min} \le d \le d_{max} \\ s_{max}, & d > d_{max} \end{cases} \quad (2)$$

Here, s represents the dynamically adjusted expansion step size; s_{min} and s_{max} denote the minimum and maximum step sizes, respectively; d represents the distance between the current node and the goal point, while d_{min} and d_{max} are the minimum and maximum distance thresholds. When the distance d between the node and the goal point is less than d_{min}, the step size is set to the minimum value s_{min} to ensure expansion accuracy. When d is larger than d_{max}, the step size is set to the maximum value s_{max} to accelerate the search. For intermediate distances, the step size is dynamically adjusted using linear interpolation.

The adaptive step size adjustment strategy enhances the flexibility of path planning by dynamically updating the expansion step size. When far from the target region, a larger step size accelerates the search and reduces the time spent on invalid expansions. As the algorithm approaches the target, the step size decreases to improve accuracy and prevent overshooting or collisions. Additionally, the strategy modifies the step size according to the environment of different nodes, further improving its adaptability to complex scenarios. This approach improves global efficiency while ensuring local precision, overcoming the limitations of fixed step sizes, and enhancing the robustness of the algorithm in various applications.

3.4. *Multi-Objective Optimization*

In path planning problems, optimizing only the path length often fails to meet practical application requirements, particularly in complex environments where path smoothness, obstacle avoidance, and kinematic characteristics are equally critical. To address this, this paper proposes a multi-objective optimization strategy that transforms the path planning problem into an optimization problem considering multiple objectives, thereby generating smoother, shorter paths that meet practical application needs.

The multi-objective path optimization model incorporates multiple objective functions to comprehensively evaluate path quality. First, to ensure the path is as short as possible, thereby reducing motion time and energy consumption, the objective function for path length minimization is defined as:

$$f_1 = \sum_{i=1}^{N-1} \| P_i - P_{i+1} \| \tag{3}$$

Here, P_i and P_{i+1} represent adjacent nodes on the path, and $\|P_i - P_{i+1}\|$ denotes the Euclidean distance between them. Second, to reduce oscillations and acceleration changes in the robot's motion, the smoothness of the path is assessed using the second-order difference. The objective function for smoothness is formulated as:

$$f_2 = \sum_{i=2}^{N-1} \left\| P_i - \frac{P_{i-1} + P_{i+1}}{2} \right\| \tag{4}$$

Here, $\left\| P_i - \frac{P_{i-1} + P_{i+1}}{2} \right\|$ denotes the Euclidean distance between node P_i and the midpoint of its neighboring nodes P_{i-1} and P_{i+1}. Additionally, to ensure the path maintains a safe distance from obstacles, an obstacle distance function is introduced. The objective is to maximize the distance between each node in the path and the nearest obstacle. The objective function is defined as:

$$f_3 = -\min_{i=1,\dots,N} d(P_i, O_j) \tag{5}$$

Here, $d(P_i, O_j)$ denotes the shortest distance between path node P_i and obstacle O_j, and the negative sign indicates the objective is to maximize this distance, thereby ensuring obstacle avoidance.

To achieve the aforementioned multi-objective optimization, this paper employs a Multi-Objective Genetic Algorithm (MO-GA) for global search and optimization. The algorithm first generates an initial path using the RRT algorithm, which is then trimmed and sampled to form the initial population. Next, for each path in the population, the values of the three objective functions—path length, smoothness, and obstacle avoidance—are calculated. Based on the Pareto optimal principle, the algorithm performs non-dominated sorting of the paths, retaining Pareto front solutions to ensure a balance among multiple objectives. Subsequently, genetic operations, including crossover and random perturbations (mutation) of path nodes, are applied to explore better solutions. The process iterates until the Pareto front of the objective functions stabilizes.

This multi-objective path optimization method not only achieves a balance between path length, smoothness, and obstacle avoidance but also significantly enhances the applicability and robustness of path planning.

3.5. *Path Smoothing*

The initial path generated by path planning often contains unnecessary sharp turns and redundant nodes due to the randomness of sampling and expansion, which negatively impacts both path length and smoothness. To address this issue, this paper employs a cubic B-spline-based smoothing method. First, redundant nodes are removed by traversing the path and eliminating nodes that do not contribute to the path's structure. Next, cubic B-splines are applied to smooth the path.

Cubic B-splines are effective interpolation methods that generate smooth, continuous curves while maintaining local control. Extra control points are inserted at the beginning and end of the path to ensure smooth transitions. The path is processed in segments, with each group consisting of four adjacent nodes generating a cubic B-spline curve. The mathematical expression of the cubic B-spline curve is as follows:

$$C(t) = \sum_{i=0}^{3} F_i(t) \cdot P_i \tag{6}$$

Here, P_i represents the control points of the path, and $F_i(t)$ denotes the cubic B-spline basis functions, given as:

$$\begin{cases} F_0(t)=\frac{1}{6}(1-t)^3, \\ F_1(t)=\frac{1}{6}\left(3t^3-6t^2+4\right), \\ F_2(t)=\frac{1}{6}\left(-3t^3+3t^2+3t+1\right), \\ F_3(t)=\frac{1}{6}t^3 \end{cases} \tag{7}$$

Here, $t\in[0,1]$ represents the curve parameter. For each segment of the path, the corresponding cubic polynomial expression is calculated, and the path is smoothed by interpolating the path points. For example, for the x, y, z coordinates of the path, the fitting formulas are:

$$\begin{cases} x(t)=a_0+a_1t+a_2t^2+a_3t^3, \\ y(t)=b_0+b_1t+b_2t^2+b_3t^3, \\ z(t)=c_0+c_1t+c_2t^2+c_3t^3 \end{cases} \tag{8}$$

Here, the coefficients a_i, b_i, c_i are calculated based on the coordinates of the control points. Finally, all cubic B-spline segments are continuously concatenated to ensure the overall continuity and smoothness of the entire path.

By utilizing this cubic B-spline-based path smoothing method, the original path is converted from a series of discrete key points into a continuous and smooth curve. This method not only enhances the geometric continuity and smoothness of the path, eliminating sharp turns, but also effectively reduces the total path length, making it more compact. Additionally, the smoothed path significantly reduces oscillations and errors in the robot's motion, greatly enhancing kinematic performance and better meeting the requirements of practical motion planning.

4. Experiment and Results Analysis

4.1. Experimental Setup and Environment

To verify the performance and effectiveness of the enhanced RRT algorithm, simulation experiments were carried out using MATLAB R2019b. The hardware setup included a laptop running a 64-bit Windows 10 operating system, equipped with an Intel® Core™ i7-11700F processor and 16GB of RAM.

The experimental environment was set in a three-dimensional space with dimensions of 100×100×100. The starting point was set at (5, 5, 5), and the goal point was set at (90, 90, 90). Multiple spherical obstacles were placed in the environment to simulate a complex path planning scenario. The algorithm parameters were set as follows: the fixed iteration step size was 5, the target bias threshold ranged from 0.1 to 1, the connection threshold was 0.8, and the maximum number of iterations was 5,000. To ensure the reliability of the results, each algorithm was run independently 100 times, and the average values were calculated.

The comparison algorithms included the traditional RRT, RRT*, and the improved RRT. The traditional RRT served as the baseline, RRT* enhanced path quality through asymptotic optimality, and the improved RRT integrated a target-biased sampling strategy, bidirectional expansion, adaptive step size adjustment, path smoothing, and multi-objective optimization.

4.2. Experimental Results

Figure 2 illustrates the path planning results of the traditional RRT, RRT*, and improved RRT algorithms. In the figure, the circles represent obstacles, the blue lines depict the random expansion process during algorithm execution, the green circle marks the starting point, the red circle marks the goal point, and the red solid line shows the final path generated by each algorithm. The path from the traditional RRT has many extra nodes, making it longer and less smooth. The RRT* algorithm makes the path better but still has some extra nodes and isn't very smooth. In contrast, the path from the improved RRT is much smoother, with a notable reduction in redundant nodes, resulting in a shorter and more efficient path.

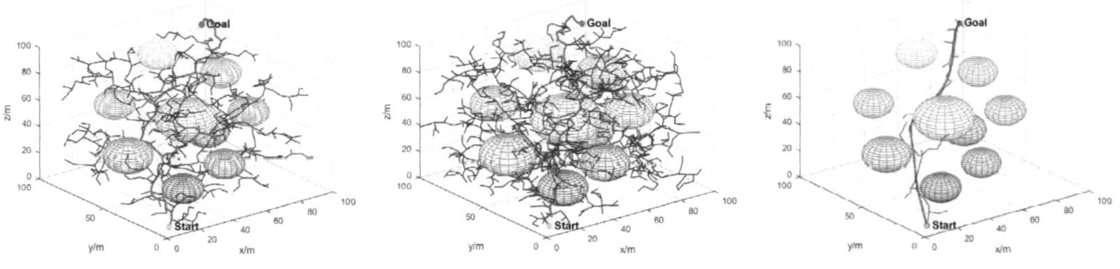

a. Traditional RRT b. RRT* c. Improved RRT

Figure 2. The path search process diagram of the three algorithms

Table 1 provides a comparison of the traditional RRT, RRT*, and enhanced RRT algorithms in terms of search time, path length, number of path nodes, and number of sampled nodes. All metrics are averaged over 100 independent runs.

Table 1. The search results of the three algorithms

Methon	Search Time (s)	Average Length (m)	Average Path Nodes	Average Sampled Nodes
Traditional RRT	2.973	223.905	45.02	1311.957
RRT*	16.550	183.540	19.95	3118.060
Improved RRT	0.187	156.735	7.12	63.520

4.3. *Result Analysis*

From the data in Table 1, it is clear that the enhanced RRT algorithm substantially decreases the number of sampling points in 3D space compared to the traditional RRT and RRT* algorithms, while generating shorter, smoother paths with fewer bends. This decrease in sampling points is mainly due to the target-biased sampling strategy and adaptive probability adjustment, while the shorter, smoother paths result from the redundant node removal strategy and cubic B-spline-based path smoothing method.

The enhanced RRT algorithm shortens the path length by 30.0% compared to RRT and by 14.6% compared to RRT*. with redundant node elimination and path smoothing being the key contributors. Moreover, the search time is decreased by 93.7% compared to RRT and 98.9% compared to RRT*, owing to the target-biased sampling strategy and adaptive step size adjustment, which narrow the search space and improve efficiency. The average number of path nodes is cut by 84.2% compared to RRT and by 64.3% compared to RRT*, while the average number of sampled nodes is reduced by 95.2% compared to RRT and 98.0% compared to RRT*, demonstrating the efficiency improvements achieved through the target-biased sampling and adaptive step size approaches.

5. Conclusion

This paper introduces an enhanced RRT algorithm to overcome the limitations of traditional RRT algorithms, such as poor guidance, slow convergence, low search efficiency, and non-smooth paths. Key innovations, including target-biased sampling strategy optimization, bidirectional expansion, adaptive step size adjustment, path smoothing, and multi-objective optimization, significantly enhance path planning performance. The improved algorithm increases search efficiency by reducing ineffective sampling and expansion time, balances global exploration and local optimization through adaptive step size adjustment, generates smoother paths using cubic B-splines, and produces more practical paths through multi-objective optimization. The experimental results show that the enhanced RRT algorithm surpasses the traditional RRT and RRT* algorithms in path length, search time, number of path nodes, and number of sampled nodes, effectively tackling path planning challenges in complex environments. This research provides a novel solution for efficient path planning with broad application prospects. Future work will explore its potential in robotic arm path planning and UAV navigation to further enhance its practicality and adaptability.

Acknowledgments

This research was funded by the Key Research and Development Program of Guangxi, (Grant No.AB24010110); The Key Research and Development Program of Guilin, (Grant No.20210206-3); The Key Research and Development Program of Guilin, (Grant No.20220113-6); And the Guangxi Science and Technology Plan Project (Grant No.AD22035141).

References

[1] LaValle S M, Kuffner J J 2001 Rapidly-exploring random trees: Progress and prospects Algorithmic and Computational Robotics, (Tokyo: University of Tokyo)
[2] Karaman S and Frazzoli E 2011 Sampling-Based Algorithms for Optimal Motion Planning The International Journal of Robotics Research vol 30 (Thousand Oaks: SAGE Publications) pp 846–894
[3] Adiyatov O and Varol H A 2017 A Novel RRT*-Based Algorithm for Motion Planning in Dynamic Environments Proc. 2017 IEEE International Conference on Mechatronics and Automation (ICMA) (Takamatsu, Japan) p 1416–1421
[4] Gammell J D, Srinivasa S S, Barfoot T D 2014 Informed RRT*: Optimal Sampling-Based Path Planning Focused via Direct Sampling of an Admissible Ellipsoidal Heuristic Proc. 2014 IEEE/RSJ International Conference on Intelligent Robots and Systems (IROS) (Chicago, IL, USA)
[5] Yin X, Chen Y, Guo W H, Yang Z C, Chen H X, Liao A, Yao D J 2025 Flexible grasping of robot arm based on improved Informed-RRT star Chinese Journal of Engineering, vol 47, no 1, pp 113–120
[6] Li Z J, Gao H B, Wang X Q 2024 Research on obstacle avoidance path planning for robotic arms using improved RRT algorithm Computer Engineering and Applications
[7] Wan N J, Hou Y L, Liang S P, Deng Y J, Zeng D X 2024 Improving the RRT Algorithm for Obstacle Avoidance in Robotic Arm Path Planning Mechanical Science and Technology for Aerospace Engineering
[8] Wu K, Yang J Z, Huang S, Wu H T, Zhang Q 2024 Efficient Path Planning of Robot Arm Using an Improved RRT Algorithm Mechanical Science and Technology for Aerospace Engineering
[9] Salzman O, Halperin D 2016 Asymptotically near-optimal RRT for fast, high-quality motion planning IEEE Transactions on Robotics vol 32, no 3, pp 473–483
[10] Otte M, Frazzoli E 2015 RRT^ X RRT X: Real-Time Motion Planning/Replanning for Environments with Unpredictable Obstacles Algorithmic Foundations of Robotics XI: Selected Contributions of the Eleventh International Workshop on the Algorithmic Foundations of Robotics, (Springer International Publishing) pp 461–478
[11] Zhang Y, Wang H, Yin M, Wang J, Hua C 2023 Bi-AM-RRT*: A fast and efficient sampling-based motion planning algorithm in dynamic environments IEEE Transactions on Intelligent Vehicles vol 9, no 1, pp 1282–1293
[12] Chen J G, Zhao Y, Xu X 2021 Improved RRT-Connect Based Path Planning Algorithm for Mobile Robots IEEE Access vol 9 (Piscataway: IEEE) pp 145988–145999

Kinematic Analysis and Printing Experiments of a Multi-DOF Robotic Arm-Based Additive Manufacturing

Kunteng Lu[1], Xingguo Han[1, 2, 3*], Xuan Liu[1], Nankang Liu[1]

[1] School of Mechanical and Electrical Engineering, Guilin University of Electronic Technology, Guilin 541004, China;

[2] Guangxi Key Laboratory of Special Engineering Equipment and Control (Guilin University of Aerospace Technology), Guilin 541004, China

[3] Key Laboratory of Special Engineering Equipment Design and Intelligent Driving Technology（Guilin University of Aerospace Technology）, Education Department of Guangxi Zhuang Autonomous Region

[*]Corresponding author's e-mail: hanxingguo2004@163.com

Abstract This study developed a multi-degree-of-freedom (DOF) robotic arm additive manufacturing(AM) device to address the limitations of traditional three-axis and five-DOF additive manufacturing path planning, thereby enables unsupported printing for curved-surface models. The research employed a structural configuration combining a six-DOF robotic arm with a two-DOF Positioning machine. The kinematics model of the additive manufacturing system of the multi-degree-of-freedom robot arm is established by using the standard Denavit-Hartenberg (D-H) parameter method, and the forward and inverse kinematics equations are deduced systematically. Experimental results demonstrate that the proposed device can efficiently complete printing tasks for simple curved-surface components, providing both hardware solutions and methodological support for multi- DOF additive manufacturing technologies.

1. Introduction

AM technology represents a paradigm shift in traditional manufacturing, integrating computer-aided design with physical processing to enable customized production. Since its inception, this technology has gained prominence due to its operational convenience, rapid prototyping capability, error-testing feasibility, cost-effectiveness, and cross-industry adaptability. With technological maturation and reduced equipment costs, 3D printing has found extensive applications across industrial production[1], biomedical engineering[2], aerospace[3], innovative education[4], and emerging consumer sectors.

Extensive research has been conducted domestically and internationally on robotic arm-based 3D printing methodologies, primarily focusing on kinematic solutions of robotic arms, 3D printing layer processing, and infill path planning. This review examines current research achievements in critical aspects affecting model forming quality and printing efficiency, particularly analyzing three key areas: robotic arm kinematic solutions, 3D printing layer processing, and infill path planning.

Numerous scholars have contributed to inverse kinematics research. Yang Q R et al.[5] developed an improved Levenberg-Marquardt algorithm addressing low precision, slow convergence, and poor generality through three strategies: nonlinear convergence factor updating, trust-region step adjustment, and iterative initial value retention. Lin Y et al.[6] addressed the inverse kinematics solutions for robots failing to meet Pieper's criterion by implementing a multi-population genetic algorithm, thereby proposing a high-precision parallel inversion algorithm applicable to general robotic manipulators. Internationally, Keating et al.[7] leveraged multi-DOF robotic flexibility for hybrid additive-subtractive manufacturing with interchangeable end-effectors. Li H et al.[8] developed a 6-DOF robotic 3D printer with offline programming for automatic trajectory generation, successfully printing precise conductive paths using conductive ink. In the domain of curved-layer slicing research, Chakraborty et al.[9] addressed the surface quality challenges inherent to conventional additive manufacturing by critically analyzing the fundamental limitations of layer-by-layer deposition mechanisms. Their methodology specifically employs curved-layer fused deposition for low-curvature surface regions, thereby significantly enhancing the geometric fidelity and surface finish of fabricated models. The conformal dynamic slicing algorithm proposed by Huang J G et al[10] can effectively alleviate the surface roughness problem of traditional curved surface printing.

Three-axis and five-DOF AM[11] systems still face significant challenges in the efficient and high-precision printing of complex curved parts. System design, kinematic modeling and experimental results are described in detail below.

2. Research principle

2.1. The principle of robot movement

The homogeneous kinematic modeling commonly employs the standard D-H parameters method, which offers advantages of fewer required parameters and simplified modeling. By establishing spatial Cartesian coordinate systems for each joint through the standard D-H parameters method, as illustrated in Figure 1, the position and orientation of the end-effector can be expressed via the end-effector coordinate system relative to the base coordinate system.

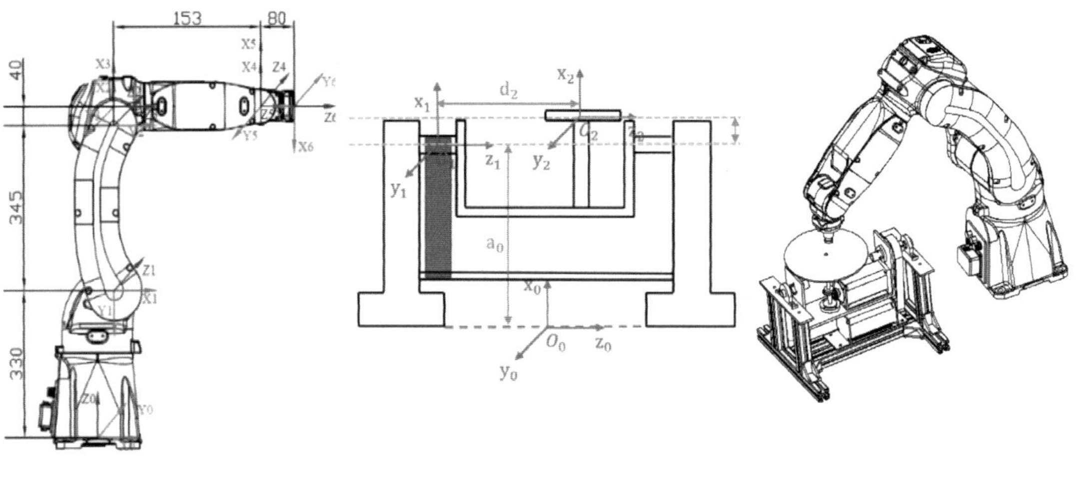

(a) (b) (c)

Figure 1. Joint size and coordinate system diagram (a)Robotic arm (b)Positioning machine (c)Structure diagram of GP8 robotic arm and positioning machine

Figure 1(a) shows the joint size and schematic diagram of GP8 robotic arm, and Figure 1(c) shows the structure diagram of GP8 robotic arm and positioning machine. The motion modeling of the manipulator and the positioner is carried out according to the standard D-H method. In Figure 1(b), the base coordinate system O_0 is located on the central ground of the base of the positioning machine, the coordinate system O_1 is located on the central axis of the axis belt of the upper rotation axis, and the coordinate system O_2 is located on the axis center of the surface of the rotating table. The D-H parameters are shown in Table 1 and Table 2:

Table 1. D-H parameter of GP8 robotic arm

Link(i)	d_i/mm	a_i/mm	α_i/mm	$\theta_i/°$	Offset/°
1	D1	L1	-90	θ_1	0
2	0	L2	0	θ_2	-90
3	0	L3	-90	θ_3	0
4	D4	0	90	θ_4	0
5	0	0	-90	θ_5	0
6	D6	0	0	θ_6	±180

Table 2. D-H parameter of Positioning machine

Link(i)	a_i/mm	$\alpha_i/°$	d_i/mm	$\theta_i/°$	Offset/°
0	0	0	0	0	0
1	a_0	-π/2	d_1	θ_1	[-90,90]
2	0	π/2	d_2	θ_2-π/2	[-360,360]

The Offset parameter documented in the table defines the linear displacement along the common normal between adjacent joint axes, along with the angular misalignment between actuator shafts measured at the robotic system's zero configuration.Where D1=330, L1=40, L2=345, L3=40, D4=153, D6=80, a_0=222, d_1=62, d_2=175, unit (mm).

2.2. *The principle of robot movement*

2.2.1. *Forward kinematics solution of 6R robotic arm*

The forward kinematics of a 6R robotic arm involves determining the end-effector's pose in the Cartesian coordinate system based on joint angles θ_i in joint space. Using translational and rotational transformations, the spatial relationship between consecutive adjacent coordinate frames can be established. By applying the Denavit-Hartenberg parameters in Equation (1), a homogeneous transformation matrix for each link is derived to describe sequential coordinate transformations.

$$ {}^{i-1}_{\ \ \ i}T = \begin{bmatrix} c\theta_i & -s\theta_i c\alpha_i & s\theta_i s\alpha_i & a_i c\alpha_i \\ s\theta_i & c\theta_i c\alpha_i & -c\theta_i s\alpha_i & a_1 s\theta_i \\ 0 & s\alpha_i & c\alpha_i & d_i \\ 0 & 0 & 0 & 1 \end{bmatrix} \quad (1) $$

In the formula, $s\theta_i$ denote $\sin\theta_i$, $c\theta_i$ denote $\cos\theta_i$.Substituting the respective link parameters yields a transformation matrix of the form shown below.

Then the expression of the homogeneous transformation matrix of the end effector relative to the base coordinate system is:

$$ {}^0_6T = \prod_{i=1}^{6} {}^{i-1}_{\ \ \ i}T(\theta_i) \quad (2) $$

The homogeneous transformation matrix of the end indicator relative to the base coordinate system can be obtained from the above equation.

2.2.2. *Positioner kinematics forward solution*

The coordinate system and the transfer matrix of each link are established according to the standard D-H method, The transformation matrix of each connecting rod is:

$$ {}_1^0T = \begin{bmatrix} c_1 & 0 & s_1 & a_0c_1 \\ s1 & 0 & -c1 & a_0s1 \\ 0 & -1 & 0 & d_1 \\ 0 & 0 & 0 & 1 \end{bmatrix}, {}_2^1T = \begin{bmatrix} s_2 & 0 & c_2 & 0 \\ -c_2 & 0 & -s_2 & 0 \\ 0 & 1 & 0 & d_2 \\ 0 & 0 & 0 & 1 \end{bmatrix} \quad (3) $$

Multiplying the two matrices in Equation (3) yields the homogeneous transformation matrix from the base coordinate system O_0 to the end-effector coordinate system O_2 of the manipulator, which constitutes the forward kinematic solution of the manipulator as expressed in Equation (4).

$$ {}_2^0T = {}_1^0T \cdot {}_2^1T = \begin{bmatrix} c_1s_2 & -c_1c_2 & s_1 & a_0c_1 + d_2s_1 \\ s_1s_2 & -s_1c_2 & -c_1 & a_0s_1 - d_2c_1 \\ -c_2 & -s_2 & 0 & d_1 \\ 0 & 0 & 0 & 1 \end{bmatrix} \quad (4) $$

2.3. *Solution of inverse kinematics of robot*

2.3.1. *Inverse kinematics of GP8 robotic arm*

The GP8 robotic arm satisfies Pieper's criterion (where the axes of the last three joints converge at the wrist center), enabling a stepwise analytical solution. The inverse kinematics for the first three joint angles (θ_1, θ_2, θ_3) can be derived from the end-effector's positional coordinates $[x, y, z]$. This process involves solving the positional inverse kinematics for the initial three joints through analytical methods.

$$ \theta_1 = arc\tan 2(y, x) - arc\tan 2(d_3, \sqrt{x^2 + y^2 - d_3^2}) \quad (5) $$

$$ \theta_2 = arc\tan 2\left(z - d_1, \sqrt{x^2 + y^2}\right) - arc\cos\left(\frac{l_2^2 - l_3^2 - r^2}{2l_2l_3}\right) \quad (6) $$

$$ \theta_3 = arc\cos\left(\frac{r^2 - l_2^2 - l_3^2}{2l_2l_3}\right) \quad (7) $$

The last three joints are inversely solved according to the attitude, and the end rotation matrix R_6^0 inversely extrapolates θ_4, θ_5, θ_6.

$$ \theta_4 = arc\tan 2(R_{32}, -R_{31}) \quad (8) $$

$$ \theta_5 = arc\tan 2(\sqrt{R_{31}^2 + R_{32}^2}, R_{33}) \quad (9) $$

$$ \theta_6 = arc\tan 2(R_{32}, -R_{31}) \quad (10) $$

$$ r = \sqrt{(x - a_1)^2 + (y - a_2)^2 + (z - d_1)^2} \quad (11) $$

2.3.2. *Inverse kinematics of Positioning machine*

When the positioning machine is in a specific orientation, the end-effector coordinates (A, B, C) can be used to analytically determine the joint angles 1 and 2. According to Equation(11), the coordinates of the positioner's end-effector in the base frame are given by $(c_1d_2 + a_0, s_1d_2, 0)$. Therefore, the following relationship can be derived:

$$ A = c_1d_2 + a_0 \quad (12) $$

$$ B = s_1d_2 \quad (13) $$

It can be solved by the above formula:

$$ \theta_1 = arc\cos(A - a_0, d_2) \quad (14) $$

$$ \theta_2 = arc\sin(B, d_2) \quad (15) $$

3. Experiment

3.1. Experimental apparatus

The experimental platform of melt deposition additive manufacturing based on the six-DOF and two-DOF displacement units of Yaskawa GP8 robotic arm was established. It includes the design of the wire feeding mechanism.

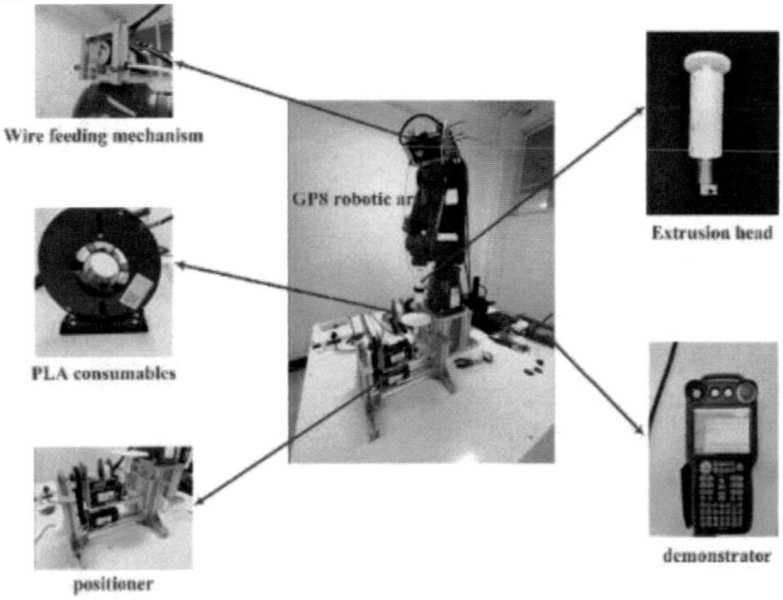

Figure 2. Experimental installation drawing

The main specifications of the AM platform is illustrated in Table 3.

Table 3. The main specifications of the AM platform

Name	Parameter
Robotic arm	Yaskawa GP8
Consumables wire feeding mechanism	SGM7J-04APK-YR11
inspection	YR1000
Consumable materials	PLA(polylactic acid)
Diameter of consumables	1.75 mm
Print the diameter of the nozzle	0.4 mm
Positioning machine	SGM7J-02APK-YR11/ SGM7J-04APK-YR11

3.2. Print experiment

To address the inherent limitation of conventional additive manufacturing (AM) systems requiring auxiliary support structures for curved-surface model fabrication, this study proposes a 6R+2R multi-degree-of-freedom AM apparatus (comprising a 6-DOF robotic arm synergistically integrated with a 2-DOF positioning mechanism), achieving support-free printing of curved structural components through coordinated motion planning.Printing experiments were conducted using a multi-DOF experimental setup based on the Yaskawa GP8 robotic arm integrated with a positioner. By utilizing the two rotational degrees of freedom of the positioner in coordination with the 6-DOF serial robotic arm, high-efficiency fabrication of simple curved surface models was achieved. The printed curved component is illustrated

Journal of Physics: Conference Series **3024** (2025) 012032 doi:10.1088/1742-6596/3024/1/012032

in Figure 3, Figure 4 shows the printed curved tubular structure. Demonstrating the system's capability in dynamic collaborative motion control for additive manufacturing applications.

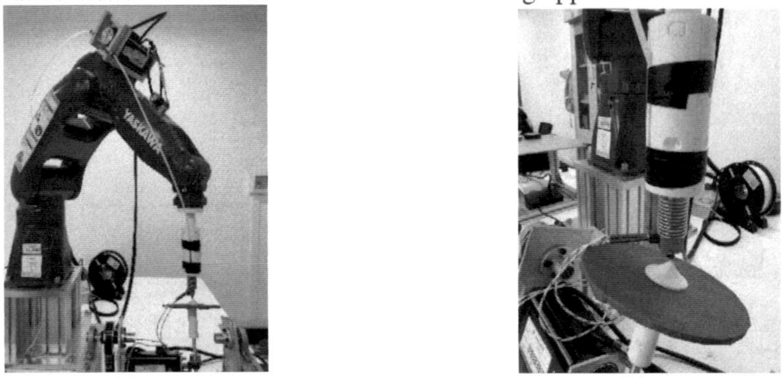

(a) (b)

Figure 3. Printing diagram of curved parts (a) Printing process diagram (b) Printing completion diagram

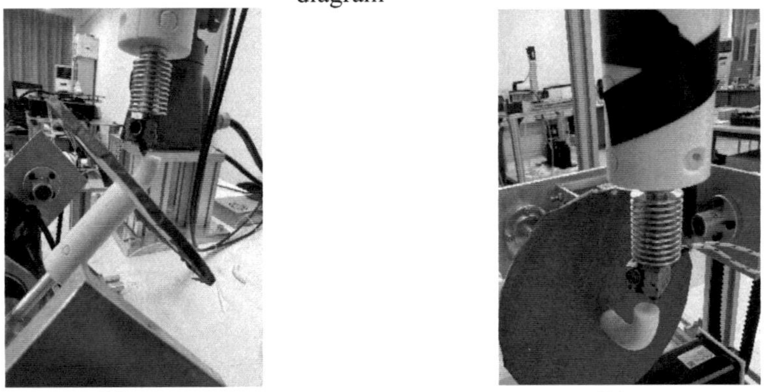

(a) (b)

Figure 4. Printing diagram of curved tubular model (a) Printing process diagram (b) Printing completion diagram

Table 4 presents a comparative analysis between the GP8 robotic system and the developed apparatus, with benchmarking metrics derived from statistically averaged printing duration and workspace utilization during fabrication of two standardized test models under identical process conditions.

Table 4. Comparison between the GP8 robotic arm and developed apparatus

Parameter	GP8 Robotic Arm	Developed Apparatus	Improvement
Workspace Volume	250cm^3	345cm^3	+38%
Print Time	120min	85min	-29%

Compared to the GP8 robotic arm-based AM system, the developed experimental setup effectively addresses the curved surface fabrication challenges inherent in the original configuration. The system enhances the machining efficiency of simple curved-surface components by implementing dual-degree-of-freedom synchronized printing through a positioning mechanism. As demonstrated in Table 3, substantial optimizations are achieved in both printing time and workspace utilization when comparing the GP8 robotic arm with the proposed apparatus. The innovation extends the robotic arm's operational workspace by 38%. The experimental results demonstrate a 29% reduction in printing time compared to the GP8 robotic arm system, while improving nozzle normal deviation issues encountered during high-curvature surface deposition processes.

4. Summary and prospect

To address the limitations of conventional three-axis and five-DOF AM systems in fabricating curved-surface components, this study focuses on the development of a novel multi-degree-of-freedom AM apparatus. The principal contributions are summarized as follows:

1.6R+2R AM experimental equipment was designed and implemented to achieve enhanced kinematic flexibility for complex curved-surface fabrication.

2.Experimental validation through curved-surface part fabrication demonstrated the system's capability to efficiently produce simple curved geometries while mitigating nozzle normal vector deviation in high-curvature regions.Future research directions will prioritize:

(1) Mechanical structure optimization to enhance dynamic stability during continuous deposition processes. Current additive manufacturing techniques are restricted to components featuring simple curved tubular geometries, with future research prioritizing the development of complex surface models.

(2) Comprehensive benchmarking against commercial multi-axis AM systems to quantitatively evaluate performance advantages in surface quality and deposition accuracy.

Project fund

This research was funded by the institution of Guangxi Nature Science Foundations under Grant 2024GXNSFAA010338 and 2023JJA160223 , Guilin Key Research and Development Project under Grant 20210101-5, GUAT Special Research on the Strategic Development of Distinctive Interdisciplinary Fields under Grant：TS2024121, and the National Natural Science Foundation of China under Grant 52365043.

References

[1] Wang X Y, Zhu L. Application and development trend of 3D printing in the field of new energy vehicle manufacturing [J]. Metal Working (Hot Working). 2022(07): pp 1-6.
[2] Zeng Y W, Liu H, Sha CL, et al. 3D Printing of Collagen-based Materials and its Application in Regeneration and Repair of Oral Tissue [J]. Leather Science and Engineering. 2023, 33(02): pp 47-54.
[3] Joshi S C, Sheikh A A. 3D printing in aerospace and its long-term sustainability[J]. Virtual and Physical Prototyping. 2015, 10(4): pp 175-185.
[4] Lu X L. Research on the Application of 3D Printing Technology in the Training of Higher Vocational Mechanical Students [J]. Times Motor. 2023(22): pp 46-48.
[5] Yang Q R, Cui Y X, Wang X L. Robot Inverse Kinematics Solving Based on Improved LM Method [J]. Modular Machine Tool & Automatic Manufacturing Technique. 2023(05): pp 94-96.
[6] Lin Y, et al. Solution of Inverse Kinematics for General Robot Manipulators Based onMultiple Population Genetic Algorithm [J]. Journal of Mechanical Engineering. 2017(03): pp 1-8.
[7] Keating S, Oxman N. Compound fabrication: A multi-functional robotic platform for digital design and fabrication[J]. Robotics and Computer-Integrated Manufacturing. 2013, 29(6): pp 439-448.
[8] Li H, Wang T, Sun J, et al. The adaptive slicing algorithm and its impact on the mechanical property and surface roughness of freeform extrusion parts[J]. Virtual and physical prototyping. 2016, 11(1): pp 27-39.
[9] Chakraborty D, Aneesh Reddy B, Roy Choudhury A. Extruder path generation for Curved Layer Fused Deposition Modeling[J]. Computer-Aided Design. 2008, 40(2): pp235-243.
[10] Huang J G, Qin Q, et al. A dynamic slicing algorithm for conformal additive manufacturing[J]. Additive Manufacturing. 2022(51).
[11] Han X G,Wu G F,Liu X , et al.Research on the Five-Axis Support-Free Additive Manufacturing Method for Overhanging Parts[J].Actuators,2025,14(2):pp99-99.

Journal of Physics: Conference Series **3024** (2025) 012033 doi:10.1088/1742-6596/3024/1/012033

Trajectory Tracking Control of Multi-axle Hybrid Steering Vehicles on Soft Terrain

Zhengkai Zhan[1], Bo Leng[1*], Wenhai Piao[1], XiHua Hu[1], Lu Xiong[1], Yishi Lu[2]

[1] School of Automotive Studies, Tongji University, Shanghai 201804, PR China

[2] Shanghai Gongji Technology Co., Ltd., Shanghai 201804, PR China

[*]Corresponding author's e-mail: lengbo@tongji.edu.cn

Abstract. Aiming at the problem of large trajectory tracking control errors for multi-axle vehicles on soft terrain, this study comprehensively considers complex wheel-terrain interaction relationships on soft terrain and develops a time-varying parameter-based trajectory tracking control algorithm for soft terrain using MPC (Model Predictive Control). Specifically, it establishes calculation methods for tire longitudinal/lateral forces on soft terrain and fits corresponding force curves. A vehicle dynamics model and trajectory tracking error model adapted to soft terrain conditions are constructed to enhance model accuracy. Considering load transfer during vehicle movement on soft terrain and the correlation between tire lateral forces and wheel loads, time-varying parameter matrices are established to accurately characterize vehicle dynamic characteristics under such conditions. By incorporating time-varying lateral force proportionality coefficients for wheel-terrain interaction, an MPC-based control algorithm is designed. Simulation tests of multi-axle vehicle trajectory tracking demonstrate the effectiveness of the proposed algorithm, showing significant improvement in control accuracy compared with conventional control methods that neglect parameter time-variation characteristics.

1. Introduction

The multi-axle hybrid steering vehicles have both the Ackermann geometric steering mode and the differential steering mode and have the advantages of high maneuverability, full wire control, and redundant safety [1]. Complex unstructured environments, especially soft terrain in the wild have become an important factor restricting the transportation efficiency of special vehicles [2]. However, how to consider the wheel-terrain interaction relationship in vehicle dynamics control and use a more accurate dynamics model for trajectory tracking control on soft terrain is a crucial issue.

It is necessary to conduct a tire force analysis for soft terrain. The semi-empirical method is an approach that combines mechanical analysis with appropriate experimental data to address wheel-terrain interaction problems, aiming to establish approximate simplified formulas for describing such interactions. Based on the theory of Ground vehicle mechanics, Jjagwe [3] developed a set of soil tool measurement systems for estimating the soil model parameters of the Bekker and Reece semi-empirical method. Alkhalifa [4] developed the tire-soil measurement method and the resulting validation data are expected to enhance terrain-mechanics modeling and support performance analysis of off-road vehicle mobility and their interactions with terrain. Han [5] established a parametrically modified wheel-terrain model, dynamic model, and motion trajectory model. Further research is required to explore how to incorporate wheel-terrain interaction relationships into vehicle dynamics control and employ more accurate dynamic models for trajectory tracking control on soft terrain.

To validate the accuracy of vehicle motion control algorithms under soft terrain conditions, Vieira D [6] developed a model based on plasticity theory and soil mechanics, employing an adjusted Burckhardt tire model combined with the Bekker-Wong wheel-soil interaction framework, and validated its accuracy for vehicle motion control across various soil types. Additionally, the grid-based LPV (Linear Parameter-Varying) synthesis [7] explicitly incorporated model uncertainties, time-varying parameters, and the effects of deformable soils. However, limited research has addressed the effects of soil parameters and the time-varying characteristics of wheel load transfer on wheel-terrain interaction relationships.

In the field of motion-tracking control methods for multi-axle vehicles, the MPC (Model Predictive Control) algorithm has been widely adopted by numerous researchers. As safety constraints, a collision-avoidance safety domain incorporating spatial safety and stability safety is established. Additionally, the proposed controller [8] integrates Active Front Steering and Direct Yaw Moment Control to enhance yaw response, thereby improving vehicle performance during collision avoidance maneuvers. To address the challenge of coordinating trajectory tracking accuracy and yaw stability for high-speed unmanned ground vehicles operating in dynamic unstructured environments, this paper proposes a learning-based sliding-mode predictive control method for integrated motion control [9]. Zhang [10] proposes a dual-steering mode based on direct yaw moment control for enhancing vehicle steering ability in complex environments. However, the hybrid steering mechanisms considered in these articles are relatively simple, and further exploration is required.

Considering comprehensively the complex wheel-terrain interaction relationship under the soft terrain, for the 8×8 independently driven vehicle with mechanically linked steering on the front two axles, this paper designs an MPC control algorithm considering parameter time-variation. It builds a trajectory-tracking control strategy for the soft terrain.

The main contributions:

1) A method for calculating the longitudinal and lateral forces of tires on soft terrain has been established, and the longitudinal and lateral force curves of tires on soft terrain have been fitted.

2) Considering the vehicle lateral load transfer during steering conditions, a parameter-time-varying matrix is established, which accurately characterizes the vehicle dynamics characteristics on soft terrain.

3) Based on the parameter-time-varying characteristic matrix, a multi-axle hybrid steering vehicle trajectory tracking control algorithm under soft terrain was designed and the algorithm's effectiveness was verified.

2. Modeling

2.1. wheel-terrain interaction relationship

The development of high-fidelity dynamic models for off-road unmanned ground vehicles fundamentally requires establishing precise wheel-terrain interaction mechanics. The Bekker model is derived from the relationship between pressure and sinkage observed in plate penetration tests. The force diagram of the wheel-terrain interaction is illustrated in Figure 1, where W represents the wheel vertical load and R denotes the wheel radius.

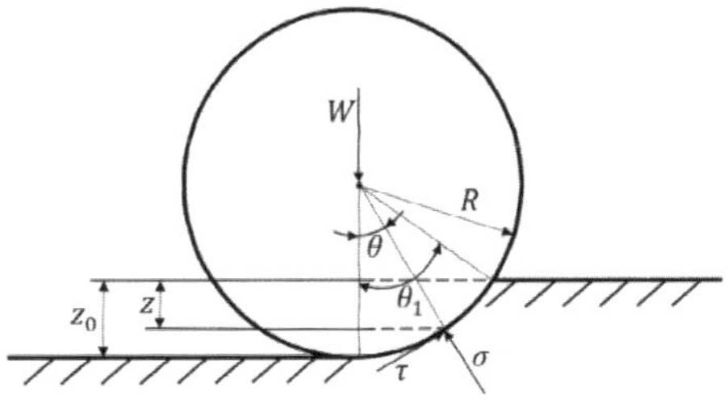

Figure 1. A rigid wheel operating on soft terrain

The normal stress distribution in the Bekker model can be described by the following equation

$$\sigma_n(\theta) = (k_c + k_\varphi b)(\frac{R}{b})^n(\cos\theta - \cos\theta_1)^n \tag{1}$$

In the above equation, $\sigma_n(\theta)$ represents the normal stress, $Z^n(\theta)$ denotes the sinkage at the contact point with an angle θ relative to the lowest contact point, k_c is the cohesive deformation modulus of the soil, k_φ is the frictional deformation coefficient of the soil, n is the sinkage exponent of the soil, b is the tread width of the tire; The approach angle θ_1 is defined as the angle between the line connecting the leading edge of the wheel-terrain contact patch to the wheel center and the vertical normal line of the terrain surface.

The shear stress distribution in the Bekker model can be determined by the following equation

$$\tau_x(\theta) = (c + \sigma_n(\theta)\tan\varphi)(1 - e^{(-\frac{j_d(\theta)}{k_d})}) \tag{2}$$

Where $\tau_x(\theta)$ is shear stress at an angle θ, φ is the soil internal friction angle, c is soil cohesion, $j_d(\theta)$ is Shear displacement, and k_d is the shear deformation modulus, shear displacement $j_d(\theta)$ can be calculated by

$$j_d(\theta) = R\left[(\theta_1 - \theta) - (1 - s_x)(\sin\theta_1 - \sin\theta)\right] \tag{3}$$

Here, s_x represents the longitudinal slip ratio. When the normal stress and shear stress distributions in the contact zone are known, integrating these over the contact area allows the calculation of the vertical force F_Z and drawbar pull F_D

$$F_Z = Rb\left[\int_0^{\theta_1}\sigma_n(\theta)\cos\theta d\theta + \int_0^{\theta_1}\tau_x(\theta)\sin\theta d\theta\right] \tag{4}$$

$$F_D = Rb\left[\int_0^{\theta_1}\tau_x(\theta)\sin\theta d\theta - \int_0^{\theta_1}\sigma_n(\theta)\cos\theta d\theta\right] \tag{5}$$

However, the equations alone are insufficient to calculate θ_1, thereby making it impossible to determine corresponding longitudinal forces such as the net traction force. To resolve this, the following formula must be derived based on force equilibrium principles

$$W = Rb\left[\int_0^{\theta_1}\sigma_n(\theta)\cos\theta d\theta + \int_0^{\theta_1}\tau_x(\theta)\sin\theta d\theta\right] \tag{6}$$

The Newton-Raphson iterative method is applied to solve the longitudinal forces in the wheel-terrain contact model. An initial guess θ_1 is assigned, and defining a residual function $f(\theta_1) = Rb\left[\int_0^{\theta_1}\sigma_n(\theta)\cos\theta d\theta + \int_0^{\theta_1}\tau_x(\theta)\sin\theta d\theta\right] - W$, the problem of solving θ_1 is transformed into finding the root of $f(\theta_1)$, The iterative process continues until convergence, with the final iterated value accepted as the numerical solution. Once this θ_1 is determined by this method, the longitudinal forces in the wheel-terrain interaction model can be fully resolved.

The lateral force can be calculated using the Janosi model, and the corresponding lateral shear stress can be expressed by the following equation

$$\tau_y(\theta) = (c + \sigma_n(\theta)\tan\varphi)(1 - e^{(-\frac{j_y(\theta)}{k_y})}) \tag{7}$$

Where lateral shear modulus $k_y = k_d$ and lateral shear displacement $j_y(\theta)$ can be calculated by

$$j_y(\theta) = R(1 - s_y)(\theta_1 - \theta)\tan\alpha \tag{8}$$

The above formula s_y represents the lateral slip rate of the wheel and α represents the side deflection angle of the wheel. Because it is a soft terrain, the wheel is regarded as a rigid wheel. The lateral force can be expressed as the following formula

$$F_y = Rb\int_0^{\theta_1}\tau_y(\theta)d\theta \tag{9}$$

To facilitate subsequent research on vehicle motion control, we need to simplify the relationship between lateral force and side deflection angle. According to the expression of lateral shear stress, we do Taylor expansion for the exponential term behind, assuming that the vehicle is turning at a small side deflection angle. At this point, we keep the first-order terms and ignore the higher-order terms, the lateral force formula can be summarized as

$$F_y = \alpha \int_0^{\theta_1} Rb(c + \sigma_n(\theta)\tan\varphi \frac{R(\theta_1 - \theta)}{k_y}) d\theta \tag{10}$$

In this way, it can be considered that under the condition of small side deflection angle, the lateral force received by the vehicle on the soft road surface is proportional to the wheel side deflection angle, and the proportional coefficient of the lateral force of the wheeling action is defined as k_{soils}, which is also affected by soil parameters and wheel load and other parameters, that is $F_y = k_{soils} \cdot \alpha$.

2.2. Dynamics model

The lateral motion and yaw motion of the vehicle are described based on the vehicle coordinate system. The schematic diagram of the monorail dynamics model of the multi-axle vehicle is shown in Figure 2. The two-degree-of-freedom monorail dynamics model of a four-axle vehicle is established, as shown in the equation.

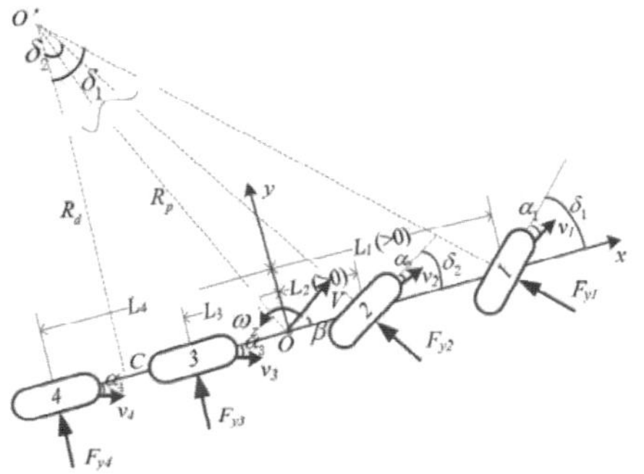

Figure 2. Two-degree-of-freedom monorail dynamics model of multi-axle vehicle

$$\begin{cases} ma_y = F_{y1}\cos\delta_1 + F_{y2}\cos\delta_2 + F_{y3} + F_{y4} \\ I_z\ddot{\varphi} = F_{y1}\cos\delta_1 l_1 + F_{y2}\cos\delta_2 l_2 - F_{y3}l_3 - F_{y4}l_4 + M_z \end{cases} \tag{11}$$

Where, m is the vehicle mass, I_Z is the moment of inertia of the vehicle yaw, $F_{yi}(i=1,2,3,4)$ is the lateral force of the tire of the four axles of the vehicle, M_Z is the additional yaw moment, δ_1, δ_2 is the rotation angle of the first and second axles, $l_i(i=1,2,3,4)$ is the distance from the center of mass to each axle, v_x、v_y is the longitudinal and lateral speed of the vehicle, and a_y、$\ddot{\varphi}$ is the lateral acceleration and yaw angle acceleration of the vehicle.

Figure 2, $\alpha_i(i=1,2,3,4)$ is the side angle of each axle of the vehicle; The lateral force of each wheel tire can be calculated by equation (10). The lateral acceleration can be expressed by

$$a_y = \dot{v}_y + v_x \cdot \dot{\varphi} \tag{12}$$

According to the geometric relationship, the expression of the side angle can be obtained as

$$\begin{aligned} \alpha_i &= \delta_i - \frac{v_y + \dot{\varphi} l_i}{v_x} (i=1,2) \\ \alpha_i &= \frac{\dot{\varphi} l_i - v_y}{v_x} (i=3,4) \end{aligned} \tag{13}$$

Based on the Ackermann steering geometry model, the kinematics model of the two-front axle steering system is established. It can be considered that there is a steering ratio between the two axle angles, then the second axle angle can be expressed as

$$\delta_2 = a_{ss}\delta_1 \tag{14}$$

Combined with the lateral force calculation formula of each axle, based on the small angle assumption, and assuming that the soil mechanical characteristics parameters are the same under the same road surface, the dynamic model is rewritten as

$$\begin{cases} \dot{v}_y = -\dfrac{4k_{soils}}{mv_x}v_y + (-v_x - \dfrac{k_{soils}(l_1+l_2-l_3-l_4)}{mv_x})\dot{\varphi} + \dfrac{k_{soils}+k_{soils}a_{ss}}{m}\delta_1 \\ \ddot{\varphi} = -\dfrac{k_{soils}(l_1+l_2-l_3-l_4)}{I_z v_x}v_y - \dfrac{k_{soils}(l^2{}_1+l^2{}_2+l^2{}_3+l^2{}_4)}{I_z v_x}\dot{\varphi} + \dfrac{k_{soils}l_1+k_{soils}l_2 a_{ss}}{I_z}\delta_1 + \dfrac{M_z}{I_z} \end{cases} \tag{15}$$

2.3. Trajectory tracking error model

To model lateral error, heading error, velocity error, and longitudinal position error, the Frenet coordinate system is introduced as shown in Figure 3.

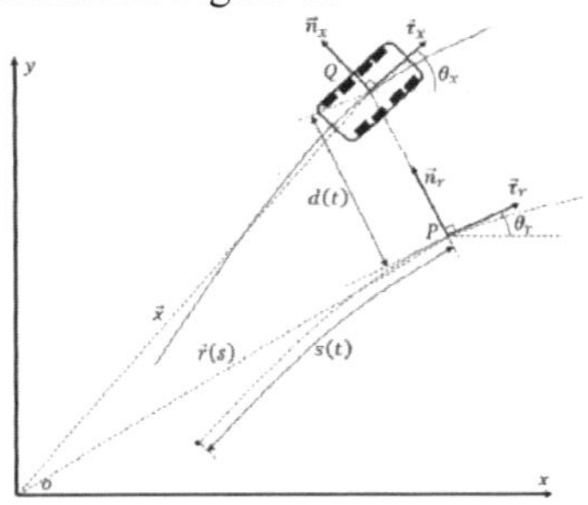

Figure 3. Frenet coordinate system and global coordinate system

To carry out lateral trajectory tracking, it is defined that e_d represents lateral error, $\dot{e}_d$ represents the first-order rate of change of lateral error, $\ddot{e}_d$ represents the second-order rate of change of lateral error, e_φ represents the heading error, $\dot{e}_\varphi$ represents the first-order rate of change of heading error, and $\ddot{e}_\varphi$ represents the second-order rate of change of heading error. d represents the lateral displacement of the vehicle in the Frenet coordinate system, $\dot{d}$ represents the lateral velocity of the vehicle in the Frenet coordinate system, and $\ddot{d}$ represents the lateral acceleration of the vehicle in the Frenet coordinate system. The above quantities can be calculated by the following formula

$$\begin{cases} e_d = d \\ \dot{e}_d = \dot{d} = v_y + v_x e_\varphi \\ \ddot{e}_d = \ddot{d} = \dot{v}_y + v_x \dot{e}_\varphi \\ e_\varphi = \varphi - \theta_r \\ \dot{e}_\varphi = \dot{\varphi} - \dot{\theta}_r \\ \ddot{e}_\varphi = \ddot{\varphi} - \ddot{\theta}_r = \ddot{\varphi} \end{cases} \tag{16}$$

Where $\dot{\theta}_r$ represents the first-order change rate of the slope of the expected trajectory, and $\ddot{\theta}_r$ represents the second-order change rate of the slope of the expected trajectory, which is generally small and negligible.

3. Control Method

The research team previously designed a yaw attitude control algorithm for differential-steered unmanned vehicles [11]. To address the impacts of tire slip during motion, static-to-dynamic friction transition in steering, and actuator torque constraints on vehicle dynamics control, a modified yaw rate virtual control variable was introduced into the reference yaw rate-differential speed control closed loop.

This led to the proposal of a dual closed-loop nested differential steering control method (Figure 4), which uses the modified yaw rate to track the reference yaw rate. The modified yaw rate is then converted into wheel reference speeds. Subsequently, the authors' proposed anti-integral-windup variable structure sliding mode control algorithm [12] is employed to track the reference wheel speeds, achieving smooth and rapid following of differential steering intentions.

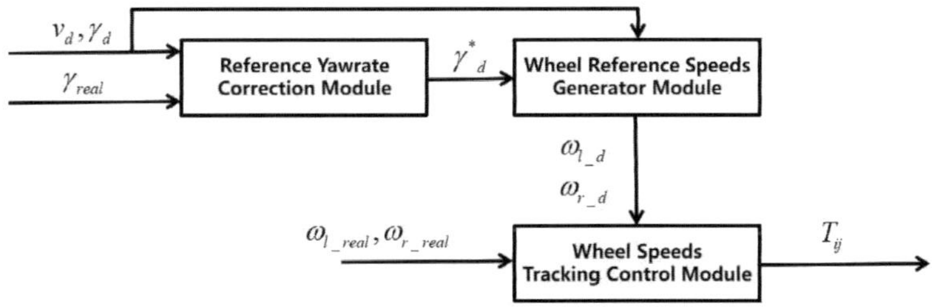

Figure 4. A dual closed-loop nested differential steering control method

The research object of this paper is a hybrid steering vehicle. Compared to pure differential steering, the hybrid steering system exhibits actuator redundancy, increased control objectives, and multi-actuator coupling issues, necessitating multi-objective optimization at the control level. Therefore, this study retains the low-level wheel state tracking algorithm and redesigns the trajectory tracking algorithm based on MPC, while coordinating the hybrid steering heterogeneous actuators. The MPC algorithm optimizes the actions of multiple actuators simultaneously via a state-space model, avoiding coupling problems inherent in traditional modular control design. It also enables real-time adjustment of control strategies through online rolling optimization to maintain system stability. To address wheel-terrain interaction mechanisms on soft terrain, this paper introduces time-varying parameter matrices to dynamically adapt to wheel-terrain interactions and load distribution variations on deformable surfaces. A multi-axle hybrid steering vehicle MPC trajectory tracking control algorithm is thus designed for soft terrain applications, and its control logic block diagram is shown in Figure 5.

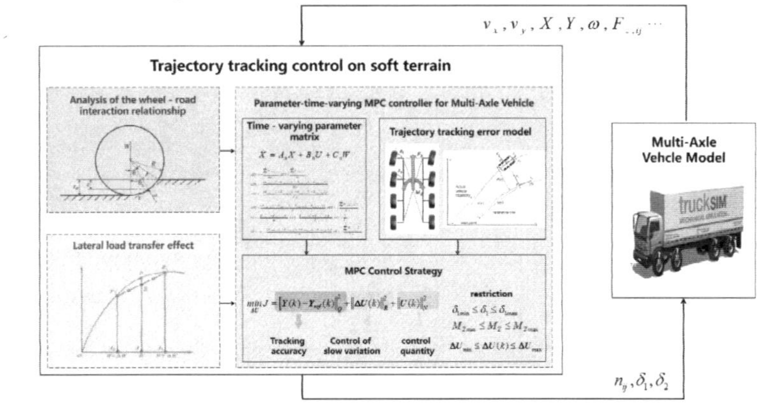

Figure 5. Logic block diagram of trajectory tracking control for multi-axle vehicles on soft terrain

3.1. State space equation

Formula.16 can be substituted into the transverse dynamics model to sort out

$$\begin{cases} \ddot{e}_d = -\frac{4k_{soils}}{mv_x}\dot{e}_d + \frac{4k_{soils}}{m}e_\varphi - \frac{k_{soils}(l_1+l_2-l_3-l_4)}{mv_x}\dot{e}_\varphi \\ +\frac{k_{soils}+k_{soils}a_{ss}}{m}\delta_1 + (-v_x - \frac{k_{soils}(l_1+l_2-l_3-l_4)}{m})\dot{\theta}_r \\ \ddot{e}_\varphi = -\frac{k_{soils}(l_1+l_2-l_3-l_4)}{I_z v_x}\dot{e}_d + \frac{k_{soils}(l_1+l_2-l_3-l_4)}{I_z}e_\varphi \\ -\frac{k_{soils}(l^2{}_1+l^2{}_2+l^2{}_3+l^2{}_4)}{I_z v_x}\dot{e}_\varphi + \frac{k_{soils}l_1+k_{soils}l_2a_{ss}}{I_z}\delta_1 \\ +\frac{M_z}{I_z} - \frac{k_{soils}(l^2{}_1+l^2{}_2+l^2{}_3+l^2{}_4)}{I_z v_x}\dot{\theta}_r \end{cases} \tag{17}$$

In trajectory tracking control, the transverse and longitudinal control of the vehicle is usually decoupled, and then controlled separately. In the governing equation of the above formula, the longitudinal velocity is taken as a known quantity to facilitate the linearization and discretization of the equation, and then the control quantity is solved.

Because of the above equation, the vector $\left[e_d, \dot{e}_d, e_\varphi, \dot{e}_\varphi\right]^T$ is selected as the state variable, $[\delta_1, M_z]^T$ as the control quantity, and $\dot{\theta}_r$ as the disturbance quantity. The above equation can be sorted into the following state space expression:

$$\dot{X} = AX + BU + DW \tag{18}$$

The lateral trajectory tracking model is not only related to the parameters of the vehicle itself, such as the distance from each axle to the center of mass, the moment of inertia of the vehicle body and the lateral force coefficient of wheel-terrain interaction.

3.2. *Time-varying parameters*

If the body rolls, the left and right wheels of each axle will increase the load on one side and reduce the load on the other side, that is the so-called load transfer. Assuming that the load transfer amounts of the front and rear of the vehicle are W_i, in the position plane of the four-axle wheels perpendicular to the longitudinal direction of the vehicle, and if no load transfer occurs between the front and rear axles at this time, the kinematics equation based on lateral motion is

$$\begin{aligned} \Delta W_1 &= \frac{a_y W_s}{d_1}\left(\frac{K_{\phi 1}h_s}{\sum_1^4 K_{\phi i} - W_s h_s} + \frac{l-l_1}{l}h_1\right) \\ \Delta W_2 &= \frac{a_y W_s}{d_2}\left(\frac{K_{\phi 2}h_s}{\sum_1^4 K_{\phi i} - W_s h_s} + \frac{l-l_2}{l}h_2\right) \\ \Delta W_3 &= \frac{a_y W_s}{d_3}\left(\frac{K_{\phi 3}h_s}{\sum_1^4 K_{\phi i} - W_s h_s} + \frac{l-l_3}{l}h_3\right) \\ \Delta W_4 &= \frac{a_y W_s}{d_4}\left(\frac{K_{\phi 4}h_s}{\sum_1^4 K_{\phi i} - W_s h_s} + \frac{l-l_4}{l}h_4\right) \end{aligned} \tag{19}$$

ΔW_i is the load transfer amount of each axle, $K_{\phi i}$ is the roll stiffness of each axle suspension, h_i is the height of the roll center of each axle from the ground, the wheelbase of each axle is d_i, W_s and h_s are the weight of the body and the distance between the center of mass of the vehicle and its roll axle.

Based on the Trucksim-Simulink co-simulation platform, steady-state rotation tests under different lateral acceleration conditions were carried out, and the soil lateral force proportional coefficient of each round was calculated in real time under different load transfer amounts, and the time-varying soil lateral force proportional coefficient matrix was obtained.

The state space equation with a constant proportional coefficient is rewritten as

$$\dot{X} = A_n X + B_n U + D_n W \tag{20}$$

In the formula,

$$A_n = \begin{pmatrix} 0 & 1 & 0 & 0 \\ 0 & a22 & a23 & a24 \\ 0 & 0 & 0 & 1 \\ 0 & a42 & a43 & a44 \end{pmatrix} B_n = \begin{pmatrix} 0 & 0 \\ b21 & 0 \\ 0 & 0 \\ b41 & b42 \end{pmatrix}, D_n = \begin{pmatrix} 0 \\ d2 \\ 0 \\ d4 \end{pmatrix}$$

$$a22 = -\frac{\sum_1^4 k_{soils,i,n}}{mv_x}, a23 = \frac{\sum_1^4 k_{soils,i,n}}{m}, a24 = -\frac{(k_{soils,1,n}l_1 + k_{soils,2,n}l_2 - k_{soils,3,n}l_3 - k_{soils,4,n}l_4)}{mv_x},$$

$$a42 = -\frac{(k_{soils,1,n}l_1 + k_{soils,2,n}l_2 - k_{soils,3,n}l_3 - k_{soils,4,n}l_4)}{I_z v_x}, a43 = \frac{(k_{soils,1,n}l_1 + k_{soils,2,n}l_2 - k_{soils,3,n}l_3 - k_{soils,4,n}l_4)}{I_z},$$

$$a44 = -\frac{\sum_1^4 k_{soils,i,n}l_i^2}{I_z v_x}, b21 = \frac{k_{soils,1,n} + k_{soils,2,n}a_{ss}}{m}, b41 = \frac{k_{soils,1,n}l_1 + k_{soils,2,n}l_2 a_{ss}}{I_z}, b42 = \frac{1}{I_z},$$

$$d2 = -v_x - \frac{(k_{soils,1,n}l_1 + k_{soils,2,n}l_2 - k_{soils,3,n}l_3 - k_{soils,4,n}l_4)}{m}, d4 = -\frac{\sum_1^4 k_{soils,i,n}l_i^2}{I_z v_x}$$

Where, $k_{soils,i,n}$ is the equivalent proportional coefficient of lateral force of wheel-terrain action of each axle under different lateral acceleration.

3.3. *MPC controller*

The discretized system model is obtained as

$$\begin{cases} x(k+1) = A_k x(k) + B_k u(k) + D_k w(k) \\ y(k) = C_k x(k) \end{cases} \tag{21}$$

To facilitate the consideration of actuator constraints and rate of change constraints, and to achieve gradual control input changes, the state variables are augmented to obtain new state variables as

$$\tilde{x}(k) = \begin{bmatrix} x(k) \\ u(k-1) \end{bmatrix} \tag{22}$$

Where $u(k) = u(k-1) + \Delta u(k)$ and $\Delta u(k)$ is the control input increment.

Therefore, Eq. (23) can be rewritten as

$$\begin{cases} \tilde{x}(k+1) = \tilde{A}_k \tilde{x}(k) + \tilde{B}_k \Delta u(k) + \tilde{D}_k w(k) \\ y(k) = \tilde{C}_k \tilde{x}(k) \end{cases} \tag{23}$$

By iterating the model, we derive the predictive model as

$$Y(k) = \Psi \tilde{x}(k) + \Theta \Delta U(k) + \Gamma \Phi(k) \tag{24}$$

where the output, control, and disturbance sequences of the system over the prediction horizon are represented as follows:

$$Y(k)=\begin{bmatrix} y(k+1) \\ y(k+2) \\ \vdots \\ y(k+N_P) \end{bmatrix} \quad \Delta U(k)=\begin{bmatrix} \Delta u(k) \\ \Delta u(k+1) \\ \vdots \\ \Delta u(k+N_C-1) \end{bmatrix} \quad \Phi(k)=\begin{bmatrix} w(k) \\ w(k+1) \\ \vdots \\ w(k+N_P-1) \end{bmatrix}$$

$$\Psi=\begin{bmatrix} \tilde{C}_k\tilde{A}_k \\ \tilde{C}_k\tilde{A}_k^2 \\ \vdots \\ \tilde{C}_k\tilde{A}_k^{N_P} \end{bmatrix} \quad \Theta=\begin{bmatrix} \tilde{C}_k\tilde{B}_k & 0 & \cdots & 0 \\ \tilde{C}_k\tilde{A}_k\tilde{B}_k & \tilde{C}_k\tilde{B}_k & \cdots & 0 \\ \vdots & \vdots & \ddots & \cdots \\ \tilde{C}_k\tilde{A}_k^{N_P-1}\tilde{B}_k & \tilde{C}_k\tilde{A}_k^{N_P-2}\tilde{B}_k & \cdots & \tilde{C}_k\tilde{A}_k^{N_P-N_C}\tilde{B}_k \end{bmatrix}$$

$$\Gamma=\begin{bmatrix} \tilde{C}_k\tilde{D}_k & 0 & \cdots & 0 \\ \tilde{C}_k\tilde{A}_k\tilde{D}_k & \tilde{C}_k\tilde{D}_k & \cdots & 0 \\ \vdots & \vdots & \ddots & \vdots \\ \tilde{C}_k\tilde{A}_k^{N_P-1}\tilde{D}_k & \tilde{C}_k\tilde{A}_k^{N_P-2}\tilde{D}_k & \cdots & \tilde{C}_k\tilde{D}_k \end{bmatrix}$$

The designed objective function is shown in the equation

$$\min_{\Delta U} J=\left\|\boldsymbol{Y}(k)-\boldsymbol{Y}_{ref}(k)\right\|_{\boldsymbol{Q}}^2+\left\|\boldsymbol{\Delta U}(k)\right\|_{\boldsymbol{R}}^2+\left\|\boldsymbol{U}(k)\right\|_{\boldsymbol{N}}^2 \tag{25}$$

In the objective function, the first term is used to represent the deviation between the output of the controlled state and the output of the reference state in the prediction time domain, that is, it reflects the tracking ability of the vehicle to the expected lateral displacement and heading Angle, and Q is the weight matrix of the state tracking target. The second term is used to indicate that in the control time domain, the control increment should be as small as possible, that is, it reflects the requirement for the control quantity to change slowly, and R is the weight matrix of the control quantity to change slowly. The third term is used to indicate that in the control time domain, the control quantity of the system should be as small as possible, and N is the weight matrix of this goal.

The control quantity of the system is a bridge Angle and direct yaw moment. To ensure the rationality of the calculation output of the controller, control increment constraint and control quantity constraint are set, as shown in the formula

$$\boldsymbol{\Delta U}_{min}(k)\le\boldsymbol{\Delta U}(k)\le\boldsymbol{\Delta U}_{max}(k) \tag{26}$$

$$\boldsymbol{U}_{min}(k)\le\boldsymbol{U}(k)\le\boldsymbol{U}_{max}(k) \tag{27}$$

3.4. Reference Wheel speed

The upper-layer parameter time-varying trajectory tracking controller calculates the front two-axle steering angles and additional yaw moment values for multi-axle vehicles to achieve trajectory tracking control. Based on prior research by the team, it was found that using speed control mode on soft terrain delivers superior performance compared to torque control mode. Therefore, this study proposes converting the desired torque of each wheel into desired wheel speed signals, which are then transmitted to the motor controller.

The torque distribution strategy based on rules is proposed in this section. The motor torque distribution strategy is shown as

$$\begin{cases} T_{il}=\dfrac{T_{des}}{8}-\dfrac{R}{\sum_1^4 B_i}M_z \\ T_{ir}=\dfrac{T_{des}}{8}+\dfrac{R}{\sum_1^4 B_i}M_z \end{cases} (i=f,r) \tag{28}$$

where, T_{il},T_{ir} are wheel motor torque for each axle. T_{des} is the desired driving torque of the vehicle. B_i and are track widths for each axle.

Considering the actuator's capacity and the limit of road adhesion coefficient on driving torque, the driving motor torques of each wheel are shown as

$$T_{ij} = min\left\{T^*_{ij}, T_{\max}, \mu F_{zij} R\right\} \quad (i = 1,2,3,4; j = l,r) \tag{29}$$

where, T^*_{ij} is the motor torque calculated by the torque distribution controller. $T_{\max}$ is the maximum torque of the motor. μ is the road adhesion coefficient and F_{zij} is the tire vertical load.

Based on the single-wheel model shown in Figure 6, the dynamic equation of wheel rotation is obtained:

$$J_w \dot{\omega} = T - F_x r_w = T - r_w F_Z \mu(\lambda) \tag{30}$$

where, J_w is the equivalent moment of inertia at the wheel, T is the wheel torque, F_x is the tire longitudinal force, F_z is the tire vertical load, and μ(λ) $\mu(\lambda)$ is the product of the normalized longitudinal slip stiffness and slip rate under the current road surface and the slip rate obtained based on the tire model.

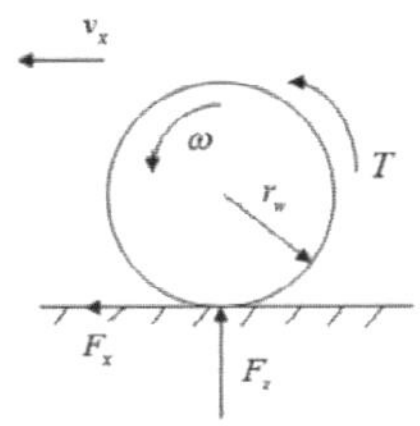

Figure 6. A dual closed-loop nested differential steering control method

4. Simulation

The vehicle parameters and soil mechanical properties are obtained as shown in Table 1, and the vehicle dynamics model is established in Trucksim. The longitudinal lateral force curve in the tire model is replaced by the longitudinal lateral force curve fitted based on the wheel-terrain action relationship, which is used to simulate the longitudinal lateral force response of the vehicle on the soft road surface. The simulation conditions were set as a longitudinal speed of 20km/h, and the test conditions were Double Lane Change Conditions. Simulink-Trucksim simulation platform was used to conduct simulation experiments.

Table 1. Vehicle parameters and soil mechanical properties

Parameter	Value	Unit	Parameter	Value	Unit
Vehicle load m	2500	Kg	cohesive deformation modulus k_c	102	$KN{\cdot}m^{-(n+1)}$
Yaw moment of inertia I_Z	1500	$Kg{\cdot}m^2$	frictional deformation coefficient k_φ	5301	$KN{\cdot}m^{-(n+2)}$
Wheel track b	1.485	m	sinkage exponent n	0.79	/
Wheel radius R	0.383	m	shear deformation modulus k_d	0.012	/
Height of center of mass h_g	0.644	m	soil cohesion c	1300	Kpa
Distance from each axle to the center of mass $l_i (i = 1,2,3,4)$	1.408/0.469/ 0.469/1.408	m	soil internal friction angle φ	31.1	deg

To verify the effectiveness of the control algorithm, the dynamic model control algorithm that considers the relationship between the wheel-terrain interaction and the control algorithm that does not consider the time-varying characteristics of the proportional coefficient of the lateral force of the wheel-earth action is compared. The simulation results are shown in Figure 7.

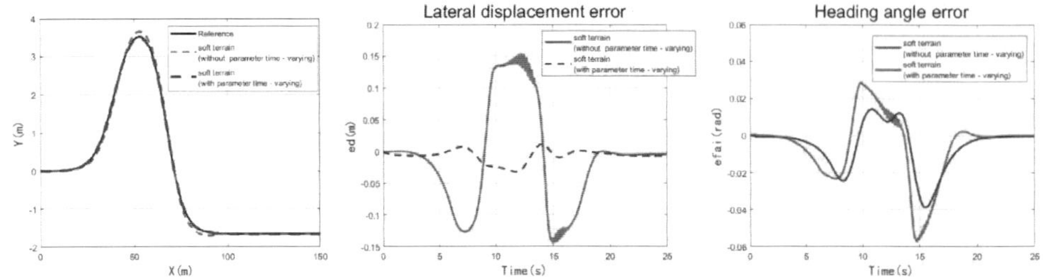

(a) (b) (c)

Figure 7. Simulation results of multi-axle vehicle trajectory tracking control

Figure 7 (a) shows the vehicle driving trajectory; Figure 7 (b) shows the lateral displacement error; Figure 7 (c) shows the heading Angle error. The simulation results show that the trajectory tracking error model that considers the time-varying parameters of the wheel-road relationship reduces the maximum lateral displacement error by 43%, The tracking accuracy of the heading angle is improved by 10.4%. The applicability and accuracy of the algorithm in soft terrain track tracking control are proved.

5. Conclusion and discussion

This study focuses on multi-axle vehicles operating on soft terrain, incorporating wheel-terrain interaction relationships to explore expressions for longitudinal and lateral forces under such conditions. A multi-axle vehicle dynamic model tailored for soft terrain is developed, with explicit consideration of the impact of load transfer on lateral force calculations across individual wheels. These analyses provide a foundation for the subsequent design of control algorithms. To address tracking accuracy degradation caused by model errors, a time-varying parameter matrix is designed for the dynamic model, enabling precise computation of soil lateral force proportionality coefficients for each wheel. Comparative experiments are conducted using a Simulink-Trucksim co-simulation platform. Simulation results demonstrate that the proposed control algorithm, which accounts for wheel-terrain interactions and time-varying characteristics on soft terrain, significantly enhances trajectory tracking accuracy. Both lateral tracking error and heading angle error outperform conventional models that neglect these time-varying properties. However, the current framework does not incorporate the intrinsic configuration of soft terrain, thus failing to fully simulate critical features such as terrain deformation and stress relaxation in practical simulations. In the future, it is necessary to improve the co-simulation platform to increase the simulation accuracy.

Acknowledgments

This work is supported by the National Natural Science Foundation of China under Grant No.52372394 and the Fundamental Research Funds for the Central Universities.

References

[1] Zhang Z, Ma X J, Liu C G 2022 Dual-steering mode based on direct yaw moment control for multi-wheel hub motor driven vehicles: Theoretical design and experimental assessment. Defence Technology, 18(1), 49-61

[2] Huang B, Fu W, Yuan Z 2024 Trajectory Tracking for Autonomous Multi-Axle Vehicles Based on Linear Quadratic Regulator No. 2024-01-5023 SAE Technical Paper

[3] Jjagwe P 2023 Modeling of tire-to-soil interaction on soft artificial soil Iowa State University

[4] Alkhalifa N 2023 Systematic measurement of 3D tire-soil contact on an artificial soil Iowa State University

[5] Han B, Dong H, Du X 2023 Research on interaction law of longitudinal/lateral slip of mountain tractor based on tire-soil mode Proceedings of the Institution of Mechanical Engineers, Part D: Journal of Automobile Engineering

[6] Vieira D, Orjuela R, Spisser M 2021 Longitudinal vehicle control based on off-road tire model

for soft soil applications IFAC-PapersOnLine 54(10): 304-309
[7] Vieira D, Vie A, Orjuela R 2023 A Robust Lateral Control Architecture for Off-Road Vehicle Guidance on Deformable Soils Electronics 12(11): 2395
[8] Yu Z, Han P, Leng B 2023 An Emergency Collision-Avoidance Controller for Distributed Drive Autonomous Vehicles Based on AFS and DYC 2023 35th Chinese Control and Decision Conference (CCDC). IEEE 175-181
[9] Liu C, Liu H 2024 Learning-based Sliding Mode Predictive Trajectory Tracking and Stability Control for Autonomous Vehicle in Unstructured Environments Journal of Mechanical Engineering 60(10): 399-412
[10] Zhang Z, Ma XJ, Liu CG, Wei S G 2022 Dual-steering mode based on direct yaw moment control for multi-wheel hub motor driven vehicles: Theoretical design and experimental assessment. Defence Technology, 18(1), 49-61
[11] Yu ZP, Zhang RX, Xiong L 2017 Dynamic control for unmanned skid-steering vehicle with conditional integrators. Journal of Mechanical Engineering, 53(14), 29-38
[12] Leng B, Xiong L, Yu ZP, Sun K, Liu M 2020. Robust variable structure anti-slip control method of a distributed drive electric vehicle IEEE Access 8 162196-162208

Journal of Physics: Conference Series **3024** (2025) 012034 doi:10.1088/1742-6596/3024/1/012034

CSEA: Confidence-guided Semantic-Enhanced Alignment for Online Test-Time Adaptation

Tengfei Zhang[1], Ya Wu[1], Yuan Huang[1*], Zhengfa Liu[2], Dequan Zeng[3]

[1]Artificial Intelligence R&D Center, CNNC Equipment Technology Research (Shanghai) Co., Ltd, Shanghai 201899, P.R. China

[2] School of Electrical and Electronic Engineering, Shanghai Institute of Technology, Shanghai 201418, P.R. China

[3] School of Mechatronics and Vehicle Engineering, East China Jiaotong University, Nanchang, Jiangxi Province, 330013, P.R. China

[*]Corresponding author's e-mail: huangyuan@cnncmail.com

Abstract. Online test-time adaptation (OTTA) is a more general method for accomplishing domain adaptation in real-world scenarios because it just needs a pre-trained source model to adapt the mini-batch target online. One popular approach is to adapt the model with empirical risk minimization using generated pseudo-labels. However, it can fail to learn optimal feature space due to the noisy pseudo-labels under domain shift. To address this restriction, we propose a novel online test-time adaptation technique called Confidence-guided Semantic-Enhanced Alignment (CSEA) that separates the mini-batch target into distinct subsets based on the pseudo-labels confidence and applies tailored constraints to each subset for reliable target structure mining. We first divide the mini-batch target data into confident and non-confident based on an online updated memory bank for each class. Then, we customize the alignment strategy to fit each subset best. Specifically, we propose a Re-weighted Local Clustering for the confident subset to learn the local semantic structure information of unlabeled target. Further, we propose a Semantic-enhanced Alignment that applied to the non-confident subset to guide the model to learn additional discriminative semantic knowledge. Extensive experiments on three OTTA benchmarks indicate effectiveness of the proposed CSEA.

1. Introduction

Unsupervised Domain Adaptation has significantly enhanced model generalization by addressing the problem of distribution shifts, considering the difference between training (source) and testing (target). Specifically, UDA assumes that both source and target domains are available but have different distributions. Learning a classifier from the source domain and adapting it to the target domain for prediction is the goal of UDA. However, such assumptions aren't always accurate in numerous real-world situations. For instance, the case where the source data are inaccessible during testing owing to privacy or business [1], [2], and the case where that target data are available at the test-time [3], [4]. To solve this issue, online test-time adaptation (OTTA) [3-5] was developed. In comparison, OTTA simply requires the model that has been trained from the source domain and adaptations it to the desired domain in an online way, allowing for a single access to each mini-batch.

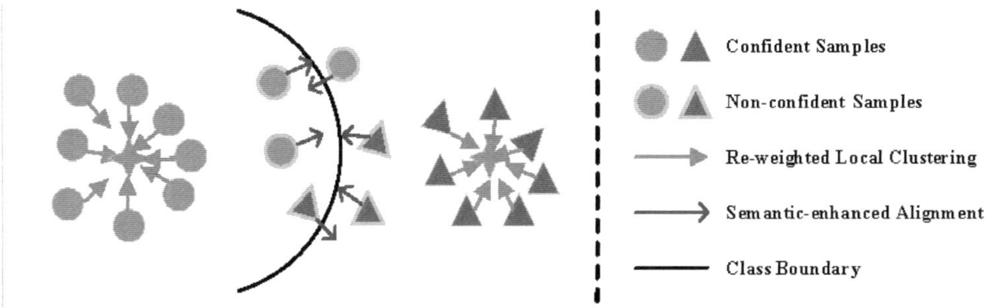

Figure 1: A toy representation of our semantic-enhanced feature alignment methodology. The mini-batch target samples are separated into two subgroups: the confident set and the non-confident set. CSEA learns the semantic-enhanced features through Re-weighted Local Clustering and Semantic-enhanced Alignment. Among them, the non-confident samples enhance the cluster's inner density and inter-separability, enabling the reliable alignment of two domains with a large shift.

One popular OTTA approach is to utilize the generated pseudo-labels at test time to fine-tune the adapted model parameters with the empirical risk minimization (ERM), e.g., entropy regularization [4]. However, under a shift, the pseudo-labels are inevitably inaccurate under a domain shift [6]. For instance, when pseudo-label noise rises with a large domain gap, resulting in a continual reduction in the test-time cumulative accuracy. To mitigate the effects of loud pseudo-labels, some OTTA methods propose to update the model by taking into account only the reliable pseudo-labels. T3A [3] proposes a pseudo-prototypical classifier that is built on a support set consisting of a reliable pseudo-labeled representation above an entropy threshold. The distance between the sample and the pseudo-prototype is used to anticipate the target data. Similar to T3A, TSD [5] filters noisy labels for prototype generation using both entropy and consistency filters.

Although the above methods work exceptionally well, they only learn discriminative knowledge from reliable samples and fail to acquire semantics of unreliable samples on clusters, which can lead to poor feature space structure. However, in high-dimensional space, the nearest-neighbor consistency of representations can be broken [7], [8]. Only a small number of neighbors will have the right labels, for example, numerous features gather in low-density locations due to a domain gap. In solution to these questions, we propose Confidence-guided Semantic-Enhanced Alignment (CSEA) (refer to Fig. 1), an efficient method that synchronously takes into account confident and non-confident samples to generate more reliable test-time unlabeled data clustering. To take advantage of the target domain structure information and accomplish flexible adaption for various data qualities, CSEA divides the min-batch into confident and non-confident, then creates various regularization techniques to acquire the structure.

In CSEA, we first divide the mini-batch target data using an entropy filter into two subsets: confident and non-confident. In contrast to the non-confident subsets, the confident subsets indicate that the samples have low-entropy predictions. The class prototypes are calculated, and the memory support sets bank is updated using the confident samples. By matching the non-confident samples with the prototypes and promoting local consistency for the confident samples, the mining of target structural information is then successfully enhanced. To improve the local semantic credibility, we specifically propose a re-weighted clustering method based on the K-nearest neighbors on the confident samples. We use a semantic-enhanced alignment loss to non-confident samples in order to aid the model in learning more discriminating semantic information. The main contributions are:

- We show a novel OTTA model CSEA that uses confidence-based subsets (confident *vs.* non-confident) to learn the enhanced semantic representation for robust online target clustering.
- We propose distinct training strategies for subsets. A re-weighted clustering technique is proposed for confident samples to strengthen the local semantic credibility, and a semantic-enhanced alignment loss for non-confident samples to guide the model towards target structure mining.
- We evaluate our proposed method CSEA on several benchmark datasets, showing that it can consistently outperform existing methods.

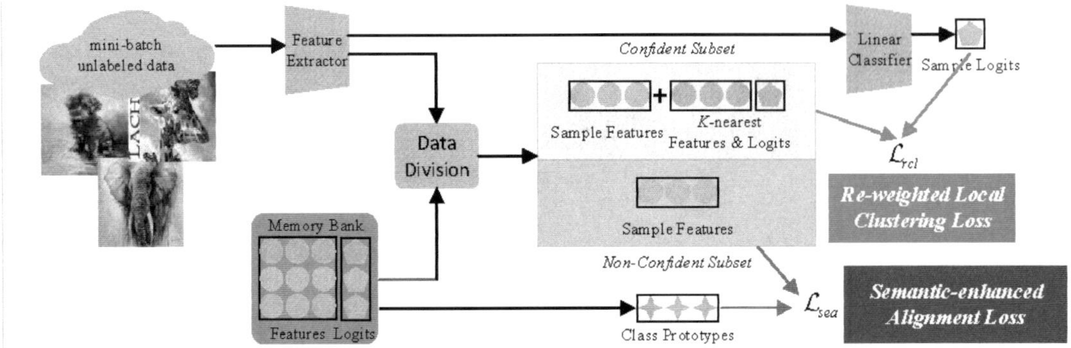

Figure 2. Overview of the proposed CSEA. Green lines denote the data stream for the confident subset and the red lines for the non-confident subset. $\mathcal{L}_{rc\ell}$: Re-weighted Local Clustering loss. $\mathcal{L}_{sea}$: Semantic-enhanced Alignment loss.

2. Related Work

With every mini-batch of unlabeled data, OTTA updates a previously trained source model online; each mini-batch is displayed just once. Many studies have been focused on the OTTA problem. Tent [5] minimizes generalization error on shifted data by using a test-time entropy minimization technique. Each sample is categorized by T3A [3] according to how far away it is from the class pseudo-prototype, where the pseudo-prototype is adjusted during test time based on a support set. To increase the effectiveness and performance on OOD samples, a sample-adaptive entropy minimization loss is developed by EATA [9] for the reliable and non-redundant samples. EATA incorporates a Fisher regularizer to further reduce the forgetting issue by preventing large model parameters from changing significantly. LAME suggests a Laplacian regularization to modify only the outputs, as opposed to the previously mentioned approach that modifies the parameters of a pre-trained model. Some works additionally make advantage of Batch Normalization Calibration [10], Entropy Minimization [11], Pseudo-labeling [5], Consistency Regularization [12]. Readers can see the recent surveys [1] for a complete list.

3. Preliminaries

The target domain samples in OTTA are only accessible online and lack labels. Adaptation aims to use a pre-trained model learned from the labeled source domain to predict the labels of the target domain. Specifically, using a well-trained model $f = g \circ h$ from the source data $\mathcal{D}_S$, we aim to adapt f to the unlabeled target domain $\mathcal{D}_T$ for an image recognition task. g refers to the feature extraction network, and h refers to the linear classifier. During adaptation/testing, the initialized f are updated using mini-batches of the online data stream $\mathcal{X}_n$ sampled from $\mathcal{D}_T$. $\mathcal{X}_n = \{\mathbf{x_1}, \mathbf{x_2}, \ldots, \mathbf{x_i}, \ldots, \mathbf{x_n}\}$ and i is the mini-batch index. $\mathcal{D}_S$ and $\mathcal{D}_T$ have different distributions. Given a target sample x_i, the feature representation can be denoted as $z_i = g(x_i)$, and the output logits of the classifier h can be referred to as $p_i = h(z_i)$. The pseudo-label can be generated using self-training-based methods [13], [14] for x_i : $\hat{y}_i = \underset{c}{\operatorname{argmax}}(p_i)^c$, where $\hat{y}_i$ is the assigned class label. $(p_i)^c$ is the c-th element of p_i.

4. Method

Fig. 2 shows the CSEA pipeline as a whole. In this section, we go over the details of this model.

4.1. Distinguishing Confident and Non-confident Samples

4.1.1. Target Domain Prototypes Generation. Following [3], [5], We keep logits p_i and feature representations z_i in a memory bank. The linear classifier's weights are used to initialize the memory bank. When the target sample arrives, the sample embedding z_i and logits p_i are added to the memory bank $\mathcal{B}$. In the mini-batch, same process is repeated for every sample. Thereafter, the updated memory bank $\mathcal{B}$ is used to compute target domain prototypes. Specifically, the memory bank $\mathcal{B}$ is updated:

$$\mathcal{B}_n^c = \begin{cases} \mathcal{B}_{n-1}^c \cup \{z_i, p_i\} & \text{if } \hat{y}_i = c \\ \mathcal{B}_{n-1}^c & \text{else} \end{cases}, \tag{1}$$

where $\mathcal{B}_n^c$ denotes the updated memory bank for the class c. n denotes the n-th mini-batch target data. With the updated memory bank, the prototype for each class could be formulated as:

$$o^c = \frac{1}{|\mathcal{B}^c|} \sum_{z \in \mathcal{B}^c} z, \tag{2}$$

where o^c is the centroid/prototype of $\mathcal{B}^c$. Eq. (1) uses all pseudo-labels to update the memory bank $\mathcal{B}$, which may be incorrect for prototype computation in Eq. (2) due to the wrong pseudo-labels. To avoid this issue, we use the prediction entropy to filter unreliable pseudo-labeled data. In particular, only the feature representations with the top-M lowest entropy are added to the memory bank for each class. And the $\mathcal{B}_n^c$ in Eq. (1) can be rewritten as:

$$\mathcal{B}_n^c = \{z \mid z \in \mathcal{B}_n^c, H(p \mid z) \le a^c\} \tag{3}$$

where a^c is the M -th lowest entropy within $\mathcal{B}_n^c$, and M is a hyperparameter. $H(p|z) = -\sum \sigma(p|z)\log_{\sigma(p|z)}$ is the prediction entropy and $\sigma(\cdot)$ is the softmax function. After that, the class prototypes in Eq. (2) are calculated based on the $\mathcal{B}_n^c$ in Eq. (3).

4.1.2. Dynamic Data Division. We employ the memory bank $\mathcal{B}$ to adaptively divide the mini-batch data in x_i into confident subsets *CON* and non-confident sets *NON*. As believed in [7], the class's learning impact may be shown in the sample-level hit rate. In order to show the learning status at the class level, our main premise is that the data division should be connected to the prediction confidence of the unlabeled data on various classes. In particular, classes are thought to struggle with acquiring local structural information if fewer samples meet a threshold of prediction confidence. Furthermore, while the model is continuously enhanced throughout training, the threshold ought to be raised gradually. With the hyperparameter M, the memory bank $\mathcal{B}$ can be served as the sample-level hit rate. When a target embedding z_i falls into a memory bank $\mathcal{B}^c$, it can be grouped into the confident subset *CON*, otherwise into the non-confident subset *NON*. Therefore, based on the memory bank $\mathcal{B}$, the mini-batch target samples can be grouped into *CON* as follows:

$$CON = \{z \mid z \in \mathcal{B}_n^c, c \in \{1, 2, \ldots, c, \ldots, C\}\}, \tag{4}$$

and the remaining mini-batch target data, *i.e.*, $\mathrm{NON} = x_n \backslash \mathrm{CON}$, are the non-confident samples. We adapt learning algorithms for *CON* and *NON* subsets in the following.

4.1.3. Test Time Self-Distillation. We add self-distillation loss after [5] to ensure the homogeneity of the current mini-batch of samples. In particular, we reduce the gap between the outputs of prototype-based and linear classifiers in the manner described below:

$$\mathcal{L}_{tsd} = -\frac{1}{|\mathbf{x}_n|} \sum_{i=1}^{|\mathbf{x}_n|} \sigma(p_i) \log\ y_i, \tag{5}$$

where $|x_n|$ is the mini-batch x_n number. y_i is the prototype-based classification:

$$y_i^c = \frac{\exp(\mathrm{sim}(z_i, o^c))}{\sum_{c'=1}^{C} \exp(\mathrm{sim}(z_i, o^{c'}))}, \tag{6}$$

where $\mathrm{sim}(z_i, o^c)$ indicates how comparable between z_i and o^c are cosine similarity.

4.2. *Confidence-Guided Enhanced Semantic Learning*

According to [7], poor feature space structure may result from applying neighbors consistency regularization. On the one hand, the noisy pseudo-labels under domain shift may cause misalignment and the formation of false clusters. The cross-label risk will also be increased since the neighbors in the low-density zones (*i.e.*, non-confident samples) break the local consistency assumption. We adapt learning algorithms to various data characteristics in order to solve these issues.

4.2.1. Re-weighted Local Clustering for Confident Samples. Motivated by the notion that the source model's target features have developed certain semantic structures, we investigate the target data's local semantic structure information using the confident subset. In particular, we promote local clustering using K-nearest neighboring features in the memory bank $\mathcal{B}$. Formally, we align the logits p_j of a confident embedding z_j to match its closest neighbors' logits in the latent space. To reduce the noisy pseudo-labels impact, we re-weigh the alignment loss:

$$\mathcal{L}_{rcl} = \frac{1}{|CON|}\sum_{j=1}^{|CON|}\sum_{m=1}^{K} wKL(p_j \,\|\, p_m), \tag{7}$$

where K represents the K-nearest neighbor set of z_j, *KL* is the Kullback-Leibler (KL) divergence, and $|CON|$ is the number of *CON*. $w = \text{sim}(\exp(z_j, z_m))$ is z_j 's the cosine similarity to its neighbors z_m.

4.2.2. Semantic-enhanced Alignment for Non-confident Samples. Unlike the confident embedding, the non-confident embeddings are situated in the low-density zone and are distant from every other point. Instead of using a signal cluster, we suggest using a semantic-enhanced loss (SEL) to align non-confident samples with all the prototypes for mining potential relationships between non-confident samples. Formally, as the prototypes are obtained by the cosine similarity (Eq. (6)), we minimize the cosine distance between the non-confident embedding and all prototypes:

$$\mathcal{L}_{sea} = \frac{1}{|NON|}\sum_{j=1}^{|NON|}\sum_{c=1}^{C} \frac{\exp(\text{sim}(z_j, o^c))}{\sum_{c'=1}^{C}\exp(\text{sim}(z_j, o^{c'}))}, \tag{8}$$

where $|NON|$ is the number of *NON*. By taking into consideration all possible semantic relationships, SEL can further guide the model to acquire more discriminative semantic knowledge. The following is a summary of the overall loss function for CSEA:

$$\mathcal{L} = \mathcal{L}_{tsd} + \lambda_0\mathcal{L}_{rcl} + \lambda_1\mathcal{L}_{sea}, \tag{9}$$

where λ_0 and λ_1 are the trade-off parameters.

5. Experiment

5.1. Experimental Setup

(i) Dataset. In a manner similar to [6], we implement the test experiments on three common domain adaptation datasets: PACS, OfficeHome, and VLCS. ***(ii) Training Setup.*** The foundation for feature extraction is made up of the ResNet-18 and ResNet-50. We split the entire source domain into 80%/20% training and validation images, adhering to the leave-one protocol for source training (the pre-trained model). We employ the Adam optimizer with $lr = 5e^{-5}$ as the learning rate. For test-time adaptation, we use the Adam optimizer with a learning rate $lr = e^{-4}$ and the batch size is 128. ***(iii) hyperparameters.*** CSEA has four primary hyperparameters: M for deciding the number of supports to restore for each class in the memory bank $\mathcal{B}$, K for selecting the number of neighbors for confident, λ_0 and λ_1 for weighting the alignment losses. For simplicity, the four primary hyperparameters are fixed for all test experiments. Specifically, we set $M = 100$, $K = 3$, $\lambda_0 = 1.0$, $\lambda_1 = 10$.

5.2. Quantitative Analysis

The performance comparison of our suggested approach, CSEA, with the baselines on three distinct datasets using the backbone ResNet-18 and ResNet-50 is displayed in Table I. The baseline values are referred to from [5]. The average outcomes of three trials using various random seeds are presented for

CSEA. From Table I, we can observe that the proposed CSEA achieves the best average performance of 76.69%/79.62% over the baselines and achieves 88.10%/91.87%, 65.90%/69.71%, and 76.07%/77.29% performance for each dataset with RestNet-18/50. Specifically, CSEA achieves +1.44%/2.09% performance gains compared to the second-best method TSD. It can be attributed to the use of non-confident samples for target semantic structure exploration. TSD enforces all the min-batch samples to remain consistent with the neighbors and ignores that the non-confident samples are low-density regions, where there is no nearest neighbor support. This explicitly demonstrates the necessity of customizing different learning strategies for confident and non-confident.

Table 1. Performance comparison (%) among different methods (best in bold).

Method	PACS	OfficeHome	VLCS	Avg(% ↑)
		ResNet-18/ ResNet-50		
ERM [15]	82.07/84.59	63.12/67.37	72.75/74.01	72.65/75.32
BN [16]	82.82/85.03	62.30/66.10	64.31/64.78	69.81/71.97
Tent [4]	84.92/87.48	63.75/67.96	67.36/69.20	72.01/74.88
SHOT-TM [17]	82.55/85.50	63.42/67.39	64.90/65.23	70.29/72.71
T3A [3]	83.50/86.04	64.25/68.29	73.03/73.98	73.59/76.10
EATA [9]	82.70/85.04	62.46/66.21	64.35/64.79	69.84/72.01
TSD [5]	87.32/89.41	64.83/68.67	73.61/74.52	75.25/77.53
ours (CSEA)	**88.10/91.87**	**65.90/69.71**	**76.07/77.29**	**76.69/79.62**

Table 2. ResNet-50 performance comparison (%) of various variations on PACS (best in bold, second in underline).

TsD	RLC	SeA	Art	Cartoon	Photo	Sketch	Avg(% ↑)
√	(baseline)		82.68	<u>89.72</u>	<u>98.62</u>	78.50	87.38
√	√		<u>93.26</u>	89.51	98.56	84.70	<u>91.51</u> (+4.13↑)
√	√	√	**93.41**	**90.06**	**98.62**	**85.37**	**91.27** (+4.19↑)

5.3. Ablation Study

Using the PACS dataset and ResNet-50, we also do an ablation research to investigate the effects of the various CSEA components. The ablation study will take place in these settings: The loss function corresponding to each component of the model is progressively added. Among the elements are: Test Time Self-Distillation (TsD), Re-weighted Local Clustering (RLC), and Semantic-enhanced Alignment (SnA). Table II presents the findings. The baseline refers to only TsD loss to train the model, which ignores the different data disparities, such as confident and non-confident. As expected, the average performance of CSEA gradually improves with the addition of the right components. In particular, applying the Re-weighted Local Clustering alignment to the confident subset results in an average accuracy improvement of +4.13% when compared to the baseline. Moreover, the semantic-enhanced alignment for non-confident can further yield additional performance improvements of +0.36%, demonstrating that the non-confident successfully mines the target semantic structure information. Interesting, we find that a small performance degradation occurs in Cartoon and Photo domains when only the RLC loss is introduced, with 89.51% *vs.* 89.72% and 98.56% *vs.* 98.62%. One possible reason is the pseudo-label noise. The best average accuracy of 91.87% and an improvement of +4.49% are obtained by integrating all those components, proving their necessity and efficacy.

6. Conclusion

In this paper, we present Confidence-guided Semantic-Enhanced Alignment (CSEA), a novel online test-time domain adaptation technique that aims to accomplish effective feature clustering while accounting for various data disparities. Based on an adaptive memory bank, CSEA divides the mini-batch data into confident and non-confident subsets. The confident subset is aligned with the K-nearest neighbors with a cosine similarity-based weight, encouraging positive clustering. Further, the non-confident subsets are aligned with all prototypes, learning the potential target structure while combating spurious clustering. The efficacy of the suggested CSEA is therefore shown by comprehensive tests on three popular OTTA benchmarks: PACS, Office-Home, and VLCS.

Acknowledgments

This paper was supported by the Ganpo Talent Support Program-Leading Academic and Technical Personnel in Major Disciplines of Jiangxi Province (Grant No. 20232BCJ23091), the National Natural Science Foundation of China (Grant No. 52462053), and the Natural Science Foundation of Jiangxi Province (Grant No. 20232BAB214092).

References

[1] J. Liang, R. He, and T. Tan, "A comprehensive survey on test-time adaptation under distribution shifts," International Journal of Computer Vision, vol. 133, no. 1, pp. 31–64, 2025.

[2] S. Qu, T. Zou, L. He, F. Ro¨hrbein, A. Knoll, G. Chen, and C. Jiang, "Lead: Learning decomposition for source-free universal domain adaptation," in Proceedings of the IEEE/CVF Conference on Computer Vision and Pattern Recognition (CVPR), June 2024, pp. 23 334–23 343.

[3] Y. Iwasawa and Y. Matsuo, "Test-time classifier adjustment module for model-agnostic domain generalization," Advances in Neural Information Processing Systems, vol. 34, pp. 2427–2440, 2021.

[4] D. Wang, E. Shelhamer, S. Liu, B. Olshausen, and T. Darrell, "Tent: Fully test-time adaptation by entropy minimization," arXiv preprint arXiv:2006.10726, 2020.

[5] S. Wang, D. Zhang, Z. Yan, J. Zhang, and R. Li, "Feature alignment and uniformity for test time adaptation," in Proceedings of the IEEE/CVF Conference on Computer Vision and Pattern Recognition (CVPR), June 2023, pp. 20 050–20 060.

[6] M. Jang, S.-Y. Chung, and H. W. Chung, "Test-time adaptation via self-training with nearest neighbor information," arXiv preprint arXiv:2207.10792, 2022.

[7] G. Wang, C. Ding, W. Tan, and M. Tan, "Decoupled prototype learning for reliable test-time adaptation," IEEE Transactions on Multimedia, 2025.

[8] S. Teng, W. Liu, L. Teng, Z. Zheng, and W. Zhang, "Joint marginal and central sample learning for domain adaptation," World Wide Web, vol. 27, no. 5, p. 55, 2024.

[9] S. Niu, J. Wu, Y. Zhang, Y. Chen, S. Zheng, P. Zhao, and M. Tan, "Efficient test-time model adaptation without forgetting," in International conference on machine learning. PMLR, 2022, pp. 16 888–16 905.

[10] S. Ioffe, "Batch renormalization: Towards reducing minibatch dependence in batch-normalized models," Advances in neural information processing systems, vol. 30, 2017.

[11] Z. Liu, G. Chen, Z. Li, Y. Kang, S. Qu, and C. Jiang, "Psdc: A prototype-based shared-dummy classifier model for open-set domain adaptation," IEEE Transactions on Cybernetics, vol. 53, no. 11, pp. 7353–7366, 2022.

[12] S. Qu, G. Chen, J. Zhang, Z. Li, W. He, and D. Tao, "General class-balanced multicentric dynamic prototype pseudo-labeling for source-free domain adaptation," International Journal of Computer Vision, pp. 1–22, 2025.

[13] L. Jiang and T. Lin, "Test-time robust personalization for federated learning," arXiv preprint arXiv:2205.10920, 2022.

[14] Q. Wang, O. Fink, L. Van Gool, and D. Dai, "Continual test-time domain adaptation," in Proceedings of the IEEE/CVF Conference on Computer Vision and Pattern Recognition (CVPR),

June 2022, pp. 7201–7211.

[15] V. N. Vapnik, "An overview of statistical learning theory," IEEE transactions on neural networks, vol. 10, no. 5, pp. 988–999, 1999.

[16] S. Schneider, E. Rusak, L. Eck, O. Bringmann, W. Brendel, and M. Bethge, "Improving robustness against common corruptions by covariate shift adaptation," Advances in neural information processing systems, vol. 33, pp. 11 539–11 551, 2020.

[17] J. Liang, D. Hu, and J. Feng, "Do we really need to access the source data? source hypothesis transfer for unsupervised domain adaptation," in International conference on machine learning. PMLR, 2020, pp. 6028–6039.

Research on the Key Technology of Ground-Air Inspection Robot for an explosive depot

Fusheng Tan[1,2], Zipeng Yu[2], Zerun Zhou[2], Liguo Zhang[1], Zhonggang Xiong[3*]

[1]Ordos Research Institute of Liaoning Technical University, Ordos 017004, China

[2]School of Electrical and Control Engineering, Liaoning Technical University, Huludao 125105, China

[3]Guangxi Key Laboratory of Special Engineering Equipment and Control, Guilin University of Aerospace Technology, Guilin 541000, China

[*]Corresponding author's e-mail: xzglsl2013@163.com

Abstract. This paper proposes an intelligent inspection system integrating an all-terrain hexapod robot and a UAV for explosive depot surveillance. The system realizes intelligent inspection in a complex terrain environment through multi-sensor fusion perception, positioning and navigation, cooperative communication between UAVs and ground robots, and other technologies.

1. Introduction

Managing dangerous explosive storage is an essential part of the safe production of blasting enterprises. Due to the strong demand for blasting operations in open-pit mines in Ordos, the number of civil explosive storage warehouses has increased. As of March 2021, Inner Mongolia has a total of 329 production coal mines, including 159 open-pit mines with a capacity of 439 million tons/year. The Ordos coal seam is shallow, and the open-pit mine accounts for a high proportion. The Heidaigou open-pit mine in Zhungeer Banner is the largest in the country. The stripping operation of an open-pit mine needs blasting, so it is very important to control the dangerous explosives, such as explosives and detonators. In 2009, the Ministry of Public Security issued the " Civil Explosives Storage Security Prevention Requirements " (GA837-2009), which made detailed provisions on the security prevention of dangerous explosives storage (civil defense, physical defense, technical defense, dog defense). Based on the concept of " intelligent blasting " proposed by academician Xuguang Wang[1], this paper introduces intelligent robot robots to the inspection of dangerous explosive storage, and discusses the inspection system of dangerous explosive storage of intelligent robots to reduce the exposure of personnel in these high-risk areas, thus reducing the risk of accidents to personnel injury. With the rapid development of robot technology, domestic oil and gas, chemical warehouses, and UHV environments have also begun to use robots for inspection. The United States uses explosion-proof robots for bomb removal and mine clearance tasks[2]. Therefore, this paper proposes a collaborative inspection system based on all-terrain robots and UAVs, which aims to improve the inspection efficiency and safety of hazardous explosive storage.

2. Research status at home and abroad

In 2022, Greek scholars proposed the use of mobile robots for warehouse inspection, focusing on the rapid identification and early warning of abnormal conditions through computer vision technology[3]; in 2021, Bao Jiatong of Yangzhou University proposed a robot experimental scheme for the safety inspection of hazardous chemical warehouses. He developed a ROS-based scheduling system using a two-wheel differential mobile chassis, a single-line lidar, and a night vision and thermal imaging camera with a pan-tilt. The scheme has done simple experiments with and without light, but it is still mainly experimental at present. The functions of dangerous goods detection and abnormal sound need to be improved[4].

Researchers in the GRASP laboratory of the University of Pennsylvania used multiple fixed-wing drones, multiple unmanned vehicles, and one airship to form a ground-air multi-robot system. The system was used to conduct urban crowd monitoring and search experiments. The situation awareness, collaborative search, recognition and positioning, and three-dimensional map construction of the ground-air multi-robot system were studied[5]. Lorenzo Rosa et al. from the University of Rome studied the multi-task coordination problem of ground-air multi-robot systems. In the designed system, the drone is equipped with a camera for environmental information perception and ground unmanned vehicle positioning, and the unmanned vehicle is used to perform tasks[6].

The high-flying team of Zhejiang University in China proposed ' Swarm of micro flying robots in the wild ' in 2022. The micro and fully autonomous aerial robot developed by the team can realize cluster flight according to the limited information provided by airborne sensors in a highly chaotic field environment. The trajectory planning algorithm on the robot can independently plan the flight path, and meet the requirements of flight efficiency, obstacle avoidance, collision avoidance between robots, dynamic feasibility, cluster coordination and so on[7].

At present, the direction of dangerous goods storage inspection robots is mainly based on the four-wheel chassis scheme of a flat road surface. With the help of a visual module, abnormal intrusion, fire point detection and response can be tried in some scenarios, which is still in its infancy.

3. Overall design of the robot

3.1. Research objectives

Hazardous explosive storage is usually located far away from key facilities and residential areas, and the building structure is mostly planned according to the original terrain, as shown in Figure 1. This kind of site selection and construction method often leads to uneven pavement inside the reservoir area, and some areas have complex terrain such as ravines, which puts forward higher requirements for the adaptability of inspection robots.

Figure 1. Civilian explosives storage area

3.2. Integral structure

Based on the special needs of the dangerous explosive storage, an intelligent inspection strategy of ground-air coordination is proposed. An AI-powered inspection system has been developed, combining an all-terrain ground robot with a UAV for comprehensive, cross-domain monitoring.

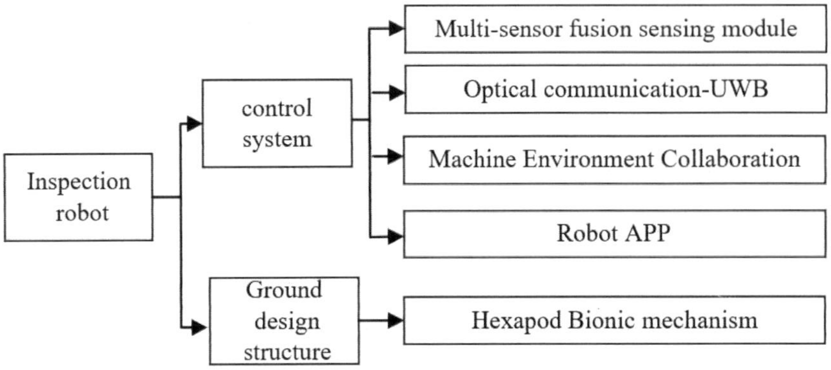

Figure 2. Whole frame

3.2.1. Base template structure. The ground robot adopts the bionic hexapod chassis design (Figure 3), adopts the hexapod bionic mechanism, and is based on the design scheme of the bird-catcher spider in the biological world. The six feet are all designed with independent suspension, which can complete the 30-degree climbing, ensure the independent driving of the foot without losing stability, and incorporate the explosion-proof scheme.

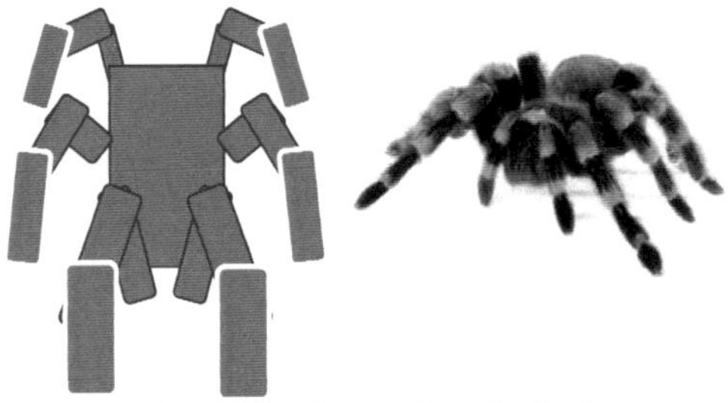

Figure 3. Bionic chassis design

3.2.2. Robot structure design. The robot chassis adopts the bionic leg-foot structure, and integrates the vehicle computer, odometer, 2D laser radar and RGBD camera. The robot is equipped with a 2-DOF pan-tilt mounted with a color night vision camera and a thermal imaging camera. The software module design is based on the ROS open-source system development, and the overall architecture adopts a B/S architecture. The hardware is shown in Figure 4.

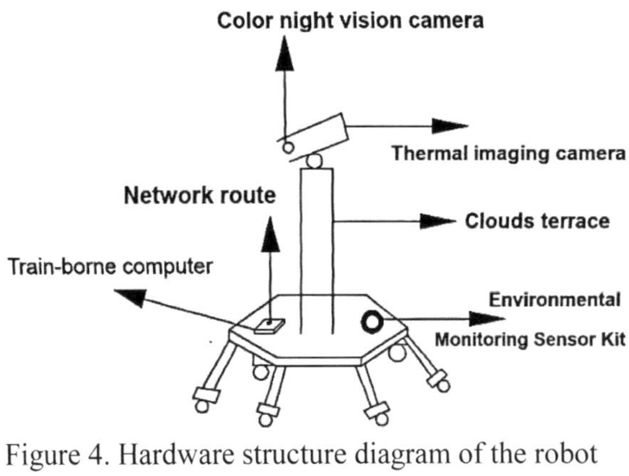

Figure 4. Hardware structure diagram of the robot

3.2.3. Ablation Study. In order to verify the contribution of the hexapod bionic leg-foot structure design, we replaced it with a traditional four-wheel chassis in the same environment. The results show that the success rate of the hexapod bionic leg-foot structure is 98% and that of the wheeled structure is 62% under slope crawling. For obstacle avoidance, the stability of the hexapod bionic leg-foot structure is lost by 3%, and the wheel is lost by 28%.

3.3. Control system design

3.3.1. Research on multi-sensor fusion sensing module. The robot integrates a gyroscope, laser positioning and navigation module, high-definition camera, etc., and can realize three-dimensional positioning, map creation, autonomous navigation, automatic obstacle avoidance and path planning. It is equipped with infrared temperature sensors, sound sensors, etc., to achieve all-weather multi-dimensional data acquisition, and through multi-sensor fusion to achieve scene anomaly detection, intrusion detection and other functions.

3.3.2. Research and development of reliable communication technology between UAV and ground mobile robot. The communication coordination between the UAV and the ground mobile robot is realized by optical communication relative networking[8]. Short-range beacon links establish connectivity, while the ground control hub orchestrates data relay across robotic and aerial platforms, guaranteeing real-time synchronization and robust interoperability for mission-critical operations. The design of the communication scheme should also meet the requirements of the minimum power of 6W in the reservoir area.

3.3.3. Research on the collaborative operation of robot and environment fusion. By the security specifications outlined in GA837-2009 for the storage of civil explosives, the facility is equipped with a range of advanced systems, including NFC patrols, fixed cameras, and infrared alarms. These tech innovations have improved hazardous explosives management and provided reliable inspection solutions for storage safety.

3.3.4. Flameproof and charging scheme design of robot and UAV. The circuit part is flameproofly designed through the aluminum metal shell, and a mobile charging seat is set up.

3.4. SLAM simulation experiment

MATLAB R2025a is used as the virtual simulation environment to construct the indoor environment model, and the binary occupied grid map is used to represent the obstacle distribution (resolution of 10 cells / m). Four symmetrically distributed rectangular obstacles are set up by the binaryOccupancyMap function to simulate the typical indoor corridor structure[9].

3.4.1. Map building. The binocular vision SLAM framework is used to generate a sparsely occupied map. By simulating the UAV's environmental exploration along the preset waypoints, the UAV's mapping strategy based on key frames is simulated. By constructing a virtual environment map and a UAV-generated map, the obstacle position error is less than 0.3 m.

3.4.2. Path planning and dynamic obstacle avoidance. The global path planning uses the Probabilistic Roadmap (PRM) algorithm to successfully plan a safe path (red trajectory as shown in Figure 5) in a complex obstacle environment, with a path length of 12.7 m.

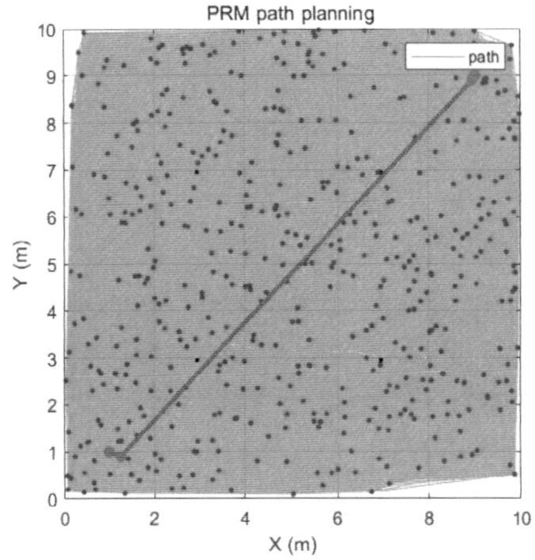

Figure 5. PRM path planning

The VFH algorithm is configured and the differential drive motion control model is used to realize dynamic obstacle avoidance. Firstly, the VFH algorithm constructs an obstacle avoidance controller based on a vector field histogram, and the differential drive model realizes the pose update. The radius of the robot is set to 0.5m, the scanning range of the simulated laser radar is set to ±90° (resolution 0.05 rad), the motion speed is 0.5m/s, and the control period is 0.1s. When the robot moves along the global path, the lidar data (lidarScan object) is input into the VFH controller in real-time to generate a safe steering angle. Fig.6 is a real-time obstacle avoidance simulation experiment. Several experiments were carried out to obtain experimental data. Experiments show that the robot can effectively avoid dynamic obstacles within a distance of 0.3m, and the success rate of obstacle avoidance reaches 96.5%, which is 40% higher than that of the traditional Bug algorithm[10].

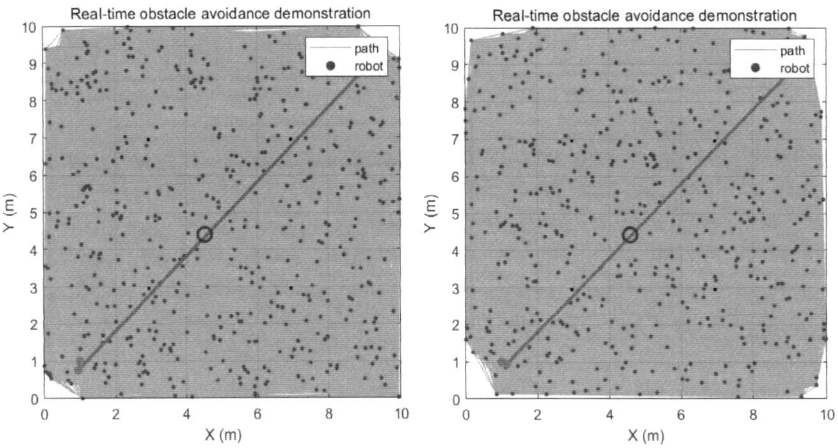

Figure 6. Real-time obstacle avoidance demonstration

3.5. *Limitations and Future Work*

The current scheme faces challenges in extreme weather: in rainy and snowy weather, LiDAR accuracy is reduced by 20% due to signal scattering; the thermal imager is less affected. And when the communication function is in heavy rain, the UWB range is reduced from 200m to 120m.

However, future work will integrate weather-resistant sensors and adaptive algorithms.

4. Conclusion

This study proposes an innovative ground-air collaborative intelligent inspection robot system, which is specifically aimed at the inspection problem of dangerous explosive storage. The system effectively solves the inspection challenges in complex environments by integrating advanced all-terrain mobile platforms, high-precision multi-sensor fusion positioning and navigation technologies, and intelligent cooperative communication mechanisms between drones and ground robots. Three core technologies are studied and innovated: one is the design of a mobile platform adapted to complex terrain, the other is the precise positioning and navigation system based on multi-source information fusion[11], and the third is the intelligent communication protocol for cooperative operation of ground-air robots. These tech innovations have improved hazardous explosives management and provided reliable inspection solutions for storage safety.

Acknowledgments
This work was funded by the Guangxi Key Laboratory of Special Engineering Equipment and Control, Guilin University of Aerospace Technology (2411KFYB04) and Erdos Emergency Rescue Intelligent Robot R&D Team (YJY-ZX-2024-A-001).

References
[1] WANG Xuguang. Intelligent Blasting. Metallurgical Industry Press. 2020.12
[2] Wang Shiqiang et al. Design of automatic inspection robot system for oil and gas station. Science, technology and engineering, 2022, 22 (36): 16124-16132
[3] Fotios K. Konstantinidis; Vasiliki Balaska etc. AROWA: An autonomous robot framework for Warehouse 4.0 health and safety inspection operations. IEEE 2022 30th Mediterranean Conference on Control and Automation (MED), pp494-499
[4] Bao Jiatong et al., design and experimental research of safety inspection robot system for hazardous chemicals warehouse. Journal of Nanjing University of Information Science and Technology, 2021, 13 (3), 311-319
[5] Fernando Cladera, Zachary Ravichandran etc. Enabling Large-scale Heterogeneous Collaboration with Opportunistic Communications, IEEE ICRA2023
[6] Lorenzo Rosa, Marco Cognetti etc. Multi-task Cooperative Control in a Heterogeneous Ground-Air Robot Team. 3rd IFAC workshop on Multivehicle Systems 2015, pp53-58
[7] XIN ZHOU, Xiangyong Wen, Fei Gao etc. Swarm of micro flying robots in the wild. May 2022. Science Robotics 7(66)
[8] Yan Dong, Zhou Nain. Design and implementation of the application system of CH-3 UAV aerogeophysical magnetic amplifier integrated station. Software, 2020,41 (10): 120-125.
[9] Lin Xiaohua, Wang Maosen, Dai Jinsong. Design of UAV Occupancy Map Construction Scheme Based on ROS. According to Journal of Ordnance and Equipment Engineering, 2021,42 (05): 215-220.
[10] Zhang Y, Abdullah A, Koppal S J, et al. ClipRover: Zero-shot Vision-Language Exploration and Target Discovery by Mobile Robots. arxiv preprint arxiv:2502.08791, 2025.
[11] Wang Dayi, Hou Bowen, Wang Jiongqi, etc. Review of state estimation methods for spacecraft autonomous navigation. Journal of Aeronautics, 2021, 42 (04): 72-89.

Multiscale Reliability-Guided Dual Attention Fusion Network for Multimodal Object Detection

Yong Wang, Hongyu Yang, and Pei-Qiu Huang*
School of Automation, Central South University, Changsha 410083, China
E-mail: pqhuang@csu.edu.cn

Abstract. Multimodal images (RGB and thermal images) can provide complementary information, making object detection more reliable and robust. However, most existing methods suffer from two issues: 1) interference from low-quality modality information in extreme environments (e.g., darkness, overexposure, and bad weather); 2) insufficient utilization of intra- and inter-modality information. To overcome these two issues, we propose a multiscale reliability-guided dual attention fusion network (MRDAN). In MRDAN, we first propose a multiscale reliability-guided module (MRGM). MRGM uses multiple parallel dilated convolution branches to calculate the reliability scores of RGB and thermal features at each level, and then uses these scores to enhance high-quality modality features and suppress low-quality modality ones. Then, we design a dual attention fusion module (DAFM). DAFM adopts parallel global and local attention units to capture long-range inter-modality contextual information and specific intra-modality information, respectively, and aggregates these two types of information for further performance improvement of object detection. Results on two benchmark datasets (FLIR and LLVIP) show that MRDAN achieves better overall performance than state-of-the-art multimodal object detection methods.

1. Introduction

Deep learning-based object detection methods [1–4] are becoming increasingly popular in many fields (such as autonomous driving). Considering that multimodal images can provide complementary information, effectively fusing multimodal images demonstrate the capability to enhance the reliability and robustness of deep learning-based object detection methods.

Currently, many multimodal object detection methods have been proposed for the fusion of RGB and thermal images. For example, a hybrid architecture integrating an intermediate fusion mechanism with an optimized decision tree algorithm was developed by Konig *et al.* [5], which can reduce potential false positive detection.

However, under harsh conditions such as darkness, overexposure, and rain and fog occlusion, low-quality modality information may interfere with multimodal fusion, resulting in poor adaptability of the multimodal object detection model.

To overcome this challenge, some methods based on CNNs have been proposed to evaluate the modal quality under modal failures in extreme environments. An *et al.* [6] used a CNN to analyse the light condition information of the original RGB image and then calculated the effective weights of the RGB and thermal modalities. However, the above methods rely only on shallow features and extract feature information in a single receptive field, which is not enough to accurately evaluate the modal quality. In addition, these CNN-based methods all employ CNNs

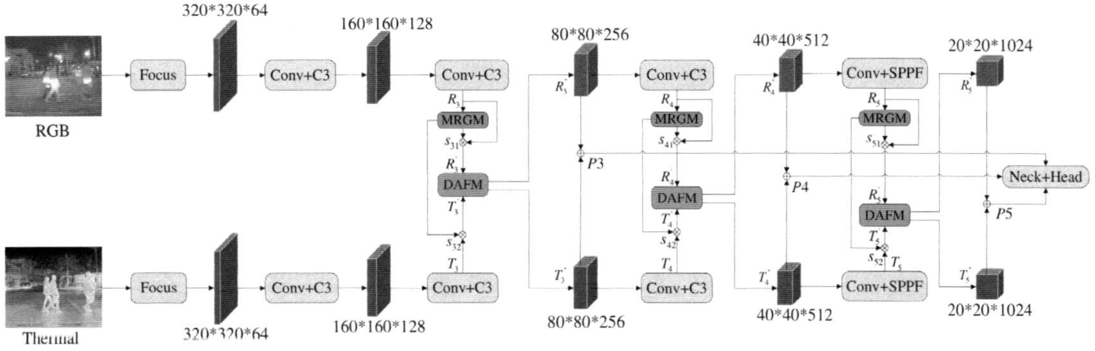

Figure 1. Architecture of the proposed MRDAN.

to learn intra-modality local features, while ignoring the utilization of inter-modality global features, resulting in insufficient modal interaction and loss of image contextual information.

To this end, several transformer-based methods have been proposed to learn long-range inter-modality contextual information. For instance, Fang *et al.* [7] proposed a transformer-based cross-modal feature fusion method, which introduces a cross-fusion transformer module within a dual-stream YOLOv5 framework to learn long-range dependencies between different modal features. Fu *et al.* [8] proposed a multimodal feature fusion method based on long-range correlation fusion, which enhances modal feature correlations through positional encoding. Lee *et al.* [?] introduced a cross-guided attention mechanism and a hierarchical transformer structure, effectively facilitating information interaction between RGB and thermal modalities and improving multiscale object detection performance. However, these transformer-based methods focus only on long-range inter-modal contextual information, while ignoring intra-modality local features, thus losing modality-specific image detail information.

In summary, existing studies on multimodal object detection have the following two issues: 1) in extreme environments, low-quality modality information can interfere with multimodal fusion and make the model exhibit poor adaptability; and 2) either CNN-based or transformer-based methods cannot capture both intra-modal local information and inter-modality global information simultaneously. To address these two issues, we propose a multiscale reliability-guided dual attention network (MRDAN) by designing a multiscale reliability-guided module (MRGM) and a dual attention fusion module (DAFM) at each level of the network. In each MRGM, we first extract reliability features of multireceptive fields by designing multiple parallel dilated convolution branches. These reliability features are then sent to the channel attention refinement unit to calculate the reliability scores of RGB and thermal features. In each DAFM, we design global and local attention units. Specifically, the global attention unit concatenates RGB and thermal features in the spatial dimension and adopts a pooling-based self-attention method to capture long-range inter-modality contextual information. The local attention unit concatenates multimodal features in the channel dimension and designs a convolution-based self-attention method to learn specific intra-modal information.

The main contributions of this study can be summarized as follows:

(i) The proposed MRGM can enhance high-quality modal features and suppress low-quality modal features by extracting multireceptive field reliability features.

(ii) The proposed DAFM with global and local attention units can capture long-range inter-modality contextual information and specific intra-modality information.

(iii) Experiments conducted on FLIR and LLVIP datasets show that MRDAN achieves better

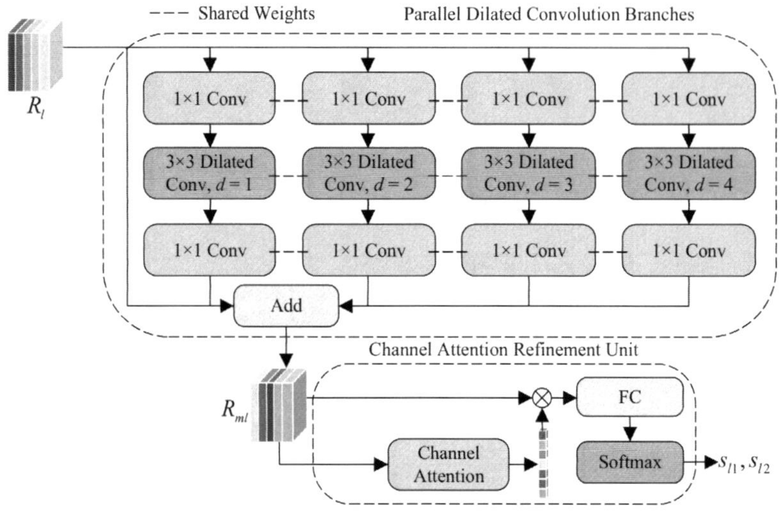

Figure 2. Structure of the proposed MRGM.

overall performance than state-of-the-art multimodal object detection methods.

2. METHODOLOGY

2.1. Architecture Overview

As illustrated in Fig 1, MRDAN adopts a two-stream YOLOv5 structure for multimodal feature extraction, including an RGB branch to extract RGB features and a thermal branch to extract thermal features. Between these two branches, we first design multiple MRGMs to evaluate the quality of modalities at different levels of the network. We denote the RGB and thermal features at the lth level of the network as R_l and T_l, respectively, where $l \in \{3, 4, 5\}$. With R_l as input, each MRGM first uses multiple parallel dilated convolution branches to extract multireceptive field reliability features, and then calculates the reliability scores of R_l and T_l through the channel attention refinement unit, denoted as s_{l1} and s_{l2}, respectively. Based on s_{l1} and s_{l2}, we can adjust the contributions of R_l and T_l on the final multimodal object detection, which generates $R_l^{'}$ and $T_l^{'}$, respectively. Following each MRGM, we further introduce DAFM, which takes $R_l^{'}$ and $T_l^{'}$ as inputs and implements global and local attention units to capture long-range inter-modality contextual information and specific intra-modality information, respectively.

2.2. MRGM

In extreme environments, low-quality modal information may deteriorate the performance of multimodal object detection. To this end, we design multiple MRGMs at different levels of the network, each containing four parallel dilated convolution branches and a channel attention refinement unit, as shown in Fig. 2. With R_l as input, four parallel branches with different dilation rates ($d \in \{1, 2, 3, 4\}$) are used to extract features from multireceptive fields. For each branch, it consists of a 1×1 convolution, a 3×3 dilated convolution, and a 1×1 convolution. Then, we add the extracted features to R_l to generate the reliability feature, denoted as R_{ml}. The above operations can be expressed as follows:

$$R_{ml} = R_l + \sum_{i=1}^{4}(conv_2(dilated\_conv_i(conv_1(R_l)))) \quad (1)$$

where $conv_1$ and $conv_2$ represent two 1×1 regular convolutions, and $dilated\_conv_i$ denotes the 3×3 dilated convolution at the ith branch.

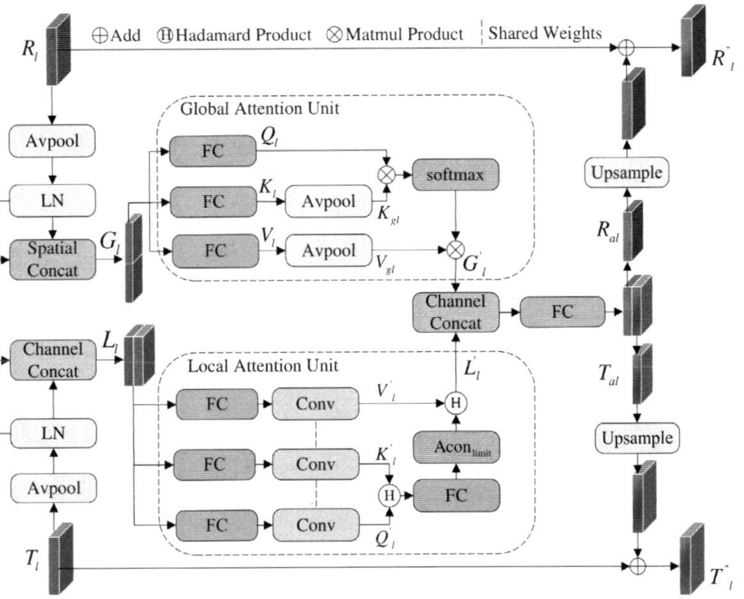

Figure 3. Structure of the proposed DAFM.

Afterward, R_{ml} is sent to the channel attention refinement unit for further feature integration and optimization. In this unit, we first calculate the channel attention weight through the classic channel attention mechanism [9]:

$$W_{cl} = CA\left(R_{ml}\right) \tag{2}$$

where CA represents the channel attention operation. Then, we implement matrix multiplication on R_{ml} and $W_{cl}^{\top}$, which applies a weighted summation of R_{ml} along the channel dimension, generating a feature with a single channel. Finally, this 1-channel feature is passed through a fully connected layer and a softmax function to produce the reliability scores of the RGB and thermal features (i.e., s_{l1} and s_{l2}), respectively. Finally, s_{l1} and s_{l2} are multiplied by R_l and T_l to obtain $R_l^{'}$ and $T_l^{'}$, respectively. The above operations can be expressed as follows:

$$s_{l1}, s_{l2} = softmax\left(FC\left(R_{ml}W_{cl}^{\top}\right)\right) \tag{3}$$

$$\begin{bmatrix} R_l^{'} \\ T_l^{'} \end{bmatrix} = \begin{bmatrix} s_{l1} & 0 \\ 0 & s_{l2} \end{bmatrix} \cdot \begin{bmatrix} R_l \\ T_l \end{bmatrix} \tag{4}$$

where FC denotes the fully-connected layer and $softmax(\cdot)$ is the softmax function. $R_l^{'}$ and $T_l^{'}$ represent the RGB and thermal features adjusted by MRGM, respectively, which will be used as the inputs of DAFM. From (4), the learned s_{l1} and s_{l2} act on R_l and T_l, which can enhance high-quality modal features and suppress low-quality modal features.

2.3. DAFM

Following each MRGM, we design DAFM with global attention and local attention units. As shown in Fig. 3, DAFM takes $R_l^{'}$ and $T_l^{'}$ as inputs. These two features are first downsampled using average pooling and then processed through layer normalization. Subsequently, the resulting two features are concatenated along the spatial and channel dimensions to generate G_l and L_l, respectively, which are then fed into the global and local attention units for further feature extraction.

2.3.1. Global Attention Unit In this unit, G_l is first mapped to Q_l, K_l, and V_l through the corresponding fully connected layers, and then K_l and V_l are average pooled to focus more on the global contextual information:

$$Q_l, K_l, V_l = FC(G_l) \tag{5}$$

$$K_{gl}, V_{gl} = Avpool(K_l), Avpool(V_l) \tag{6}$$

where $Avpool$ denotes the average pooling operation; Q_l, K_l, and V_l are the query, key, and value, respectively; and K_{gl} and V_{gl} represent the global key and value after pooling, respectively. The standard self-attention mechanism [10] is then applied as follows:

$$G_l^{'} = softmax\left(\frac{Q_l K_{gl}^{\top}}{\sqrt{d}}\right) V_{gl} \tag{7}$$

where d is the dimension of Q_l and $G_l^{'}$ represents the feature after the global attention-based information interaction.

2.3.2. Local Attention Unit In this unit, L_l is similarly mapped to its query, key, and value through the corresponding fully connected layers, and then they are processed by respective convolution layers with shared weights for local feature aggregation. As a result, the following three local features are obtained:

$$Q_l^{'}, K_l^{'}, V_l^{'} = Conv_{shared}(FC(L_l)) \tag{8}$$

Subsequently, the Hadamard product of $Q_l^{'}$ and $K_l^{'}$ is computed, followed by a linear transformation through a fully-connected layer. The resulting feature is passed through the capped version of ACON function [11] to obtain the local-aware weight, whose values are in the range of $[-1, 1]$. Finally, we perform the Hadamard product of the local-aware weight and V_l for local feature enhancement:

$$L_l^{'} = ACON_{limit}(\frac{FC(Q_l^{'} \odot K_l^{'})}{\sqrt{d^{'}}}) \odot V_l^{'} \tag{9}$$

where $L_l^{'}$ represents the feature after the local attention-based interaction, $d^{'}$ is the dimension of $Q_l^{'}$, and $\odot$ denotes the Hadamard product. Note that $ACON_{limit}$ represents the capped version of ACON function, which can provide a stronger nonlinear representation capability:

$$ACON_{limit} = \begin{cases} -1, & ACON_{limit} < -1 \\ ACON, & other \\ 1, & ACON_{limit} > 1 \end{cases} \tag{10}$$

where $ACON$ indicates the original ACON function.

After the above two units, $G_l^{'}$ and $L_l^{'}$ are concatenated along the channel dimension, transformed by the fully-connected layer, and separated according to the original channels to generate R_{al} and T_{al}:

$$R_{al}, T_{al} = FC(Concat(G_l^{'}, L_l^{'})) \tag{11}$$

These two features are then upsampled and skip-connected with their initial inputs to produce the outputs of DAFM:

$$R_l^{''}, T_l^{''} = R_l^{'} + Upsample(R_{al}), T_l^{'} + Upsample(T_{al}) \tag{12}$$

where $Upsample$ is an interpolation-based upsampling function. In addition, $R_l^{''}$ and $T_l^{''}$ represent the attention-enhanced RGB and thermal features, respectively, which are sent back to their own branches. As a result, $R_l^{''}$ and $T_l^{''}$ contain the captured long-range inter-modality contextual information and specific intra-modality information, achieving effective complement of the RGB and thermal modality information.

Journal of Physics: Conference Series **3024** (2025) 012036 doi:10.1088/1742-6596/3024/1/012036

Table 1. Results on the FLIR Dataset

Model	Modality	mAP_{50}	mAP
SSD [2]	RGB	52.2	21.8
YOLOV3 [3]	RGB	58.3	25.7
YOLOV5	RGB	67.8	31.8
YOLOV7 [4]	RGB	70.2	31.6
Faster RCNN [12]	RGB	64.9	28.9
SSD [2]	T	65.5	29.6
YOLOV3 [3]	T	73.6	36.8
YOLOV5	T	73.9	39.5
YOLOV7 [4]	T	75.6	38.2
Faster RCNN [12]	T	74.4	37.6
GAFF [13]	RGB+T	72.9	37.5
CFT [7]	RGB+T	78.7	40.2
LRAF-Net [8]	RGB+T	80.5	42.8
CAMDet [14]	RGB+T	77.1	38.4
CFR_3 [15]	RGB+T	72.4	/
CSSA [16]	RGB+T	79.2	41.3
Ours	RGB+T	**83.9**	**45.5**

3. Experiments

3.1. Datasets

We used two public datasets to evaluate our method: FLIR and LLVIP datasets. The FLIR dataset contains images captured from the perspective of vehicles on the road, with target types including vehicles, pedestrians, and bicycles. It contains a total of 5142 registered image pairs, of which 4129 pairs were used for training, and 1013 pairs were used for testing. The LLVIP dataset is used for pedestrian detection in low-light scenes, with target types all being pedestrians. It contains a total of 15488 registered image pairs, of which 12025 pairs were used for training, and 3463 pairs were used for testing.

3.2. Implementation Details

Our model used the SGD optimizer with an initial learning rate of 0.01, a momentum of 0.937, a batch size of 16, and a weight decay of 5e-4. In order to make a fair comparison with previous works, mAP_{50} and mAP were selected as the evaluation metrics. mAP_{50} is the mean of all the average precision (AP) values for all categories when the intersection over union (IOU) is 0.5. mAP is the mean of a group of average AP values ranging from mAP_{50} to mAP_{95}, sampled at a step size of 0.05 for IOU in the interval $[0.5, 0.95]$.

3.3. Comparison with Existing Methods

3.3.1. FLIR Dataset Table 1 presents the mAP_{50} and mAP values of different methods on the FLIR dataset, where "RGB"/"T"/"RGB+T" indicates the RGB modality, the thermal modality, or a combination of RGB and thermal modalities. It can be seen that MRDAN outperforms the compared methods in terms of the two metrics. Specifically, MRDAN outperforms the very recent method (LRAF-Net [8]) by 3.4% and 2.7% in terms of mAP_{50} and mAP, respectively.

Table 2. Results on the LLVIP Dataset

Model	Modality	mAP_{50}	mAP
SSD [2]	RGB	82.6	39.8
YOLOV3 [3]	RGB	85.9	43.3
YOLOV5	RGB	90.8	50.0
YOLOV7 [4]	RGB	91.9	51.2
Faster RCNN [12]	RGB	91.4	49.2
SSD [2]	T	90.2	53.5
YOLOV3 [3]	T	89.7	52.8
YOLOV5	T	94.6	61.9
YOLOV7 [4]	T	96.0	63.9
Faster RCNN [12]	T	96.1	61.1
CFT [7]	RGB+T	97.5	63.6
LRAF-Net [8]	RGB+T	**97.9**	66.3
CSSA [16]	RGB+T	94.3	59.2
Ours	RGB+T	97.5	**66.5**

Figure 4. Visualized detection results of the baseline and MRDAN. The red boxes indicate the missed detection.

3.3.2. LLVIP Dataset From Table 2, the performance of MRDAN is better than that of most competitors. For the mAP_{50} metric, MRDAN is second to LDAFNet [8], which may be due to LDAFNet's larger number of network parameters. Furthermore, we visualized the object detection results of the baseline and MRDAN in Fig. 4. Herein, the baseline represents MRDAN without MRGM and DAFM, as shown in the second row of Table 3. Compared with the baseline, our method has better performance in detecting small objects and under poor light conditions.

3.4. Ablation Study

To analyze the effectiveness of MRGM and DAFM, we conducted ablation studies. Table 3 shows the ablation results on the FLIR and LLVIP datasets.

3.4.1. Effectiveness of MRGM We compared the results of MRDAN and MRDAN without MRGM. From Table 3, the use of MRGM improves the mAP_{50} and mAP metrics by 2.8%

Table 3. Results of the Ablation Studies

MRGM	DAFM	Modality	FLIR mAP_{50}	FLIR mAP	LLVIP mAP_{50}	LLVIP mAP
		RGB+T	75.5	37.6	95.2	62.4
✓		RGB+T	79.5	39.2	96.1	63.0
	✓	RGB+T	81.1	42.9	97.4	64.8
✓	✓	RGB+T	**83.9**	**45.5**	**97.5**	**66.5**

and 2.6% on the FLIR dataset, respectively. On the LLVIP dataset, it improves the mAP_{50} and mAP metrics by 0.1% and 0.7%, respectively. It is due to the fact that the FLIR and LLVIP datasets contain extreme environments such as darkness and overexposure. MRGM is able to alleviate the interference of low-quality modality information in extreme environments by adjusting modal features.

3.4.2. Effectiveness of DAFM We then compared the results of MRDAN and MRDAN without DAFM. From Table 3, the use of DAFM improves the mAP_{50} and mAP metrics by 4.4% and 6.3% on the FLIR dataset, respectively. On the LLVIP dataset, it improves the mAP_{50} and mAP metrics by 1.4% and 3.5%, respectively. The above results demonstrate that DAFM is beneficial to multimodal object detection performance.

4. CONCLUSION

In this paper, we proposed a new multimodal object detection network called MRDAN by designing two novel modules, i.e., MRGM and DAFM. On the one hand, MRGM could enhance high-quality modal features and suppress low-quality modal features by extracting multireceptive field reliability features. On the other hand, DAFM with global and local attention units could capture long-range inter-modality contextual information and specific intra-modality information. Results on the FLIR and LLVIP datasets demonstrated that MRDAN achieved better overall performance than state-of-the-art multimodal object detection methods.

References

[1] Girshick R 2015 Fast R-CNN *Proceedings of the IEEE International Conference on Computer Vision* pp 1440–1448
[2] Womg A, Shafiee M J, Li F and Chwyl B 2018 Tiny SSD: A tiny single-shot detection deep convolutional neural network for real-time embedded object detection *2018 15th Conference on Computer and Robot Vision (CRV)* (IEEE) pp 95–101
[3] Redmon J and Farhadi A 2018 *arXiv preprint arXiv:1804.02767*
[4] Wang C Y, Bochkovskiy A and Liao H Y M 2023 YOLOv7: Trainable bag-of-freebies sets new state-of-the-art for real-time object detectors *Proceedings of the IEEE/CVF Conference on Computer Vision and Pattern Recognition (CVPR)* pp 7464–7475
[5] Konig D, Adam M, Jarvers C, Layher G, Neumann H and Teutsch M 2017 Fully convolutional region proposal networks for multispectral person detection *Proceedings of the IEEE Conference on Computer Vision and Pattern Recognition Workshops* pp 49–56
[6] An Z, Liu C and Han Y 2022 *IEEE Signal Processing Letters* **29** 2562–2566
[7] Qingyun F, Dapeng H and Zhaokui W 2021 *arXiv preprint arXiv:2111.00273*
[8] Fu H, Wang S, Duan P, Xiao C, Dian R, Li S and Li Z 2024 *IEEE Transactions on Neural Networks and Learning Systems* **35** 13232–13245
[9] Hu J, Shen L and Sun G 2018 Squeeze-and-excitation networks *Proceedings of the IEEE Conference on Computer Vision and Pattern Recognition (CVPR)* pp 7132–7141
[10] Vaswani A, Shazeer N, Parmar N, Uszkoreit J, Jones L, Gomez A N, Kaiser Ł and Polosukhin I 2017 *Advances in Neural Information Processing Systems* **30** 1–11

[11] Ma N, Zhang X, Liu M and Sun J 2021 Activate or not: Learning customized activation *Proceedings of the IEEE/CVF Conference on Computer Vision and Pattern Recognition (CVPR)* pp 8032–8042
[12] Ren S, He K, Girshick R and Sun J 2016 *IEEE Transactions on Pattern Analysis and Machine Intelligence* **39** 1137–1149
[13] Zhang H, Fromont E, Lefèvre S and Avignon B 2021 Guided attentive feature fusion for multispectral pedestrian detection *Proceedings of the IEEE/CVF Winter Conference on Applications of Computer Vision* pp 72–80
[14] Jang J, Lee J and Paik J 2025 CAMDet: Condition-adaptive multispectral object detection using a visible-thermal translation model *ICASSP 2025 - 2025 IEEE International Conference on Acoustics, Speech and Signal Processing (ICASSP)* pp 1–5
[15] Zhang H, Fromont E, Lefevre S and Avignon B 2020 Multispectral fusion for object detection with cyclic fuse-and-refine blocks *2020 IEEE International Conference on Image Processing (ICIP)* (IEEE) pp 276–280
[16] Cao Y, Bin J, Hamari J, Blasch E and Liu Z 2023 Multimodal object detection by channel switching and spatial attention *Proceedings of the IEEE/CVF Conference on Computer Vision and Pattern Recognition (CVPR)* pp 403–411

Journal of Physics: Conference Series **3024** (2025) 012037 doi:10.1088/1742-6596/3024/1/012037

Improved YOLOv8 Algorithm for Indoor Object Detection

Jie Wang[1,2][0009-0003-9534-3945], **Hong Zhao**[1,2*][0000-0003-4521-7281], **Yuxi Huang**[1,2][0009-0009-6942-5967], **Baotong Zhang**[1,2][0009-0002-4007-2634]

[1] School of Mechanical and Control Engineering, Guilin University of Technology, Guilin 541000, China

[2] Key Laboratory of Advanced Manufacturing and Automation Technology(Guilin University of Technology), Education Department of Guangxi Zhuang Autonomous Region, Guilin, 541006, China.

*Corresponding author's e-mail: zhaohong@glut.edu.cn

Abstract. To address the challenges of missed detections and false positives in indoor object detection tasks caused by target clustering and occlusion, this study proposes an advanced indoor object detection algorithm, SRSE-YOLOv8. A novel indoor object detection algorithm, SRSE-YOLOv8, is proposed to address missed and false detections caused by object clustering and occlusion in indoor scenes. First, an additional smaller detection head is introduced to enhance the capture of small objects and improve feature extraction capabilities. Then, RFAConv is incorporated to achieve input-adaptive kernel adjustment through multi-scale receptive field learning and spatial-channel attention coordination, thereby enhancing the model's ability to handle complex images. Next, the Slimneck structure is applied in the neck network to reduce model complexity while maintaining accuracy. Finally, the EMA attention mechanism is introduced to filter out background noise interference and enhance feature responses in target regions. The proposed algorithm achieves an average precision of 94.9% on a self-constructed indoor waste dataset, representing a 3.5% improvement over the original YOLOv8n model. This approach effectively enhances indoor waste detection accuracy while meeting real-time detection speed requirements.

1. Introduction

Currently, waste classification methods in China remain relatively underdeveloped, primarily relying on manual sorting in garbage collection stations. This approach significantly increases the workload of sanitation workers and results in low sorting efficiency. The emergence of indoor service robots offers a potential solution by enabling waste classification at the source, freeing people from tedious household tasks. Moreover, with an aging population and improvements in social welfare systems, these robots have broad application prospects. However, existing datasets for indoor waste classification are still incomplete.

Deep learning-based object detection technology performs excellently in recognizing medium and large-sized objects. However, in real-world production and daily life scenarios, small objects far outnumber large ones, making small object detection particularly important. Detecting small objects in indoor environments presents numerous challenges, including lighting variations, object occlusion, scale differences, and complex backgrounds. Firstly, unstable lighting conditions, especially shadows in low-light or dark environments, can blur object boundaries and reduce detection accuracy. Secondly, small indoor objects often appear in clusters, making them prone to occlusion, which increases the risk of missed detections. Additionally, when an object's color or texture is similar to the background, the model may struggle to distinguish between them, severely affecting detection performance. Another major challenge is the dataset issue. Existing object detection datasets primarily focus on general objects and lack high-quality annotated data specifically for small objects in indoor environments, which may limit the model's generalization ability.

In summary, this paper proposes an improved YOLOv8 indoor object detection algorithm, the main contributions of this paper are as follows:

• Introducing a smaller detection head to enhance the extraction of detailed target features.

• Introducing RFAConv in the backbone network to adjust branch weights, focus on multi-scale features, and enhance the model's ability to process complex images.

• Employing the Slimneck structure in the neck network to reduce model complexity while maintaining accuracy.

• Integrating the EMA attention mechanism to strengthen responses to key regions, suppress background noise interference, and further improve the model's robustness.

2. Related Work

To study waste classification, Yang and his colleagues at Stanford University created the publicly available GarbageNet dataset [1]. Yang applied the Support Vector Machine method to this dataset, achieving an accuracy of only 63%. Satvilkar [2] used a Random Forest algorithm for classification, attaining an accuracy of 62.61%. Besides some non-public datasets, Mittal et al. [3] developed the GINI dataset, which contains 2,561 images, and achieved an accuracy of 87.69% using the GarbNet model. Boshra Khalili et al. [4] proposed SOD-YOLOv8 for small object detection. By introducing an additional small object detection head and integrating an attention module to enhance multi-path fusion, they further improved the detection capability for small objects. They also proposed PIOU for medium-quality anchor boxes. In the final test on the VisDrone2019 dataset[5], the model achieved an mAP of 45.1%. Zhao Yawen et al. [6] integrated the ParNetAttention module into the C2f module of the backbone to enhance feature extraction. They also incorporated CARAFE into the neck network to strengthen the focus on local information. Ultimately, their approach achieved a mAP@50 of 65.9%.

Commonly used waste classification datasets include Trash [7] and TACO [8]. The Trash dataset primarily features images with white or simple backgrounds, overlooking complex indoor scenarios such as multi-object occlusion, varying lighting conditions, and densely distributed small objects. This limitation makes it difficult for models to adapt to real-world indoor small-object detection tasks. Additionally, the dataset contains an insufficient number of small-object samples, restricting the model’s ability to learn fine-grained features. The TACO dataset mainly focuses on outdoor environments such as beaches and forests, lacking the distribution characteristics of indoor waste, making it unsuitable for indoor small-object detection tasks. The SUN397 dataset is designed for scene classification, prioritizing overall scene diversity rather than the distribution of small objects. Indoor environments often involve densely arranged and occluded objects, yet this dataset lacks optimized annotations for such scenarios.

3. Methodology

The structure of the improved YOLOv8n model is shown in Figure 1. In the backbone network, RFAConv replaces the original convolution, enhancing the ability to extract multi-scale features. The EMA attention mechanism is also introduced to strengthen the extraction of local information in feature maps. A new small-object detection head with a 4× downsampling rate is added to improve the detection performance for small targets. Furthermore, the Slim-neck structure is employed in the neck network, where GSConv replaces the standard convolution of the original network, and VOVGSCSP is used to replace all C2f modules in the neck. These modifications not only reduce the model size but also enhance detection accuracy. To facilitate further discussions on the improved model, it is referred to as SRSE-YOLOv8.

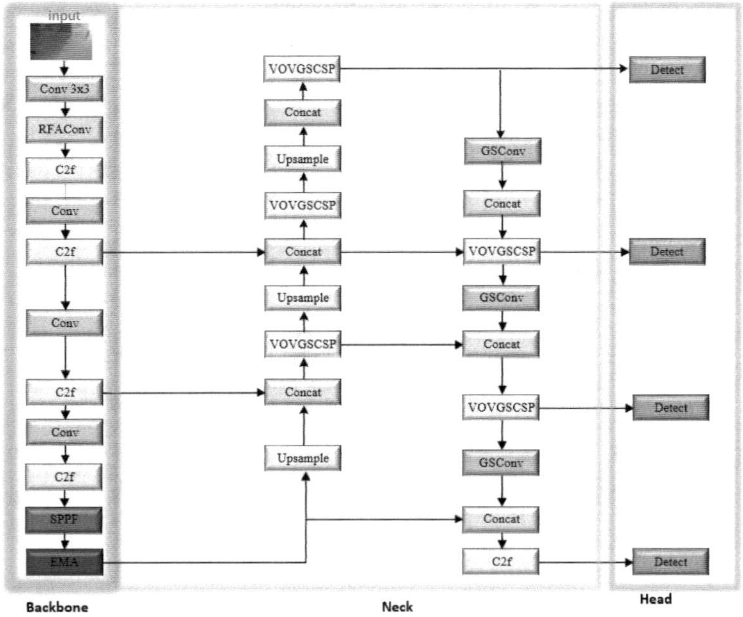

Figure 1. SRSE-YOLOv8 structure diagram

3.1. Adding a smaller detection head

Shallow feature maps have a relatively small local receptive field due to the lower depth of stacked convolutional layers. As a result, they rely more on local texture information from neighboring pixels but have limited capability for global context modeling of small objects. In indoor environments, small waste items are often accumulated in corners and edges, where they are more likely to be occluded. Additionally, lighting conditions may be uneven, with shadows or reflections affecting visibility.

Feature maps obtained through 8× downsampling may lead to insufficient target pixel coverage, making it difficult to detect small objects accurately. An additional small-object detection head with 4× downsampling is introduced to construct a 160×160 feature map to address this issue. This approach significantly enhances the model's ability to capture intricate details of small objects, thereby boosting detection precision. By offering more refined feature representations, the model gains improved capability to differentiate objects from complex backgrounds. This strengthens its resilience to occlusions and reflective surfaces and effectively lowers the likelihood of missed detections.

3.2. RFAConv module

RFAConv[10] is a dynamic, receptive field adaptive convolution that adjusts the convolution kernel based on input dependency through multi-scale receptive field learning and spatial-channel attention collaboration. In YOLOv8, traditional convolution layers have fixed kernel sizes, which introduce certain limitations. For example, while large kernels are beneficial for tasks requiring global information, they increase the parameters and computational complexity, potentially introducing more noise and reducing model accuracy. On the other hand, small kernels capture fine-grained local features but may struggle to cover the complete structure of large objects.

To address these issues, RFAConv introduces a receptive field attention mechanism, effectively solving the parameter-sharing limitation of traditional convolution operations. Additionally, its multi-branch structure dynamically adjusts the weights of each branch according to different inputs, enabling the model to focus on features of various scales. This enhances the model's capability to process complex images. The spatial variation process of the receptive field is illustrated in Fig. 2 a.

3.3. Slim-neck

Traditional Depthwise Separable Convolution[11] achieves lightweight computation using depth and pointwise convolution. However, DPConv processes each channel independently, lacking cross-channel information interaction, making preserving fine-grained features difficult.GSConv[12] addresses this issue by incorporating a parallel architecture of grouped convolution and standard convolution, combined with channel shuffling, to reduce computational cost. Meanwhile, the standard convolution branch can extract global features, compensating for the information loss in DSC.

In Fig. 2 b, the grouped convolution branch splits the input channels into two equal groups, applying a 3×3 convolution to each group, resulting in an output of C2/2 channels. Simultaneously, the standard convolution branch applies a 1×1 convolution to the first 50% of the input channels, producing C2/2 channels. Finally, the channels are randomly shuffled to enhance cross-group feature interaction, improving the model's representational capacity.

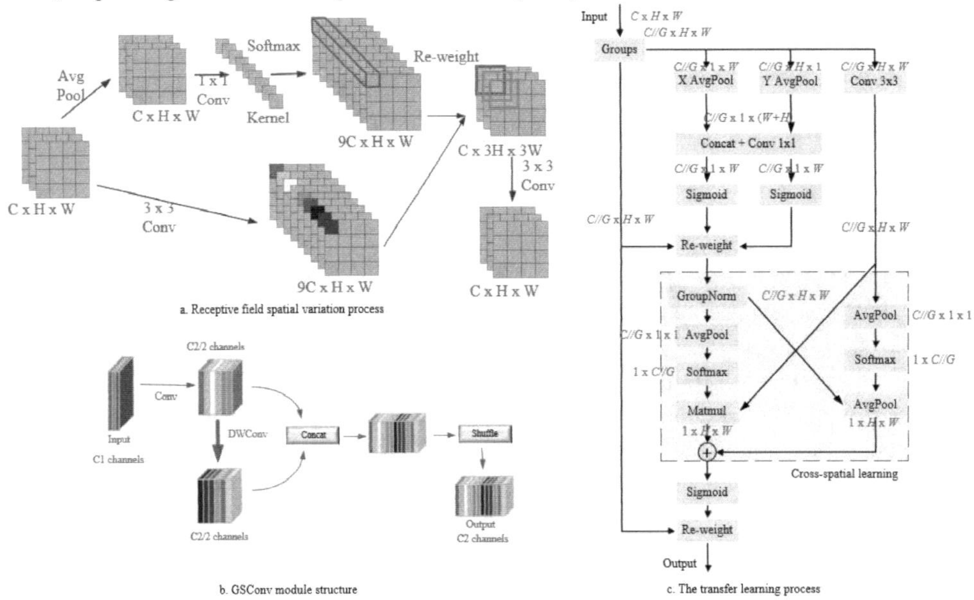

Figure 2. a. Receptive field spatial variation process,b. GSConv module structure and c. The transfer learning process

3.4. EMA attention mechanism

The backbone network of YOLOv8 (CSPDarknet) extracts features through multi-level downsampling. However, traditional convolution operations suffer from feature response homogenization, meaning they treat all regions equally. This can lead to the loss of fine details in small objects. Therefore, an attention mechanism is needed to make feature selection more intelligent, enabling the network to have a "visual focus" capability.

Objects may be occluded or resemble the background in real-world scenarios, leading to false positives or missed detections. The attention mechanism enhances feature responses in target regions, reducing these errors. As a lightweight module, EMA[13] improves the detection of small objects without significantly increasing computational complexity. The process is shown in Fig. 2 c.

4. Expremental Results And Analysis

4.1. Dataset

Some public datasets fail to adequately meet indoor waste recognition requirements, so image collection was carried out using methods such as ordinary cameras and smartphone photography. The collected dataset of common indoor waste was organized according to the PASCAL VOC 2007 standard framework, establishing three core directories: ImageSets, JPEGImages, and Annotations.

(a)

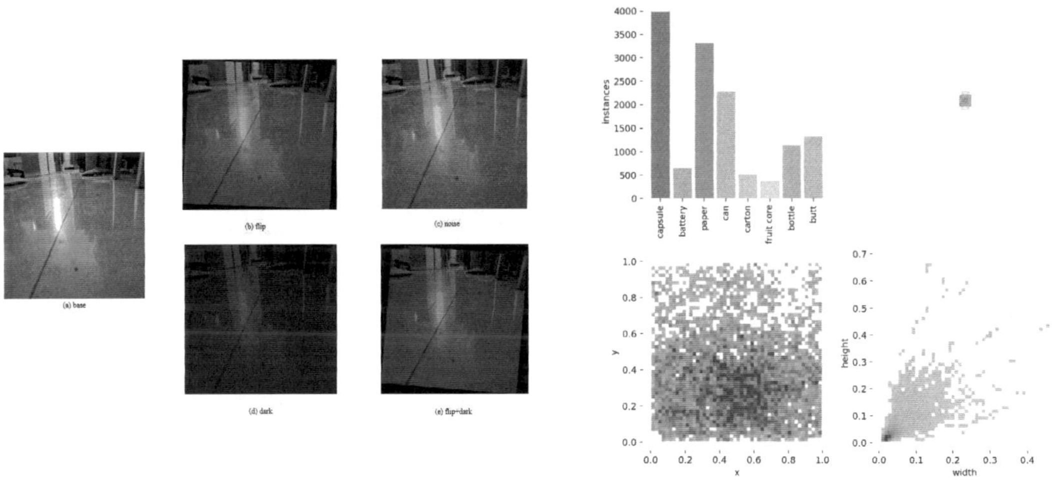

(b) (c)

Figure 3.(a) LabelImg; (b) Augmentation; (c) Target distribution

The ImageSets folder contains partition files, such as training, validation, and test sets. The JPEGImages directory stores the original image files, named in a continuous numerical sequence. The Annotations folder holds XML annotation files, which follow the same naming convention as the images, ensuring a one-to-one correspondence between images and their annotations. Figure 3(a) shows that these XML files were generated using the LabelImg professional annotation tool, where targets in the images were manually labeled.

The experiment selected eight common types of waste from daily life for analysis, including capsule, paper, can, carton, fruit core, bottle, butt, and battery. Objects such as fruit core, capsule, and butt are representative of small object recognition tasks. These objects are small and occupy fewer pixels in the images, posing a more significant challenge to the detection model and helping to improve its robustness. Objects like cans and bottles may have similar shapes but different materials, with varying reflections and textures, which enhances the dataset's diversity and improves the model's generalization ability.

In Figure 3(b), the dataset was expanded to 3,604 images through data augmentation techniques such as rotation, noise addition, light intensity adjustment, and combined transformations to meet the requirements for model training. Finally, Figure 3(c) illustrates the target distribution of the dataset, providing a clear overview of its composition.

4.2. *Analysis of results*

4.2.1. Ablation experiment. To isolate the contribution of each component, we conduct comprehensive ablation studies under controlled experimental settings. The experiment utilized the controlled variable method to evaluate the independent effects and combined impact of the four proposed enhancement modules. Using YOLOv8n as the baseline model, the mAP@50 reached 91.4%. After adding the small-object detection head (Smaller head), mAP@50 increased by 1.7%. Incorporating RFAConv further improved mAP@50 by 1.1%. While adding SlimNeck slightly reduced accuracy, it effectively decreased model complexity. The inclusion of the EMA attention mechanism led to an additional 1.2% improvement in mAP@50. The results of the ablation study indicate that Smallerhead and RFAConv form the core framework for accuracy enhancement. Without Smallerhead, the accuracy improved compared to the baseline but still lagged significantly behind the Smallerhead-integrated model.EMA provided attention enhancement, while SlimNeck controlled computational expansion. Ultimately, these improvements resulted in a 3.5% increase in mAP@50 and a 2.6% increase in mAP@50:95, validating the effectiveness of the proposed modifications.

Table 1. Ablation experiment

Smallerhead	RFAConv	slimneck	EMA	mAP0.5(%)	mAP0.5:0.95(%)	GFLOPs
-	-	-	-	91.4	60.8	8.1
√	-	-	-	93.1	61.9	12.2
-	√	-	-	92.5	61.3	8.2
-	-	√	-	91.1	59.7	7.3
-	-	-	√	92.6	61.4	8.1
√	√	-	-	94.3	62.7	12.3
√	-	√	-	94.1	62.4	11.1
√	-	-	√	94.0	62.2	12.2
-	√	√	-	92.7	60.8	7.4
-	√	√	√	92.8	60.7	7.4
√	√	√	-	94.5	62.9	11.2
√	√	√	√	94.9	63.4	11.2

4.2.2. Comparative experiments. To demonstrate the enhanced performance and effectiveness of the improved algorithm proposed in this paper, experiments were conducted under the same conditions, using the same dataset for comparison with the YOLO series algorithms. The YOLOv8-MSS model enhances its performance in small object detection and anti-interference capabilities by introducing an additional small object detection head and incorporating lightweight modules, MLCA and SENetV2, into the network. As shown in Table 2, the improved SRSE-YOLOv8 model achieves higher precision than other models. Although the model size has increased, the improvement in precision compensates for this drawback. The enhanced model also achieves the highest F1 score, indicating more reliable object detection with reduced missed detections and false positives.

Table 2. Comparative experiments

Baseline	P/%	R/%	mAP0.5(%)	mAP0.5:0.95(%)	GFLOPs	F1(%)
YOLOv5s	93.3	84	90.8	58.6	7.1	86
YOLOv6	91.8	81.4	87	56.8	11.8	81
YOLOv7-tiny	92.1	81	88.4	57.2	13.2	82
YOLOv8n	93.3	87.2	91.4	60.8	8.1	90
YOLOv8-MSS[14]	93.7	90.6	94.3	62.5	12.3	91
SRSE-YOLOv8	94.4	90.9	94.9	63.4	11.2	92

4.2.3. Generalization experiment. Additional tests were conducted on other publicly available small-object datasets to verify the model's robustness further. The VisDrone2019 dataset was gathered by a research team from Tianjin University using drones in multiple cities across China. This dataset includes real-world interference factors such as intense lighting, haze, motion blur, and object occlusion. Small objects account for 60% of the dataset, making it a widely used benchmark for small-object detection.

As shown in Table 3, the detection performance for all target objects improved to varying degrees. From the drone's perspective, Pedestrians, People, Tricycles, and Motors are classified as small objects. The mAP@50 for these categories increased by 10%, 9.6%, 3.9%, and 4.7%, respectively. On average, the SRSE-YOLOv8 model achieved a 4.4% improvement, effectively demonstrating its effectiveness in small-object recognition.

Table 3. Generalization experiment(%)

Method	Pedestrain	People	Bicycle	Car	Van	Trunk	Tricycle	Awning	Bus	Motor	mAP@50
Base	35.8	27.6	24.1	76.1	38.7	29.1	22.2	11.4	47.4	37.5	35
Ours	45.8	37.2	28.4	79.3	43.0	30.5	26.1	13.1	48.5	42.2	39.4

4.3. Test results and analysis

Fig. 4 (b1) shows a densely packed object, with paper and cigarette butts under the chair in a dark area and the legs of the chair causing occlusion. This increases the difficulty of detection, resulting in significant false negatives. In (b2), the cigarette butt is partially occluded by the paper in front of it, leading to a missed detection as well. From (c1) and (c2), it is clear that the improved model not only enhanced the detection accuracy but also successfully identified the missed cigarette butt and paper. (b3) shows a cardboard box obstructing the bottle, causing the detection of two bottles and incorrectly classifying a nearby stain as a capsule. In (b4), the rust on the stool is mistakenly detected as a can. Compared with (c3) and (c4), the SRSE-YOLOv8 model demonstrates a strong ability to resist interference.

Figure 4. Visualization analysis

5. Conclusion

This paper proposes an indoor object detection and recognition method based on YOLOv8. The method enhances the capture of fine-grained features by adding a small object detection head and introduces RFAConv to focus on features at different scales, improving the handling of complex images. The neck network utilizes the Slimneck structure to reduce model complexity while maintaining high accuracy. Lastly, the EMA attention mechanism is added to filter background noise interference, further improving small object detection capability. The improved model demonstrates strong resistance to interference and effectively enhances detection accuracy for indoor objects.

References

[1] Babaei, A.A., Alavi, N., Goudarzi, G., et al. (2015). Household recycling knowledge, attitudes and practices towards solid waste management. Resources, Conservation & Recycling, 102, 94-100.

[2] Krizhevsky, A., Sutskever, I., Hinton, G.E. (2017). ImageNet classification with deep convolutional neural networks. Communications of the ACM, 60(6), 84-90.

[3] Srivastava, N., Hinton, G., Krizhevsky, A., et al. (2014). Dropout: a simple way to prevent neural networks from overfitting. The Journal of Machine Learning Research, 15(1), 1929-1958.

[4] Khalili, B., Smyth, A.W. (2024). SOD-YOLOv8—Enhancing YOLOv8 for Small Object Detection in Aerial Imagery and Traffic Scenes. Sensors, 24, 6209.

[5] Du, D., et al. (2019). VisDrone-DET2019: The vision meets drone object detection in image challenge results. Proceedings of the IEEE/CVF International Conference on Computer Vision Workshops.

[6] Xiao, J., Ehinger, A.K., Hays, J., et al. (2016). SUN Database: Exploring a Large Collection of Scene Categories. International Journal of Computer Vision, 119(1), 3-22.

[7] Yang, M., Thung, G. (2016). Classification of trash for recyclability status. CS229 Project Report, 2016(1), 3.

[8] Proenca, F., Simoes, P. (2020). TACO: Trash Annotations in Context for Litter Detection. arXiv, 06975.

[9] Lin, T.Y., Dollár, P., Girshick, R., He, K., Hariharan, B., Belongie, S. (2017). Feature Pyramid Networks for Object Detection. IEEE Computer Society.

[10] Zhang, X., Liu, C., Yang, D., et al. (2023). RFAConv: Innovating spatial attention and standard convolutional operation. arXiv preprint arXiv:2304.03198.

[11] Schick, C. (2009). Differential scanning calorimetry (DSC) of semicrystalline polymers. Analytical and Bioanalytical Chemistry, 395(6), 1589-1611.

[12] Li, H.L., et al. (2022). Slim-neck by GSConv: A better design paradigm of detector architectures for autonomous vehicles. arXiv preprint arXiv:2206.02424, 10.

[13] Ouyang, D.L., He, S., Zhang, G.Z., et al. (2023). Efficient multi-scale attention module with cross-spatial learning. 2023 IEEE International Conference on Acoustics, Speech and Signal Processing (ICASSP). New York: IEEE, 1-5.

[14] Wang, J., Zhao, H. (2024). Improved YOLOv8 Algorithm for Water Surface Object Detection. Sensors, 24(15), 5059.

Multilayer Type-2 Fuzzy Q-Learning-Based Energy Management for Fuel Cell Hybrid Mobile Robot System with Varying Temperature

Yunlong Wang[1,*], Yongfu Wang[1]

[1] School of Mechanical Engineering and Automation, Northeastern University, Shenyang, Liaoning Province, 110819, PR China

E-mail: wangyl@mail.neu.edu.cn

Abstract. Fuel cell hybrid mobile robots emerge as a promising solution for enhancing manufacturing productivity while enabling eco-friendly industrial operations. Nevertheless, their dynamic power performance critically depends on coordinated management of hybrid energy systems under fluctuating environmental conditions. This investigation develops a multilayer type-2 fuzzy Q-Learning (T2-FQL) energy management strategy (EMS) to achieve autonomous power allocation optimization across diverse temperature environments. The research methodology encompasses four key aspects: First, we establish a comprehensive mathematical framework characterizing the fuel cell-battery hybrid architecture in mobile robotic systems. Subsequent the relation between temperature with critical system performance parameters is analysed. To simultaneously address multi-objective optimization requirements in economy, efficiency, and durability, we innovatively integrate fuzzy logic systems (FLS) with Markov decision processes (MDP) through a multilayer T2-FQL architecture. This synergistic approach effectively enhances Q-learning generalization capabilities while maintaining operational scenario adaptability. Finally, hardware-in-the-loop (HIL) validation demonstrates the proposed EMS's superior optimization performance, particularly in varying temperature conditions.

1. Introduction

The ongoing technological revolution in industrial automation has transformed manufacturing paradigms, creating unprecedented demands for mobile robots with extended-range material transport capabilities [1]. Proton exchange membrane fuel cells (PEMFCs) have emerged as a promising alternative, particularly given global concerns over depleting fossil fuel reserves and urgent climate change mitigation needs through CO_2 emission reduction. These devices generate electricity through efficient electrochemical conversion of hydrogen and oxygen [3]. As an energy carrier for PEMFCs, hydrogen offers distinct advantages including exceptional power density, low-temperature operation, rapid activation characteristics, near-silent functioning, and zero direct carbon emissions [4]. However, technical barriers remain, particularly regarding PEMFCs' sluggish dynamic response caused by intrinsic electrochemical latency and limited load-following capabilities – critical limitations impeding commercial deployment in dynamic robotic applications [5]. To overcome these constraints, researchers propose hybrid power architectures combining PEMFCs with lithium battery buffers, creating synergistic energy systems that satisfy instantaneous power demands during complex maneuvers. Such integration not only reduces fuel cell stress but also improves overall energy utilization efficiency and system durability through intelligent energy management strategies (EMS).

The advancement of machine learning has spurred significant exploration of reinforcement learning (RL) for EMS design. For instance, Q-learning-based frameworks have been implemented to achieve real-time power optimization in fuel cell hybrid system [7], while multilayer perceptron architectures have been integrated with fuzzy controllers to enhance fuel efficiency and service durability through driving pattern recognition and power allocation in fuel cell/ultracapacitor systems [8]. Further extensions include multi-source energy systems managed by RL strategies, which optimize both fuel cell efficiency and energy consumption economy [9]. Nevertheless, conventional Q-learning approaches suffer from computational inefficiency in real-time applications due to high-dimensional discrete state and action spaces [10]. Comparative studies reveal that double Q-learning algorithms outperform traditional counterparts in optimization performance [11].

Despite these advancements, critical challenges persist. Hybrid mobile robot energy systems are inherently susceptible to multi-scenario temperature perturbations arising from dynamic environments, which compromise model accuracy and induce solution drift during optimization [12]. This positions parameter uncertainty as a pivotal concern in multi-energy power allocation, effectively framing it as an uncertain optimization problem. Additionally, although DRL achieves superior optimization, its dependency on deep neural networks introduces latency constraints that challenge real-time applicability.

To this end, this study aims to address the critical challenge of dynamic power allocation in fuel cell hybrid mobile robots under varying temperature conditions. By integrating type-2 fuzzy logic with Q-learning, we propose a novel EMS to achieve robust optimization of system efficiency, hydrogen economy, and component durability across diverse thermal environments. This paper begins by modeling the fuel cell mobile robot system. The relation between maximum efficiency point performance and the uncertain parameter is analyzed. To achieve the power distribution, the multilayer type-2 fuzzy Q-learning (T2-FQL) algorithm is designed in the EMS scheme. Meanwhile, the concept of temperature scenario is introduced and integrated into the idea of the Markov decision process (MDP) to enhance the generalization. This paper is organized as follows: Section 2 provides a detailed mathematical model of the fuel cell and battery hybrid mobile robot system. Section 3 outlines the multilayer T2-FQL for EMS scheme design. Section 4 presents the hardware-in-the-loop (HIL) experiment results, and Section 5 concludes with a summary of the key findings.

2. System modeling of hybrid system

2.1. Fuel cell system modeling

The PEMFC operates through a sophisticated electrochemical conversion process and the fundamental energy conversion mechanism generates electrical current through the multi-stage process. PEMFC generates electricity and water through an electrochemical reaction between hydrogen gas from a hydrogen tank at the anode and oxygen from a compressor at the cathode, which in turn supplies power to the load. The overall voltage of the fuel cell can be expressed using the Nernst equation as follows:

$$V_{total} = 1.23 - 8.5 \times 10^{-4} \times (T_{FC} - 298) + 4.31 \times 10^{-5} \times T_{FC} \left[\ln(P_{H_2}) + 0.5\ln(P_{O_2})\right] \quad (1)$$

where T_{FC} is the fuel cell temperature. P_{H_2} and P_{O_2} are the hydrogen partial pressure and oxygen partial pressure. Taking these loss voltages into account, the operating voltage of a single fuel cell V_{fc} and total power P_{stack} can be expressed as follows:

$$V_{fc} = V_{total} - V_{act} - V_{ohm} - V_{conc} \quad (2)$$

$$P_{stack} = iV_{fc}A_{fc}N_{cell} \quad (3)$$

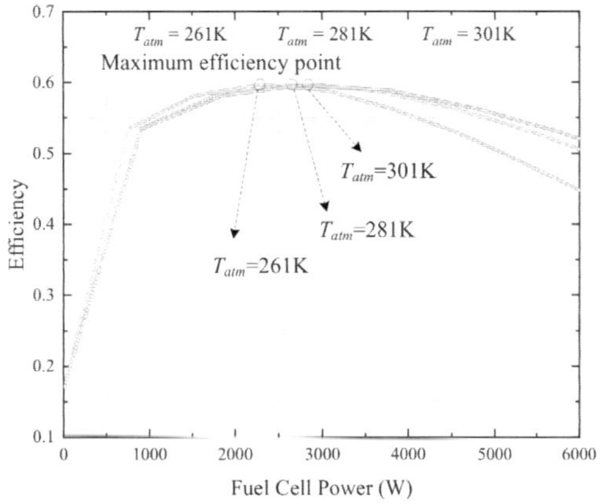

Fig. 1. Fuel cell efficiency curve at different temperatures.

where V_{act}, V_{ohm} and V_{conc} are the activation loss, Ohmic loss and concentration loss, respectively. i represents the current density. N_{cell} is the number of fuel cell. The majority of parasitic power is attributed to the power consumption of the compressor and the power of the compressor P_{cp} can be determined using the following equation:

$$\begin{aligned} P_{cp} &= \tau_{cp}\omega_{cp} \\ \tau_{cp} &= \frac{C_p T_{atm}}{\omega_{cp}\eta_m\eta_{cp}}\left[\left(\frac{p_{ca}}{p_{in}}\right)^{\frac{\gamma_a-1}{\gamma_a}} - 1\right] m_a \\ m_a &= 3.57\times10^{-7}\lambda_{O_2} i A_{fc} n_{cell} \end{aligned} \tag{4}$$

where T_{atm} represents the atmospheric temperature. p_{ca} and p_{in} represent the cathode pressure and inlet air pressure of compressor. λ_{O_2} is the oxygen excess ratio. The consumed hydrogen mass can be calculated by:

$$m_{H_2} = \int_0^t \frac{P_{net}(\tau)}{\eta_{FC}\cdot \mathrm{LHV}} \tag{5}$$

$$P_{net} = P_{stack} - P_{cp} \tag{6}$$

where P_{net} is the net power. LHV represents the low heating value of hydrogen. η_{FC} is the efficiency of fuel cell.

To elucidate the operational point drift phenomenon in PEMFCs, Fig. 1 presents a comparative analysis of efficiency-power characteristics under varying environmental conditions. The quantitative analysis reveals that the maximum efficiency operating point exhibits significant dependence on ambient temperature fluctuations. At 261K, the PEMFC achieves peak efficiency (η=61.2%) at 2500W output power. However, when ambient temperature increases to 298K, the maximum efficiency operating point shifts to 2750W with η=58.7%, demonstrating a 6.8% power requirement increase for optimal efficiency maintenance.

2.2. Battery system modeling

Lithium-ion batteries serve dual critical functions in hybrid energy systems: providing primary power supply and enabling regenerative energy recovery during braking phases. To maintain DC bus voltage stability, the battery pack is interfaced with the DC bus through a bidirectional

DC-DC converter, with its terminal voltage governed by the following dynamic model:

$$V_{bat} = V_{oc}(\text{SOC}) - I_{bat}R_{bat}(\text{SOC}) \tag{7}$$

where V_{oc} is the open-circuit voltage. R_{bat} is the internal resistance.

The current and power of battery can be represented as follows:

$$\begin{aligned} I_{bat} &= \frac{V_{oc}(\text{SOC}) - \sqrt{V_{oc}(\text{SOC})^2 - 4R_{bat}(\text{SOC})P_{bat}}}{2R_{bat}(\text{SOC})} \\ P_{bat} &= V_{bat}I_{bat} \end{aligned} \tag{8}$$

where the V_{oc} and R_{bat} can be represented as the functions of SOC. The SOC of the battery represents the remaining capacity, which is defined as follows:

$$\text{SOC}(t) = \text{SOC}_{ini} - \int_0^t \frac{\eta_{bat}I_{bat}(\tau)}{Q_{bat}}d\tau \tag{9}$$

where η_{bat} is the battery efficiency.

To assess the durability index of the battery, the battery state-of-health (SOH) is utilized as a metric, which is defined as follows:

$$\text{SOH}(t) = \text{SOH}_{ini} - \int_0^t \frac{|I_{bat}(\tau)|}{2N_{EoL}(c_{rate})Q_{bat}}d\tau \tag{10}$$

where $N_{EoL}(c_{rate})$ represents the total number of cycles.

3. Multilayer type-2 fuzzy Q-learning energy management strategy for hybrid mobile robot system

3.1. Theoretical Analysis of RL-based EMS

The purpose of performance optimization is to find the optimal fuel cell power so that the comprehensive performance can approach the ideal performance as much as possible, which can be expressed as follows:

$$\begin{cases} \max\limits_{P_{fc}} F(P_{fc}(t)) = \Sigma_{n=1}^{N}\Sigma_{t=0}^{T}\alpha_n f_n(t) \\ P_{fc}(t+1) = f\left(s(t), \Delta P_{fc}(t)\right) \end{cases} \tag{11}$$

where $F(\cdot)$ and $f_n(\cdot)$ are the comprehensive performance and specific performance. α_i is the weight coefficient. $s(t)$ is the current state. $P_{fc}(t)$ is the reference fuel cell power. The dynamic process of RL can be formulated as a MDP, denoted with a data tuple $\langle s_t, a_t, r_t, s_{t+1}\rangle$, where s_t, a_t, r_t represent the current state, action, reward and future state, respectively. s_{t+1} represents the future state.

In traditional Q-learning scheme, the updated principle is based on the following expression:

$$\text{New } Q\left(s_t, a_t\right) = Q\left(s_t, a_t\right) + \alpha\left[r_t + \gamma \max_a Q\left(s_{t+1}, a\right) - Q\left(s_t, a_t\right)\right] \tag{12}$$

where s_t and a_t represent the discrete values for states and actions in Q table. Increasing the number of discrete points can increase the possibility of selecting the best performance points, but on the other hand, it will increase the dimension of the system and the computational complexity.

To solve the problem of excessive discrete Q table dimensions, the FLS is used to compute the continuous action u_t:

$$u_t = \frac{\sum_{l=1}^{M_s}\left(\left(\prod_{i=1}^{n_s}\mu_s^{F_i^l}(s_i)\right)\cdot a_t^l\right)}{\sum_{l=1}^{M_s}\left(\prod_{i=1}^{n_s}\mu_s^{F_i^l}(s_i)\right)} = \sum_{l=1}^{M_s}\Phi_s^l a_t^l \tag{13}$$

where M_s represents the number of rules and n_s represents the number of states. μ_s represents the degree of membership for states. a_t represents the discrete action.

Conventional Q-learning suffers from high-dimensional discrete state-action spaces, the proposed framework reduces complexity through continuous state representation in Eq. 13. By combing with fuzzy Q-learning (FQL) method, the MDP of EMS can be established as the following process:

$$\begin{cases} \max R(s_t) = \sum_{t=0}^{T}\gamma^t r_t \\ s_t = [P_{req,t}, \mathrm{SOC}_t] \\ u_t = \triangle P_{fc,t} = \sum_{l=1}^{M_s}\Phi_s^l a_t^l \\ r_t = F(P_{fc,t}) \\ P_{fc,t+1} = P_{fc,t} + \triangle P_{fc,t} \end{cases} \tag{14}$$

As revealed by the temperature sensitivity analysis in Section 2, the conventional strategy Eq. 14 demonstrates scenario-specific optimization capabilities, achieving peak performance exclusively under fixed operating conditions. However, its efficacy degrades significantly when confronted with dynamic parametric variations, necessitating computationally intensive re-optimization for each new scenario. Besides, the methodology enhances Q-learning generalization through the integration of a FLS, which maps continuous state-action spaces into interpretable linguistic rules. To overcome these limitations and enable robust cross-scenario generalization, this work proposes a multilayer T2-FQL framework:

1) For simplicity, singleton fuzzification is applied. We use product inference and obtain the following firing interval $\tilde{A}^j$ of each rule:

$$\tilde{A}^l : \begin{cases} \tilde{A}^l(\boldsymbol{s}) \equiv [\underline{a}^l(\boldsymbol{s}), \bar{a}^l(\boldsymbol{s})] \\ \underline{a}^l(\boldsymbol{s}) \equiv \prod_{i=1}^{n_s}\underline{\mu}_{\tilde{A}_i}(s_i) \\ \bar{a}^l(\boldsymbol{s}) \equiv \prod_{i=1}^{n_s}\bar{\mu}_{\tilde{A}_i}(s_i) \end{cases} \tag{15}$$

2) Compute the following interval weighted average:

$$Y_{\mathrm{TSK}}(\boldsymbol{s}) = 1/\sum_{l=1}^{M_s}\left[a_t^l\tilde{A}^l/\sum_{l=1}^{M_s}\tilde{A}^l\right] = 1/[y_l(\boldsymbol{s}), y_r(\boldsymbol{s})] \tag{16}$$

The computations of $y_l(\boldsymbol{s})$ and $y_r(\boldsymbol{s})$ are performed by using KM algorithms that are applied to:

$$\begin{cases} y_l(\boldsymbol{s}) = \min_{\forall a_l\in[\underline{a}^l(s),\bar{a}^l(s)]}\sum_{l=1}^{M_s}a_t^l a_l^l/\sum_{l=1}^{M_s}a_l^l \\ y_r(\boldsymbol{s}) = \max_{\forall a_l^r\in[\underline{a}^l(s),\bar{a}^l(s)]}\sum_{l=1}^{M_s}a_t^l a_l^r/\sum_{l=1}^{M_s}a_l^r \end{cases} \tag{17}$$

3) Compute the defuzzified output as continuous variable:

$$u_t = \frac{1}{2}[y_l(\boldsymbol{s}) + y_r(\boldsymbol{s})] \tag{18}$$

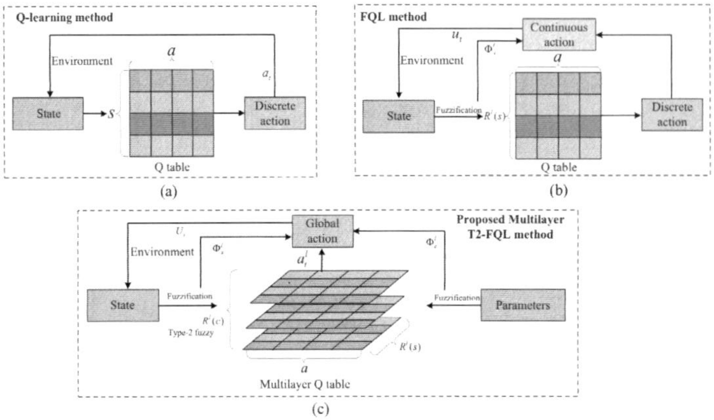

Fig. 2. Different optimization schemes. (a) Q-learning method. (b) FQL method. (c) Multilayer T2-FQL method.

4) Compute the global output at different environment parameters:

$$U_t = \frac{\sum_{l=1}^{M_c}\left(\left(\prod_{i=1}^{n_c}\mu_c^{F_l^l}(c_i)\right)\cdot u_t^l\right)}{\sum_{l=1}^{M_c}\left(\prod_{i=1}^{n_c}\mu_c^{F_i^l}(c_i)\right)} = \sum_{l=1}^{M_c}\Phi_c^l u_t^l \tag{19}$$

The interactive process between agent and environment in multi-scenario temperature can be implemented and the generalization ability can be improved. The different schemes for the Q-learning, FQL and proposed multilayer T2-FQL are compared in Fig. 2.

3.2. Description of multilayer T2-FQL-based EMS Scheme

Different from the traditional Q-learning method, the proposed multilayer T2-FQL can find the continuous optimal global action in different multi-scenario temperature.

Consider the fuel cell mobile robot as an intelligent agent, and design the three elements of RL: state space, action space and reward function.

1) State space: To capture the dynamic characteristics of the system across different condition, we utilize the required power and SOC values of battery as state variables, which are defined as: $s_t = \{P_{req}(t), SOC(t)\}$. By incorporating these values into the state, we aim to improve the optimization performance of agent.

2) Action space:

This paper focus on the power distribution of hybrid power system to obtain the optimal system performance. Therefore, we choose the variation value of fuel cell power as the action variables, which are defined as: $a_t = \{\Delta P_{fc,t}\}$. The remaining power is replenished by lithium battery.

3) Reward function:

Based on the above analysis, the EMS problem for mobile robot can be transformed into the markov process, with the optimization controller viewed as the agent, and overall mobile robot various power sources regarded as the environment. The agent interacts with the environment to acquire control strategies that achieve power flows and improved performance including the power flows index, lower hydrogen cost, higher efficiency, appropriate SOC and SOH. Thus, the optimizing problem can be included to achieving the minimum cost function at each time step:

$$J_{\text{total}} = J_{\text{per}} + J_{\text{H}_2} + J_{\text{Eff}} + J_{\text{SOC}} + J_{\text{SOH}} \tag{20}$$

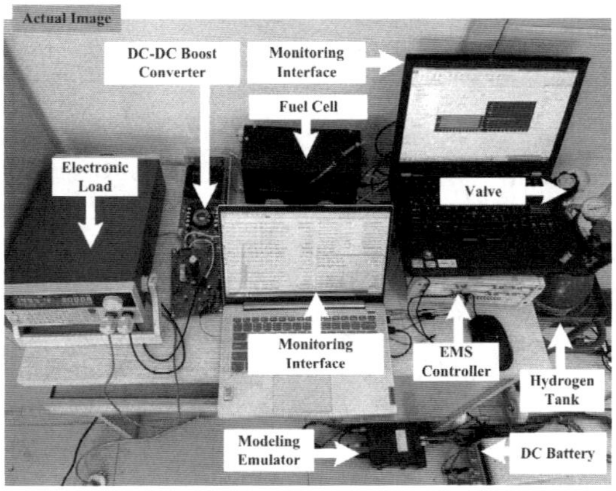

Fig. 3. HIL test platform.

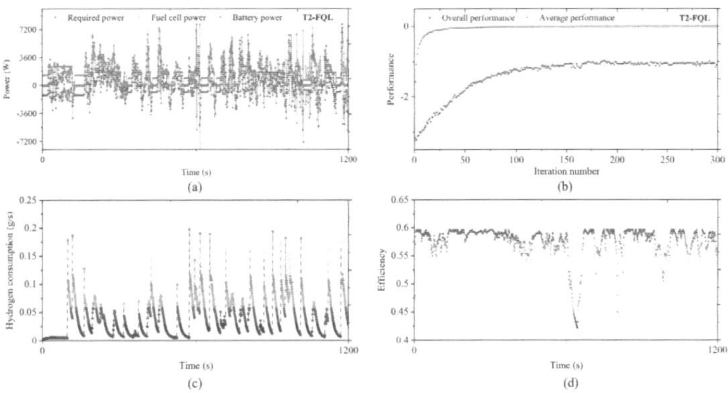

Fig. 4. Performance evolution of T2-FQL method.

4. Results and discussion

A HIL experimental platform was established to validate the effectiveness of the proposed multilayer FQL method. The experimental setup comprises the following key components: a fuel cell system, programmable electronic load, real-time ES910 simulator, ES930 module, dSPACE controller, and auxiliary instrumentation. The schematic diagram of the implemented experimental architecture is presented in Fig. 3.

A power segment is selected as the working condition of the hybrid power system and trained under different temperatures and pressures. The results are shown in Fig. 4. Fig. 4(a) shows the required power of the hybrid power system, the power distribution of the fuel cell and the battery. It can be seen that the fuel cell can better follow the required power, and the remaining insufficient power is supplemented by the battery. Fig. 4(b) shows the evolution of the reward value in the iterative process, and it can be seen that the reward value gradually increases and the final region is stable. Fig. 4(c) and 4(d) shows the performance of each performance of the system battery, and it can be seen that at low power requirements, the fuel cell power value is lower, and therefore the hydrogen consumption is also less. In addition, the efficiency of the fuel cell in the middle and high power working stage is always in the high efficiency range. The battery SOC was also maintained relatively well, with little difference from the initial value.

5. Conclusion

This article proposes the multilayer T2-FQL-based EMS control for a fuel cell hybrid power system that can automatically adjust the power distribution in a multi-scenario temperatures. Based on the system modeling of a hybrid system in different working conditions and the analysis of system performance, the RL algorithm is introduced into the EMS process and the optimization mechanism including the states, reward and action is proposed. Different multi-scenario parameter conditions are randomly injected for the training process to enhance the generalization performance, and the comparison results with FQL and traditional Q-learning algorithms verify the advantages of the proposed multilayer T2-FQL method. In the HIL experiment tests with different environment parameters are carried out and the results demonstrated that the multilayer T2-FQL method achieves better total performance, which could be applied as the reference for the power optimization of hybrid power applications.

Acknowledgments

This work was supported by the National Natural Science Foundation of China under Grant 52405535, and the Natural Science Foundation of Liaoning Province under Grant No. 2023-BSBA-101, and Postdoctoral Science Foundation, Northeastern University under Grant 20240103, and the Fundamental Research Funds for the Central Universities of China under Grant N25ZJL016.

References

[1] X. Lü, R. Deng, C. Chen, Y. Wu, R. Meng, and L. Long, "Performance optimization of fuel cell hybrid power robot based on power demand prediction and model evaluation," *Appl. Energy*, vol. 316, Jun. 2022, Art. no. 119087.

[2] J. Xie and T. Yao, "Quantified assessment of internal short-circuit state for 18650 batteries using an extreme learning machine based pseudo-distributed model", *IEEE Trans. Transp. Electrif.*, vol. 7, no. 3, pp. 1303–1313, Sep. 2021.

[3] O. Erdinc and M. Uzunoglu, "Recent trends in PEM fuel cell-powered hybrid systems: Investigation of application areas, design architectures and energy management approaches", *Renew. Sustain. Energy Rev*, vol. 14, no. 9, pp. 2874–2884, Dec. 2010.

[4] M.U. Farooq, A. Eizad, and H.K. Bae, "Power solutions for autonomous mobile robots: A survey," *Robot. Auton. Syst.*, vol. 159, Jan. 2023, Art. no. 104285.

[5] L. Xu, Y. Huangfu, R. Ma, R. Xie, Z. Song, D. Zhao, Y. Yang, Y. Wang adn L. Xu, "A comprehensive review on fuel cell UAV key technologies: propulsion system, management strategy, and design procedure," *IEEE Trans. Transp. Electrif.*, vol. 8, no. 4, pp. 4118–4139, Dec. 2022.

[6] R. Zhang, J. Tao, and H. Zhou, "Fuzzy optimal energy management for fuel cell and supercapacitor systems using neural network based driving pattern recognition," *IEEE Trans. Fuzzy Syst.*, vol. 27, no. 1, pp. 45–57, Jan. 2019.

[7] H. Sun, Z. Fu, F. Tao, L. Zhu, and P. Shi, "Data-driven reinforcement-learning-based hierarchical energy management strategy for fuel cell/battery/ultracapacitor hybrid electric vehicles," *IEEE Trans. Fuzzy Syst.*, vol. 455, Mar. 2020, Art. no. 227964.

[8] Q. Li, W. Chen, Z. Liu, M. Li, and L. Ma, "Development of energy management system based on a power sharing strategy for a fuel cell-battery-supercapacitor hybrid tramway," *J. Power Sources*, vol. 279, pp. 267–280, Apr. 2015.

[9] D. Yang, L. Wang, K. Yu, and J. Liang, "A reinforcement learning-based energy management strategy for fuel cell hybrid vehicle considering real-time velocity prediction," *Energy Convers. Manage.*, vol. 274, Dec. 2022, Art. no. 116453.

[10] Y. Zhang, R. Ma, D. Zhao, Y. Huangfu, and W. Liu, "A novel energy management strategy based on dual reward function Q-learning for fuel cell hybrid electric vehicle," *IEEE Trans. Ind. Electron.*, vol. 69, no. 2, pp. 1537–1547, Feb. 2022.

[11] Y. Wang, Y. Wang, and P. Li, "Adaptive hierarchical energy management strategy for fuel cell mobile robot hybrid power system based on working condition recognition," *Renew. Energy*, vol. 237, Oct. 2024, Art. no. 121628.

Vision-Guided Intelligent Robotic System for In-core Thermocouple Maintenance in Nuclear Reactors

Tengfei Zhang[1], Weiguo Han[2], Zhihui Zhang[2], Yufan Jiang[2], Zhengfa Liu[3], Dequan Zeng[4*]

[1] Artificial Intelligence R&D Center, CNNC Equipment Technology Research (Shanghai) Co., Ltd, Shanghai 201899, P.R. China

[2] First Maintenance Division, China National Nuclear Power Operation Management Co., Ltd, Jiaxing, Zhejiang Province, 314300, P.R. China

[3] School of Electrical and Electronic Engineering, Shanghai Institute of Technology, Shanghai 201418, P.R. China

[4] School of Mechatronics and Vehicle Engineering, East China Jiaotong University, Nanchang, Jiangxi Province, 330013, P.R. China

[*]Corresponding author's e-mail: 3340@ecjtu.edu.cn

Abstract. The Maintenance of reactor core thermocouples, especially online welding operations following component replacement, poses critical challenges, including its stringent precision requirements, specialized welding processes, complex implementation conditions, and limited workspace. In response to these challenges, this paper proposes an online intelligent welding robot system. The system is primarily composed of 5 parts: a three-stage high-precision visual perception system, a six-axis robotic arm with adaptive tracking, a dedicated laser welding system, a control and interactive center, and auxiliary equipment. The research achieves an overall positioning error of $\leq$0.1 mm at a variable distant and an operational time within 3 minutes, which can ensure automated, standardized, and remotely controlled high-precision welding repair. Preliminary validation on a dedicated simulator platform on 1:1 scale to the real scene confirms its effectiveness, marking it as the first complete system in the domestic nuclear power sector for automated repair of reactor core thermocouples. This technology introduces a new method for the intelligent operation and maintenance of critical nuclear power plant equipment and can be extended to other high-precision repair applications.

1. Introduction

In the operation and maintenance of nuclear power plants, repairing reactor core thermocouples is both a critical and highly challenging task. Traditional repair methods require the complete removal of the thermocouple from the reactor, a process that demands operators with specialized qualifications and is accompanied by high costs, significant radiation exposure, and considerable safety risks. Field inspections have revealed that most failures originate from connection issues at the circuit connectors; in many cases, simply replacing the connectors outside the reactor core and performing localized welding is sufficient to restore functionality. However, the domestic welding process for thermocouples has yet to mature, facing several challenges, including:

1. **Detection and Positioning Challenges:** The welding equipment must be capable of accurately navigating from an external position over 50 cm away into the confined area where the thermocouple connectors are installed.
2. **High-Precision Recognition Requirements:** It is essential to clearly identify key components, such as the thermocouple connection sheath and wires, and to precisely segment their boundaries to provide accurate six-degree-of-freedom (6DOF) information, which in turn guides the subsequent positioning of the driving mechanism for vertical welding.
3. **Welding Precision Demands:** Given that the thermocouple wire has a diameter of only 0.8 mm, the welding process demands extremely high precision; the overall spatial error must be $\leq$0.2 mm.
4. **Stringent Welding Process Requirements:** Nuclear power plants impose rigorous standards on the welding process, requiring a filler-free operation, tailored melt pool and weld position control for different materials, and weld integrity that meets specialized testing standards.
5. **Operational Environment Constraints:** Limited workspace and unique installation conditions necessitate that the equipment be designed to be both lightweight and compact.

Furthermore, existing high-precision visual recognition technologies have their limitations. Although fixed-focus cameras can achieve micron-level accuracy, they are not directly applicable when the initial position is uncertain and the detection distance exceeds 50 cm. In contrast, typical non-fixed-focus 3D cameras exhibit errors at the millimeter level, which are inadequate for accurately positioning thermocouple wires with a diameter of 0.8 mm. Coupled with the strict technical requirements of nuclear power operations, conventional methods such as electric welding, tungsten inert gas welding, and resistance welding are unsuitable. While laser welding shows promise, its bulky configuration, the need for fine-tuning of melt pool parameters, and the requirement for remote wireless operation to suit high-risk environments further complicate its application. To date, a complete solution addressing these issues is absent in domestic practice. To overcome these challenges, this paper presents an innovative online repair robotic system guided by high-precision vision, which enables intelligent welding repair that meets on-site requirements (refer to Fig. 1). This system is the first advanced device in the domestic nuclear power sector to satisfy the relevant standards and has been comprehensively validated on a simulation experimental platform. Its main contributions include:

Long-Range High-Precision Vision Guidance System: By integrating the advantages of fixed-focus and non-fixed-focus 3D cameras, a three-stage high-precision visual recognition algorithm is proposed to acquire 6DOF high-precision information of the welding point.

Adaptive Tracking Motion Control Method: Incorporates deep learning-based object detection techniques to optimize the trajectory of the robotic arm.

Welding Process Optimization: Enhances welding success through parameter tuning and inert gas shielding; employs a design that separates the welding gun head from the laser generator to reduce the operational footprint, and includes a coaxial camera for real-time welding quality monitoring.

Modular Mechanical Structure Design: Facilitates easy assembly and disassembly, significantly reducing setup time and enhancing practicality.

Remote Wireless Control Welding System: Supports an intelligent repair process via web or app, validated on certain nuclear power station simulation platform, ultimately achieving an overall positioning error of < 0.1 mm and an operation time of <3 minutes.

2. Related Work

Recent advances in nuclear welding technology have paved the way for improved manufacturing and repair processes. LI Xiang and LIU Huan [1] explored the localization and intelligent development of welding in the nuclear industry, offering valuable insights into reducing dependence on imported technology. The New Nuclear Manufacturing (NNUMAN) Programme [2] provided a comprehensive review of automated welding processes for nuclear applications, while studies on resistance, electron, and laser-beam welding of zirconium alloys [3] have further clarified the challenges faced in nuclear welding. In addition, early work by Kobayashi et al. [4] demonstrated the potential of automatic welding techniques for nuclear power plants in enhancing welding precision and safety. However,

Journal of Physics: Conference Series **3024** (2025) 012039 doi:10.1088/1742-6596/3024/1/012039

despite these advancements, automated welding repair of reactor core thermocouples remains challenging due to the need for high-precision positioning in confined spaces and strict process control. This underscores the urgent demand for innovative, vision-guided online repair systems.

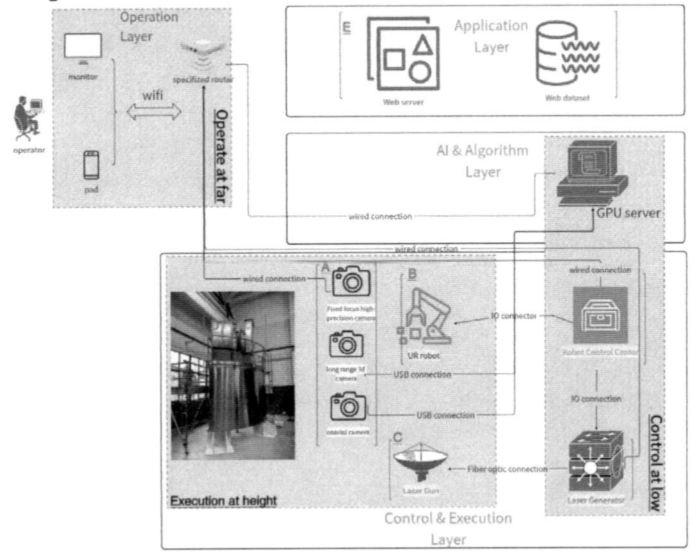

Figure 1. Overall of whole robotic system. The main subsystems are labeled with red capital letters.

Figure 2. crosshair in Coaxial Camera

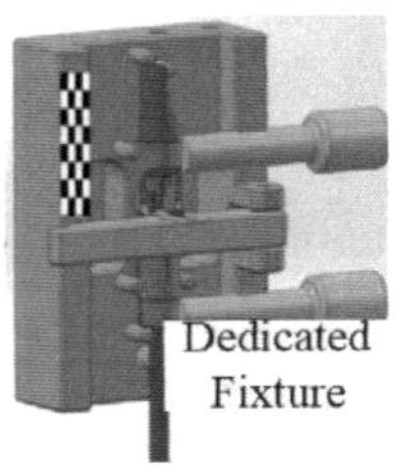

Figure 3. Dedicated Fixture with pose information

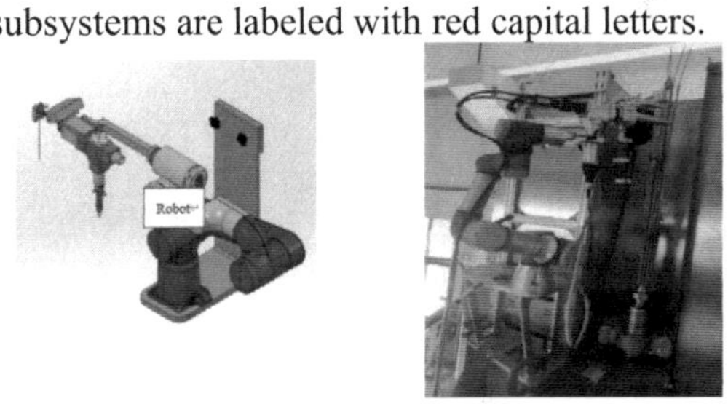

Figure 4. Design of robot with cameras, laser gun and fixed bracket

3. Method

3.1. Visual Guidance System

3.1.1. Three-Stage High-Precision Visual Localization Algorithm. **Stage 1: Long-Range Coarse Localization and Target Detection:** During the long-range localization phase (working distance >50 cm), a non-telecentric 3D camera (millimeter-level depth error) combined with an enhanced YOLOv8 [5] detection framework achieves rapid coarse localization of thermocouple targets.

Robust Target Detection: A lightweight YOLOv8-based model outputs target bounding boxes $B = [x_{min}, y_{min}, x_{max}, y_{max}]$.

Depth-to-Position Mapping: Using the 3D camera's depth API, the bounding box center pixel coordinates $(u_c, v_c) = ((x_{min} + x_{max})/2, (y_{min} + y_{max})/2)$ are converted to 3D coordinates $P_c \in R^3$ in the camera coordinate system:

$$P_c = d \cdot \mathbf{K}^{-1} \begin{bmatrix} u_c \\ v_c \\ 1 \end{bmatrix}, K = \begin{bmatrix} f_x & 0 & c_x \\ 0 & f_y & c_y \\ 0 & 0 & 1 \end{bmatrix} \quad (1)$$

where d is the depth value, **K** is the camera intrinsic matrix, f_x, f_y are focal lengths, and (c_x, c_y) denotes the principal point.

Stage 2: Short-Range High-Precision Segmentation and 3D Localization: When the target enters the field of view of the telecentric high-precision camera (working distance: 10 cm ±5 mm, micron-level positioning error), the system switches to a UPerNet [6] semantic segmentation network for sub-pixel feature extraction:

Fine-Grained Semantic Segmentation: Utilizing a Swin Transformer backbone, UPerNet outputs pixel-level masks $M \in \{0,1\}^{H \times W}$ with classes including Background, Terminal, Wire, and Calibration Board. This enhances multi-scale feature capture via Atrous Spatial Pyramid Pooling [7].

Post-Processing and Geometric Centroid Calculation: a) Morphological Refinement: Apply closing operations (3×3 circular kernel) to fill small gaps and fractures. **b) Convex Hull Extraction:** Compute the minimal enclosing convex hull H using Andrew's monotone chain algorithm to ensure contour continuity. **c) Centroid Localization:** Calculate geometric moments $M_{pq} = \sum_{(x,y) \in H} x^p y^q$ via OpenCV. The centroid coordinates (x_g, y_g) are:$x_g = \frac{M10}{M00}$, $y_g = \frac{M01}{M00}$.

3D Coordinate Mapping: Invoke the 3D camera API to map centroid pixel coordinates (x_g, y_g) to 3D space, generating precise welding point coordinates $P_c = [X_w, Y_w, Z_w]^T$.

Stage 3: Coaxial Camera for Weld Quality Monitoring: A coaxial camera with customized aperture settings is deployed to overcome angular occlusion caused by the welding torch. It aligns the camera's crosshair with the welding point through real-time calibration, enabling continuous observation of weld quality (e.g., molten pool geometry and defects) during the welding process.

3.1.2. Thermocouple Pose Estimation Algorithm with Dedicated Fixture: PnP-Based Pose Solving Method. A dedicated fixture securely holds the thermocouple wire, ensuring that both the wire and the fixture maintain identical pose states. A calibration chessboard is mounted on the fixture, enabling the use of a monocular camera and the Perspective-n-Point (PnP) algorithm to estimate the chessboard's pose. The detail can be seen at Fig 3.

Coordinate System Definition and Problem Formulation:

1. World Coordinate System: Defined by the calibration chessboard on the fixture, with the origin at the top-left corner and the Z-axis perpendicular to the chessboard plane. The 3D coordinates of chessboard corners $\{X_i^W = (x_i, y_i, 0)\}$ are known (grid size: 2.5 mm).

2. Camera Projection Model: Let C and I denote the camera and image coordinate systems, respectively. The perspective projection relationship for corners is:

$$s \begin{bmatrix} u_i \\ v_i \\ 1 \end{bmatrix} = \mathbf{K}[\mathbf{R} \mid \mathbf{t}] \begin{bmatrix} X_i^W \\ 1 \end{bmatrix}, \mathbf{R} \in SO(3), \mathbf{t} \in R^3 \tag{2}$$

where s is a scale factor, and **R**, **t** represent the rotation matrix and translation vector from the world to camera coordinate system.

Robust PnP Solving: The OpenCV solvePnPRansac [8] algorithm is employed to compute **R** and **t**. That is to say, if we know the some chessboard corners in 3D conordinates and corresponding points in 2D camera picture, we can get **R**, **t** by using solvePnPRansac. The impact related to pose estimation is relatively small compared to 3D space coordinates. Using this method, a preliminary rough estimate shows an error of less than 5 degrees, which basically meets actual needs.

Robot Base Coordinate Transformation: Using hand-eye calibration, the camera-to-robot base transformation $\{\boldsymbol{R_{cb}}, \boldsymbol{t_{cb}}\}$ is pre-calibrated. The thermocouple's pose in the base coordinate system B:

$$\boldsymbol{R_b} = \boldsymbol{R_{cb}} \boldsymbol{R_c}, \quad \boldsymbol{t_b} = \boldsymbol{R_{cb}} \boldsymbol{t_c} + \boldsymbol{t_{cb}} \tag{3}$$

This transformation bridges vision-based localization with robotic kinematics, ensuring 6DOF welding precision.

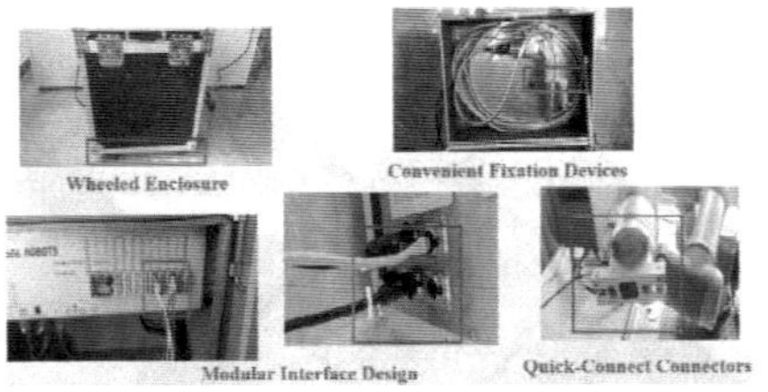

Figure 5. Engineering Optimization

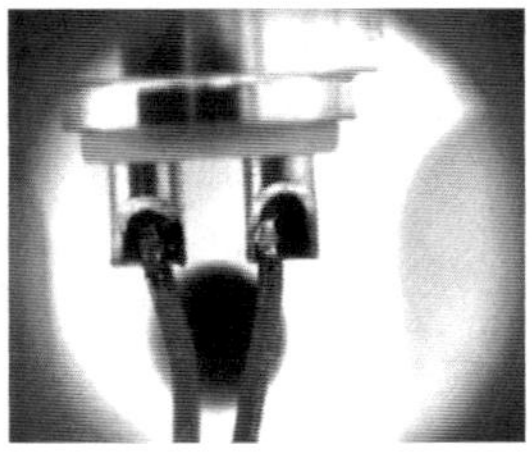
Figure 6. Welding effect in Coaxial Camera

3.2. Automatic Positioning Mechanism

Robotic Arm Motion Control: The UR5 6-DOF collaborative robotic arm (payload: 3 kg) was selected for its compact design and high repeatability (±0.03 mm). Control was implemented through ROS-based API integration, enabling Ethernet/IP communication for trajectory planning and synchronized digital I/O operations.

Multi-Camera Coordinate System Switching: A hierarchical coordinate transformation strategy was developed for three operational phases:

Stage 1: Eye-to-hand calibration achieved <3 mm positioning errors using Tsai-Lenz algorithm[9]. Robotic end-effector setting is not strict at this stage.

Stage 2: The robotic end-effector was constrained to the telecentric camera's focal plane (depth-of-field: ±5 mm), ensuring 4K imaging clarity for subpixel feature extraction.

Stage 3: Through iterative experimental validation, the robotic end-effector was positioned at 10cm from the welding torch tip, empirically determined as the optimal standoff distance to balance weld quality assurance and real-time visual monitoring precision.

Synchronized Laser Activation: A hardwired I/O interface (24V DC optocoupled) connected the robot controller to the laser generator. TTL pulse signals triggered laser emission precisely at waypoints, synchronized with robot motion through cross-compiled state machines. This hardware-level synchronization reduced timing jitter to <0.1ms, critical for repeatable weld seam formation.

3.3. Welding System

(i) ***Welding Equipment:*** The system utilizes a YLM ytterbium-doped fiber laser and is controlled via a PLC. A dedicated debugging interface allows for flexible adjustment of process parameters, and the system supports both hard-wired and remote limited operation for switching. ***(ii) Welding Parameter Tuning:*** For specific materials, the melt pool parameters and welding distance are optimized to achieve the best welding performance. ***(iii) Welding Process Protection:*** An auxiliary argon gas supply is integrated to protect the welding process. "Pre-activation" refers to turning on the argon gas in advance so that the protective atmosphere reaches the welding area before the process begins, effectively preventing oxidation of the workpiece; "post-delay" refers to delaying the shutdown of the gas supply after the laser is activated to ensure adequate protection of the weld pool. ***(iv) Lightweight Improvements:*** The welding gun is equipped with a camera and features a hollow, optimized structural design to reduce weight. Additionally, the laser generator is separated from the gun head, with only the gun head positioned on the operating table, significantly reducing the operating weight and enhancing responsiveness and flexibility. The total weight of the gun and cameras are < 2.6kg.

3.4. Supporting Tools and optimal engineering design

Bracket Design: A multi-level, modular lightweight bracket system is implemented to accommodate various installation heights and usage requirements. This design ensures stable mounting of modules across different levels, while providing excellent structural stability and ease of use.

Engineering Optimization: ***(i) Wheeled Enclosure:*** A wheeled enclosure is designed to facilitate easy transportation and efficient on-site deployment. ***(ii) Convenient Fixation Devices:*** Easy-to-install and remove fixation devices are employed to ensure that all components remain secure and stable during operation. ***(iii) Modular Interface Design:*** The system features integrated and bundled modular interfaces for IO, signals, power, and network cables, streamlining connectivity and maintenance through standardized design. ***(iv) Quick-Connect Connectors:*** Quick-connect connectors are provided for the welding gun and robotic arm, enabling rapid assembly and disassembly to enhance operational efficiency on-site.

Table 1. Target Detection Quantitative Results

Model	AP (IoU=0.50:0.95)	AR (IoU=0.50:0.95)
YOLOX[10]	65.0%	70.3%
YOLOv8	72.1%	75.0%
PPYOLOE[11]	68.0%	73.2%

Table 2. Semantic Segmentation Results (%)

Model	aAcc	mIoU	mAcc
UPerNet	92.2	80.5	84.3
Segformer[12]	91.5	78.1	82.0
Segmenter Mask[13]	90.8	77.9	81.5

Table 3. Welding parameters optimization

No.	Current (%)	Pulse Width (ms)	Penetration	Pull Test	Resistance (Ω)	Remarks
1	50	5	No	Yes	0.1	Excellent welding, recommended
2	50	6	Yes, with spatter	Yes	0.1	Slightly high power
3	40	5	No	No, not welded	∞	Detaches with slight pull
4	60	5	Yes	Yes	0.1	Severe penetration, backside scorched
5	70	5	Yes	No	0.1	Thermocouple wire mostly melted, easy to break with lateral pull

Note: In welding processes, the Pulse Repetition Frequency (PRF) is typically set at 1Hz based on empirical parameters. The welding distance is generally chosen at the focal position, where the coaxial fixed-focus camera should achieve clear imaging. Through multiple measurements of the weld spot size, the focal distance has been preliminarily determined to be 13.6 millimeters. After establishing the welding distance, optimization adjustments are primarily made to the current and pulse width.

Table 4. Welding effect

Test Number	Final Error (mm)	Welding Quality	Welding Successful	Conductivity (%)	Time
1	0.03	Excellent	Yes	100	2min30s
2	0.03	Good	Yes	100	2min15s
3	0.06	Good	Yes	100	2min30s
4	0.06	Acceptable	Yes	100	2min40s
5	0.06	Acceptable	Yes	100	2min50s
6	0.00	Acceptable	Yes	100	2min10s
7	0.03	Acceptable	Yes	100	2min10s
8	0.00	Acceptable	Yes	100	2min10s
9	0.00	Acceptable	Yes	100	2min30s
10	0.09	Acceptable	Yes	100	2min30s

Note: The welding conditions and quality are based on on-site judgement by professional engineers, and effect example is shown in Fig. 6.

Table 5. Method comparation

Experiment	Components Included	Overall Positioning Error (mm)	Welding Surface Condition
Baseline	Target Detection Only	∞	Typically unable to weld thermocouples
Model A	Target Detection + Semantic Segmentation (without Pose Estimation)	0.6	Sometime welding points are very close to the boundary,
Model B	Model A + PnP Pose Estimation (without welding process protection and welding parameters optimization)	0.12	Overoxidation occurred after some time
Full Model	Model B + Welding Process Optimization	0.09	Acceptable

3.5. Human-Machine Interaction System.

(i) Hardware Control System: Comprising a server, a PLC-controlled laser generator, a robotic arm control box, and wireless communication modules, this system ensures efficient and coordinated operation of all components. ***(ii) Wireless Control Methods:*** Supports remote control via both a web platform and a mobile app, enabling users to monitor and operate the system in real time.

4. Experiment

4.1. Experimental Setup

(i) Overall Accuracy Experiment: On the simulation platform, measurements were taken at 10 different positions on various thermocouples. The error from the initial position to the final welding position was recorded (in mm) by controlling the robotic arm movement (0.03 mm increments) and visually confirming the alignment. Ten measurements were recorded to evaluate the actual welding position error. ***(ii) Welding Process Experiment:*** By varying key parameters such as welding gun current, pulse width, light emission frequency, and the working distance of the welding gun head, 5 welding parameters optimization experiments were conducted. Welding quality was evaluated by visual inspection and conductivity measurements, while the welding position offset was recorded. ***(iii) Implementation Details.*** The experimental environment was established both in the laboratory and on a dedicated simulation platform to validate the automated reactor core thermocouple welding repair system. The setup includes a 3D camera for long-range target detection, a high-precision fixed-focus camera for precise localization, and a dedicated computing unit running deep learning algorithms. This environment simulates various operating conditions of nuclear power plants, such as different lighting, distances, occlusions, and installation heights.

4.2. Quantitative Analysis

Table 1 compares object detection models, where YOLOv8 achieves the best performance (AP 72.1%, AR 75.0%), followed by PPYOLOE, while YOLOX performs the lowest. This highlights YOLOv8's superior accuracy and recall, we choose yolov8 as our detection model choice. Table 2 presents semantic segmentation results, showing UPerNet outperforming others with mIoU 80.5% and mAcc 84.3%, benefiting from multi-scale feature fusion. So UPerNet is chosen as the main model for semantic segmentation process. Table 3 evaluates welding parameters, with 50% current , 5ms pulse width and 1Hz PRF , providing optimal welding quality. Higher currents cause excessive penetration, while lower currents result in weak bonding.Table 4 shows welding effects, with all tests achieving 100% conductivity and positioning errors within 0.09mm. The best welding quality is observed with robot arm motion minimal deviation (0.03mm error). Table 5 compares various methods, where the full model (target detection + segmentation + pose estimation + welding optimization) achieves the best accuracy (0.09mm) and acceptable welding surface, demonstrating its effectiveness.

4.3. Ablation Study

To evaluate the impact of different components in our method, we conduct an ablation study based on Tables 1–5. We progressively integrate key components, including target detection, semantic segmentation, pose estimation, and welding process optimization, to assess their contributions to overall performance. Additionally, from Table 3 and Table 4, the optimized welding parameters (50% current, 5ms pulse width) yield the best welding quality, preventing excessive penetration or weak bonding. The final welding tests confirm high conductivity (100%) and minimal positioning errors (0.03mm in optimal cases). From Table 5, the baseline model (target detection only) fails to achieve effective welding, highlighting the need for additional processing. Adding semantic segmentation (Model A) reduces positioning error to 0.6mm, but occasional misalignment occurs. Incorporating PnP pose estimation (Model B) further improves accuracy to 0.12mm, though overoxidation remains an issue. The full model, which includes welding process optimization, achieves the best accuracy (0.09mm) and an acceptable welding surface, demonstrating the necessity of each component. Overall, the results validate the effectiveness of integrating detection, segmentation, pose estimation, and welding optimization, with each component progressively improving accuracy and welding quality.

5. Conclusion

In this paper, we propose the first domestic system for integrated welding detection and optimization, combining target detection, semantic segmentation, pose estimation, and welding process optimization to enhance accuracy and reliability. The full model reduces positioning error to <0.1 mm , controls operation time within 3 mins and ensures stable welding quality. Ablation studies confirm the necessity of each component, with optimized welding parameters further improving consistency. This is the first domestic solution for high-precision welding in this direction, and it holds significant potential for future industrial applications, offering a promising outlook for widespread adoption.

References

[1] Xiang L I, Huan L I U. Exploration on Promoting the Localization and Intelligent Development of Welding in Nuclear Industry[J].Southen Energy Construction, 2020, 7(4): 93-97.J. Liang, R. He, and T. Tan, "A comprehensive survey on test-time adaptation under distribution shifts," International Journal of Computer Vision, vol. 133, no. 1, pp. 31–64, 2025.
[2] Smith M C, Vasileiou A N, Rathod D W, et al. A review of welding research within the new nuclear manufacturing (NNUMAN) programme[C]//Pressure Vessels and Piping Conference. American Society of Mechanical Engineers, 2017, 58004: V06BT06A072.
[3] Slobodyan M. Resistance, electron-and laser-beam welding of zirconium alloys for nuclear applications: a review[J]. Nuclear Engineering and Technology, 2021, 53(4): 1049-1078.
[4] Kobayashi M, Koakutsu A, Kawahara W. Automatic welding techniques for nuclear power plants[J]. Genshiryoku Kogyo (Nuclear Engineering), 1996, 42.
[5] Reis D, Kupec J, Hong J, et al. Real-time flying object detection with YOLOv8[J]. arXiv preprint arXiv:2305.09972, 2023.
[6] Xiao T, Liu Y, Zhou B, et al. Unified Perceptual Parsing for Scene Understanding with Pyramid Pooling[J]. European Conference on Computer Vision (ECCV), 2018.
[7] Chen L C, Papandreou G, Schroff F, Adam H. Rethinking Atrous Convolution for Semantic Image Segmentation[J]. arXiv preprint arXiv:1706.05587, 2017.
[8] Bradski G. OpenCV Library[J]. Dr. Dobb's Journal of Software Tools, 2000. (solvePnPRansac is part of OpenCV)
[9] Tsai R Y, Lenz R K. A New Technique for Fully Autonomous and Efficient 3D Robotics Hand/Eye Calibration[J]. IEEE Transactions on Robotics and Automation, 1989, 5(3): 345-358.
[10] Ge Z, Liu S, Wang F, et al. YOLOX: Exceeding YOLO Series in 2021[J]. arXiv preprint arXiv:2107.08430, 2021.
[11] Xu P, Zhang C, Wang J, et al. PP-YOLOE: An Evolved Version of YOLO[J]. arXiv preprint arXiv:2203.16250, 2022.
[12] Xie E, Wang W, Yu Z, et al. SegFormer: Simple and Efficient Design for Semantic Segmentation with Transformers[J]. Advances in Neural Information Processing Systems (NeurIPS), 2021.
[13] Strudel R, Garcia R, Laptev I, Schmid C. Segmenter: Transformer for Semantic Segmentation[J]. Proceedings of the IEEE/CVF International Conference on Computer Vision (ICCV), 2021: 7262-7272.

Research on Sea Surface Gimbal Control Strategy Based on LESO Observation PID

Jizhen Li[2], Jichi Yan[1 2*], Mingfan Chen[3]

[1] Key Laboratory of Advanced Manufacturing and Automation Technology (Guilin University of Technology), Guilin, Guangxi Zhuang Autonomous Region, 541006, PR China

[2] School of Mechanical and Control Engineering, Guilin University of Technology, Guilin, Guangxi Zhuang Autonomous Region, 541006, PR China

[3] School of Mechatronic Engineering and Automation, Shanghai University, Shanghai, 200444, PR China

*Corresponding author's e-mail: 18577339537@163.com

Abstract. Aiming at the problems of poor anti-disturbance effect and slow response speed of the wave gimbal based on traditional PID algorithm on the sea surface, a PID algorithm based on linear extended state observer (LESO) is designed. This algorithm aims to achieve stable attitude control and enhance the gimbal's anti-disturbance capabilities. The unknown perturbations of the system are estimated and compensated by LESO in real time; in order to solve the problem of difficult measurement due to multiple disturbances and high coupling of angles as well as angular velocities, the perturbation compensation link is introduced on the basis of the traditional serial PID control strategy. The verification is carried out on the actual built sea surface gimbal. The proposed control strategy shows significant superiority over the traditional PID control strategy in both internal and external disturbance experiments. This reflects the real-time estimation of the system error. It can be seen that the proposed algorithm is more resistant to external perturbations and has a higher response speed compared to the traditional PID.

1. Introduction

The stability of various marine equipment cannot be guaranteed due to the influence of waves. Therefore many scholars at home and abroad have analysed and studied it. The following two options are currently in common use for stabilisation:

(1) Adjusting the length of the platform's motion strut chain through mechanical structure to reduce or eliminate the impact of waves on marine equipment.

(2) Motor control to counteract wave fluctuation, so as to stabilise the platform and minimise equipment shaking.

Stewart D.[1] designed the parallel mechanism gimbal according to (1). Parallel mechanism with its unique characteristics of simple structure, large load bearing capacity and high accuracy of movement is more and more people's attention. Parallel platform technology as the frontier technology of the discipline has been studied by many scholars in recent years.[2][3] For example, Professor Huang Zhen of Yanshan University in China established the dynamic model of Stewart by using the Lagrange method.

Qianwen Duan et al [4] Aytaç Altan et al [5] Franz Königseder et al [6], on the other hand, designed the gimbal according to (2). The microcontroller unit (MCU) uses the position sensor to obtain the current attitude data and makes the gimbal stable by controlling the motors for position correction.

For the research on the control strategy of gimbal, various control algorithms have been proposed by researchers and scholars from various countries. G.P. Liu[7] used PID control algorithm. PID does not need to model the system accurately, but the control algorithm through pure feedback has some limitations in the requirements of high positioning accuracy and rapidity[8]. PID control algorithm has poor robustness and insufficient anti-jamming ability on sea surface gimbal. Literature[9] uses sliding-mode adaptive control (SMC), which has improved control but is more dependent on model accuracy. Literature [10][11] proposes active disturbance rejection control (ADRC) approach to estimate the internal and external disturbances of the system. However, the parameter tuning required by this algorithm is cumbersome. The literature[12] summarises the principles of parameter tuning for second-order nonlinear dilated state observers. Aiming at the problem of ADRC's difficulty in tuning parameters, the linear[13] active disturbance rejection control (LADRC) theory has been proposed in the literature. Since then the bandwidth method has gradually evolved into the mainstream method of LADRC parameter tuning. The literature[14] gives a method for determining the parameters of a linearly expanding state observer, but the rectification of the active disturbance rejection control algorithm remains difficult. The principal innovation of this research lies in the development of a cascade PID architecture integrated with LESO, which fundamentally transcends the inherent limitations of conventional LADRC and standard PID configurations. This novel structure strategically eliminates the complex Tracking Differentiator (TD) and State Error Feedback (SEF) modules characteristic of traditional LADRC implementations.

Different control strategies are used in the experiment to compare the control performance, and the verification can be obtained that the controller has strong robust performance and control effect, and has high value for engineering applications.

2. Structural Principle of PID Controller Based on LESO Observation

2.1. Mathematical Modelling of Stepper Motors

In this paper, an independently designed and built sea surface gimbal is used as a research object. As shown in Figure 1, it mainly consists of attitude sensor, interfernce stage, controlled stage, stepper motor, disturbance servo, fixed platform and microcontroller. The data from the position sensor is parsed by the microcontroller, the data is processed by the control algorithm, and the target position data is output to the stepper motor.

In order to ensure accurate control of the stepper motor and avoid the phenomenon of lost steps, this experimental platform adopts position closed-loop control in the bottom layer of the stepper motor drive control.

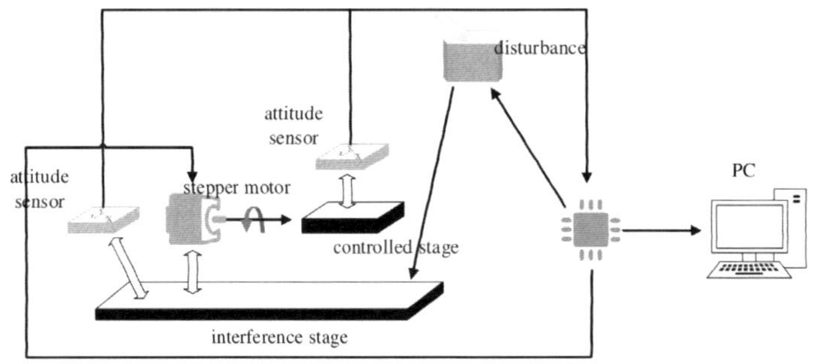

Figure 1. Sea Surface Gimbal Mock-Ups

The motor is single-phase excitation with no external load on the winding arm. By simplifying the model, a stepper motor with single-phase excitation has only one phase of excitation at any given moment. The equations of motion of the motor are linearised and then subjected to the Rasch transform and substituted for the initial value 0 to obtain:

$$G(s)=\frac{\theta_0(s)}{\theta_i(s)}=\frac{i_A^2Z_r^2L}{2Js^2+2Bs+i_A^2Z_r^2L} \tag{1}$$

Where: J is the total rotational inertia of the motor rotor and winding arm; B is the speed proportionality factor; T_L is the load torque.

The motor model used is 42ZDT40A motor and the motor parameters are shown in Table 1.

Table 1. 42ZDT40A motor parameters

Parameters	Rated Current /A	Zr/pc	L/mH	J/(g-cm2)	B/(N/(m-s-1))
Numerical Value	1.4	50	4.2	38	0.3

From equation (4) the transfer function is given as:

$$G(s)=\frac{\theta_0(s)}{\theta_i(s)}=\frac{270}{s^2+0.008s+270} \tag{2}$$

2.2. Principles of LESO

To address the problem of difficult ADRC parameter tuning, Professor Gao proposed the Linear Auto-Disturbance Control (LADRC), which relates the ADRC parameters to the frequencies of the controller and the observer, and transforms the parameter tuning problem into a bandwidth adjustment problem. The Linear Expanded State Observer (LESO) belongs to the core part of the Auto-Disturbance Control Algorithm (ADRC), and its main function is to estimate in real time the internal states in the system as well as the external perturbation. Core advantages reside in active disturbance compensation, low model dependency, smooth control output, and simplified deployment, making this architecture particularly suited for industrial environments with significant model uncertainties and frequent disturbance occurrences. Take the stepper motor as the control target and construct its state equation.

The above dynamic equation for stepper motor can be written as:

$$\begin{cases}\dot{x}=\omega\\ \dot{\omega}=\dfrac{N_r\psi_m i_q}{2J}-B\omega-TL-\dfrac{T_e}{J}sin(2N_r x)+h\end{cases} \tag{3}$$

Where x is the mechanical position; ψ_m is the maximum magnetic flux; h is the internal and external perturbations of the system; y is the output of the system; $f=-B\omega-TL-\frac{T_e}{J}\sin(2N_r x)$ represents the dynamics of the controlled object, for the sake of narrative convenience, it is abbreviated as f, assumed that $x_1=x$, $x_2=\omega$, $x_3=h+f+bi_q-b_0i_{qref}$ formally introduced into the 'dilation state', to meet the following formula:

$$\begin{cases}\dot{x}_1=x_2\\ \dot{x}_2=x_3+bu\\ \dot{x}_3=\omega(t)\end{cases} \tag{4}$$

where, $\omega(t)=\dot{f}(x,\omega)$, where b_0 is an estimate of b .

The dilated state estimate of the system state variable x is obtained as z , and the dilated state observer for the system is established as:

$$\begin{cases} e_1 = x_1 - z_1 \\ z_1 = z_2 - log_1(e_1) \\ z_2 = z_3 + bu - log_2(e_1) \\ z_3 = -log_3(e_1) \end{cases} \tag{5}$$

Taking $g_i(e_1) = e_1$, i.e., using the Linear Expansion State Observer (LESO).

Real-time estimation of internal and external perturbations of the system is performed using LESO as a way to enhance the stability and adaptability of the system. In the actual deployment, the controller structure diagram is shown in Figure 2.

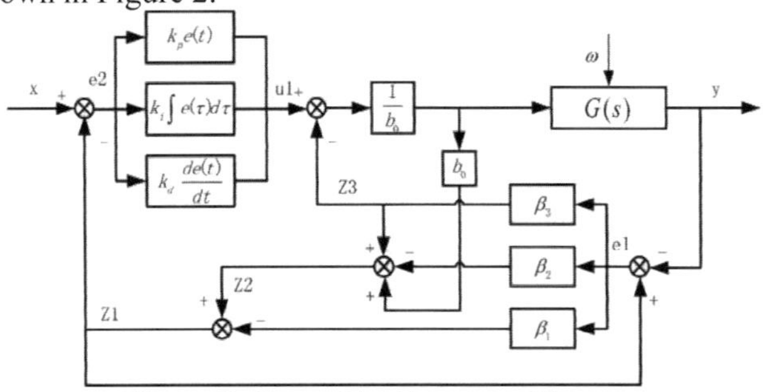

Figure 2. Controller Structure Diagram

The above state equation describes the working principle of the controller based on LESO observation of single neural adaptive PID. In this way, the states and total disturbances estimated by the LESO are used to adjust the output of the PID controller, thus improving the robustness and control accuracy of the system.

3. System Construction and Implementation

3.1. Sea Surface Gimbal System Framework

The design of the experimental platform is shown in Figure 3. The system takes angular displacement as well as angular velocity as system inputs, and outputs position data to the stepper motor driver through the MCU to control the amount of pulses driven by the stepper motor. During operation, two six-axis attitude sensors collect the angles of the interfering platform and the controlled platform respectively, and input the collected data into the MCU as real-time feedback. All of the above parts are mounted on a fixed platform, the stepper motor is placed in the interfering platform, following the rotation of the controlled platform; the interfering platform is fixed to the servo rudder, and the controlled platform is fixed to the output shaft of the stepper motor. All the above platforms rotate along the same axis.

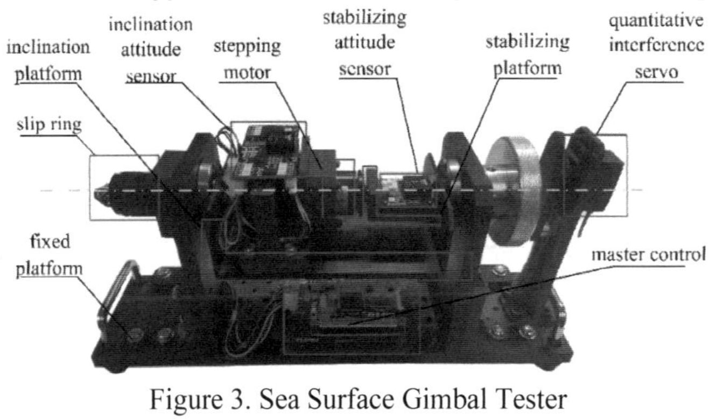

Figure 3. Sea Surface Gimbal Tester

3.2. System Hardware Design

The STM32F103ZET6 chip is used as the MCU, the highest 72MHz working frequency can guarantee the real-time control of the stepper motor, use its internal two USART interface to collect the position sensor, output a PWM to the interference servo, and control the stepper motor through another USART interface, and design the RS232 interface to communicate position data with the upper computer and display the position data to the upper computer. The RS232 interface is designed to communicate with the host computer for position data and display the position data in the host computer. The HWT606 six-axis attitude sensor is selected, and the three-axis angle data are solved by the Kalman filter fusion algorithm. In the circuit design of the attitude sensor, in order to avoid the role of the electrostatic field within the electronic components to generate high voltage, and more than a certain threshold resulting in breakdown phenomenon, so the use of ESD protection technology to protect the attitude sensor. The direct-drive topology is implemented in the motor actuation system to preserve signal integrity throughout the power transmission chain. The general design of the system hardware circuit is shown in Figure 4.

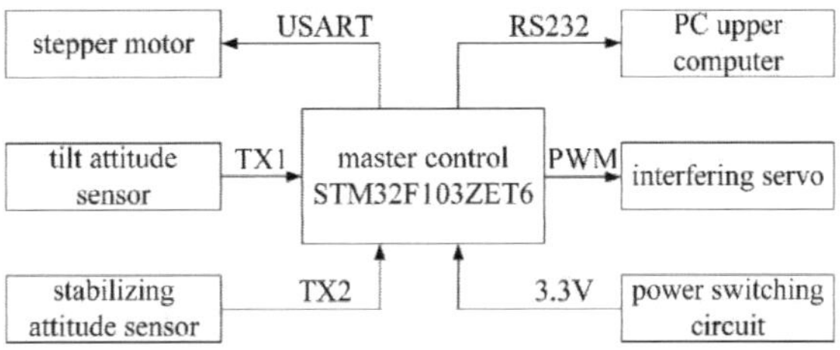

Figure 4. System Hardware Circuit Diagram

3.3. System Software Design

As shown in Figure 5, the system first initialises various peripherals such as clock, GPIO, USART, RS232, etc. The main program takes the time-slice polling method. the MAIN function loops the procedure of controlling the stepping motor as well as the bit position data transfer with the upper computer. Two timer interrupts are set to execute the timer tasks by judging the flag bits, and the position data collection and control algorithm are written into the timer tasks respectively. The timer configuration will be precisely calibrated to satisfy the experimental frequency requirements.

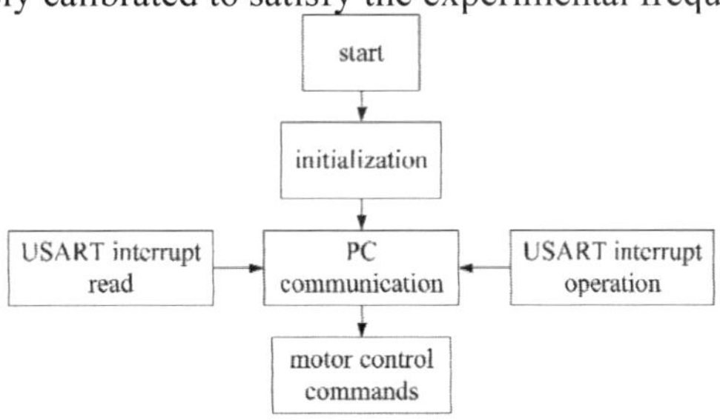

Figure 5. System Main Programme Flowchart

4. Experimental Tests and Analysis of Results

On the basis of completing the system hardware and software design, the sea surface gimbal interference experiment is carried out to verify. The sea surface gimbal experiment is shown in Figure 6. The LESO and PID parameters are designed as shown in Table 2.

Journal of Physics: Conference Series **3024** (2025) 012040 doi:10.1088/1742-6596/3024/1/012040

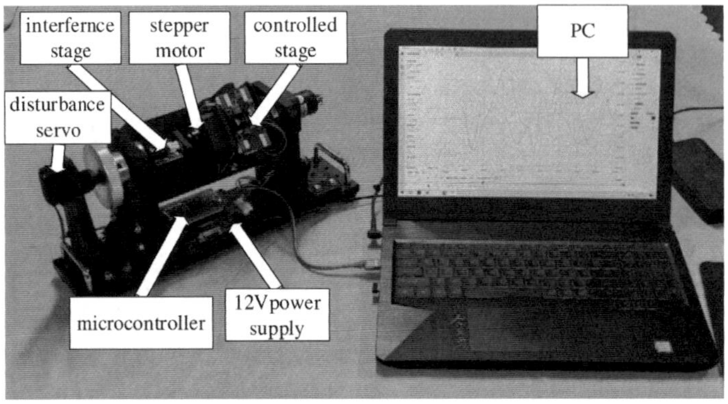

Figure 6. Sea Surface Gimbal Experiment

Table 2. Controller Parameter

b_0	h	r	ω_0	ω_c	K_{p1}	K_p	K_i	K_d
7000	0.007	2000	90	10	0.7	1.9	0.003	0.002

where k_{p1} is the outer loop PID parameter.

4.1. External 110° Large Single Disturbance

The jamming servo is initially positioned at -50° and then rotated to 60° to produce a large 110° rotation of the jamming stage, which is used to simulate a sea surface gimbal experiencing extreme ups and downs in the waves. Comparison of the experimental diagrams are as follows.

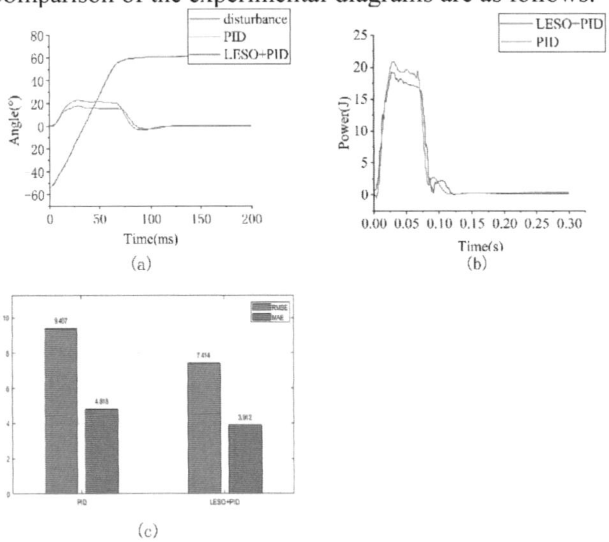

Figure 7. 110°Single Disturbance Angle Tracking Plot

The effect of each control strategy is shown in Table 3 below.

Table 3. Indicators under 110° Large Single Disturbance Case

	Peak Value	RMSE	MAE	Power
LESO + PID	17.99°	7.414	3.912	1.28J
PID	22.93°	9.407	4.818	1.31J

4.2. External 1Hz Frequency Disturbance

The dominant spectral energy of oceanic waves primarily resides within 0.05-0.3 Hz bandwidth. However, laboratory replication of such low-frequency wave dynamics encounters fundamental constraints arising from actuator response characteristics and temporal scaling limitations. To resolve this, elevates disturbance frequency to 1Hz regime. This approach effectively excites system dynamics within constrained experimental durations while circumventing nonlinear distortion phenomena in servo actuators induced by friction thresholds at ultra-low frequencies. The jamming servo generates a sinusoidal wave at a frequency of 1Hz, which causes the jamming gimbal to rotate back and forth within an amplitude of 76°, which is used to simulate a sea surface gimbal in waves. The comparison test diagram is shown below.

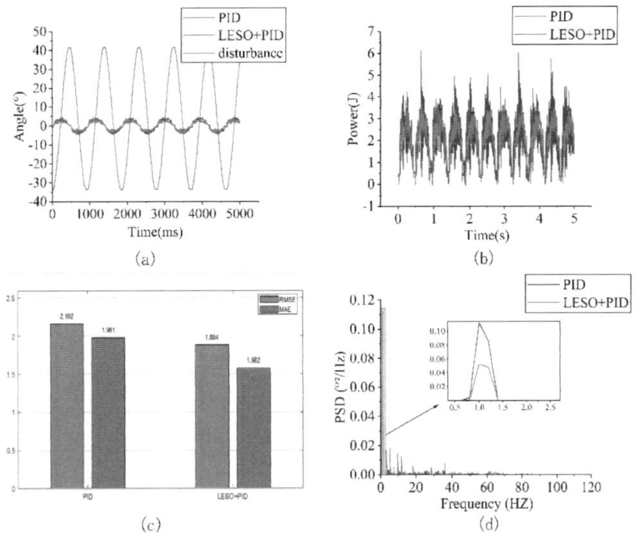

Figure 8. 1Hz Continuous Disturbance Angle Tracking Plot

The effect of each control strategy is shown in Table 4 below.

Table 4. Indicators under 1Hz Continuous Disturbance Case

	Peak Value	RMSE	MAE	Power
LESO + PID	7.91°	1.999	1.820	7.91J
PID	9.49°	2.822	2.560	9.91J

4.3. *Internal 30° Single Disturbance*

The MCU changes the control target from 0° to 30°, causing the controlled console to rapidly rotate by 30°. In this way the situation where the device needs to change the control angle is simulated. The comparison test chart is shown below.

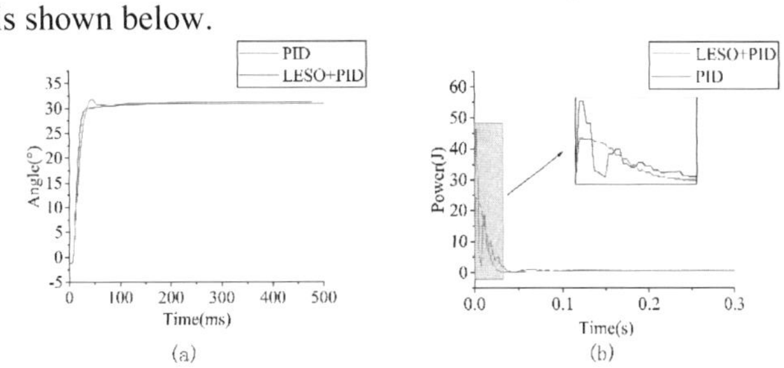

Figure 9. Internal Disturbance Angle Tracking Plot

The effect of each control strategy is shown in Table 5 below.

Table 5. Indicators for Internal Disturbance Scenarios

	Maximum Deviation	Timing of Adjustments	Power
LESO + PID	5.7%	38ms	0.75J
PID	0%	39ms	0.63J

5. Conclude

In this paper, a control strategy for LESO observation PID is proposed with the aim of improving the robustness as well as the immunity of the PID control system. It is verified by using it on a sea surface gimbal that is actually built. After experimental verification:

The PID control strategy of LESO observation proposed in this paper can use LESO to quickly estimate the internal and external perturbations of the system and compensate for them, which greatly reduces the maximum tracking error of the experimental gimbal, significantly improves the system's robustness and interference resistance, and reduces the difficulty of parameter tuning, so it can be conveniently deployed for gimbal control.

Acknowledgments

2022 Guilin Innovation Platform and Talent Plan, Contract No.: 20220124-2.

References

[1] Stewart D. 1965 A Platform with Six Degrees of Freedom (Proceedings of the Institution of Mechanical Engineers) pp 371-386.
[2] Zhen Huang, Lingfu Kong, Yuefa Fang 1997 Theory and Control of Parallel Robot Mechanism (Mechanical Industry Press)
[3] Li chen 2004 Establishment of a complete dynamic model for 6-DOF parallel robot on Stewart platform (Journal of Yanshan University) pp 228-232.
[4] Qianwen Duan, Xi Zhou, Qiunong He, Xinglong Chen, Wansheng Liu, Yao Mao 2021 Pointing control design based on the PID type-III control loop for two-axis gimbal systems (Sensors and Actuators A: Physical) pp 0924-4247
[5] Aytaç Altan, Rıfat Hacıoğlu 2020 Model predictive control of three-axis gimbal system mounted on UAV for real-time target tracking under external disturbances(Mechanical Systems and Signal Processing) pp 0888-3270
[6] Franz Königseder, Wolfgang Kemmetmüller, Andreas Kugi 2017 Attitude control strategy for a camera stabilization platform (Mechatronics) pp 0957-4158
[7] G.P. Liu, S. Daley 2001 Optimal-tuning PID control for industrial systems (Control Engineering Practice) pp 0967-0661
[8] Hognyu Lu, Gangshi Jin, Lizhi Liu. 2018 Research on servo control strategy of phototracking rotary table (Laser and Infrared) pp 503-508
[9] Kim H, Ahn H, Chung Y, et al. 2023 Quadrotor Position and Attitude Tracking Using Advanced Second-Order Sliding Mode Control for Disturbance (Mathematics)
[10] Tushar B, Sumit A, Rajan C, et al. 2023 Enhanced ADRC for sinusoidal trajectory tracking of an upper limb robotic rehabilitation exoskeleton (International Journal of Dynamics and Control) pp 2424-2436
[11] Li T, Sun S, Li M, et al. 2023 Intelligent ADRC - based inertia control for offshore wind farm (IET Generation, Transmission & Distribution) pp 530-541
[12] Jingqing Han 2002 From PID technology to 'self-immobilisation control' (Control Engineering) pp 13-18
[13] GAO Z 2003 Scaling and bandwidth-parameterization based controller tuning (Proceedings of the American Control Conference). pp 4989– 4996
[14] Jingqing Han 2008 Self-immunity control technology--control technology for estimating and compensating uncertainties (National Defence Industry Press) p 209

Journal of Physics: Conference Series **3024** (2025) 012041 doi:10.1088/1742-6596/3024/1/012041

Simulation Study of Trajectory Tracking of Picking Robotic Arm Based on Inverse Dynamics and Fuzzy PID Algorithm

Deqing Liu[1,2], Zhonggang Xiong*[1,2], Xintao Liu[2], Jinyun Cai[1,2], Dandan Wu[1,2], jiabao Li[1,2], Shuguang Wang[1,2]

[1] School of Mechanical and Electrical Engineering, Guilin University of Electronic Technology, Guilin, 541004

[2] Guangxi Key Laboratory of Special Engineering Equipment and Control, (Guilin University of Aerospace Technology), Guilin, 541004

*Corresponding author's e-mail: xzglsl2013@163.com

Abstract. Aiming at the problem of low efficiency and poor precision of the picking robotic arm, a trajectory tracking control strategy of the robotic arm based on a fuzzy PID algorithm is proposed. Firstly, through the kinematic analysis of the picking robotic arm, the robotic arm is converted from the working space to the joint space; secondly, the inverse dynamics model of the robotic arm is constructed using the Lagrangian method as the feed-forward compensation module; base on which the fuzzy rules are introduced to design the fuzzy PID controller, and finally the fuzzy PID controller is verified through Matlab/Simulink simulation. The simulation shows that compared with the traditional PID controller, the tracking error of the designed fuzzy PID controller is reduced by 80%, and the convergence speed is improved, which provides reliable technical support for enhancing the trajectory tracking control accuracy of the mechanical arm for picking agricultural products.

1. Introduction

At present, most fruits and vegetables are picked manually, not only the labor cost is high, but the picking efficiency is low[1]. The use of robotic arm as a human hand to pick crops is a meaningful way to realize harvest automation and smart agriculture[2]; in the picking process, the requirements of robotic arm can be fast, accurate, and stable three aspects, to replace the human hand, the so-called "fast" is to require that the robotic arm can be a very short reaction time into the working state; The so-called "fast" is to require that the robotic arm can enter the working state with a very short reaction time; "accurate" requires that the robotic arm can move along the planned trajectory; "stable" requires that the robotic arm can be in the process of operation, the joints run smoothly, no jamming, no shaking.

Robotic arms applied in agricultural picking can significantly improve picking efficiency[3][4]. Scholars have proposed many kinds of control algorithms to improve the operating accuracy of robotic arms. For example, sliding film control, neural network and sliding film combined control, sliding film and fuzzy combined control, enhanced learning control, adaptive robust control, and other methods. Among them, the slip film control has insensitivity to perturbation, so the control method combined with it is used more in the trajectory tracking of the robotic arm.Sun Guofa[5] et al, for the problem of modeling error as well as perturbation of the robotic arm in practical applications, designed an improved terminal sliding film control method, by estimating the total perturbation and error, adopting a nonlinear gain function, realizing the convergence of the perturbation and error, and effectively suppressing the jittery vibration of the sliding film control. Finally, the control method is validated by Matlab simulation.Jia Hua[6] et al, proposed a control method based on the combination of neural network and slip film control for the problem of modeling inaccuracy and external interference affecting trajectory tracking of the robotic arm. The model is reconstructed by using a neural network approximation dynamics model and online adjustment of weights by the slip film to eliminate the error effect. Finally, through Simulink simulation, it is concluded that the maximum error of the joint is reduced by 83.7%. Wang Sanxiu[7] et al, proposed an adaptive robust control method for the problem of interference and model uncertainty. Robust control is utilized to eliminate the effects of interference and uncertainty, and simulation validation shows that the control method has better anti-interference ability and higher convergence of trajectory tracking error.

The control methods proposed by the above scholars, there is a certain degree of modeling, validation, application and other aspects of difficulty, the form of the designed controller is more complex, not easy to popularize. Industrial control of the robotic arm is still more traditional PID control algorithms, but the lack of robustness, this paper proposes the use of fuzzy PID trajectory tracking control algorithms, adding fuzzy rules for the traditional PID. Taking the knuckle card Zu5 six-degree-of-freedom robotic arm as the picking robotic arm, firstly, the kinematic analysis of the robotic arm is carried out to realize the joint transformation, and the simulation is verified by using Matlab. Then Lagrangian method is used to construct its inverse dynamics model to do feed-forward compensation, followed by adding fuzzy rules to the traditional PID to reduce the trajectory error, and finally,the results show that the designed fuzzy PID controller based on inverse dynamics feedforward can improve the tracking accuracy and tracking efficiency.

2. Picking robot arm modeling

2.1. Kinematic modeling

In this paper, the knuckle card Zu5 six-degree-of-freedom collaborative robotic arm is taken as the research model, the first three degrees of freedom are used for the adjustment of the picking spatial position, and the last three degrees of freedom are used for the adjustment of the picking attitude. Positive kinematics model to establish the design base, the position of each joint coordinate system, first using the improved D-H parameter method for the establishment of the coordinate system of this picking robotic arm. As shown in Figure 1 is the improved joint coordinate diagram of Zu5s.

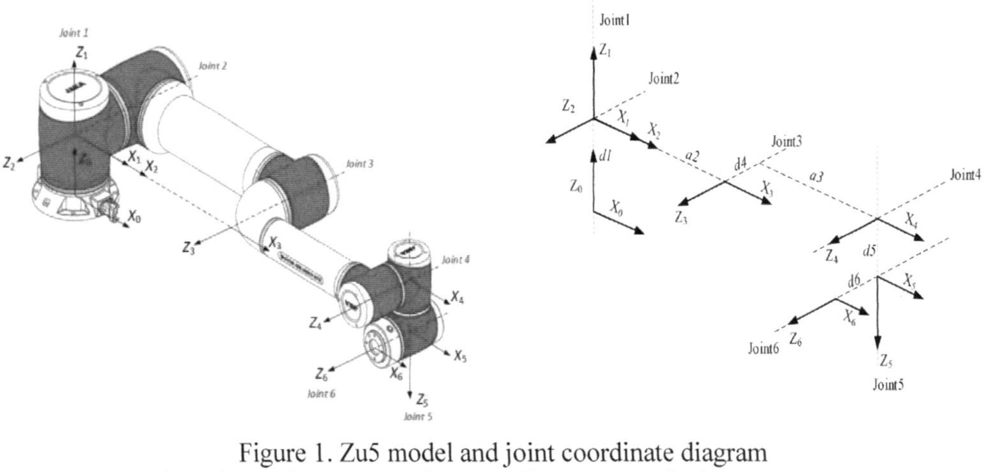

Figure 1. Zu5 model and joint coordinate diagram

The transformation of coordinate system i to coordinate system k+1:

$$ {}^{k-1}_{k}T = {}^{k-1}_{R}T \bullet {}^{R}_{Q}T \bullet {}^{Q}_{P}T \bullet {}^{P}_{k}T \tag{1} $$

The conversion formula is

$$ {}^{k-1}_{k}T = \begin{bmatrix} \cos\theta_k & -\sin\theta_k & 0 & a_{k-1} \\ \sin\theta_k \cos\alpha_{k-1} & \cos\theta_k \cos\alpha_{k-1} & -\sin\alpha_{k-1} & -d_k \sin\alpha_{k-1} \\ \sin\theta_k \sin\alpha_{k-1} & \cos\theta_k \sin\alpha_{k-1} & \cos\alpha_{k-1} & d_k \cos\alpha_{k-1} \\ 0 & 0 & 0 & 1 \end{bmatrix} \tag{2} $$

Eq. 2, α_k a_k θ_k d_k, and denote the connecting rod angle of rotation, connecting rod length, joint angle, and connecting rod deflection, respectively. According to the method of determining each parameter of the robotic arm and using the improved D-H method to derive the parameter information of the Zu5 robotic arm.

Table 1. Parameters of Zu5 Robotic Arm Improved D-H

Joint k	a_{k-1} /mm	α_{k-1} /°	d_k /mm	θ_k /°
1	0	0	120.15	θ_1
2	0	90	0	θ_2
3	430	0	0	θ_3
4	368.5	0	-114	θ_4
5	0	90	113.5	θ_5
6	0	-90	107	θ_6

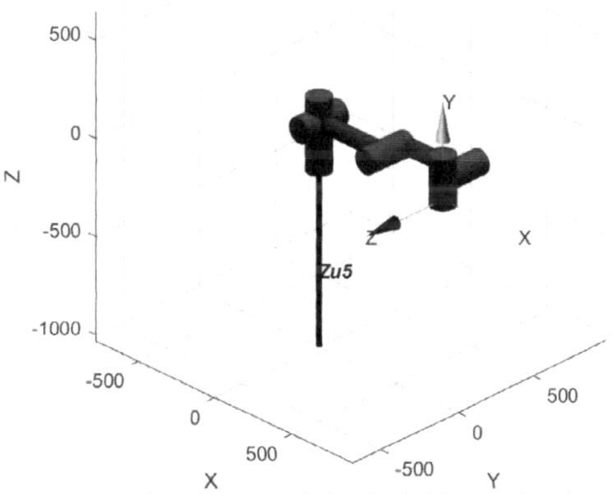

Figure 2. Validation model of picking robotic arm

Matlab is utilized to validate the simulation, as shown in Fig. 2 for the positive kinematics simulation model of the robotic arm, which is consistent with the joint model, and the information of the constructed improved D-H parameters is not incorrect.

2.2. *Kinetic Modeling*

According to the kinematics of the robotic arm, the end trajectory of the robotic arm, is realized by changing the angle of the joints In order to make the robotic arm follow with high accuracy, its dynamics model is also needed. The Lagrangian method is used to establish the kinetic model of the picking robotic arm.

$$ L - P = K \tag{3} $$

$$F_k + \frac{\partial L}{\partial q_k} = \frac{d}{dt}\frac{\partial L}{\partial \dot{q}_k} \tag{4}$$

Where: L denotes the Lagrangian function; K denotes the total kinetic energy of the picking robotic arm in J; P denotes the total potential energy of the picking robotic arm in J; F_k denotes the joint driving torque in N*M; $\dot{q}_k$ denotes the angular velocity of each joint in rad/s; and q_k denotes the angle of each joint in rad.

The dynamics equations of the final picking robotic arm are derived from Eqs. (3) and (4):

$$\tau = M(q)\ddot{q} + C(q,\dot{q})\dot{q} + G(q) \tag{5}$$

Where: $M(q)$ denotes the matrix of inertia of the arm; $C(q,\dot{q})$ denotes the matrix of Kurtosis and centrifugal forces of the arm; $G(q)$ denotes the matrix of gravitational forces of the arm; τ denotes the moments applied to the joints, in units of N*M. q、$\dot{q}$、$\ddot{q}$ denotes the angular vector, angular velocity vector, and angular acceleration vector of the arm, in units of rad, rad/s, rad/s2.

3. Picking robot arm fuzzy PID controller design

3.1. *PID control principle*

The traditional PID control is a classical control method, which P denotes the proportional link, I denotes the integral link, and D denotes the differential link. The precise control of the system output is realized by dynamically adjusting the control quantities to achieve the effect of improving the system performance, stability and adaptability.The output equation of the PID controller is:

$$u(t) = K_p * e(t) + K_i * \int_0^t e(\tau)d\tau + K_d * \frac{de(t)}{dt} \tag{6}$$

Where: Kp, Ki, Kd are denoted as proportional, integral, and differential coefficients, respectively; u(t) denotes the output of the PID controller; e(t) is denoted as the systematic deviation between the input value and the output value.

Although the traditional PID control is simple and easy to understand, but it is more difficult to adjust the parameters, often based on the trial-and-error method of adjustment, it is difficult to find the optimal parameters for nonlinear or complex systems. Need to let the traditional PID and can only control ideas combined, can do online adaptive adjustment of control parameters, such as fuzzy PID control.

3.2. *Principles of fuzzy control*

For the trajectory tracking problem of the picking robotic arm, the joint angle error and the error change rate of trajectory tracking are taken as input variables, which are defuzzified and converted into fuzzy values; according to the created fuzzy rules, the relationship between the input quantities and the output quantities is described; reasoning is carried out to determine the fuzzy set of the output variables; and finally, the output variables are converted into precise values[8]. The schematic diagram of the fuzzy controller is shown in Fig. 3.

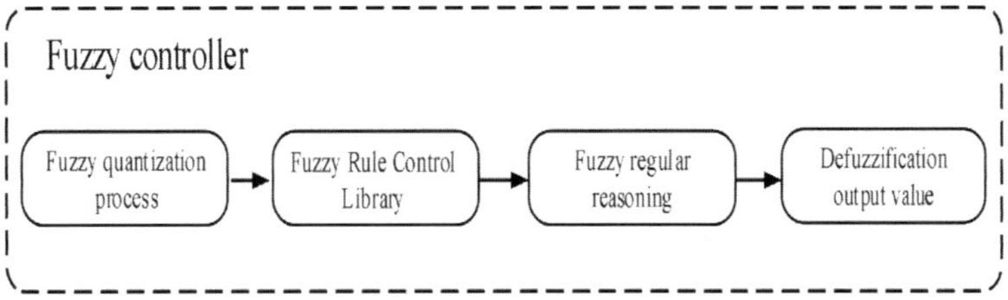

Figure 3. Schematic diagram of fuzzy controller

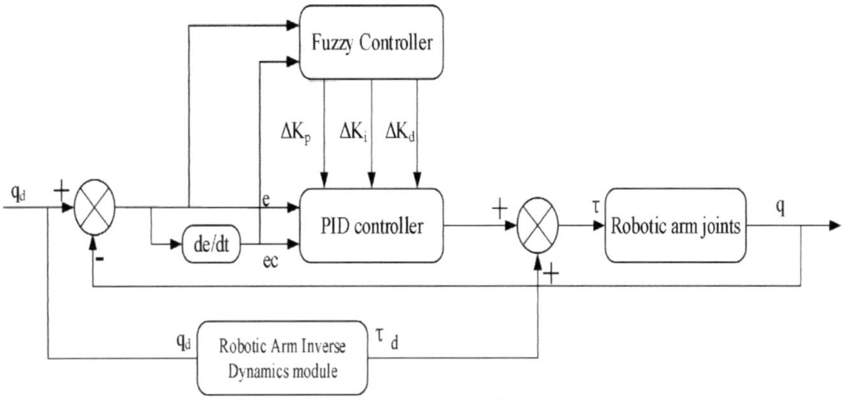

Figure 4. Schematic diagram of inverse dynamics feedforward compensation of robotic arm with fuzzy PID controller

3.3. *Fuzzy PID controller design*

3.3.1. *Fuzzy PID controller principle*

Fuzzy PID controller is based on the PID control algorithm, the error e and the rate of change of the error $\dot{e}$ as input , ∇K_p、∇K_i、∇K_d as output, fuzzy reasoning through fuzzy rules to meet, at each time of the rectification requirements.A schematic of the model based on inverse dynamics and fuzzy pid control is shown in Fig. 4

The final output of the fuzzy PID controller is:

$$\begin{aligned} KP &= K_p + \nabla K_p \\ KI &= K_i + \nabla K_i \\ KD &= K_d + \nabla K_d \end{aligned} \tag{7}$$

3.3.2. *Fuzzy rule determination*

Fuzzy rules are the core of fuzzy controllers where the selection of linguistic variables and affiliation functions is the key to fuzzy rules.The robotic arm joint angle error e and the rate of change of the error as input linguistic variables $\dot{e}$ as output linguistic variables.and defined five fuzzy subsets.whose fuzzy subsets are e , $\dot{e}$ = {NB, NS, ZE, PS, PB}.The subsets represent negative large, negative small, zero, positive small, and positive large, respectively, and can accurately, quickly, and adaptively capture the dynamic changes in the control process. Since the robotic arm has multiple uncertainty data, such as modeling inaccuracy and external disturbances.The joint Gaussian-type affiliation function (Gauss2mf), which is capable of handling large amounts of uncertainty data and allows for fine control of the change in affiliation, is chosen, and its expression is given as:

$$\mu(x) = \exp(-\frac{(x-a)^2}{2\sigma_1^{\ 2}}) \bullet \exp(-\frac{(x-b)^2}{2\sigma_2^{\ 2}}) \tag{8}$$

where: a and b are the means of the Gaussian distribution; σ_1, σ_2 are the standard deviations.

The Gauss2mf used in this paper is shown in Fig. 5.

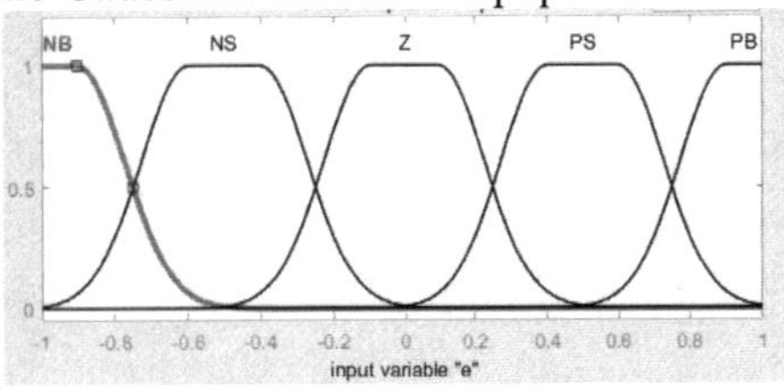

(a) Error e affiliation function

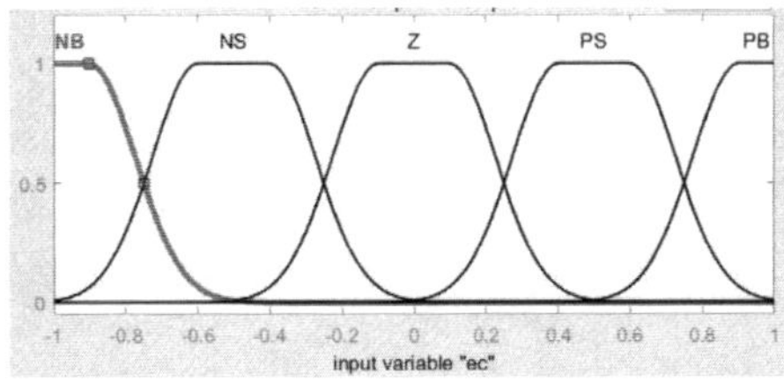

(b) Error change rate $\dot{e}$ affiliation function

Figure 5. shows the affiliation function of the joint error and the derivative of the joint error.

Finally, A fuzzy rule is a set of conditional statements used to define the relationship between input variables and output variables, usually expressed in the form of "if-then". In order to reduce the joint error of the robotic arm and improve the trajectory tracking accuracy and the stability of the system. Then the generated control rules are shown in Fig. 6.

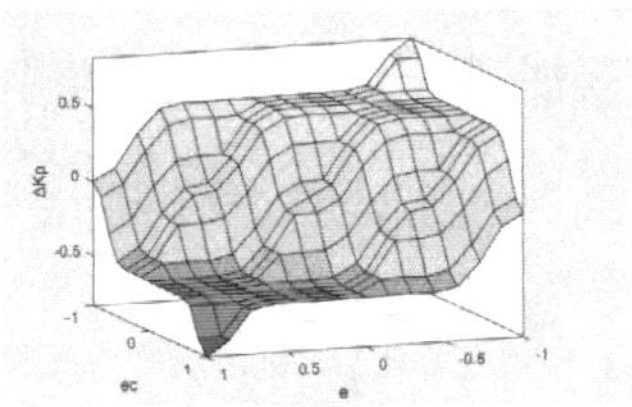

(a)Fuzzy control rules for ∇Kp

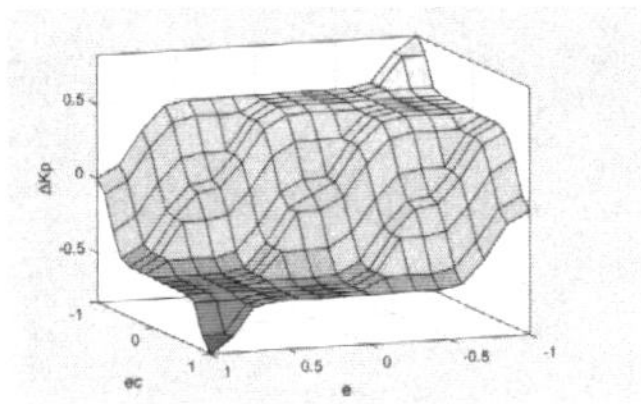

(b)Fuzzy control rules for ∇Ki

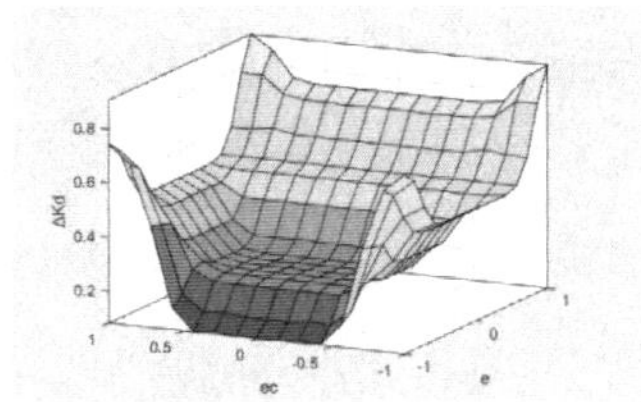

(c)Fuzzy control rules for ∇Kd

Figure 6. the fuzzy rule map

The fuzzy control libraries about ∇Kp , ∇Ki , ∇Kd and are constructed based on the domain experts' experience, knowledge, and understanding of the system behavior shown in Figure 6.

4. Simulation analysis

In order to verify that the proposed inverse dynamics and fuzzy PID based controller can have high accuracy and speed in controlling the picking Zu5 robotic arm. Matlab/simulink is used to set up the knuckle card Zu5 robotic arm model. Each joint of the robotic arm is set as moment input and joint angle output mode. And in this way, the fuzzy PID controller is built, and the setting time is 10s anyway. the first and second joints are selected for consideration. The trajectory tracking effect diagrams of joint 1 and joint 2 under PID control are shown in Figs.7 and 8.

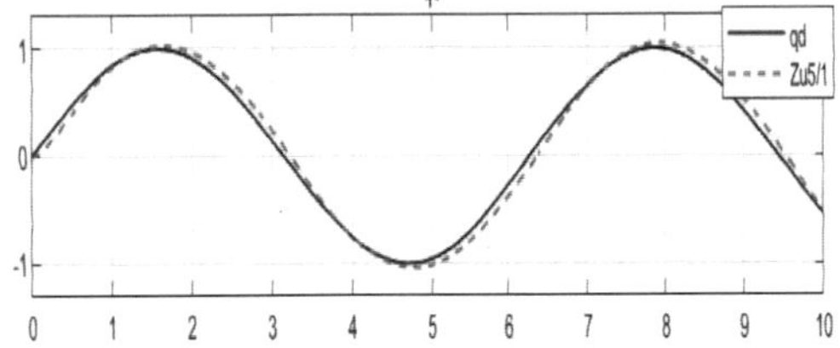

Figure 7. Trajectory tracking of Joint 1

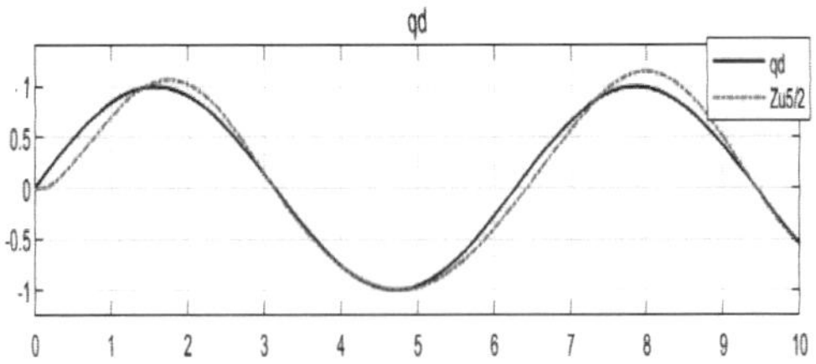

Figure 8. Trajectory tracking of Joint 2

From Figures 7 and 8, it can be seen that when the sinusoidal function trajectory is given to joint 1 and joint 2, the actual joint trajectory (the red line in the figure) and the desired trajectory (the black line in the figure) of the picking robotic arm under the traditional PID control are in some error, and the tracking effect is not ideal.

From Figure 8 according to the tracking error graph, the traditional PID control its control effect is not ideal, the error has not converged to 0, and the error fluctuation amplitude is large. For this reason, fuzzy rules and the inverse dynamics model of the robotic arm are added to the traditional PID controller to improve the tracking effect. As shown in Figures 9 and 10.

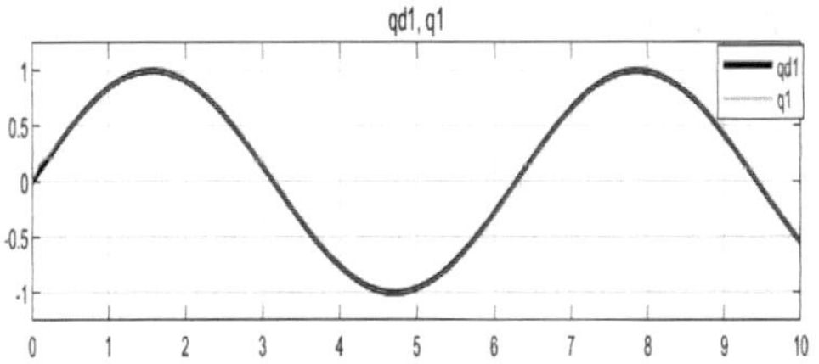

Figure 9. Trajectory tracking of joint 1 under fuzzy PID control

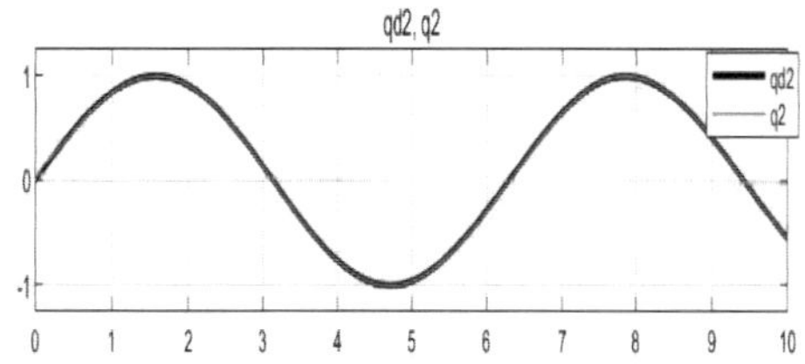

Figure 10. Trajectory tracking of joint 2 under fuzzy PID control

It is clear from Figs. 9 and 10 that after adding fuzzy rules and inverse dynamics modeling to the conventional PID controller, the actual motion angles of joint 1 and joint 2 follow closely with the expected motion angles. The tracking errors of the two joints are shown in Fig. 11.

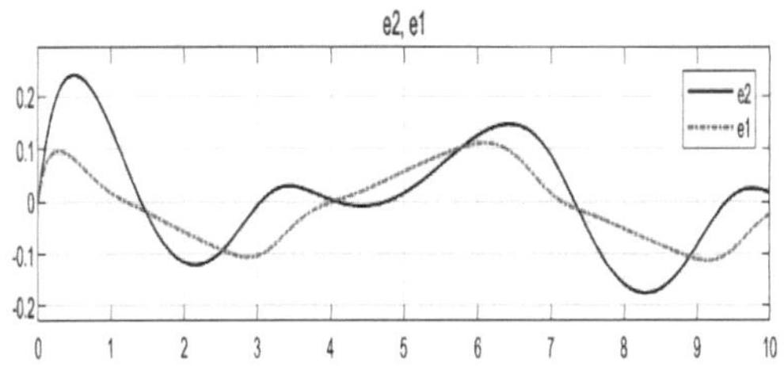

Figure 11. Plot of joint tracking error

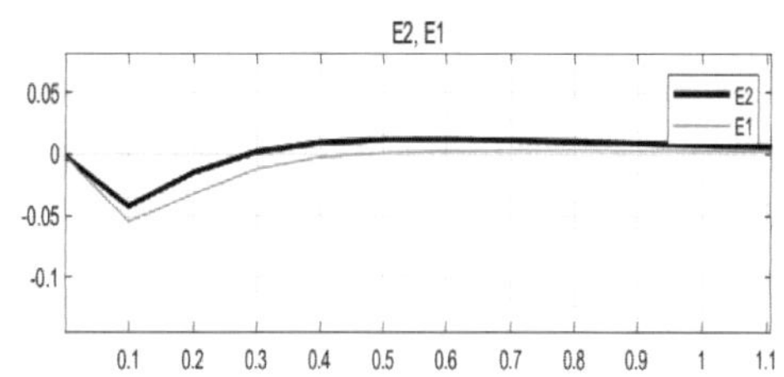

Figure 12. Error profile of two joints under fuzzy PID control

From Fig. 12, it can be seen that its joint tracking error is almost converged to 0 at 0.3 s. The maximum error of joint tracking without adding fuzzy rules is 0.25; the maximum error of trajectory tracking of PID controller and inverse dynamics model with adding fuzzy rules is 0.05; the tracking error is reduced by 80%.

5. Conclusion

In order to improve the accuracy of the picking robotic arm in the process of operation, a trajectory tracking algorithm that adds fuzzy rules and an inverse dynamics model for traditional PID controller to do feed-forward compensation is proposed with the knuckle-card Zu5 collaborative robotic arm as the object of the study, and the following conclusions are drawn:

1. A kinematic and dynamic model of the picking robot arm was developed to analyze the relationship between the joint angle and the end position of the arm, the attitude and the joint moment..

2. The inverse dynamics model and fuzzy rules are introduced, and through simulation, it is concluded that not only can the joints move smoothly, but also the tracking error is reduced by 80% compared with the traditional PID controller. The effectiveness of the control method is proved. It provides a concise and feasible program for the robot joint control problem in industrial, agricultural and medical fields. In addition, more advanced control strategies with better tracking effect can also be used, and this part will be continued to be studied in depth as a follow-up content.

Acknowledgments
Funding: National Natural Science Foundation of China(62241304); enterprise-commissioned lateral project (KH201812507); Guangxi special engineering equipment and control key laboratory director subject (SEEC24ZZR08)

Author's information: Xiong Zhonggang(1986-), Male, Changde, Hunan, Ph.D., Professor, Master Supervisor.

References

[1] ZHANG Weimin, DONG Hongzheng, CHEN Wenqing. Structural design and simulation analysis of end-effector of citrus picking robot[J]. Journal of Yangzhou University (Agriculture and Life Science Edition), 2019, 40(4): 113-118.
[2] ZHANG Wei, MU Huiqun, GUO Jian. Application of fuzzy algorithm in trajectory control of cucumber picking robot[J]. Agricultural Mechanization Research, 2021, 43(3): 230-235.
[3] ZHAO Kaixuan, WANG Shubo, LI Dongwu. Sliding mode control of robotic arm based on improved convergence law[J]. Control Engineering, 2022, 29(12): 2221-2226.
[4] XIONG Zhonggang, YE Zhenhuan, HE Juan, et al. Intelligent tracking control of small agricultural machinery path based on immune fuzzy PID[J]. Robotics, 2015, 37(2): 212-223.
[5] SUN Guofa, ZHAO Erquan, ZHANG Guoju, et al. Non-singular fast terminal sliding mode control of robotic arm based on disturbance observer compensation[J]. Control Theory and Applications, 2022, 39(8): 1506-1515.
[6] JIA Hua, LIU Yanjun, WANG Yu, et al. Neural network adaptive terminal sliding mode control of a six-axis robotic arm[J]. Journal of Xi'an Jiaotong University, 2022, 56(11): 21-30.
[7] WANG Sanxiu, YU Li, XU Jianming, et al. Adaptive robust trajectory tracking control of robotic arm[J]. Control Engineering, 2015, 22(2): 241-245.
[8] Zhou Mengcheng, Jin Xiaolong. Trajectory tracking control of a six-degree-of-freedom robotic arm based on fuzzy sliding mode switching gain[J]. Journal of Chongqing Technology and Business University (Natural Science Edition), 2024: 1-9.

Sampling System of Salt Lake UAV Based on Coverage Path Planning

Song Li*, **XinZheng Huang, Yan Zhang, Gong Lu**

Intelligent System Laboratory of Qinghai University,Xining, Qinghai Province 810000, P. R. China

*Corresponding author's e-mail: 2021990072@qhu.edu.cn

Abstract. The aim is to address the problems of complex terrain and high sampling frequency of Salt Lake, such as high intensity, significant error, low efficiency, and high risk of traditional manual sampling. In this paper, a salt lake UAV sampling system is designed to cover path planning based on the advantages of UAV mobility, high security, and manpower saving. The system first covers the salt lake with a circular area. It automatically plans the sampling points based on the sampling radius information set by a Salt Lake user, then uses an intelligent algorithm for path planning, and finally obtains the lowest energy loss through different speed planning strategies to complete the whole Salt Lake sampling. The experimental simulation results show that the system realizes low-power Automatic full coverage Salt Lake sampling under different algorithms and different speed planning strategies, improves the sampling efficiency, reduces the sampling cost, and is expected to contribute to the development of the Salt Lake industry.

1. Introduction

Salt Lake usually refers to the water body with the highest concentration in the natural water body with a salt content of more than 50 g/L [1]. China is rich in salt lake resources. With the acceleration of the construction of a world-class Salt Lake industrial base, Salt Lake water quality sampling is of great significance for the analysis and utilization of Salt Lake resources [2,3]. However, in the current water quality sampling system [4,5,6], online monitoring equipment and manual collection in key areas are mostly used, with limited scope of operation, poor mobility, and low sampling efficiency. While the water quality parameters of the salt lake change with time and space factors, it is difficult to obtain the water conditions, and the salt lake has a large area and complex terrain. Therefore, this paper proposes a system using UAV to carry sampling equipment for the salt lake. However, the landscape of the salt lake is vast, the endurance capacity of UAVs is limited, and the coverage area of sampling should be guaranteed. Therefore, reasonable coverage path planning is particularly important.

The path coverage problem of UAVs refers to that in a given target area, UAV carries airborne sensors to scan and cover the target area [7]. In the fields of agricultural monitoring, UAV aerial photography, and so on, most of them are faced with regular plane areas. The grid map method and region decomposition method can be used for environmental modeling. In the face of mixed integer programming and mixed integer linear programming, there are many path-planning algorithms, including classical algorithms, graph search heuristic algorithms, and meta-heuristic algorithms [8, 9,

Journal of Physics: Conference Series **3024** (2025) 012042 doi:10.1088/1742-6596/3024/1/012042

10]. The salt lake has a large area, but it belongs to the regular plane area. It needs to establish more sampling points as much as possible while taking into account the endurance ability of UAVs to obtain more comprehensive information about the water quality parameters of the salt lake. Therefore, taking Chaka Salt Lake in Qinghai, China as an example, this paper uses the circular region decomposition method to divide the salt lake into multiple circular regions and cover the whole Salt Lake as much as possible. According to the sampling requirements, it can adjust the coverage radius of the circular region, to determine the sampling coordinate points, and then use the classical algorithm for path planning. However, to complete the sampling of the whole Salt Lake, UAVs should have sufficient endurance ability. The endurance ability of UAVs is not only related to path planning but also related to the speed planning strategy in the flight process. Different speed planning strategies will produce different energy losses, This will affect the endurance capability of UAVs.Therefore, this paper proposes to use the S-curve planning method, which is a commonly used speed planning method. By ensuring the continuity of acceleration and speed, the speed is changed by fixed acceleration, to achieve the purpose of smooth control of UAV speed. To sum up, to meet these challenges, this paper makes the following contributions: 1) this paper designs a salt lake sampling device based on UAV, which can realize the automatic sampling of UAV. 2) The circular domain decomposition method is proposed to realize the UAV coverage path planning problem and realize the full coverage of the shortest path sampling points. 3) The S-curve velocity planning method is proposed to ensure the endurance of UAVs during sampling. It is expected to contribute to the development of salt lake industry through this study.

2. Overall scheme design

The Salt Lake UAV sampling device mainly includes UAV and sampling devices. The UAV part mainly completes the coverage path planning and speed planning, while the sampling device part mainly completes the water sample extraction work when the UAV hovers. In this paper, a small UAV sampling device is designed and built. The overall design framework of UAV sampling is shown in Figure 1. The UAV coordinates with GPS and obtains the current altitude information of the UAV, but the GPS altitude detection accuracy is not high. Therefore, when the UAV lands for sampling, it measures the altitude in real time through two high-precision detection devices, ultrasonic and near-infrared, for water quality sampling. When the UAV starts, it receives the coordinate information to be sampled from the ground station through the wireless receiver and calculates the optimal path through the path planning algorithm. After the calculation is completed, the UAV starts to fly to the coordinate point and ensures the maximization of energy consumption utilization through speed control to reduce energy loss. After reaching the sampling point, it starts height detection for a slow descent. After reaching the precise height from the water surface, the auxiliary rod descends so that the water sampling pipeline extends into the water for water sample extraction through the micropump. After sampling, it flies to the second sampling point and circulates in turn. After completing the last sampling point, the UAV automatically returns to complete all sampling work.

The traditional Nansen water sampler can only sample a single time and has a large volume, which is obviously not suitable for the UAV sampling system with many sampling points in the salt lake. At the same time, in order to ensure that the water quality of each sampling point in the salt lake does not interfere with each other during the sampling process, a purging process is required. Therefore, according to the characteristics of Salt Lake sampling, a sampling device as shown in Figure 2 is designed. The flexible hose is used as the waterway and water sampling bag for water sample storage. The water sample is extracted by the micro pump and purged by the purge pump. The water inlet, waterway, water storage system and purging part are isolated by the solenoid valve. The water quality sampling process is mainly divided into two processes: micro pump pumping water body and purge pump purging. When the UAV slowly drops to a fixed height after reaching the sampling point, the water inlet reaches a fixed water level, the solenoid valve of the water inlet pipeline opens, the corresponding micro pump pumps water body and stores the water sample in the corresponding sampling bag. In this process, other micro pumps and solenoid valves are closed. After the extraction is completed, the solenoid valve at the end of the micro pump is closed, the solenoid valve of the purge pump is opened, and the purge pump is purged. After the purging is completed, all solenoid valves are closed to complete the sampling at the sampling point.

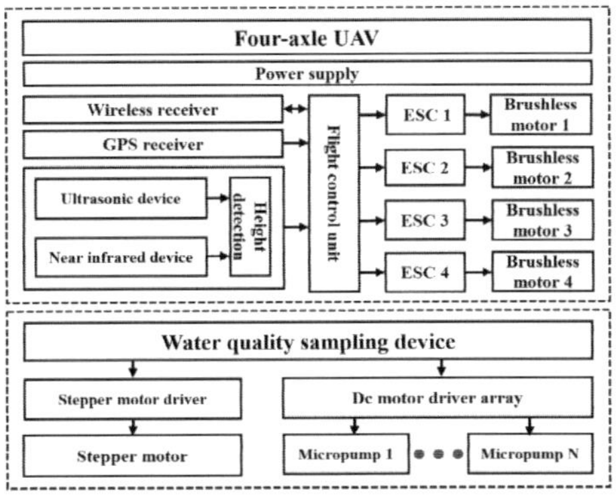

Figure 1. Overall design framework

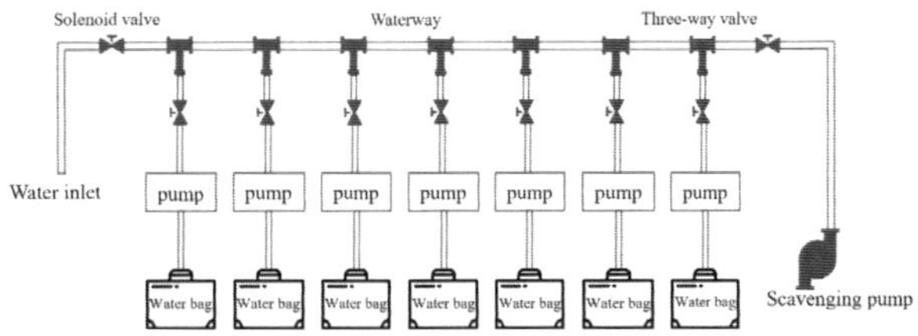

Figure 2. Sampling device design block diagram

3. Coverage path planning simulation

The problem of coverage path planning of water quality sampling in Salt Lake can be described as follows: firstly, the sampling points of UAV cover the whole of Salt Lake as much as possible, so that the sampling is representative; secondly, UAV multi-point sampling and finally return to the starting point. The determination of the sampling sequence of each sampling point is very important and determines the flight path and flight distance of the UAV. If the flight distance of a UAV is the shortest, it can shorten the flight time, significantly improve the sampling efficiency, and collect more samples at the same time. At the same time, due to the reduction of flight distance, energy consumption is reduced, battery-powered UAV endurance is extended, and more samples can be completed by one charge. Taking Chaka Salt Lake in Qinghai, China as an example, the circular region decomposition method is used to decompose the salt lake into several circular regions with the same radius for full coverage. The sampling radius can be set to 1.4km, 1.1km, and 0.8km respectively for full coverage sampling. Three sets of sampling points with different numbers, namely 15, 25, and 38, are obtained. The take-off point is set to 36.71 degrees north latitude and 99.18 degrees east longitude. The traveling salesman problem with 16, 26, and 39 points is obtained. The simulated annealing algorithm, ant colony algorithm, and genetic algorithm are used for optimal path planning, respectively.

When the sampling radius is 1.4km, the number of sampling points is 16, and the simulated annealing algorithm is used for coverage path planning. The results are shown in Figure 3. The scale of 16 sampling points is limited, and the initial temperature is set to 100. For simulated annealing at 16 sampling points, the number of points is less, and the annealing coefficient is 0.99. The termination temperature is 0.00001. The results of path planning covered by the 16 sampling point ant colony algorithm are shown in Figure 4. In the ant colony algorithm for the 16-point traveling salesman problem, too few ants may not be able to fully explore the solution space, and too many will increase the computational complexity. Set it to 32, the pheromone importance factor is set to 5, the importance of the initiation function is set to 5, and the pheromone volatilization coefficient is set to 0.5. The coverage path planning results of the 16 sampling points genetic algorithm are shown in Figure 5, and the population size is set to 100. The crossover probability is set to 0.8, the mutation probability is set to 0.5, and the number of iterations is 100.

Journal of Physics: Conference Series **3024** (2025) 012042 doi:10.1088/1742-6596/3024/1/012042

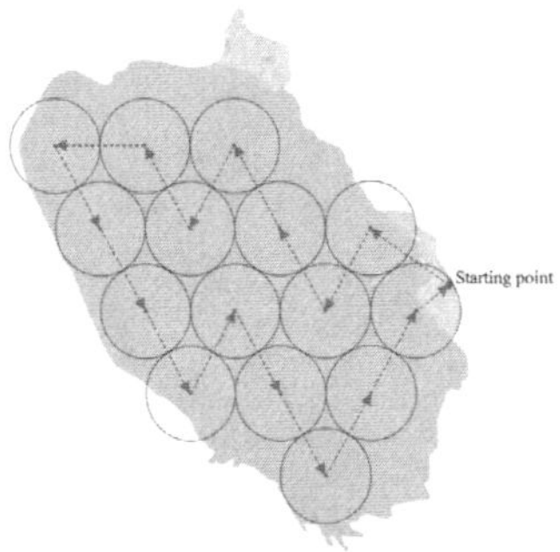

Figure 3. Simulated Annealing Algorithm

Figure 4. Ant Colony Algorithm

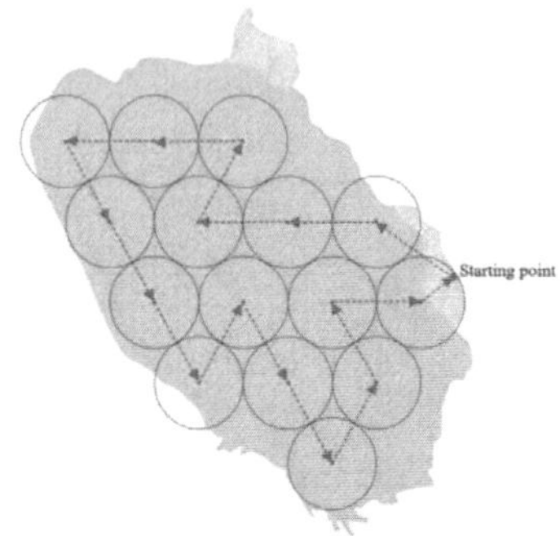

Figure 5. Genetic Algorithm

The performance indicators of path planning covered by the three algorithms are shown in Table 1. The path planning obtained by simulated annealing operation at 16 sampling points is shown in Figure 1. It takes 0.06s and the calculation distance is 44.92km; The operation time of the ant colony algorithm is 0.33s, and the operation distance is 44.72km; The operation time of the genetic algorithm is 0.26s, and the operation distance is 44.16km. Compared with the former two, the genetic algorithm is the best.

Table 1. Performance Indicators of Coverage Path Planning for Three Algorithms at 16 Sampling Points

Performance Indicator	Simulated Annealing Algorithm	Ant Colony Algorithm	Genetic Algorithm
Calculation Results	44.92km	44.72km	44.16km
Computation Time	0.06s	0.33s	0.26s

In the path planning with 26 sampling points, reset the parameters of each algorithm as the number of points increases. The initial temperature of the simulated annealing algorithm is 1000, and the cooling coefficient is 0.99. The path optimization results are shown in Figure 6; The number of ants in the ant colony algorithm is set to 52, and the path optimization result is shown in Figure 7; The number of iterations of the genetic algorithm is 150, and the path optimization results are shown in Figure 8. The path planning and comparison results of the three algorithms are shown in Table 2. At 26 o'clock, the simulated annealing operation took 0.22s and the calculated distance was 60.32km; The operation time of the ant colony algorithm is 2.14s, and the operation distance is 63.04km; The operation time of the genetic algorithm is 0.50s, and the operation distance is 57.72km. Compared with the former two, the genetic algorithm is the best.

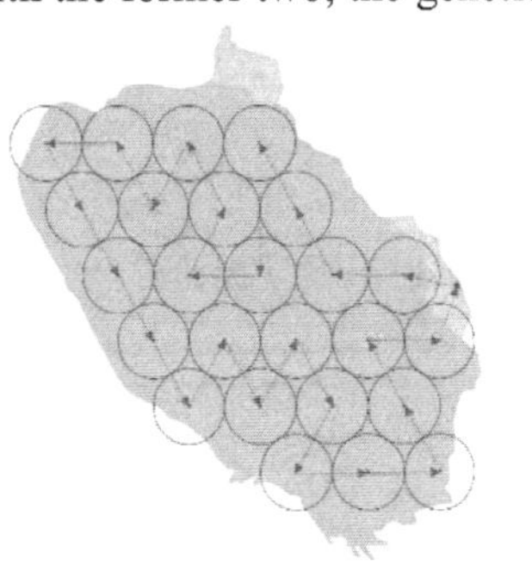

Figure 6. Simulated Annealing

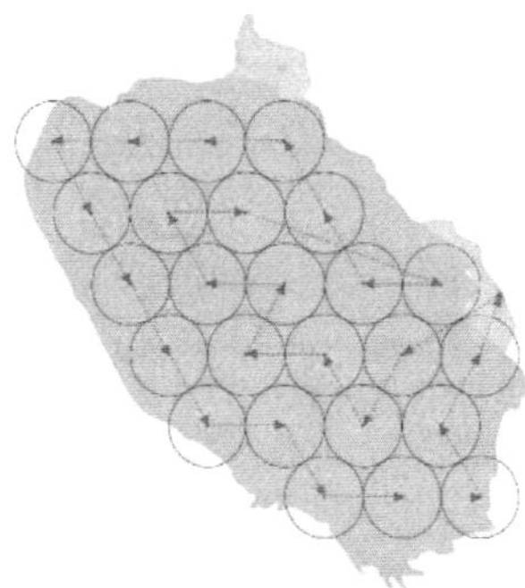

Figure 7. Ant Colony Algorithm

Figure 8. Genetic Algorithm

Algorithm

Table 2. Performance Indicators of Coverage Path Planning for Three Algorithms at 26 Sampling Points

Performance Indicator	Simulated Annealing Algorithm	Ant Colony Algorithm	Genetic Algorithm
Calculation Results	60.32km	63.04km	57.72km
Computation Time	0.22s	2.14s	0.50s

In the path planning with 39 sampling points, reset the parameters of each algorithm as the number of points increases again. The initial temperature of the simulated annealing algorithm is 1500, the number of ants in the ant colony algorithm is 78, and the number of iterations of the genetic algorithm is 300. The path planning and comparison results of the three algorithms are shown in Table 3. At 39, simulated annealing takes 0.32s and the calculation distance is 78.61km; The operation time of the ant colony algorithm is 7.22s, and the operation distance is kilometers; The operation time of the genetic algorithm is 1.37s, and the operation distance is 68.75km. Compared with the former two, the genetic algorithm is the best.

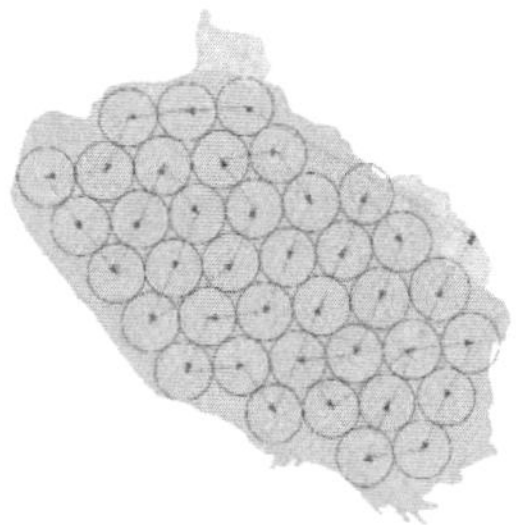

Figure 9. Simulated Annealing Algorithm

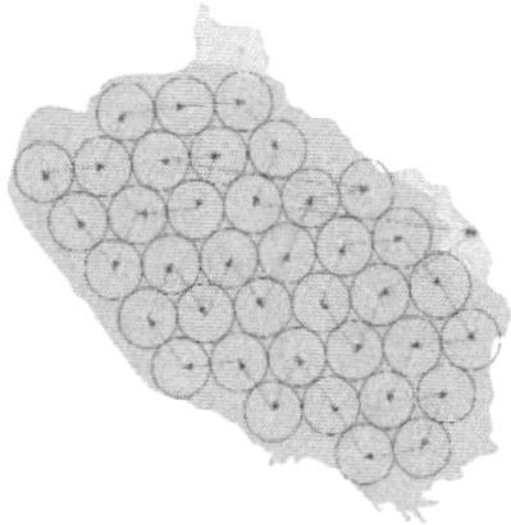

Figure 10. Ant Colony Algorithm

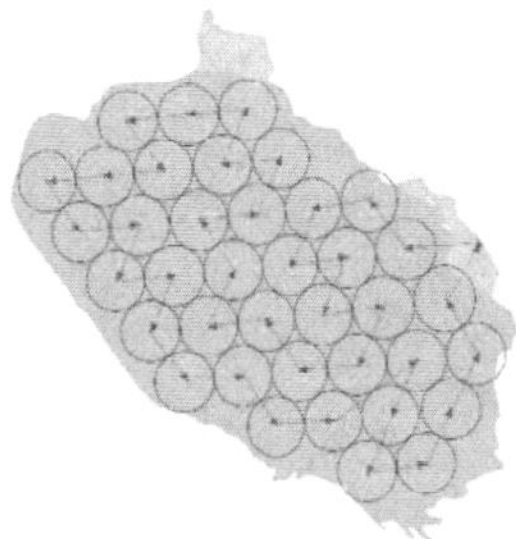

Figure 11. Genetic Algorithm

Table 3. Performance Indicators of Coverage Path Planning for Three Algorithms at 39 Sampling Points

Performance Indicator	Simulated Annealing Algorithm	Ant Colony Algorithm	Genetic Algorithm
Calculation Results	78.61km	79.83km	68.75km
Computation Time	0.32s	7.22s	1.37s

Among the path planning tasks with different points, the three algorithms have their performance. The simulated annealing algorithm has certain advantages in the calculation speed, but in the shortest path optimization, the results of the genetic algorithm are the best, followed by the ant colony algorithm when the number of points is small. The operation distance of 16 sampling points is 44.72km, and the operation time is 0.33s. However, with the increase of points, the operation time increases significantly. The operation time of 26 sampling points is 2.14s, and that of 39 points is 7.22s. The calculation efficiency is greatly affected by the number of points, and the operation result is poor. When the number of points is small, the simulated annealing algorithm has the best result in operation time. In the 16 sampling point path planning, it only takes 0.06s, and can quickly give a feasible path planning scheme with less calculation. This is because the simulated annealing algorithm can converge to a better solution in a short time when dealing with relatively simple problems. Although the calculated path distance is 44.92km, which is slightly longer than the genetic algorithm's 44.16km, the difference is not very large, and the genetic algorithm has high requirements for the computing power of the chip. When the computing power of the UAV control chip is low, the simulated annealing algorithm has certain advantages. With the increase in the number of points, the genetic algorithm shows better comprehensive performance. The genetic algorithm searches for the optimal solution through iteration. When the number of points increases, the number of iterations can be adjusted accordingly. In this process, it can better balance the path optimization and calculation time. It is better than the simulated annealing algorithm in the path optimization effect, and the calculation time is significantly shorter than 7.22s of the ant colony algorithm.

4. Speed planning simulation

When the speed of the UAV is not smooth, there will be a large energy loss. During the acceleration process, the motor needs to output a large amount of power to overcome the inertia of the UAV and make the speed of the UAV increase rapidly. A large acceleration means that a larger force is required, which requires the motor to consume more power. At the same time, during the deceleration process, it means that a larger force is required to reduce the speed of the UAV rapidly, which also needs to consume a lot of energy. On the contrary, the power system of UAVs with smooth speed can run with relatively stable power output during flight. Since the speed does not need to be changed frequently and significantly, the motor does not need to provide high power in a short time to accelerate or decelerate through energy consumption. It can maintain flight in a relatively continuous and stable state, and the UAV can use electric energy more effectively with smooth speed, to achieve more energy-saving effects.

The trapezoidal curve speed planning has a linear change in speed during acceleration and deceleration, and the acceleration will change suddenly. The S-curve planning algorithm has excellent performance in motion stability. Its speed and acceleration change continuously during acceleration and deceleration, which effectively avoids the speed jump and acceleration mutation. The smooth transition of acceleration greatly reduces the impact force on the system. The S-curve planning is more stable than the t-curve speed planning. UAV is mainly affected by air resistance during flight. When the UAV flies at a certain speed, the air resistance is as shown in Formula 1:

$$F_d = \frac{1}{2} C_d \rho A v^2 \tag{1}$$

F_d is the air resistance, C_d is the resistance coefficient, it is related to the shape, surface roughness and other factors of the UAV, ρ is the air density, its size is related to the altitude, air temperature, air pressure and other meteorological conditions, A is the projection area of the UAV in the air flow direction, v is the speed of the UAV relative to the air. Ignoring the hovering power of the UAV, the flight power of the UAV can be expressed by multiplying the flight resistance by the flight speed. The energy consumption of the UAV is equal to the integration of power and time, as shown in formula 2.

$$P = F_d v,\quad E = \int_0^{T} P dt \tag{2}$$

The T-shaped smoothing algorithm is shown in Figure 12. In this algorithm, the distance between the two sampling points is l, the translational flight speed of the UAV starts from zero, the translational flight acceleration is a constant value a_0, when it accelerates to the maximum speed v_{max}, it keeps the speed constant, then decelerates with constant acceleration $-a_0$, and when it reaches another sampling point, the speed drops to zero, so that the time of the constant speed phase is T_u, the flight time of the whole phase is T_d, and the flight energy consumption of the whole process is the integration of the power of each phase with the corresponding time.Because the total distance is one, the integral of the whole process speed versus time, that is, the area of the trapezoid, is the total distance, as shown in Formula 3, then the total energy consumption of the t-smoothing algorithm can be deduced from Formula 1, formula 2 and formula 3, as shown in formula 4.

$$\frac{(T_u+T_d)v_{max}}{2} = l \tag{3}$$

$$E = \frac{l C_d \rho A v_{max}^2 + C_d \rho A v_{max}^3 T_u}{4} \tag{4}$$

When T_u is equal to zero, the energy consumption of the UAV is the least, that is, when the UAV accelerates to the maximum value, it decelerates evenly to zero immediately, and when the speed is zero, the UAV just reaches the destination. Therefore, the energy consumption is shown in formula 5. The flight time of the whole process is shown in formula 6.

$$E = \frac{lC_d\rho A v_{max}^2}{4} \tag{5}$$

$$T_d = \frac{2l}{v_{max}} \tag{6}$$

The S-curve planning algorithm is shown in Figure 13. In this algorithm, the distance between the two sampling points is l, the horizontal speed and acceleration of the UAV start from zero, the acceleration of the UAV is constant j, the acceleration is maintained at constant acceleration to accelerate, and then the acceleration movement of the UAV is maintained at constant $-j$ acceleration to reduce the acceleration reaches the maximum speed v_{max}, the flight time is constant T_m, and then the deceleration movement of the UAV is performed at constant acceleration and the acceleration is increased at constant acceleration. When reaching the destination, the speed is zero, and the acceleration is also zero. The energy consumption of the whole flight process consists of five stages: accelerated motion with increased acceleration, accelerated motion with reduced acceleration, uniform motion, deceleration motion with reduced acceleration, and deceleration motion with increased acceleration. The *v-t* image of the process is symmetrical, and the time calculation formula of the first, second, fourth and fifth stages is shown in formula (7):

$$\sqrt{\frac{v_{max}}{j}} \tag{7}$$

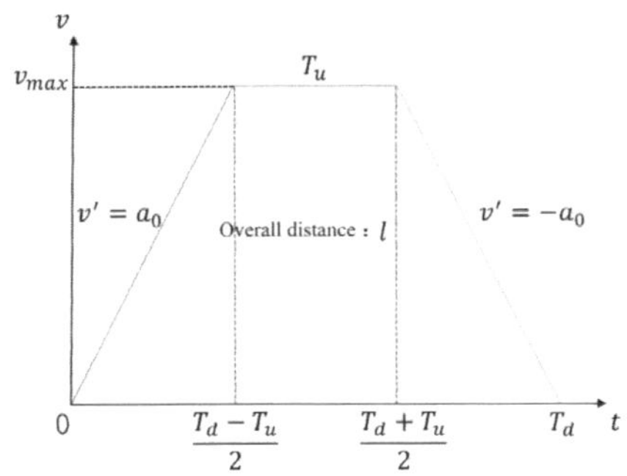

Figure 12. T-smoothing algorithm

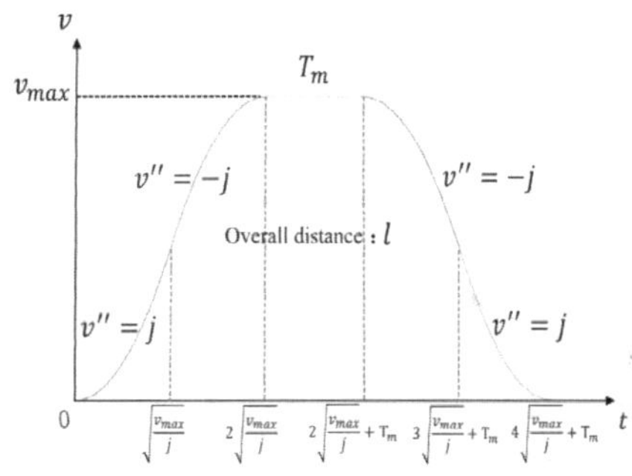

Figure 13. S-curve planning algorithm

The total distance of S-curve planning is one, and the speed expression of the first segment is shown in formula 8, the initial speed and acceleration of the second segment are the end speed $v_{max}/2$ and acceleration $\sqrt{jv_{max}}$ of the first segment, and the acceleration is −j, so the speed expression of the second segment is shown in formula 9, and the third segment is a constant speed flight segment.

$$v = \frac{j}{2}t^2 \tag{8}$$

$$v = \frac{v_{max}}{2} + \sqrt{jv_{max}}\left(t - \sqrt{\frac{v_{max}}{j}}\right) + \frac{-j}{2}\left(t - \sqrt{\frac{v_{max}}{j}}\right)^2 \tag{9}$$

The energy consumption of the first stage is the integration of the power of this segment with time. From Formula 1, formula 2 and formula 6, the energy consumption of this segment can be deduced as shown in formula 10. The energy consumption of the fifth segment is equal to that of the first segment.

$$E = \frac{C_d\rho A v_{max}^3\sqrt{v_{max}}}{112\sqrt{j}} \tag{10}$$

From Formula 1, formula 2 and formula 7, it can be deduced that the energy consumption of the second segment is as shown in formula 11, the energy consumption of the fourth segment is equal to that of the second segment, and the third segment is a constant speed flight phase, and its energy consumption is as shown in formula 12:

$$E = \frac{177 C_d \rho A v_{max}^3 \sqrt{v_{max}}}{560\sqrt{j}} \tag{11}$$

$$E = \frac{1}{2} C_d \rho A v_{max}^3 T_m \tag{12}$$

The third stage is the stage of uniform velocity, in which the fourth and fifth segments are symmetrical with the first and second segments, and the distance is equal to the integral of velocity versus time. After calculation, formula 13 is obtained.

$$2v_{max}\sqrt{\frac{v_{max}}{j}} + v_{max}T_m = l \tag{13}$$

From formula 12 and formula 13, the formula for calculating the energy consumption in the uniform speed stage is derived as shown in formula (14), while the total energy is the sum of the energy of five segments, and the calculation formula is shown in formula (15). It can be seen that the smaller the segment is j, the minimum energy loss is when the segment T_m is zero. Therefore, the relationship between and can be derived as shown in formula 16, the minimum total energy consumption in the whole process is calculated as shown in formula 17, and the flight time in the whole process is shown in formula 18.

$$E = \frac{1}{2} C_d \rho A v_{max}^2 l - C_d \rho A v_{max}^3 \sqrt{\frac{v_{max}}{j}}, \tag{14}$$

$$E = \frac{1}{2} C_d \rho A v_{max}^2 l - \frac{49}{140} C_d \rho A v_{max}^3 \sqrt{\frac{v_{max}}{j}} \tag{15}$$

$$2v_{max}\sqrt{\frac{v_{max}}{j}} = l \tag{16}$$

$$E = \frac{91 C_d \rho A v_{max}^2 l}{280} \tag{17}$$

$$T_d = \frac{2l}{v_{max}} \tag{18}$$

Taking the Chaka Salt Lake area as an example, according to the standard atmospheric model, the altitude of the Chaka Salt Lake is about 3.05km, and the air density in this area can be calculated to be about $0.907 kg/m^3$. The drag coefficient and reference area are different when different UAVs sample. When the UAV surface is smooth, the drag coefficient C_d can be taken as 0.2, and the reference area A is set as 0.5m^2. If the maximum flight speed of the UAV in the salt lake sampling mission cannot exceed 10m/s, then v_{max}=10. The total distances of 16 points, 26 points and 39 points are calculated by genetic algorithm to be 44.16km, 57.72km and 68.75km respectively. Based on T-shaped algorithm and S-curve algorithm, the total energy consumption and total time of are shown in Table 4.

Table 4. Performance Indicators of Different Smoothing Algorithms

Algorithm Type	Energy Consumption at 16 Points	Overall Time	Energy Consumption at 26 Points	Overall Time	Energy Consumption at 39 Points	Overall Time
T-shape	27.81Wh	2.45h	36.36Wh	3.21h	43.30Wh	3.82h
S-curve	36.15Wh	2.45h	47.26Wh	3.21h	56.29Wh	3.82h

The comparative analysis shows that the total time consumption of the two algorithms is the same, and the energy consumption of S-curve planning algorithm is slightly higher than that of T-shape algorithm, but the acceleration of S-curve changes constantly, which is more stable than that of T-shape algorithm. In the water quality sampling with high requirements for stability, the comprehensive performance of S-curve algorithm is more superior, so it is more suitable for the UAV Salt Lake sampling system designed in this paper.

5. Conclusion

With the rapid construction of the world-class Salt Lake industrial base, the traditional Salt Lake sampling has been gradually eliminated, and a new type of Salt Lake sampling mode is in urgent need. Therefore, this paper proposes a salt lake sampling system based on UAVs. The system uses the coverage path planning strategy of circular region decomposition, and the sampling points maximize the coverage of the entire Salt Lake area. The genetic algorithm optimizes the shortest path, and the S-curve speed planning algorithm makes the salt lake UAV sampling more energy-saving and low power consumption, to improve the endurance and stability of UAV sampling, and further contribute to the development of the salt lake industry.

Acknowledgments

Special thanks should go to my friends who have put considerable time and effort into their comments on the draft.

References

[1] Song P S, Li W, Sun B. 2011 Progress in the development and utilization of salt lake resources. Journal of inorganic chemistry, Vol.27, pp: 801-815
[2] Zhou Y, Li L J, Wu Z J. 2013 Development and comprehensive utilization of salt lake resources in Qinghai. chemical progress. Vol.25, pp: 1613-1624
[3] Yang C. 2024 Present situation of salt-lake lithium resources and research progress of lithium extraction technology. Journal of Engineering Science.Vol.46,pp:1659–1670.
[4] Jiang B, Wu X R, Ding J, 2016 Comparison of calculation methods for thermocline depth in the South China Sea. Ocean bulletin. Vol35, pp: 64-73
[5] Song Y Z, Lan H, Jia W J, 2014 Optimal design of electronically controlled multi bottl e water sampler system. Ocean technology, Vol33,pp: 41-46
[6] Lu C H, Dong L, Fu Z K. 2020 Design of water quality sampling system for unmanne d monitoring ship based on LabVIEW. laboratory research and exploration, Vol39 pp: 68-71
[7] Sun W C, Luo Z H, Shi J M.2024 Overview of UAV coverage path planning methods. control theory and application, pp: 1-21
[8] Tan Y, Li S, Liu H, 2021 Automatic inspection data collection of building surface based on BIM and UAV. Automation in Construction, pp:131: 103881
[9] Cho S W, Park H J, Lee H, 2021 Coverage path planning for multiple unmanned aerialvehicles in maritime search and rescue operations.Computers & Industrial Engineering, p 07612
[10] Campo L V, Ledezmal A, Corrales J C. 2020 Optimization of coverage mission for lightweight unmanned aerial vehicles applied in crop data acquisition. Expert Systems with Applications, p 113227

Journal of Physics: Conference Series **3024** (2025) 012043 doi:10.1088/1742-6596/3024/1/012043

Control and Simulation of Flying Robotic Arms

Liang Xuan[1*], Haowei Yu[1], Tongjin Sun[2], Shaowei Teng[1]

[1] SuHubei Engineering Research Center for Intelligent Detection and Control of Specialized Equipment Jianghan University, No. 8 Sanjiaohu Road, Wuhan,China

[2] Yancheng City product quality supervision and inspection institute, Yancheng, China

[*]Corresponding author's e-mail: xuanl@jhun.edu.cn

Abstract. Traditional manual high-altitude operations have problems of safety risks and low efficiency. However, the combination of unmanned aircraft and mechanical arms can improve efficiency and ensure safety. Flight disturbances will affect the precise control of the mechanical arms. This paper proposes a compensation algorithm based on fuzzy PID control to deal with the motion disturbances of the flight platform. Fuzzy PID control combines the advantages of fuzzy control and classical PID control, and can handle the uncertainty and nonlinearity of the system, and real-time compensate for the disturbances of the flight platform. This innovation can solve the problems of accuracy decline and response delay that traditional PID control may encounter when dealing with disturbances or nonlinear systems. According to the deviation data of the end position of the mechanical arm in the simulation results, it can be concluded that after compensation control, the error of the end position is significantly reduced compared with traditional PID control, and the control accuracy reaches within 0.05mm, and the steady-state error is less than 0.1mm after compensation. This research provides technical support for the application of flight mechanical arms in wind turbine group detection, and provides new ideas for the design of control systems for unmanned aircraft platforms carrying mechanical arms. The simulation results show that this control strategy significantly improves the stability and accuracy of the end position of the mechanical arm, especially under the condition of flight platform disturbances, it can effectively maintain the end position of the mechanical arm at the target position.

1. Introduction

With the development of wind power industry, the threat of lightning disaster facing wind power units is gradually increasing, and the detection of lightning protection devices becomes particularly important.[1-3]Traditional detection methods rely on manual high-altitude operations, which not only pose high safety risks and labor intensity, but also require time and effort.In order to improve detection efficiency and ensure safety, the technology of combining UAV platform and robot arm has become the focus of research.By installing a customized robotic arm at the bottom of the drone and remotely controlling it for lightning protection testing, it can not only avoid the risk of high-altitude operations, but also significantly shorten the detection time.[4]The flying robotic arm system refers to a new type of aerial robot that uses drones as the flying platform and robotic arms as the operating devices. [5]This technology not only improves the safety of wind turbines, but also promotes the development of intelligent and automated inspection in the wind power industry.In order to ensure the coordinated operation of the robot arm and the UAV, a special control system needs to be designed to achieve accurate operation.

In the research of robotic arm control, many scholars have made certain progress in different fields.Based on the calculated torque control and PD control, Mathew[6] proposed a design method for the position tracking controller of the mobile robot arm, including the feedforward control ring and the feedback control ring, which can effectively deal with external disturbance and achieve accurate tracking of the reference trajectory.Han research focuses on the robust control of a rigid and flexible coupled planar two-link manipulator, and adopts a sliding mode controller to achieve the goals of joint Angle control, flexible vibration suppression and anti-interference.[7]In addition, according to the characteristics of the multi-degree of freedom robotic arm system, Yang designed an adaptive controller with real-time gravity compensation and motion constraint guarantee functions, which eliminated static errors by introducing nonlinear auxiliary terms.[8]In order to ensure the reliability of the flying robot arm, this paper decides to adopt the fuzzy PID control strategy for the five-DOF robot arm to complete the control simulation of the five-DOF robot arm carried by the UAV.

2. Disturbance analysis of flying robot arm

The UAV is in a hovering state when it carries a mechanical arm for fixed-point clamping, and it is in a state of constant disturbance due to wind and other environmental factors.The UAV has six degrees of freedom of motion, and also has six degrees of freedom of disturbance.In this paper, six degrees of freedom platform is used to simulate the motion of UAV in the air.

$$U = Ri + L\frac{di}{dt} + E \tag{1}$$

It is used to represent the motion parameters of the UAV in the world coordinate system.The UAV motion signals need to be mapped to the motion parameters of the 6-DOF platform.The coordinate system of the UAV is shown in Figure 1.

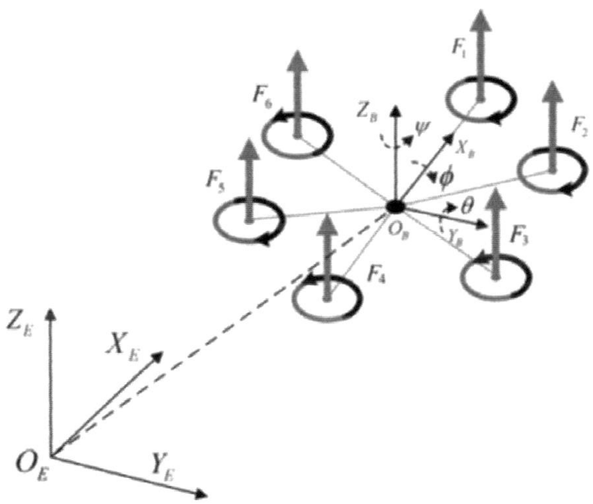

Figure 1. UAV Coordinate System

The pose deviation of the robot arm refers to the difference between the actual position and the expected position of the actuator at the end of the robot arm due to various factors when the robot arm performs the task.The common factors that affect the posture deviation of the robot arm are as follows.

a)Position Deviation

The position deviation is mainly affected by environmental factors.When the flying robot arm flies to the height, there will be environmental effects such as wind speed.The flying robot arm may be displaced from the set position, resulting in the actual position of the actuator on the mechanical alarm end being inconsistent with the expected position, which will affect the positioning accuracy of the robot arm.

b)Attitude Deviation

Attitude deviation is a common disturbance of flying robotic arms.It is mainly caused by the coupling effect between the robot arm and the UAV when it is working, resulting in rotation and other disturbances between the flying robot arm and the set position.The disturbance of the base may cause the rotation or tilt of the base of the robot arm, resulting in the attitude deviation of the end-actuator of the robot arm, which will affect the attitude control and path planning of the robot arm.

As shown in Figure 2(a), when the rotor UAV produces distance displacement relative to the expected position, the end fixture of the robot arm also produces distance deviation relative to the expected position.As shown in Figure 2(b), when the UAV rotates around itself, the end of the robotic arm produces a spatial pose change.Both kinds of deviation will lead to the pose deviation of the robot arm in the execution of work.

(a)UAV Position Disturbance (b)UAV Attitude Disturbance

Figure 2. Disturbance of the UAV to the Robotic Arm

In order to stabilize the flying robot arm and meet the actual engineering requirements, it is necessary to compensate the position change and attitude change of the end of the flying robot arm.The drone itself carries a control system and a sensor system, the sensor provides feedback to the control system, and the controller can better adjust the attitude of the robot arm.

3. Flying robotic arm control

Because the robot arm base is installed at the bottom of the UAV, the disturbance of the UAV will have a great influence on the pose control of the end of the manipulator.Compare the fuzzy PID control method, traditional PID control and sliding mode control among others. In terms of control accuracy, the fuzzy PID control can effectively deal with complex systems and disturbance environments by adaptively adjusting PID parameters, thereby improving control accuracy. Especially in nonlinear systems, it performs better. However, traditional PID control is difficult to cope with changes in system parameters and its accuracy is relatively fixed. Sliding mode control also has advantages in terms of accuracy, but the switching process may cause "jerk" phenomenon, affecting stability. In terms of robustness, the fuzzy PID control has strong adaptability by adjusting PID parameters in real time under uncertainty and external disturbances. The robustness of traditional PID is poor and it is easily affected by changes in system parameters and disturbances. Sliding mode control has strong robustness when facing external disturbances, but the switching frequency may cause fluctuations in control signals. In terms of stability, the fuzzy PID control maintains system stability by flexibly adjusting control rules, but complex rules may affect global stability. The stability of traditional PID control is poor and it requires frequent debugging. Sliding mode control has strong stability in dynamic systems, but the switching frequency may lead to jitter. In summary, the fuzzy PID controller is adopted for compensation control.

3.1. The joint position control model was established

The control of the robot arm is to use the difference between the expected Angle of the robot arm and the actual Angle of the feedback, and then output the torque.The joint motion of the robot arm needs to be driven by the joint motor.The joint model construction should take into account the dynamic characteristics of the motor, mechanical transmission components and other influencing factors, so as to provide conditions for the subsequent joint control simulation.

Kirchhoff voltage balance equation[9]:

$$U = Ri + L\frac{di}{dt} + E \tag{2}$$

$$E=K_e w \tag{3}$$

Torque balance equation of motor:

$$T_e = K_m A \tag{4}$$

$$T_e = J\frac{d\omega}{dt} + B\omega + T_l \tag{5}$$

$$\theta_m = \omega t \tag{6}$$

The transfer function between the angular displacement of the joint at the load end is obtained:

$$\theta_m(s) = \frac{\omega(s)}{s} \tag{7}$$

$$G_1(s) = \frac{\omega(s)}{U(s)} = \frac{K_m}{(Ls+R)(Js+B)+K_E K_M} \tag{8}$$

$$G_2(s) = \frac{\theta_m(s)}{U(s)} = \frac{K_m}{s\left((Ls+R)(Js+B)+K_E K_M\right)} \tag{9}$$

3.2. Fuzzy PID Controller Design

Fuzzy control includes three parts: fuzzy inference, fuzzy reasoning and fuzzy resolution.[10]In the fuzzification stage, the input variables are transformed from the precise domain to the fuzzy domain.Its purpose is to enable fuzzy logic system to deal with fuzzy input, so as to better deal with the uncertainty and fuzziness of the system.Fuzzy inference refers to the process of generating fuzzy output according to fuzzy input and a set of fuzzy rules.In the fuzzy inference stage, the system will deduce the fuzzy input according to the defined fuzzy rules to determine the fuzzy set of the output.[11]Finally, in the defuzzification stage, it converts the fuzzy output from fuzzy reasoning into the deterministic actual output.In the defuzzification stage, the system needs to map the fuzzy output to the deterministic output space in some way in order to actually execute the corresponding control action.The key is to choose the input variables reasonably, design fuzzy rules and determine the method to solve the ambiguity, so as to achieve the precise control and optimize the performance of the system.

Fuzzy PID control is a method that combines fuzzy control and PID control.[12]It makes use of the adaptive ability of fuzzy control system to fuzzy and nonlinear systems, while retaining the stability and reliability of classical PID control.Fuzzy control and PID control are combined to complement each other.Its principle is shown in the figure 3.

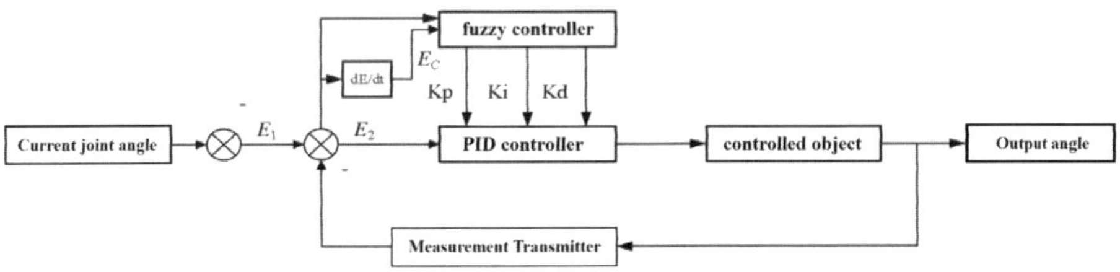

Figure 3. Fuzzy PID Control Schematic

The fuzzy PID controller is designed and simulated by Matlab.Given that the target trajectory is a step signal, the established fuzzy rules, discourse domain and membership function are loaded into the fuzzy controller.The simulation structure of fuzzy PID controller is shown in the figure 4.

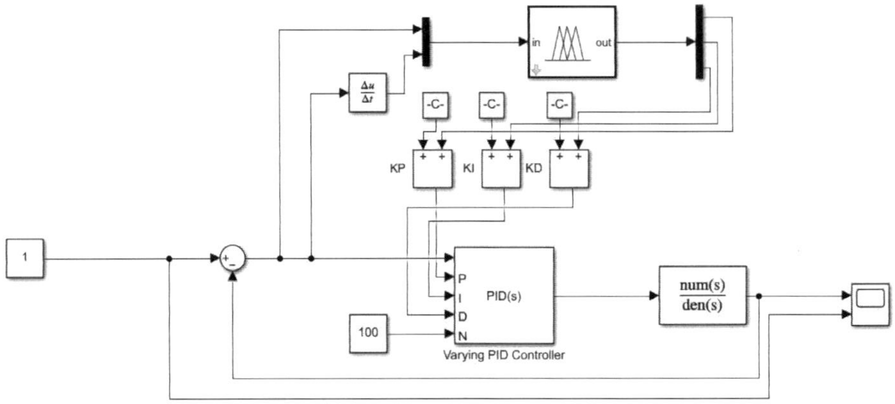

Figure 4. Fuzzy PID Control Simulation Model is Established

In actual drone scenarios, environmental factors such as wind speed and air turbulence usually exhibit periodic or oscillatory characteristics. Sinusoidal waves can effectively simulate periodic disturbance sources by adjusting their amplitudes and frequencies. Therefore, using sinusoidal signals to simulate disturbances can accurately reflect the dynamic changes of these disturbances.Add a sinusoidal signal with a amplitude of 0.3 on the 5th as a continuous interference signal.The simulation results are shown in the figure 5.

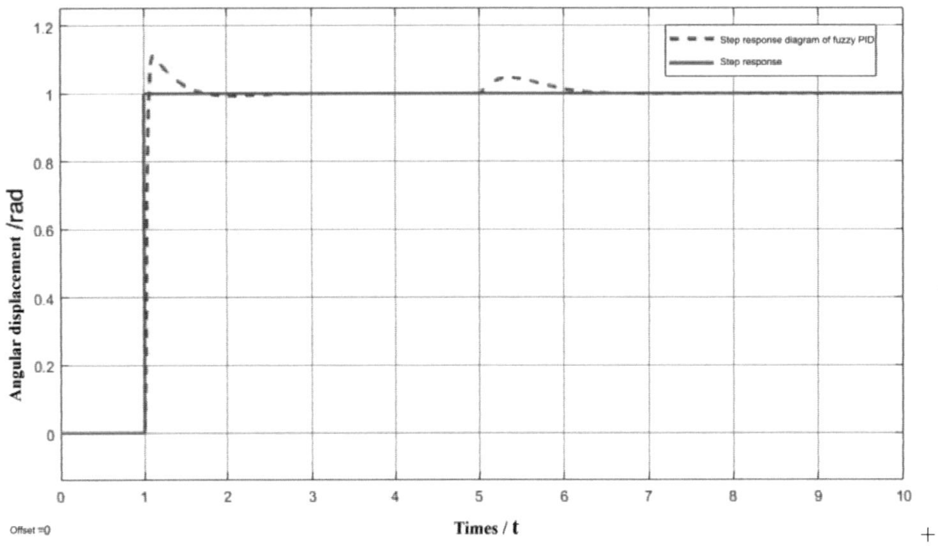

Figure 5. Fuzzy PID Control Simulation Results

The fuzzy PID position control system can achieve the preset value in the disturbance environment quickly after optimization.According to the simulation results, when the interference signal is added, it only takes 1.3s to reach the target position.According to the changing law of the position, velocity and acceleration of the end position of the flying robot arm in the global coordinate system under the condition of the movement of the flying platform.Aiming at the influence of UAV motion on robotic arm, a motion compensation control strategy is proposed for the end position of robotic arm in the floating state of UAV.The position compensation of each joint at the end of the manipulator is realized in the floating state of the flying platform.

In order to compensate the position of the end of the robot arm, the position and attitude of the flying platform must be detected first.The attitude ultrasonic sensor of the flight platform can be used to measure the elevating position of the UAV.The attitude sensor and ultrasonic sensor are jointly installed in the central position of the UAV, which can measure the pitch, roll and heave equipose information of the UAV in real time.The position of the sensor installation relative to the entire flight robotic arm system will cause pose deviation.So the sensor coordinate system is given as $\{b_s\}$.The error between the center of the flying platform $\{b_b\}$ and the sensor coordinate system $\{b_s\}$ is represented by the secondary transformation matrix T_s^b.

θ_r, θ_p, z_h

Roll angle θ_r , pitch angle θ_p and position coordinate z_h data can be obtained through sensor measurement.can be obtained from the sensor measurement data. The homogeneous transformation matrix T_n^s from the sensor coordinate system $\{b_s\}$ to the initial central coordinate system $\{b_n\}$ of the flight platform can be expressed as:

$$T_s^n = T(z_h)R(\theta_p)R(\theta_r) \tag{10}$$

The homogeneous transformation matrix of the global coordinate system $\{b_g\}$ in the end coordinate system of the manipulator is obtained with respect to T_t^g by the measured values of the sensor.

$$T_t^g = T_n^g\left(T(z_h)R(\theta_p)R(\theta_r)\right)T_b^s T_r^b T_t^r \tag{11}$$

When the flight platform is moving, the accurate position and attitude of the end of the robot arm in the global coordinate system can be determined according to the above detection data using the sensor. Achieve accurate control and positioning of the robot arm.

At the same time, the homogeneous change matrix of the end fixture coordinate system $\{b_t\}$ relative to its base system $\{b_g\}$ can be solved as:

$$T_t^r = \left(T_n^g\left(T(z_h)R(\theta_p)R(\theta_r)\right)T_b^s T_r^b\right)^{-1} T_t^g \tag{12}$$

Through the above formula, the position of the end of the robot arm $\{b_t\}$ is converted from the global coordinate $\{b_g\}$ to the position of the robot arm under the base coordinate $\{b_g\}$. The deviation between the actual position of the end of the robot arm and the target position is obtained, and then the Angle of each joint is adjusted to compensate by the inverse kinematics of the robot arm. Its control structure frame is shown in the figure 6.

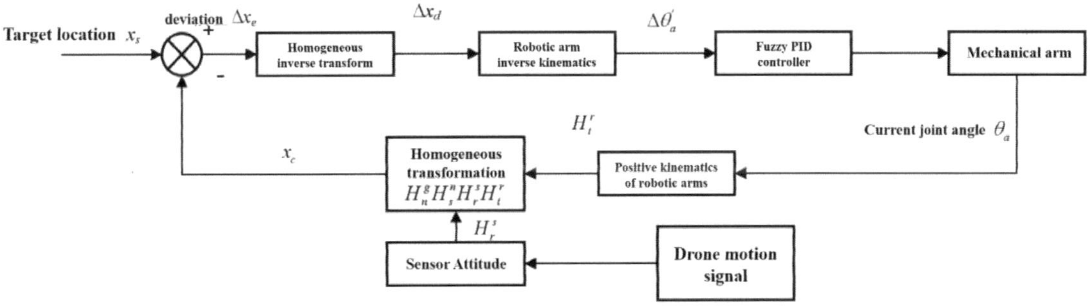

Figure 6. Compensation Control Principle Diagram of Flying Robot Arm

The motion attitude information of the platform is collected by the attitude sensor, and the homogeneous transformation from the base coordinate system to the sensor coordinate system is solved. The current Angle value θ_a of each joint is obtained through the manipulator control system. Combined with the forward and inverse homogeneous transformation matrix of the manipulator and related formulas, the position x_c of the end effector of the manipulator in the world coordinate system is calculated. In each control cycle, the target set point x_s of the manipulator's end effector is determined. The deviation Δx_e of the end position of the manipulator in the global coordinate system can be calculated.

$$\Delta x_e = x_s - x_c \tag{13}$$

The compensation formula of the end position of the robot arm marked with the base of the robot arm can be converted as follows:

$$\begin{bmatrix} \Delta x_d \\ 1 \end{bmatrix} = \left(T_n^g \left(T\left(z_{\text{heave}}\right) R\left(\theta_{\text{pitch}}\right) R\left(\theta_{\text{roll}}\right) \right) T_b^s T_r^b \right)^{-1} T_t^g \begin{bmatrix} \Delta x_e \\ 1 \end{bmatrix} \tag{14}$$

The current position of the mechanical arm in its base system $\{b_r\}$ is set as p_o.

$$p_t = p_0 + \Delta x_d \tag{15}$$

Angle adjustment θ_a' can be obtained from the end position p_t according to the inverse kinematics of the robot arm, and then the control quantity θ_a' is input to the robot arm controller to control the robot arm for motion compensation control.

4. Build simulation environment and verify compensation control algorithm

4.1. Simulation Environment Construction

The simulation environment is built based on SolidWorks and Matlab platform.Firstly, the rotation position of each joint of the manipulator is marked with the coordinate system through SolidWorks, and then each part is derived into x_t format.Each joint part of the manipulator was imported by solid module of Simulink simulation platform.The coordinate system was calibrated at each rotating joint and Rigid Transform module was used to assemble the parts.Then the simulation model of the robot arm under Simulink as shown in Figure 7 can be obtained.

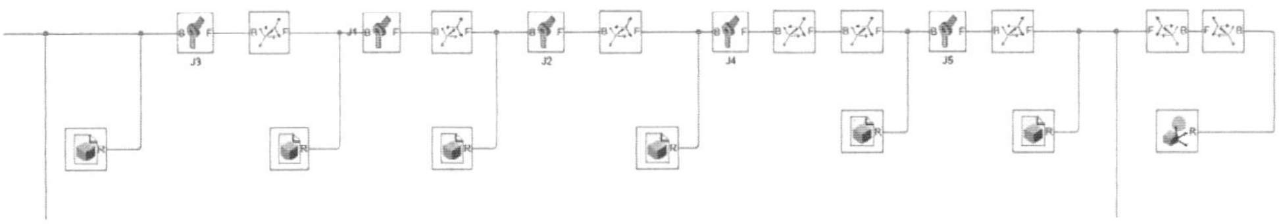

Figure 7. Robot Arm Simulation Model

Parameters of the joint of the mechanical arm can be adjusted by clicking the joint component in Simulink, as shown in Figure 8.

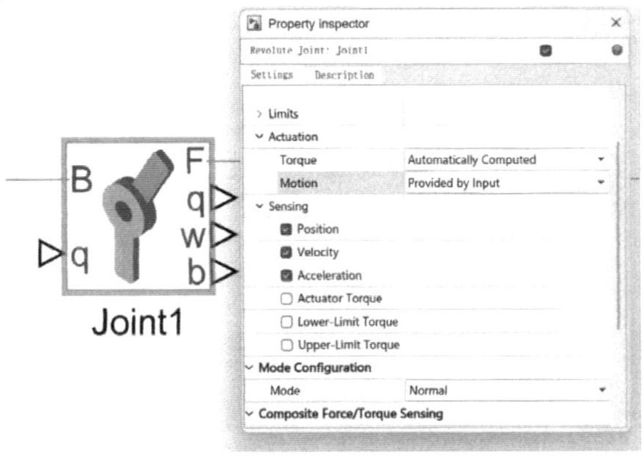

Figure 8. Robot Arm Joint Attribute Setting

In order to simulate the 6-DOF motion of the flight platform, the Bushing Joint module in Simscape is used to simulate the rotation and translation of the flight platform in the world coordinate system, and the modules are successively added between the flight platform and the world coordinate system.As shown in Figure 9.

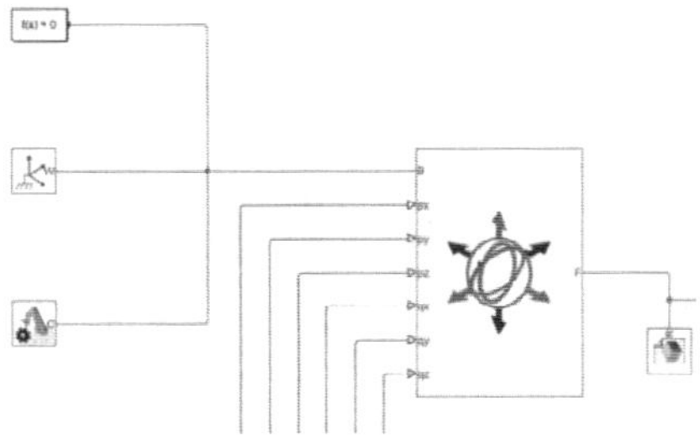

Figure 9. Flying Mobile Platform

On the basis of the above model, the manipulator kinematics module, the joint position compensation algorithm module, the manipulator end trajectory generation module, the joint controller module and the flight platform disturbance signal module are added.The simulation model of the built flight robotic arm is shown in Figure 10.

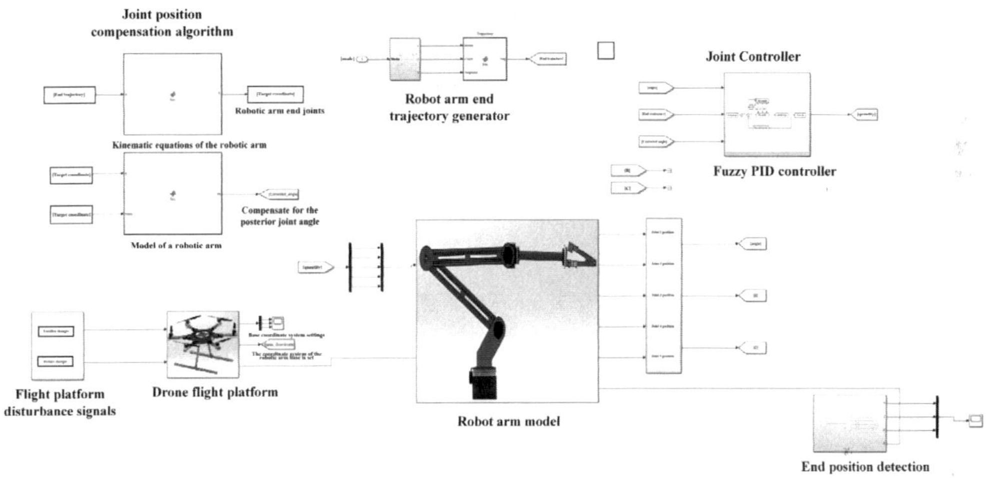

Figure 10. Overall Simulation Model of Flying Robot Arm.

4.2. *Simulation of End Compensation Control Strategy*

In order to verify the correctness of the model solution, the initial position of the joint Angle $\theta_1,\theta_2,\theta_3,\theta_4,\theta_5$ of the manipulator arm was set as [-pi/6,-pi/6, pi/2,pi/2].In order to simulate the movement of the flight platform, the movement direction and attitude $q=\left(x_0,y_0,z_0,\phi_0,\theta_0,\psi_0\right)^T$ of the platform are set.Given by sinusoidal signal, its motion curve generation formula is as follows:

$$y = A\sin(2\pi ft+\sigma\frac{\pi}{180}) \tag{16}$$

The amplitude, frequency and phase parameters of sinusoidal signal are shown in Table 1.

Table 1. Table of Amplitude, Frequency and Phase Parameters of Sine Signal

Argument	Amplitude A(m)	Frequency f(Hz)	Phase σ(deg)
x_0	0.01	0.1	0
y_0	0.02	0.1	45
z_0	0.03	0.1	90

ϕ_0	5*pi/180	0.1	0
θ_0	5*pi/180	0.15	90
ψ_0	5*pi/180	0.2	120

By comparing the simulation results without joint compensation and the relevant joint compensation, the end position compensation effect of the fuzzy PID controller designed in this paper can be analyzed.As shown in Figure 11, the blue line is the curve without joint compensation control end position, the red line is the joint compensation control curve, and the green line is the target position of the robot arm end.Through forward kinematics analysis of the manipulator arm, it can be obtained that the initial position of the end is x=664mm, y=-389mm, z=-501mm.The initial position was kept unchanged during simulation, and the simulation time was 40s.Observe the change of the position of the end of the mechanical arm, and the results are as follows:

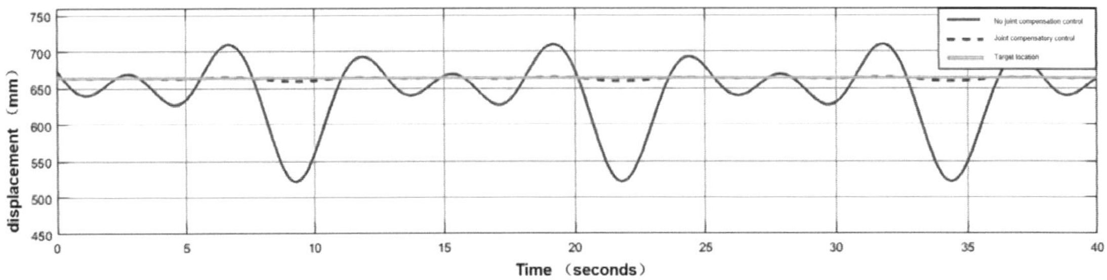

(a)Position Change Curve of The End of Mechanical Arm in X Direction

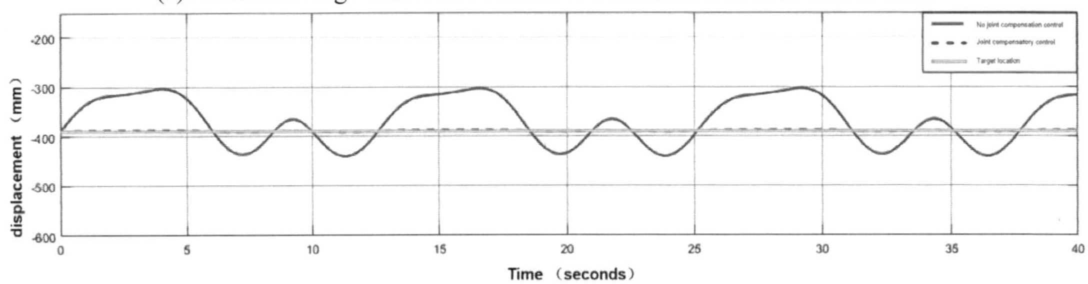

(b)Position Change Curve of The End of Mechanical Arm in Y Direction

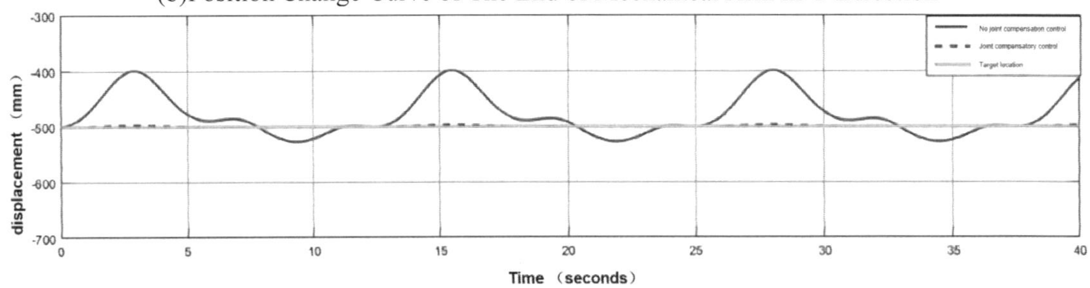

(c)Position Change Curve of The End of Mechanical Arm in Z Direction

Figure 11. Position Curve of The End of The Manipulator Arm in X, Y and Z Directions

As can be seen from the above figure, if the Angle of the robotic arm is not compensated and adjusted during the movement of the flying platform, the end position of the robotic arm will cause great deviation.After the joint compensation control algorithm, the coordinates of the end position X, Y and Z of the manipulator are stabilized at the initial target position (664mm, -389mm, -501mm).The correctness of the compensation control algorithm in this paper is verified.

5. Build simulation environment and verify compensation control algorithm

In this paper,the control strategy of the flying robot arm on the UAV platform is studied, focusing on the effect of flight disturbance on the position of the end of the robot arm. A compensation control system based on fuzzy PID control is proposed to compensate the end deviation caused by UAV movement. Combining the advantages of fuzzy control and traditional PID control, the algorithm can dynamically adjust the control parameters, and significantly improve the control accuracy of the end position of the manipulator in the face of external disturbance and uncertainty. It can be seen from the simulation results that the compensated error is greatly reduced and the control accuracy reaches a small error range, which improves the stability of the manipulator in a complex environment. The research provides technical support for the efficient and safe detection of wind turbines carried by the robot arm on the UAV platform, and provides a new idea for the design of the control system of the flight robot arm.

In the future, more types of sensors, such as vision sensors or lidar, can be introduced to improve the control system's ability to perceive environmental changes, while further adjusting the control algorithm to maintain high performance in extreme or uncertain environments. Establishing a more efficient cooperative control mechanism between the flight platform and the robot arm will better improve the stability and response speed of the entire system. To achieve seamless docking and precise coordination between the flight platform and the robotic arm.

Acknowledgments
This study was supported by Hubei Provincial Engineering Research Center for Intelligent Detection and Control of Special Equipment, and the control and simulation of the flight manipulator arm for wind power lightning protection detection was the research on the remote detection system of large-scale in-service wind turbine blades (No. KJ2023067) and the research on the acupuncture robot system based on the digital Chinese medicine model (No. 2024BCB010) project provides the technical basis.

References

[1] LI Guoqing. The new power system "Drawing a blueprint to the end" Wartsila ushered in a key window [J]. Electric Age,2021,(12):18-23.
[2] Tan Xihui. Research and Application Analysis of lightning protection Technology for Wind Turbine [J]. Encyclopedia Forum Electronic Magazine,2023(19):13-15.
[3] Gorg A, Mohammed A, Abdelali A E. Map Optimization Fuzzy Logic Framework in Wind Turbine Site Selection with Application to the USA Wind Farms[J]. Energies, 2021,14(19):6127-6127.
[4] Zhang Yibo, Xu Bin, Fan Wei. Research progress and prospect of intelligent aerial Work Robot [J]. Artificial Intelligence, 2021,(04):6-16.
[5] LIU Chao, ZHANG Guangyu, Xiong Anbin. Review of research on robotic arms for rotor flight [J]. High-tech Communications, 2019,30(07):735-747. (in Chinese)
[6] Mathew S S, Jisha V R. Tracking control of a mobile manipulator with external torque disturbances using computed torque control[C]//2020 IEEE 17th India council international conference (INDICON). IEEE, 2020: 1-7.
[7] Han F, Jia Y. Sliding mode boundary control for a planar two-link rigid-flexible manipulator with input disturbances[J]. International Journal of Control, Automation and Systems, 2020, 18(2): 351-362.
[8] Yang T, Sun N, Fang Y, et al. New adaptive control methods for $ n $-Link robot manipulators with online gravity compensation: Design and experiments[J]. IEEE Transactions on Industrial Electronics, 2021, 69(1): 539-548.
[9] Liu Zhiquan, Wei Qingqing, Wang Yaobing. Establishment of Joint Fine Dynamics Model and Joint Torque Control for Space Manipulator [J]. Journal of Astronautics,2014,35(6):663-668. (in Chinese)

[10] Narayan J, Dwivedy S K. Towards neuro-fuzzy compensated PID control of lower extremity exoskeleton system for passive gait rehabilitation[J]. IETE Journal of Research, 2023, 69(2): 778-795.
[11] Borase R P, Maghade D K, Sondkar S Y, et al. A review of PID control, tuning methods and applications[J]. International Journal of Dynamics and Control, 2021, 9: 818-827.
[12] Lu Kai, Wang Ting. Experimental study on anti-roll Control of End Spinner of Lifting Manipulator [J]. Mechanical and Electrical Engineering,2022,39(12):1776-1783.

AUTHOR INDEX